Analytical Biogeography

AN INTEGRATED APPROACH TO THE STUDY OF ANIMAL AND PLANT DISTRIBUTIONS

Analytical Biogeography

AN INTEGRATED APPROACH TO THE STUDY OF ANIMAL AND PLANT DISTRIBUTIONS

Edited by

A.A. MYERS

and

P.S. GILLER

CHAPMAN & HALL

London · New York · Tokyo · Melbourne · Madras

Published by Chapman & Hall, 2–6 Boundary Row, London SE1 8HN

Chapman & Hall, 2–6 Boundary Row, London SE1 8HN, UK

Chapman & Hall, 29 West 35th Street, New York NY10001, USA

Chapman & Hall Japan, Thomson Publishing Japan, Hirakawacho Nemoto Building, 7F, 1-7-11 Hirakawa-cho, Chiyoda-ku, Tokyo 102, Japan

Chapman & Hall Australia, Thomas Nelson Australia, 102 Dodds Street, South Melbourne, Victoria 3205, Australia

Chapman & Hall India, R. Seshadri, 32 Second Main Road, CIT East, Madras 600 035, India

First edition 1988
Reprinted as a paperback 1990, 1991

Typeset in 10/12pt Palatino by Acorn Bookwork, Salisbury, Wiltshire
Printed in Great Britain by St Edmundsbury Press Ltd, Bury St Edmunds, Suffolk

ISBN 0 412 40050 2

A catalogue record for this book is available from the British Library

Library of Congress Cataloging-in-Publication data
Analytical biogeography : an integrated approach to the study of animal and plant distributions / editors, A. A. Myers and P. S. Giller.
p. cm.
Bibliography : p.
Includes index.
ISBN 0-412-28260-7
ISBN 0-412-40050-2 (pbk)
1. Biogeography. I. Myers, Alan A. II. Giller, Paul S.
QH84.A53 1988
574.9—dc19 88-307
CIP

Printed on permanent acid-free text paper, manufactured in accordance with the proposed ANSI/NISO Z 39.48–199X and ANSI Z 39.48–1984

Contents

Contributors

Nick Barton Department of Genetics and Biometry, University College London, London, UK

Jim Brown Department of Biology, University of New Mexico, Albuquerque, New Mexico, USA

Lars Brundin Sektionen för Entomologi Naturhistoriska, Riksmuseet, Stockholm, Sweden

Robin Craw Entomology Division, DSIR, Auckland, New Zealand

Paul Giller Department of Zoology, University College, Cork, Ireland

Chris Humphries Department of Botany, British Museum, (Natural History), London, UK

Pauline Ladiges Botany School, University of Melbourne, Parkville, Victoria, Australia

John Lynch School of Life Sciences, The University of Nebraska, Lincoln, Nebraska, USA

Jack Major Department of Botany, University of California, Davis, California, USA

Larry Marshall Institute of Human Origins, Berkeley, California, USA

Alan Myers Department of Zoology, University College, Cork, Ireland

Peter Parsons Department of Genetics and Human Variation, La Trobe University, Bundoora, Victoria, Australia

Marco Roos Vakgroep Bijzondere Plantkunde, Rijksuniversiteit, Utrecht, The Netherlands

Brian Rosen Department of Palaeontology, British Museum, (Natural History), London, UK

Amy Schoener Institute of Environmental Studies, University of Washington, Seattle, Washington, USA

Thomas Schoener Department of Zoology, University of California, Davis, California, USA

Mark Williamson Department of Biology, University of York, York, UK

Rino Zandee Institute of Theoretical Biology, University of Leiden, The Netherlands

Preface

Biogeography may be defined simply as the study of the geographical distribution of organisms, but this simple definition hides the great complexity of the subject. Biogeography transcends classical subject areas and involves a range of scientific disciplines that includes geography, geology and biology. Not surprisingly, therefore, it means rather different things to different people. Historically, the study of biogeography has been concentrated into compartments at separate points along a spatio-temporal gradient. At one end of the gradient, ecological biogeography is concerned with ecological processes occurring over short temporal and small spatial scales, whilst at the other end, historical biogeography is concerned with evolutionary processes over millions of years on a large, often global scale. Between these end points lies a third major compartment concerned with the profound effects of Pleistocene glaciations and how these have affected the distribution of recent organisms. Within each of these compartments along the scale gradient, a large number of theories, hypotheses and models have been proposed in an attempt to explain the present and past biotic distribution patterns. To a large extent, these compartments of the subject have been non-interactive, which is understandable from the different interests and backgrounds of the various researchers. Nevertheless, the distributions of organisms across the globe cannot be fully understood without a knowledge of the full spectrum of ecological and historical processes.

There are no degrees in biogeography and today's biogeographers are primarily born out of some other discipline. As a subject, biogeography is not generally introduced to students until third level, and even then it is rarely taught in its own right but included as a component of a 'unit' within classical subject areas. This is largely reflected in the increasing number of books that have appeared in recent years under the title of *Biogeography*. They range from biological or ecological primers for geographers, through limited scope works on ecology, dispersal biogeography or vicariance biogeography to (a very few) well structured reviews of the field.

The present volume originated from our own feelings of despair at the lack of available texts which attempted to draw attention to major areas of controversy, highlight areas of possible agreement and synthesis and integrate, in an unbiased way, the multitudinous branches of the subject. Each research area has something to learn from another, but to suggest that integration of the various schools of biogeographical research is as yet attainable would be highly optimistic. Integration is however clearly desirable if we are to conceptualize the joint roles of ecological and historical processes in the formation of biogeographic patterns. To this end, we have brought together researchers in ecological, post-Pleistocene and historical biogeography from the fields of botany, ecology, genetics, geology, palaeontology and zoology to provide a springboard towards such integration.

Ecological biogeographers often profess that they do not understand the various divisions and controversies amongst historical biogeographers, whilst the latter often refute the significance of ecology to historical biogeography. To overcome such problems, the brief we set each contributor was to lay out clearly, and without confusing jargon, the principles and visions of their own research school. Whilst we have taken pains to eliminate infighting between competing schools and accentuate integration, the reader will become aware that this has not occurred without dissent. Where entrenched views are held, we have asked contributors to state clearly their own views without attempting to demonstrate the superiority of these by the destruction of competing views. Readers can then evaluate the competing schools themselves.

Inevitably, with a subject so diverse, the arrangement of sections and selection of chapters may not be to the liking of all readers. Further chapters could usefully have been added but would have made the volume unacceptably large. Within the constraints of size, we have tried to assemble a balanced review of the various current fields of biogeographic research. The volume has been divided into a number of sections. Section 1 is an overview, setting out the perspectives of the subject of biogeography. Section 2 examines patterns, since it is the documentation of organism distribution patterns, which is the starting point of all analytical biogeography. Section 3 deals with processes, an understanding of which is the goal of the science of biogeography, and Section 4 looks at the methods of reconstruction of the events during the history of life, which have led to present day and past distribution patterns. We have attempted to provide a balanced background to the subject area at the start of each section and also to highlight the main areas of controversy.

We hope that the volume will be a vehicle for promoting synthesis and a basis for future integration across the broad field of biogeography. We wish to record our thanks to all the authors and acknowledge their willingness to conform to the overall plan envisaged by us for the work, despite the constraints which this placed upon them. We are grateful to them for their cheerful acceptance of the various changes to their texts, which we requested for the sake of conformity. We thank them also for the various constructive comments which they made to improve the contributions and the overall structure of the volume. Finally, we thank Alan Crowden, formerly of Chapman and Hall for his continuous support and encouragement throughout the project.

ALAN MYERS and PAUL GILLER
Cork, 1988

Preface to the paperback edition

Two years have elapsed since this book first appeared in hardback, and it has, during this period, received many positive and favourable reviews. It is hoped that publication of this paperback edition will help the book to reach an even wider audience, particularly final year undergraduate and post-graduate students. The opportunity has been taken to make a few minor corrections but otherwise this edition is unchanged from the original.

ALAN MYERS and PAUL GILLER
Cork, 1990

PART I

Biogeographic perspectives

CHAPTER 1

Process, pattern and scale in biogeography

A. A. MYERS and P. S. GILLER

1.1 INTRODUCTION

Patterns of distribution of organisms over the surface of the globe can be demonstrated to be non-random. Once defined, this non-randomicity requires explanation in terms of process or processes, before a reconstruction of the events of biotic history that led to present day species distributions can be achieved. Biogeography has evolved along the two avenues of pattern definition and process identification, principally from systematics, ecology and palaeontology. Biogeography is not, however, a unified subject and this is due in part to these diverse origins. Publications in biogeography occur in rather different journals, each with its own clientele, further aggravating the lack of overlap between disciplines. Biogeographers will always have different interests born out of their different backgrounds. They include botanists, ecologists, geneticists, geographers, geologists, palaeontologists, taxonomists and zoologists. It is not surprising, therefore, that researchers in these diverse fields tend often to talk past, rather than to each other. This interdisciplinary nature has led to fragmentation rather than cooperation which has been to the detriment of the subject as a whole. To progress, biogeography must attempt to integrate these divergent interests and determine how speciation, adaptation, extinction and ecological processes interact with one another and with geology and climate to produce distributional patterns in the world's biota through time.

The origins of biogeography lie in systematics and have been attributed to Buffon (1761) who stated that the Old World and New World had no mammal species in common. This led to the framing of Buffon's Law which postulated that different regions of the globe, though sharing the same conditions, were inhabited by different species of animal and plant. Linnaeus, the founder of modern systematics, postulated (Linnaeus, 1781) that the world's biota originated from a single land mass, 'paradise', the only land not immersed by the sea, and then spread out to colonize other areas far and near as more primordial land emerged from the sea. This 'centre of origin/dispersal' hypothesis still forms a basic tenet, albeit in modified form, of much biogeographic narrative. Later Candolle (1820) distinguished between historical biogeography (his botanical geography) and ecological biogeography (his botanical topography). These twin spearheads of biogeography have, to a large extent, progressed along independent often diverging pathways through to modern times. The difference between them is largely one of scale based on temporal (short-term to evolutionary), spatial (local to global) and taxonomic (species populations to higher taxa) differences. We will return to this on p. 10.

1.2 PROCESSES

The processes implicated in forming biogeographic patterns, whether abiotic or biotic, may not necessarily be specifically biogeographic, and indeed it is questionable whether biogeographic processes *per se* exist. Large scale abiotic processes include tectonic movements of plates, eustatic changes in sea level and changes in climate and oceanic circulation. These processes will almost always operate in concert, since climate will be affected by movements of continents and changes in ocean circulation; climate will influence eustasy through glacial-interglacial periodicity; tectonic movements will affect ocean currents and so on. On a more local scale, abiotic disturbances such as fire, hurricanes, floods and volcanic eruptions also influence distribution patterns in many types of habitat. The biotic processes are both evolutionary (including adaptation, speciation and extinction) and ecological (including varied biotic interactions such as predation and competition and 'dispersal'). Indeed 'dispersal' might have a claim to being a truly biogeographic process although establishment of a population seeded by propagules after a dispersal event is clearly an ecological process. The term 'dispersal' (*sensu lato*) is retained throughout to cover all types of geographical translocation by individuals or taxa. Dispersal, as used here, incorporates several distinct biogeographic processes, which have been surrounded by much semantics and confusion. They require precise definition at the outset to avoid the misunderstanding that the confused terminology has engendered in the past.

Movements away from parents are a normal part of the life cycle of all organisms and theoretically the need to avoid competition among relatives can maintain a dispersive phase even in a homogeneous environment (Hamilton and May, 1977). Many writers have distinguished this type of 'dispersal' from that which involves the movement of propagules across uninhabitable territory (i.e. barrier crossing). Platnick (1976) has used the term 'dispersion' for the former and 'dispersal' for the latter. However Schaffer (1977) has pointed out that 'dispersion' is regularly used by ecologists to mean 'pattern and location of items that have dispersed' and has suggested the alternative term 'range expansion'. Pielou (1979a) has used the term diffusion for the process of range expansion and 'jump dispersal' for barrier crossing or chance dispersal. Baker (1978) makes a distinction between 'non-accidental' and 'accidental' dispersal. Whatever the semantics, all these writers have clearly identified two distinct processes. The terms 'range expansion' (natural movements away from parents) and 'jump dispersal' (barrier crossing) are the most unambiguous and are adopted here. A third process, 'range shift' can be recognized: this results from range expansion coupled with differential extinction in parts of an organism's former range (Fig. 1.1). Range expansion and jump dispersal are opposite probabilistic conditions of a tendency for young organisms to move away from parents. Most will settle within the existing range, a few may jump beyond it. Range shift, by contrast, is not probabilistic but is mostly the consequence both of barriers (biotic or abiotic) and environmental conditions changing through time. 'Range contraction' may also occur and leads to isolation (refugia) and ultimately to extinction.

Two revolutionary process-oriented theories have had major implications for biogeography. On the one hand, MacArthur and Wilson's (1963) equilibrium theory of insular zoogeography has strongly coloured approaches to ecological biogeography, putting forward a framework within which ecological processes and biogeographical patterns (see for example the collection of papers in *Oikos*, **31**, (1983)) and ecological theories and palaeontological data (Gould, 1981; Hoffman, 1985a) could be merged. On the other hand, the emerging science of plate tectonics, which arose from the works of Snider-Pellegrini (1858) but more influentially Wegener (1929), has had important consequences for historical biogeography. The recognition of an unstable earth has stimulated hypotheses that biota have been split (vicariated) into isolated populations by the production of barriers to dispersal resulting from tectonic activity, principally plate movements and associated orogeny. Vicariance, of course, is not only a result of plate movements, since eustatic changes in sea level, shifting climatic belts and zones, and ecological processes can also vicariate populations. The influence of many of these different processes forms the basis of any analysis of patterns and is discussed further in the following section.

1.3 PATTERN ANALYSIS

The determination of species distribution patterns is a starting point for all biogeographic analysis. Such patterns are amenable to analysis without any assumptions of underlying processes, or they can be used

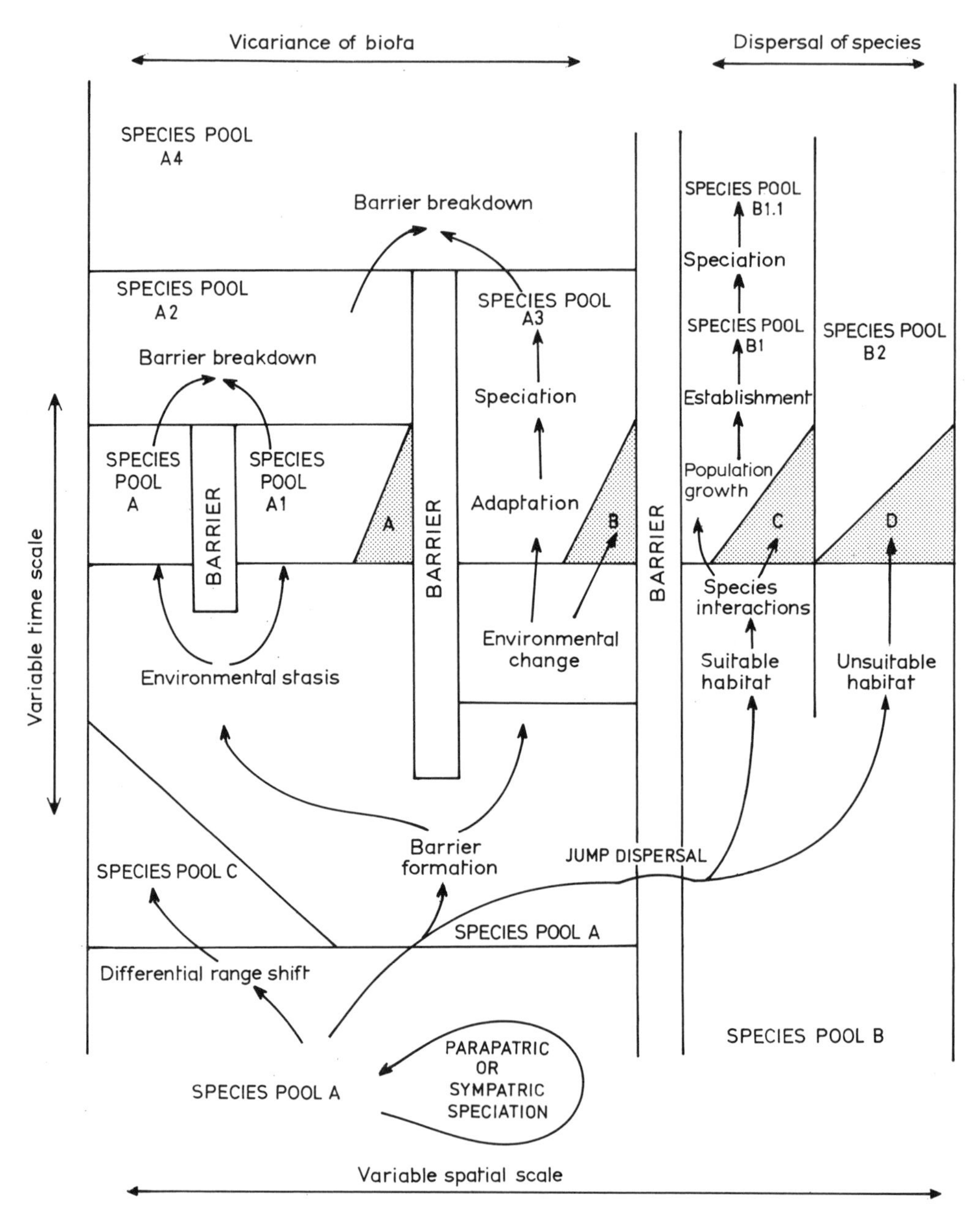

Figure 1.1 Possible processes leading to change in distribution patterns. Shaded areas = extinction. A = Extinction due to non-viable population size; B = Extinction through failure to adapt; C= Extinction through failure to establish; D = Extinction due to unsuitable habitat.

to test (in the sense of falsify) hypotheses of process. Both historical and ecological biogeographers have used observed patterns of organism distribution indirectly to test hypothesized processes (hypothetico-deduction). In addition, both have employed hypothetico-inductive methods, inferring processes from empirical data (species distribution); these include earth processes like vicariance and biotic processes such as dispersal, species interactions and disturbance events. Only ecological biogeographers have the option of testing the potential role and importance of some of these processes experimentally (Chapter 15).

Because the twin spearheads of ecological and historical biogeography have progressed largely independently, we shall examine these two approaches separately and then see how placement on scale gradients of space and time can lead to some reconciliation between the two avenues of biogeographic research.

1.3.1 Historical biogeography

Historical biogeography attempts to reconstruct the sequences of origin, dispersal and extinction of taxa and explain how geological events such as continental drift and Pleistocene glaciations have shaped present day biotic distribution patterns. Historical biogeography has however become split into two camps, dispersal and vicariance biogeography. Dispersal biogeography follows the premise that from a centre of origin, a species undergoes jump dispersal across pre-existing barriers (Fig. 1.1). Dispersal biogeographers are therefore concerned in the main with the dispersal of species, involving movements of the organisms themselves. Present day biota of an area results from the sum of the potential dispersability of individual lineages. Vicariance biogeography by contrast is concerned with discovering the commonality of the observed distribution patterns shown by unrelated taxa which suggests a common and simultaneous process. Hypotheses are based on the assumption that the biota of an area is split up (vicariated) by the emergence of barriers through, for example, continental drift, which separate parts of a once continuous biota and alter the proximity of populations of organisms irrespective of their own movements. Vicariance biogeographic hypotheses are formulated to try to piece together the fragments and allow determination of the sequence of disjunction of the various parts of the once continuous biota and allow inferences to be made of the earth processes that led to its fragmentation. Vicariance biogeography has itself become polarized by the adoption of two philosophically different approaches to biogeographic reconstruction. One approach is based on the philosophy of Croizat (1952, 1958) and is termed Panbiogeography. Croizat formulated a method of reconstructing distribution histories which commenced by drawing lines on a map (tracks) which connected the known distributions of related taxa in different areas. When two or more tracks of unrelated taxa coincided, he combined them into a 'generalized' track. The massed tracks were assumed to indicate the pre-existence of widely distributed ancestral biota which had vicariated as a result of the formation at some time in the past, of a barrier (vicariant event). The plotting of numerous generalized tracks revealed certain areas where many of the tracks converged. Croizat (1952) termed these areas 'gates' and later (Croizat, 1967) 'centres of dispersal'. They are assumed to represent tectonic convergence where fragments of two or more ancestral biotic and geologic worlds meet and interdigitate in space and time (Craw, 1979, 1982).

In the panbiogeographic method phylogenetic relationships of taxa have a subordinate position, indeed, panbiogeographic hypotheses are said to be able to allow inferences of phylogeny to be made. By contrast, the other (cladistic) method of vicariance biogeography is firmly rooted in phylogenetic taxonomy. Careful phylogenetic reconstruction using the cladistic techniques of Hennig (1966) is an essential prerequisite of any biotic reconstruction through cladistic vicariance biogeography. Relationships between areas inhabited by endemic species are determined on the basis of shared derived taxa (equivalent to shared derived character states in taxonomy). Hypotheses are based upon the three taxon-three area paradigm, i.e. given three taxa in three areas the two areas which share the taxa with the more recent common ancestor will be the more recently vicariated. Such hypotheses are based solely on groups with endemic representatives in each of the three areas, since non-endemic taxa have no useful information content in this context. To the panbiogeographer, in contrast, both endemic and non-endemic taxa have information content. Both panbiogeographers and cladistic biogeographers consider that since jump dispersal is a stochastic process, it can only result in a random pattern rather than the non-random patterns documented. They therefore accept only post-barrier dispersal as important. A counter argument is that while dispersal is stochastic, polarized phenomena (such as winds and currents) and differential survival (owing to variable establishment success) may result in a non-random pattern which is indistinguishable in practice from that resulting from vicariance.

These various divisions amongst historical biogeographers have led to cluttering of the literature with confusing jargon, personal attacks and self-serving statements which have done much to undermine the cause of historical biogeography, particularly in the eyes of ecological biogeographers. Arguments have gone back and forth on increasingly narrow issues (McCoy and Heck, 1983). Dispersal and vicariance are not, however, mutually exclusive and the two processes have probably contributed jointly to biogeographic patterns the world over.

Both dispersal and vicariance mediated distribution patterns are clearly modified over time by evolutionary and ecological factors. This was evident more than a century ago when Wallace (1880) suggested ". . . the distribution of the various species and groups of living things over the earths surface and their aggregation in definite assemblages in certain areas is the direct result and outcome of . . . firstly, the constant tendency of all organisms to increase in numbers and to occupy a wider area, and their various powers of dispersion and migration through which when unchecked they are enabled to spread widely over the globe; and secondly, those laws of evolution and extinction which determine the manner in which groups of organisms arise and grow, reach their maximum and then dwindle away . . .". The importance to biogeography of the evolutionary processes of adaptation and speciation driven by environmental change, and of subsequent differential extinction of poorly or non-adapted species, is clear. These processes lead to changes in the species pool, which in turn leads to a change in distribution patterns (Fig. 1.1, species pools A/A3).

The evolutionary history of species is also critical to historical biogeography. Distributional data from clearly defined and taxonomically well researched monophyletic groups (in the Hennigian sense – groups in which the included species have an ancestor common only to themselves), are the building blocks of historical biogeographic analysis. It is an unfortunate fact that well founded phylogenetic classifications are the exception, and debate over phylogenetic methods is rampant. When reviewing distributions of many unrelated taxa, monophyly must often be taken on trust, since the reviewer will never be sufficiently competent in all groups to be able to assess the 'quality' of the phylogenetic classifications used: in many cases the classification will not be phylogenetic at all. When a small taxonomic group is used as a model, the reviewer will usually be an expert in the particular taxon under study and will have developed a satisfactory phylogenetic classification. However, small monophyletic groups may have a unique history.

1.3.2 Ecological biogeography

At a different temporal and spatial level, ecological biogeography, working largely with extant species, tries to account for distribution patterns in terms of the interactions between the organisms and their physical and biotic environment now and in the recent past. On one scale, ecologists attempt to identify those processes that limit the distribution of a species population and maintain species diversity. On a larger scale they seek to explain such phenomena as latitudinal and other species richness gradients, and island colonization patterns. Island biogeography theory has played a large part in the development of some of these ideas over the past two decades. Natural biogeographic experiments have also helped to elucidate the link between process (speciation, extinction and immigration) and pattern (species richness and equilibrium diversity based on, for example, area and isolation). Hypotheses formulated from such studies have recently been applied to the analysis of mainland communities and habitat islands (e.g. mountain tops) and continental patterns (May, 1981; Giller, 1984). The adoption of rigorous statistical tests to differentiate observed patterns from random null hypotheses and the recent popularity of the laboratory and field experimental approaches have both greatly increased our understanding of the relevant ecological processes (Strong *et al.*, 1984b; Diamond and Case, 1986; Gee and Giller, 1987).

From an ecological point of view, dispersal is thought of simply as an adaptive part of the life history of an organism. An understanding of ecological processes following dispersal and vicariance events helps to explain local and intracontinental distribution patterns at the species level (Fig. 1.1). For example, colonization and establishment, following jump dispersal, can only occur provided suitable habitats are found across the barrier by a minimum sized propagule and providing that interspecific interactions do not drive the colonist to extinction (e.g. through predation or competition). Vicariance events can also lead to extinction of species, unless a viable population size exists in the vicariated biota, and the species richness is not too large in relation to the area of the vicariated habitat. Ecological interactions are of paramount importance in the formation of a biota following the breakdown of a barrier and the merging of two previously separated biota (Fig. 1.1, species pools A2, A4).

Ecological phenomena can also explain the great parallels between community structure of different continental areas of similar climate and topography. They cannot, however, explain the great difference in the taxonomic composition which produces such structure nor intercontinental distribution patterns at higher taxonomic levels.

1.4 SCALE

On the basis of the foregoing discussion, the historical and ecological approaches to biogeography might be viewed as incompatible, due to differences in scale. Firstly, on a temporal scale, i.e. short-term ecological periods (over tens of years) studying extant or recently extant species, to long term evolutionary periods limited only by the origination time of the group(s) under study. Secondly, on a spatial scale, i.e. from local (within habitat) intracontinental patterns, to global intercontinental patterns and thirdly, on a taxonomic scale, i.e. from distributions of species populations to those of species and higher taxa. Despite these scale differences, Vuilleumier and Simberloff (1980) have rightly drawn attention to the fact that the distinction between ecological and dispersal biogeography is hazy, since both use principles of ecology, and dispersal forms a very important aspect of island biogeography theory. In addition, both dispersal and ecological approaches explain distribution in terms of process and some processes can be investigated experimentally. Vicariance biogeographers are however concerned principally, though not exclusively, with pattern, since past processes can never be known, only hypothesized. However, the role of extant ecological processes in the formation of these patterns following vicariance events should not be underestimated. The joint roles of ecological and historical processes in determining present and past distributions of taxa have been acknowledged in hypotheses formulated by researchers attempting to explain modern distributions with reference to the Pleistocene glaciations. The effects of these glaciations were probably felt worldwide through changes in climate and eustatic changes in sea level. Glaciations have been implicated in creating refugia (and hence probably accelerating speciation, leading to increased diversity) and have important consequences for island biogeography in relation to effects on dispersal potential and species richness patterns as postulated by island biogeography theory. Because of the recency of glaciations, post-Pleistocene biogeographers tend, perhaps more than most others, to combine elements of ecological biogeography (resource availability, competition, population dynamics, predator–prey interactions and species–area relations) and historical biogeography (climate, inter-island distance, donor-recipient relations, vicariance and refugia) into their hypotheses.

The ultimate goal of biogeography is to explain the distribution of organisms over the surface of the world and hypotheses must therefore be based upon a sound knowledge of animal and plant population distributions. The scale and detail of such distributions used in analyses depends largely on the questions being asked and the nature of the answers sought. Past distributions may be known at a coarse level (from fossils or pollen), but they more often than not have to be inferred by hypotheses of vicariance or dispersal. Historical biogeographers must therefore generally be satisfied with large scale, incomplete distributional data which are nevertheless sufficient for testing historical hypotheses. Finding an organism in, for example, North America, is the important event when studying North Atlantic disjunction. The fact that it occurs in say some, but not all, eastern seaboard regions is largely irrelevant. On the other hand, species distributions are not static; species cross barriers (jump dispersal), colonize new areas from peripheral areas of their distributional range (range expansion) and shift ranges through a combination of range expansion and contraction and differential extinction (range shift) (Fig. 1.1). These dispersal processes and establishment in new areas may largely be governed by ecological processes. Ecological studies of the minutiae of distribution patterns (largely at the population level) are important as a means of understanding the underlying processes causing these changes in species distribution. In such studies, the precise limits to species distributions are required and these can be examined without paying any attention to historical factors. Thus as the spatial and temporal frames of analysis increase, so the delimiting lines of the distribution pattern become more and more blurred in time and space, and the required precision of measurement decreases.

It is these differences in scale of time and space which have largely contributed to the fragmentation of biogeography. To understand why a particular species is found in a specific place one requires knowledge of the species' adaptations and ecological interactions with the environment (i.e. factors limiting distribution). To understand why it occurs in particular areas of oceans or land masses, on the other hand, may require knowledge of recent climatic and geological history of the area. To understand why particular taxa are found only in disjunct areas requires knowledge of the evolutionary history, both of the taxa and the area, coupled ideally with some information on the past distribution of the taxa or their ancestors.

It is clear, therefore, that the distributions of organisms across the globe cannot be fully explained or understood without a knowledge of the full spectrum of ecological and historical processes. A reversal of the hitherto divergent paths taken by biogeographers working on different scales is clearly needed. Integration is desirable if we are to recognize the joint roles of ecological and historical processes in the formation of biogeographic patterns. One step in this direction is for each school of researchers to state their position clearly and positively, to enable members of the other schools to appreciate the value of their work. This would, hopefully, provide a springboard for a more holistic and all-embracing perception and study of biogeography.

PART II

Biogeographic patterns

Introduction

The raw material of biogeography is the distribution of species in space and time. Lists of organisms in different areas must be accumulated before one can proceed to the study of the more intractible problems of how they got there. However, to quote from MacArthur (1972), ''To do science is to search for repeated patterns, not simply to accumulate facts.'' The existence of repeated non-random patterns in species distributions implies the operation of some general causal processes and from an understanding of these one can work towards a reconstruction of the history of life that can explain the present day distribution of species. However, there are serious questions that need to be addressed about the very nature of biogeographical patterns, and there is even ambiguity over the term 'pattern'. An important point raised by several contributors to this book is that success in pattern recognition seems to require that we do not use all the apparently relevant data. Latitudinal diversity gradients (Chapter 3) and species–area relationships (Chapter 4) are examples. How do we decide which data are useful? Any satisfactory explanation of the patterns must encompass both supporting and conflicting data. Rosen (Chapter 2) sets out the philosophical background to these problems with respect to historical patterns in particular, but the ideas are relevant in all areas of biogeography.

PRIMARY PATTERNS

The study of patterns is primarily descriptive and the documentation of patterns is relatively easy. Observations and rigorous quantitive studies are followed by comparisons among features of different natural systems or between natural systems and theoretical models. These can lead to the delineation of species associations in replicated physical environments, along natural environmental gradients or along statistically-derived multifactorial gradients. There may, hopefully, be a relatively limited number of general patterns to be discerned. A major problem in both biogeography and community ecology has been to distinguish determinate patterns from random ones. The patterns may be real but simply a statistical property of the system rather than the end point of the action of some deterministic processes. Were all observed patterns of this random type, there would be no real controlling processes and we would be unable to understand and predict. Acceptance of some pattern-seeking and theoretical studies over the past two decades has certainly been too enthusiastic and uncritical (May, 1984). In reaction to this, it is now becoming fashionable to test for the existence of patterns of both biogeographic distributions and species compositions of communities against null or neutral hypotheses, or models that involve only stochastic processes to account for the observed pattern. When comparative studies of natural systems yield patterns beyond those in the neutral models, one can be more confident of their existence. Brown (Chapter 3) points out that it is relatively easy to document highly non-random geographic patterns in species richness and there are many problems with the use and derivation of the null models currently employed (Southwood, 1987; and see Chapter 9). Nevertheless, the use of such techniques has been valuable in introducing a degree of rigour to pattern analysis which has hitherto been absent, even though they have been largely confined to the analysis of insular distribution patterns (Brown and Gibson, 1983; Strong *et al.*, 1984b).

Spatial distribution patterns of species are observed and documented on a range of scales. Early naturalists discovered distinctly different assemblages of species around the world, and as data were gathered on these patterns, six major biogeographic realms became recognizable. Each realm was separated by a major physical or climatic barrier to dispersal and there was generally a high degree of taxonomic consistency at the familial and generic levels within regions, but marked differences from one region to another. Although there are now disagreements with respect to the exact boundaries, there is broad agreement on the usefulness of recognizing these regions (Pianka, 1983). Finer scale partitioning of the globe into floristic regions or provinces has been based on the distribution of endemic taxa as discussed in Chapter 5. At an even finer level, distribution patterns can be documented by habitat and finally microhabitat, e.g. plant distributions on soil types (Chapter 5). Such patterns are ultimately the result of physical variation in the environment and the species adaptations to it. The adaptations in turn can lead to recognizable geographical patterns in morphological traits, such as those formulated in Bergmanns, Allens and Gloger's rules (Lane and Marshall, 1981). Species can themselves impose patterns on other species, e.g. through species interactions (Chapter 9), modification of the habitat through bioturbation, or space occupancy by sessile or colonial species. If, as is believed by many community ecologists, assembly rules govern the structure of most communities, then these could also lead to recognizable patterns of both composition and functioning (see May, 1986 for a recent review) which in turn will affect the distribution of species amongst different communities. Recognition of such patterns does depend on a stricter framework of definitions for the various levels of species assemblages and for the ecological community itself than has hitherto been the case. Such a framework has recently been proposed by Giller and Gee (1987).

Temporal patterns in the composition of the biota of an area can also be documented, such as patterns of increase in species richness following a major disturbance (Chapter 3) and widespread evolutionary increases in species diversity (Chapter 8). Extinction patterns, such as evolutionary relays (successive, abrupt or gradual replacement of taxa occupying the same adaptive zone) and periodicity of mass extinctions can also be identified (Chapter 8). As mentioned earlier, there is a philosophical problem in defining what constitute historical patterns and Rosen develops the theme of a biogeographical system approach which attempts to identify processes involved in the maintenance, change and origination of species distribution patterns in general, and historical patterns in particular (Chapter 2).

SECONDARY PATTERNS

At a primary level, the patterns outlined above are not of course absolute in the sense that they are only inferred from the collection data of the taxa concerned and from the fact that the distribution of the taxa themselves is not static. This creates its own set of problems and biases (Chapters 8, 11 and 14). At a secondary level, i.e. from the analysis of primary distribution data of many species, the term pattern can be used to describe species richness or nested sets of endemicities over various scales and ultimately the biogeographic realms mentioned above. In Chapter 3, Brown describes the major features of geographic variation in the present day diversity of species and attempts to evaluate the various hypotheses proposed to account for these patterns. Most of the patterns of diversity can be classified as geographic gradients, as the number of species varies relatively continuously with geographic variation in physical features such as latitude, elevation, aridity, salinity and water depth. However, as Brown makes abundantly clear, the natural world is never constant, and any attempt to explain the spatial gradients in diversity must also consider the magnitudes of the most recent disturbances to the system, the time since these events and the dynamics of extinction, colonization and speciation. Brown and Gibson (1983) have pointed out that the history of life has resulted from a series of unique events, so taxa are products of unpredictable histories. Thus both historical events and ecological processes must be part of any complete explanation of patterns of species diversity. In this light, diversity patterns are used in some palaeogeographical reconstruction methodologies (Chapter 14) where interpretation of extant patterns are used to infer historical events from diversity patterns of fossils, e.g. latitudinal patterns can sometimes give distance constraints on hypothesized movement of biotas.

Species diversity patterns are concerned mainly with assemblages of species at a particular point in space, but the distribution of specific taxa provides patterns which also require a causal explanation and can in turn help unravel what has happened in the past. No species of animal or plant is truly cosmopolitan, although some are very widely distributed on a global scale. Such species may be eurytopic with broad local, as well as broad global, distributions, but others may be stenotopic with patchy local distributions within their broad global distributions. Species with restricted ranges are termed endemics, are generally stenotopic with narrow ecological requirements and are often morphologically specialized. Endemics are of two kinds. They may be the result of *in situ* speciation (neoendemics) or may be relicts of species once more widely distributed but which have since become extinct elsewhere (palaeoendemics). Patterns of endemism, like species diversity patterns, are controlled by both historical and ecological processes. In the short term, endemism is largely governed by ecological processes such as food availability, predator–prey interactions and competition, coupled with extant physical factors such as temperature, precipitation and soil type. The primary determinants of endemism are, however, the historical processes of plate tectonics and associated eustatic changes in sea level, and major climatic fluctuations which all isolate or reassort distributions. Clearly, endemism is also a matter of scale and how endemism is explained depends upon the scale applied. If a species is limited in its distribution to two islands, it is endemic to neither but endemic to the two islands together. Endemism can therefore be viewed as a series of nested sets, the final nest incorporating the whole world which has one hundred per cent endemicity. We can study endemism at one level in relation to ecophysiological processes and adaptation as reviewed in Chapters 5 and 6. At a historical level however and particularly in the case of the cladistic biogeographic method, in which only endemic taxa provide information (Chapter 12), the historical relationship of geographic areas is determined largely on the basis of shared endemic taxa or sister taxa.

Rarely is the nature of the endemic taxa considered. Are they old taxa gradually becoming extinct through the effects of changing climatic or competitive conditions (palaeoendemics), or new taxa gradually expanding their range (neoendemics)? The biogeographic implications are quite different. Two or more areas, which share taxa endemic to them alone, are generally assumed to be more closely related (more recently vicariated) to each other than they are to other areas. If the endemic taxa concerned are neoendemics, then the assumption is valid. If the endemics are palaeoendemics, the assumption may not be correct. The taxa may once have been more widely distributed, but through differential extinction may now occur in some, but not all of the areas in which they once occurred. The areas now sharing the endemic taxa will not then necessarily be the most recently vicariated. Insufficient attention has been paid to this problem and progress will be made only through a better understanding of the nature of endemics.

As mentioned above, endemism results from a number of interrelated factors. It has long been recognized that isolation, both temporal (the greater the age of isolation the greater the time for speciation) and spatial (the greater the distance between gene pools the lower the genetic interchange) is important in producing endemics. There is also, as pointed out in Chapter 5, a broad relationship between the number of endemics in an area and the size of the area. This, however, may simply be an expression of the relationship between the number of species in an area and the size of the area (the species–area relationship) which is discussed in Chapter 4. Major (Chapter 5) also notes that endemicity generally follows the same latitudinal trend as species richness and indeed may account, to a greater or lesser extent, for the overall global diversity.

TERTIARY PATTERNS

Species–area relationships can be considered as tertiary level patterns, which describe the relationship between secondary level data, i.e. contemporary species richness, and non-biotic data or size of area harbouring the species. As discussed in Chapter 4, the centre of distribution and the spread around the centre are different for each species investigated and one would, in general, expect larger areas to contain more species than smaller ones. But the species–area relationship is not viewed simply as a reflection of these different distribution patterns, *per se*, but as a real result of some higher level, non-random processes that lead to predictable relationships between species number and area and impinge on the structure of ecological communities. The species–area theory has also been used as a link between ecology and geology (plate tectonics and eustasy), in order to provide an insight into historical patterns of species diversity and extinctions (Chapter 8). For example, in the Permian crisis, half the invertebrate families are thought to have become extinct, most noticeably amongst shallow water marine biotas. At this time the continents coalesced into Pangea, thus decreasing the area of continental shelf through a reduction in the periphery of continents and in sea level (Gould, 1981). The ecological-historical combination is again highlighted and Williamson (Chapter 4) points out that the importance of dispersal, evolution and environmental heterogeneity on the patterns of biogeographic distribution all come together in the species–area effect. The species–area effect is also viewed by Marshall (Chapter 8) as the most important unifying concept in understanding extinction dynamics because it is applicable to and apparent in both living and fossil biotas.

PATTERN-GENERATED HYPOTHESES

Patterns are often explained by reference to inferred processes (constructing logical and realistic scenarios of how patterns and process might be linked – hypothetico-inductive) or by deduction (through the acceptance of an underlying process of a theoretical model as the causal explanation of the pattern – hypothetico-deductive). The problems of these approaches, their relation to model building and testing and the influence of our perception and use of the approaches on our understanding of species distribution patterns is addressed in Chapter 2. Much of the study of biogeographic patterns, especially historical ones, has necessarily been descriptive or narrative and the inherent problems in this scientific approach have been widely discussed. Modelling and hypothetico-deductive approaches are likewise not without their problems, especially when the assumptions take on the role of fact for higher level models without adequate testing and when the patterns derived from the models become transformed into processes. These aspects of modelling are also explored in Chapter 2. To the historical biogeographer, the repetition of similar patterns of distribution by unrelated taxa is highly significant, since it suggests a common process, or processes. The sequence of historical events leading to the observed common pattern may be hypothesized after an examination of the distribution of sister taxa of monophyletic groups, i.e. the areas sharing the most closely related, and hence most recently evolved, taxa, will be the most recently vicariated (separated). The analysis of patterns and the significance of endemism and phylogeny in the reconstruction of historical processes is the subject of Part IV of this volume.

In biogeography, the experimental approach to process derivation is largely limited to island biogeography (Chapter 15) and to the extension of community ecology concepts into an understanding of species distributions on largely local scales (Chapter 9). This can lead to direct testing of the significance of different processes, but we will leave further discussion of this to Part IV of the book.

THE FUTURE

Knowledge of the distributions of most organisms is weak, as is knowledge of the faunas and floras of many regions of the globe. In this respect, current biogeographic hypotheses are built around very incomplete basic data and we are perhaps trying to run before we can walk. Further documentation of distribution patterns is therefore badly needed. At a more inclusive level, patterns of species richness and species-area relationships are relatively well documented, and further documentation of such data is unlikely to add substantially to our understanding of the patterns. On the other hand, identification of incongruent richness and area patterns is likely to be more instructive in that it will aid the development of robust explanations for the more general patterns themselves. Whilst the species richness and species–area phenomena are very important components in a proximate explanation of species distributions, it is doubtful whether any major breakthrough in our understanding of the ultimate (historic) biogeographic processes will result from these research areas.

A better understanding of the nature of endemicity is highly desirable, since endemism is seen as fundamental to the testing of several methods of biogeographic reconstruction. Endemism results from a constellation of different factors and different endemics (palaeoendemics, neoendemics) tell us different stories. In the same way that an understanding of the phylogeny of taxa allows us to perceive the sequence of historical processes, so an understanding of the age of endemics will provide greater precision in historical reconstruction. To date, analysis of endemicity rarely addresses this temporal aspect.

CHAPTER 2

Biogeographic patterns: a perceptual overview

B. R. ROSEN

2.1 INTRODUCTION

To understand biogeographical patterns one must first ask questions about the nature of patterns themselves. This in turn requires a consideration of the aims, approaches and methods of biogeography. The need for such a discussion reflects the fact that biogeography lacks cohesion and consistency (Chapter 1). There is too little sound knowledge of biogeographical processes, reflected by disparity about what can be assumed or should be investigated, a lack of integrated models of the biogeographical system and little recognition of process levels within such a system. These problems have been present in biogeography for a long time but more recently the subject has been shaken by outright controversy, and even unpleasant disagreement and bitterness about approaches and methods (e.g. some of the contributions in the volume edited by Nelson and Rosen, 1981; and see Ferris, 1980 for an interesting summary of this meeting). This controversy may well testify to the healthiness of the subject (Brown and Gibson, 1983), but more promisingly, it may presage major changes:

"The proliferation of competing articulations, the willingness to try anything, the expression of explicit discontent, the recourse to philosophy and to debate over fundamentals, all these are symptoms of a transition from normal to extraordinary research." (Kuhn, 1970, p. 91).

What will emerge in biogeography when the dust settles?

Certainly a proper integration of biogeographical concepts is overdue, and for this reason I make a case here for considering biogeographical processes, as part of a single system, and modelling this system rationally. However, no attempt to clarify issues in biogeography should result merely in redefining its existing schisms or opening the way for new ones. An integrated approach should go beyond a well-meaning amalgamation of different approaches to the development of a balanced programme of assessing all possible biogeographical processes in a critical way.

There is also the matter of deciding what is meant by 'historical', because historical factors are not confined to what is usually covered by 'historical biogeography'. Firstly, 'historical' can refer either to evolutionary processes or to geological processes, or both, while for some biogeographers it has simply meant the documentation of biotas through geological time. Secondly, although most biogeographers see the categories of historical and ecological biogeography as the primary subdivisions of the subject, ecology and history are not themselves mutually exclusive. Thus many palaeobiological authors have attempted to relate the history of distributional changes to changes in ecological conditions. Thirdly, where does 'palaeobiogeography' fit in?

2.2 PATTERNS

2.2.1 Reality and representation

To gauge from Eldredge's (1981) discussion of patterns in biogeography, there is ambiguity in the biogeographical use of the word pattern. Firstly, it seems to refer to what we believe is absolute reality or truth, like an actual sequence of events, a real configuration of geographical features or the true, concrete nature of the earth as it may have been at some particular time in the past. These kinds of patterns evidently exist whether or not we have discovered them yet. Secondly, and much more widely, the word is used to refer to characteristic (often mathematical) features of a set of results, such as the trends and repetitions in a data set, the shapes of graphical plots, or a statistical concept like an average. Patterns in this sense are derivations, analogues, syntheses or generalizations based on initial observations.

When concrete reality (i.e. pattern in the first sense) is not directly observable, we often use the second kind of pattern to infer the first. In historical biogeography, for example, we cannot map continents, oceans and organisms as they were in the past using direct visual evidence. 'Reality' is actually an inference. Although we may believe that we are approaching reality through refinement of methods and accumulated understanding, it is unlikely that we shall ever know what events actually took place in the past. For this reason, I adopt the view that the second kinds of patterns in historical biogeography must be hypothetical representations of what may have happened, though we carry out our work in the belief that some of these derived patterns are also valid statements of what really happened; that is, they also "capture the (real) pattern" (Eldredge, 1981). Throughout the remainder of this chapter, I use 'pattern' in this representational sense.

The problem of having to infer reality from a data pattern applies not only to reconstructing the past. Even on the modern earth, there are distributions like those of deep sea organisms that we cannot observe directly, and where we must again use inference. More importantly, in the context of abstract aims, such as when we test ideas or seek possible correlations between different parameters, we must again use representational patterns. We might, for instance, take (relatively) concrete data like species lists for different localities and look for possible correlations between diversity (species richness) and latitude using graphical plots or tabulations. We would then go on to observe 'patterns' in these results, which are an abstract derivation from the original concrete information. Patterns are 'ways of seeing' raw data.

2.2.2 Patterns as perceptions

The idea that there is a distinction between fundamental truth and our representation or understanding of it, is like the distinction made by many psychologists, epistemologists and neurobiologists between the real world and the way we perceive it (Abercrombie, 1960; Young, 1971; Gregory, 1981). According to this view, perceptions are not necessarily the truth in a fundamental sense because, leaving aside sensory malfunction, we unavoidably make observations in the light of presuppositions and pre-formulated questions, no matter how vague or subconscious these might be. In this way, we (and many other animals) build up a probabilistic model of the world.

Science can be regarded as an elaborate, conscious adaptation of this modelling trait, formalized and made as explicit as possible in terms of aims, approaches, procedures, methods, interpretations, hypotheses and publications. The perceptual process applies to all stages of an investigation, even including our initial observations. Moreover, Gregory (1981) has argued that perceptions are basically hypotheses and that science thus consists of chains of hypotheses. Everything, in short, is unavoidably filtered through the neurological limitations of our brains and the experiences (first or second hand) stored within them. Unfortunately, from a scientific standpoint, this also has the effect that we reject or overlook information which seems at the time to be irrelevant, without even being aware of it. Comfort (1987) has pointed out that even the sense of seriality (which includes 'time') is generated within themselves by organisms, whereas the universe is a space-time entity. This must eventually have serious implications for the meaning of historical patterns.

It follows from this perceptual view of science that biogeographical patterns are not, in the fundamental sense, the 'truth'. To regard patterns as perceptions but also appeal for them to have no bias (Eldredge, 1981) therefore seems to be asking the impossible. We can perhaps aim only to be as aware of, and as explicit as possible about, the influences on all the various stages of a scientific investigation. I should add that it does not follow from the foregoing discussion that I believe that all patterns are mere artifacts without scientific meaning; only that some of the problems in biogeography have probably arisen from misunderstandings about the perceptual, or hypothetical, nature of patterns.

2.2.3 Objectivity

If this view, that the kind of patterns that scientists derive from their data are formalized perceptions, is correct, our observations, results and conclusions do not exist in an objective vacuum. We should expect our patterns to reflect the kinds of questions we ask in the first place as well as the way in which we formulate these questions. Although some of these background ideas may be explicit, others are passed on, often tacitly, through the generations of a particular research school or scientific sub-community, until they eventually stumble on an obvious anomaly (cf. Kuhn (1970) on paradigms). It is easy to find numerous instances of this in biogeography. In Kuhn's view of 'normal science', the dominating background to problem and method formulation in science is a prevailing scientific paradigm, while numerous other authors have identified social and cultural factors that affect the aims, methods and results of science. It follows that in biogeography we should be alert to the hidden implications in choice of method and approach.

The perceptual argument however may seem exaggerated when the application of formalized methods (like statistics) appears to produce patterns that show no bias towards a particular idea. This objectivity is illusory however. Though there are probably many reasons for this, I can only briefly suggest here the following. Firstly, some methods, like the use of averages, can serve so many different purposes, and are so widespread in science and our daily lives, that they do not obviously suggest bias to any one particular question. Built-in bias can then only be revealed by careful analysis of an individual study. Secondly, a pattern, once obtained and communicated, is available for use by others, though they will not necessarily use it to answer the same questions as the original investigator. A particular pattern may even come to be used almost as if it was, itself, a factual observation. The context of the original method, and the perceptual nature of the original pattern, may then be overlooked altogether. Thirdly, in presenting results, an investigator may simply not have explained or considered the relationship between aims, methods and resulting patterns, and a method, on its own, may then seem independent and objective. In any case, the task of spelling out in detail the history, context and theory behind every project in every publication would be overwhelming and repetitive. Instead, scientists resort to shorthand phrases encoded in 'the vocabulary of the consensus' of a particular research school (Ziman, 1968).

The factor of one pattern serving many purposes is particularly relevant to biogeography. What might an 'historical pattern' be when, for example, in a whole series of publications, the same patterns of diversity and geological age in tropical marine biotas have been taken to be factual and interpreted as either historical, or ecological, or both (Newell, 1971; Stehli and Wells, 1971; B. Rosen, 1975; McCoy and Heck, 1976; Valentine, 1984a; Rosen, 1984; McManus, 1985; Potts, 1985)? Osman and Whitlatch (1978) noted a similar situation throughout the literature on spatial and temporal patterns of species diversity.

2.2.4 Summary: a perceptual framework for pattern analysis in biogeography

I take patterns to be perceptual, not factual, and we cannot therefore consider patterns in isolation from background suppositions. 'Historical patterns' must be thought of as shorthand for either (1) patterns which we think are historical regardless of original context, or (2) patterns obtained in the search for historical processes and events regardless of whether they really are historical. Since patterns are derived from methods, and methods are developed to serve particular aims in the context of a particular approach, I have used this chain, in reverse order, to determine the structure of the rest of this chapter and its follow-up in Chapter 14.

I next discuss approaches to biogeography in the context of different ideas about scientific methods, and then attempt to clarify some aspects of aims in biogeography. I use the conclusions about perception, patterns and scientific method to advocate the development of an integrated model of the biogeographical system, and conclude with a survey of the main ideas that might be included in it, and discuss the different levels on which the various processes operate. I use the idea of levels to help define what is 'historical'. In Chapter 14, I discuss the application of such a model to the historical problem of inferring geological events from organic distributions, particularly from fossils, and review relevant methods and the kinds of pattern they generate.

2.3 APPROACHES TO BIOGEOGRAPHY

2.3.1 Explanatory or investigative?

In the course of discussing how more rigour might be brought into historical biogeography, cladistic authors have recognized a number of different approaches to the subject, in particular the 'descriptive', 'narrative', and 'analytical' phases of Ball (1975), and the 'investigative' and 'explanatory' approaches of Patterson (1983). The descriptive approach refers simply to the collection and documentation of distributional data, clearly a necessary task. In the historical biogeographical literature there now seems to be agreement that observations are subject to perceptual filtering (as above), so the main controversial issues lie in other areas.

Investigative biogeography "seeks pattern in order to discover processes", whereas explanatory biogeography "seeks to explain pattern by process" (Patterson, 1983). The problem with explanatory biogeography, as its critics see it, is that it invokes processes, so abdicating the opportunity to investigate processes properly. Invoked processes are assumed from the outset and all observations are woven into narratives that are based on these assumptions. Its critics argue that this approach is therefore inherently incapable of discovering anything fundamentally new. This is said to be the underlying reason for the 'stagnation' which has apparently afflicted systematics and historical biogeography for so long. Furthermore, narrative biogeography invariably accommodates inconvenient new facts and anomalies by *ad hoc* modifications of an earlier narrative, or special pleading. Narratives are therefore also inelegant and non-parsimonious.

Critics also draw attention to two more particular aspects of narrative historical biogeography. Firstly, that much of it has rested heavily on the prior assumptions of dispersal (in a broad but often unspecified sense) and of evolution by natural selection. Secondly, that it has also placed a logically untenable emphasis on fossil evidence (Patterson, 1981a; see also 1981b for relevant discussion of the phylogenetic significance of fossils). I discuss this further in Chapter 14. As it happens, the more general criticisms of the narrative approach stand in their own right regardless of the particular processes invoked. It is also possible to recognize the limitations of fossil evidence and still use a narrative approach.

The investigative approach in systematics and biogeography seems to refer in particular to the use of hypothetico-deductive methods, as emphasized in numerous papers in the journal *Systematic Zoology*, over the last decade or so. Such methods are apparently free of (or attempt to minimize) prior assumptions about process, concentrating only on the search for testable hypotheses and the patterns that might test these hypotheses. Its advocates emphasize their 'Popperian' standpoint, that is, the ideals of formulating hypotheses that are falsifiable, and deliberately setting out to refute, rather than to corroborate, them. Narrative biogeography is dismissed as consisting of 'scenarios' and 'Just-so stories'. In this interpretation, narratives cannot usefully contribute to scientific progress because they are not refutable. Instead, they only generate rhetoric between rival schools, unresolvable differences of opinion and *ad hoc* accretion to existing beliefs. Ultimately this approach is authoritarian.

Central to these criticisms is the issue of induction. Critics say that narratives are inductionist because they are based on uncritical amassing of facts. It follows that in these critics' eyes, induction has nothing to offer scientifically.

2.3.2 Must good biogeography be Popperian?

I agree that narrative science depends on authority rather than testing, and that hypothetico-deductive methods are a basic part of scientific methodology. Moreover, cladistic methods are surely the most successful rational means found so far of organizing and analysing systematic data, and of making systematics explicit. The problem posed by advocates of Popperian principles is not so much what they advocate positively, as what they reject. There are other views about what constitutes induction, and about what induction can achieve, than those just mentioned, and there is also another way of looking at the scientific value of narratives. In general, authentic and successful science seems to proceed on a much broader front than hypothetico-deduction alone. In particular, I ask:

(1) Is hypothetico-deduction really the only way we should consciously proceed in biogeography?
(2) Is it really true that cladistic biogeography makes no assumptions about process, and that it does not engage in 'explanation'?
(3) Is there a valid place for assumptions about processes?
(4) Do narratives really have nothing to contribute to biogeography?

In the comments that follow, I have referred to the Popperian approach largely through the way in which historical biogeographers and cladists have adopted it, and through the discussion of Popper's ideas by a number of his critics.

Hypothetico-deduction and induction

Popper's arguments have been countered by various writers, but on the subject of his rejection of induction, Gregory, (1974, 1981) has provided a particularly clear and stimulating alternative view. This also closely complements Young (1971) and others' neurobiological, homeostatic and psychological concepts of learning. Gregory's conclusions are that deduction works within an inductive framework and that all fundamental discoveries in science are made by induction. Discoveries are perceptual, hypothetical and probabilistic, not objective, and provide the premises for deduction. Gregory rejects Popper's view that the subjectivity of induction invalidates its conclusions, arguing that whatever the demands of pure logic, the reality of learning and understanding in general, and of science in particular, is that it is ultimately inductive. Science works pragmatically on what is probably true, rather than by searching only for what is absolutely true.

Schuh (1981, p. 234), echoing numerous cladistic and vicariance biogeographers, claims that vicariance biogeography represents "the hypothetico-deductive/analytic approach" and says that this "seeks to discover and test empirically some general patterns that may be true for all of life." In cladistic biogeography, these general patterns are tested not by deduction, but by congruence, which surely can be none other than induction by enumeration. What is a 'general pattern' if it is not inductive?

Assumptions, narratives and models

The hypothetico-deductive approach, or at any rate, its adoption by cladistic biogeographers, does not seem to be as free of process assumptions or models as many authors seem to believe. From Gregory's arguments, it is clear that premises in hypothetico-deduction must carry some assumptions of process; Hull (1979, 1980) has actually explored this aspect of cladistic methods. We therefore need to be clear on what is good and bad about assumptions. They are unhelpful when they are inextricably combined with fact and interpretation and not explicit, as in most narrative accounts; but this does not make the assumption of process bad in itself. Assumptions are the basis of modelling, and modelling is a fundamental part of science.

As it would seem to be impossible to collect data and infer patterns without reference to some kind of model, we should acknowledge the value of modelling, though at the same time we must try to make our models internally consistent, modifying or rejecting them as quickly and open-mindedly as possible in the light of further investigation. Some of this investigation may take the form of testing, whether by induction or deduction, and some will consist of explorations and improvements to the theoretical side of the model.

Narratives are very limited in their scientific value and seem to be attempts to model processes in a confusing and non-rigorous way. Serious models, ideas and hypotheses are embedded in much of narrative biogeography and they can be extracted and developed in a more investigative framework. In a perceptual view of science, narratives do not represent non-science, but should be viewed as an inevitable and integral part of science. There is probably a cycle in which more precise models grow out of narratives, and the models then generate hypotheses for testing by both deduction and induction (compare Ball's (1975) 'phases'). Too little knowledge of processes probably leads to an overloading of assumptions. Scenarios and models dominate when there is deadlock in trying to bring coherence to theory, observation, pattern and process – a state of affairs which has certainly afflicted historical biogeography.

Explanations

Even in an investigative context, there is a need for explanation. Investigation and explanation are not in themselves mutually exclusive. It is the way in which explanation is introduced that matters. At some point in any investigation, we usually offer an interpretation. This generally amounts to a re-affirmation or modification of an existing model of the processes and events that may have given rise to our observed patterns. At best this is couched in terms of a testable idea.

2.4 AIMS OF BIOGEOGRAPHY: A QUESTION OF LEVELS

2.4.1 Pure and applied biogeography

Biogeography is 'pure' if it is seeking to discover what causes distributions, and 'applied' when distributional data are used to investigate other processes. (Note that applied in this sense does not necessarily refer just to economic or commercial aspects.) Unfortunately there is risk of circularity between pure and applied aims because the processes in each case, whether real or envisaged, are conceptually interlocked. Thus one author might explain a distribution in terms of dispersal (pure), but another uses dispersal to make geological inferences about ancient land bridges (applied). Until we have a major breakthrough in our knowledge of biogeographical processes, this is bound to remain a recurrent difficulty.

Pure biogeography is concerned with discovering the processes which cause distributions and, to go by other authors' ideas (Valentine and Jablonski, 1982), we might expect these to amount to something more than ecological, evolutionary or geological processes on their own. It may be helpful to conceive of a biogeographical system which could be modelled in various ways, and I explore this further in Section 2.5.

Applied aims usually concern ecology, evolutionary biology and geology. Disregarding the actual validity or success of particular investigations, the following examples convey the nature of applied biogeography. In an ecological context, biogeographers have used geographical patterns of diversity to infer the ecological role of factors like temperature, water depth, primary productivity or levels of radiant energy (Rosen, 1981, 1984; Valentine, 1984b; Chapter 3). In evolutionary biology, biogeographers have been concerned with questions such as whether levels of biotic provinciality influence diversity and extinctions (dynamic palaeogeography). In a geological context, the most common biogeographical interest has been inference of past continental configurations and tectonic events from organic distributions (Chapter 14).

The same problem of assumptions that exists between pure and applied biogeography also exists within applied biogeography. It is difficult to avoid circularity because in investigating one applied area, we seem to have to make major assumptions about the processes in another area, and the same kind of distributional data comes to serve different applied aims by simply switching our assumptions. For instance, we may assume that evolution is in part the result of geological events and use evolutionary (phylogenetic) patterns to infer geological events and processes, or we may assume that geologists' reconstructions of past events are basically correct and use them to infer evolutionary processes.

This second approach ('dynamic palaeogeography') has been reviewed by Jablonski *et al.*, (1985). Its scope includes: changes in provinces through time leading to changes in taxonomic richness, originations of new taxa, extinctions of other taxa and differential survival of clades; and changes in the geographical ranges of species and in their population sizes. Note that two levels of palaeogeographical assumption are used in this kind of study, one general and one particular. The general one is that the major features of the earth's surface are mobile. The second level of assumption concerns actual reconstructions of the earth's features at a particular geological time. There is now little reason to doubt the first assumption but, in the second kind of assumption, what if the reconstructions are wrong? Certainty about them varies according to how far back in time we go, mainly because independent geophysical evidence becomes less reliable.

2.4.2 Ecological and historical biogeography: a question of levels

Numerous authors have stressed the importance of distinguishing between ecological and historical biogeography. This seems to be a straightforward distinction in applied biogeography, where one can define ecological and historical aims according to one's own notion of these subjects. But in pure biogeography, where we do not yet know enough about the causes of distributions, commitment to one or other category effectively pre-selects reasons for distributions. It is remarkable that, notwithstanding this, biogeographers' explanations have usually been couched in terms of one of these categories, with little or no reference to or acceptance of the other possibility. And yet, both ecology and history must have influenced the distribution of organisms as emphasized by Humphries and Parenti's simple point (1986) that the mountain floras on two different continents may resemble each other ecologically in their physical aspect and habitat, but as the species themselves are different on each continent, ecology alone cannot explain this phenomenon.

The persistence of the divide between ecological and historical biogeography probably reflects the predominance of narrative rather than analytical methods in biogeography because narratives allow authors to cast their explanations in terms of rival beliefs rather than rigorous inferences. On the other hand, when analytical methods are used in biogeography, the patterns obtained from them may be neither wholly historical nor wholly ecological, and considerable refinement, testing and reasoning is needed if the effects of these processes are to be distinguished (e.g. see discussions in Humphries and Parenti, 1986; Rosen and Smith, 1988; Chapter 14).

To help distinguish between ecological and historical processes, it is useful to start from a more neutral position with a general model which incorporates both. We can suppose that a biogeographical system exists as an entity, comparable with, say, a physiological system. We do not yet know much about the processes involved, but we can envisage that these processes operate at different levels. I base this idea on Valentine's (1973) use of biological 'levels', Riedl's (1978) ideas about hierarchies in living systems, and on Young's (1971) conception of evolutionary processes as just one extreme of the homeostatic continuum which characterizes living things. Blondel (1986) has also developed this idea of biogeographical levels.

In studying a very different biological system, that of brain function, Rose (1976) also discusses levels of function. He points out the need always to be aware of levels within systems, and to ensure that aims, investigations and interpretations are consistent with respect to a chosen level. Cracraft (1981) has made a similar point with regard to the study of evolutionary processes. In these subjects, failure to recognize levels has led to misunderstandings, non-explanations, and apparent contradictions, all of which have often obscured valid points and masked real progress. This state of affairs is all too familiar in biogeography. A concerted attempt to perceive biogeographical processes as a whole, but at the same time to be aware of levels within the biogeographical investigation when studying a particular problem, would undoubtedly help to clarify and unify biogeography.

Three important examples of level are taxonomic rank, time scales and spatial scales. By analogy with mechanics, in which vectors can be resolved along any axis we choose, we can similarly resolve biogeographical processes on any scale we choose and direct our researches accordingly. Levels are discussed further in Chapter 1 and Section 2.6.

In drawing the traditional distinction between ecological and historical processes, biogeographers have been resolving the continuum of all the various possible processes into two groups, an apparent difference between which is their scale: the scales in time and space of historical processes are usually taken to be greater than those of ecological processes. But 'history', unlike ecology, does not by itself refer to processes, and ecology does not by itself convey any particular time scale. On semantic grounds alone therefore, historical and ecological biogeography cannot be mutually exclusive categories. This is still true even when we consider the actual processes usually regarded as part of historical biogeography, because there is considerable overlap between ecological and historical biogeography when the scales of time and space of their processes are compared (Section 2.6).

2.4.3 Palaeobiogeography and historical biogeography

'Historical biogeography' nearly always refers to:

(1) The study of how organisms have changed their distributions through geological time.
(2) Historical explanations of the distributions of organisms, generally with respect to evolutionary and geological processes; especially dispersal, centres of origin, refuges and vicariance.
(3) The use of organic distributions to infer the distribution of geographical features in the geological past (palaeogeography), and associated geological events.

The first of these aims is essentially descriptive, though the common procedure of mapping past distributions on palaeogeographical reconstructions incorporates a conjectural element which risks circularity with the third aim (Section 14.2.2). The first two aims are 'pure'. Patterson (1983) has discussed them and pointed out the potential conflict between them. The third one is 'applied', and seems to have developed from a very different applied problem, i.e. trying to understand the biogeographical constraints on zonal stratigraphy and stratigraphical correlation. In this field, the ideal of being able to find age-diagnostic fossils that are distributed worldwide can never be realised because no organism occurs in all environments and all regions at any one time. The potential value of being able to use the restricted distribution of organisms to infer palaeoenvironmental conditions, and hence assist with reconstructing earth history, must have been grasped at an early stage. Palaeobiogeography must have been a stratigraphical 'spin-off'.

A study of literature with explicitly palaeobiogeographic titles (Hallam, 1973; Ross, 1976a) shows that there is a very considerable overlap of its aims with those of 'historical biogeography'. So why have there been two subject terms, implying two different sets of biogeographical aims? The short answer is fossils. The palaeobiogeographical equivalent of the second aim of historical biogeography for example, is the explanation of fossil distributions. Historical biogeography is usually preoccupied with explaining present day distributions, and uses fossils only in so far as they are thought to have a bearing on these distributions.

In the broadest sense, of course, palaeobiogeography should just be a category of biogeography in general, or of historical biogeography in particular, and this is how I prefer to see these headings used, but biogeography's customary neocentric or actuocentric emphasis invites use of the prefix 'neo-' to distinguish it from palaeobiogeography (compare palaeontology and neontology). Having said this, I think the traditional pathways followed by neo- and palaeobiogeographers have already diverged far enough in their content and methods, if not in their aims, to the disadvantage of both. They should not be further entrenched by terms that emphasize their differences just when more integration is needed. I therefore make use of the 'neo/palaeo' distinction only in the context of discussing biogeographical ideas and approaches.

From a neocentric biogeographer's point of view, the further back in geological time one goes for evidence, the more tenuous must become its value for understanding present distributions, except in the broadest, descriptive, 'life-through-time' sense. In any case, the quality of distributional evidence also becomes poorer as one takes older geological horizons. Since palaeobiogeographers seek to explain distributions at various given times in the past, their aims have a different emphasis, giving equal importance to fossil distributions, however far back in time their chosen geological horizon lies (though of course an individual Jurassic-orientated palaeobiogeographer, say, will tend to take a Jurassico-centric view, and probably regard the Cambrian as having as little relevance as might a neocentric historical biogeographer).

In the case of the third aim, that of the palaeogeographical use of organic distributions, there is no difference at all between historical biogeography and palaeobiogeography. Rather, differences of approach and method exist, because palaeobiogeography has largely been descriptive and narrative. Palaeontologists themselves have expressed their unease about the methods and conclusions of palaeobiogeography to date (Gray and Boucot, 1979b; Jablonski *et al.*, 1985; Rosen and Smith, 1988). I discuss this further in Chapter 14, in which I also consider the particular difficulties that arise from the nature of the fossil record, and the criticisms that these have generated.

2.5 PURE BIOGEOGRAPHY: THE BIOGEOGRAPHICAL SYSTEM

2.5.1 Models

Our view of biogeography has been sharpened by the critics' appeal for the use of analytical methods. On the other hand, it follows from the patterns-as-perceptions argument that we also work from some kind of model, consciously or otherwise, so we should recognize this by making explicit models of the biogeographical system. We need to do this in order to focus our observations, and to provide a base for formulating hypotheses.

Many biogeographical models lie submerged in the narrative biogeographical literature. More explicit biogeographical models have to date consisted either of relatively informal conceptualizations (Vermeij, 1978; Erwin, 1981; Valentine and Jablonski, 1982; Rosen, 1984), or of more formal and elaborate schemes which address only a part of the biogeographical system. The most stimulating example of a formalized mathematical model of this kind is the equilibrium biogeography of MacArthur and Wilson (1967) but in its initial conception at least, it is largely ecological. (Its evolutionary component has been on selection and adaptation rather than speciation models *per se*.) There are also good examples of formalized biogeographical schemes in the 'systems' approach to ecological biogeography, generated mostly by 'physical' geographers (Furley and Newey, 1983). What have scarcely yet been attempted in biogeography are models that integrate the main three groups of possible processes, maintenance, distributional change and originations (see Sections 2.5.3, 2.5.4 and 2.5.5), and that also take into account the different possible levels within the whole system.

The rest of this section reviews the various implicit and explicit models in the biogeographical literature by bringing them together, firstly according to the kinds of process they emphasize, and secondly by referring them to their appropriate level within the biogeographical system. My aim is to demonstrate the range of ideas about the nature of the biogeographical system as conceived by biogeographers to date, and to use this to show which patterns might be regarded as historical. The complete list of ideas (Table 2.1) should be seen, at this stage, more as an agenda for modelling than as a fully conceived model. Mention of particular ideas is not meant to indicate that I personally prefer or accept them, nor is it a comment on their rigour, testability or general acceptance. I have not attempted to discuss inconsistencies between the various ideas, nor to make more than passing critical comments on them, even though some have received telling criticisms. I have offered critical comments on the methods and patterns used to reconstruct earth history in Chapter 14.

2.5.2 Biogeographical processes

Earlier, I drew a distinction between pure and applied biogeography. The idea of pure biogeography is consistent with the idea of a biogeographical system. Both imply that there are processes that we can regard as distinctly biogeographical, rather than as extended aspects of ecology, evolutionary biology or geology. We might rephrase the pure aims of biogeography, not as the explanation of distributions (Brown and Gibson, 1983), which implies that we already have a good idea what the underlying processes are, but as the search for the causes of distributions, i.e. the 'how, when and why' of distributions (Jablonski *et al.*, 1985).

What kinds of processes might account for distributions? Patterson (1983) identified four kinds of explanations used by biogeographers to date. Two are biological (dispersal, evolution), and two are physical (geographical change, climatic change). All are primarily historical, but the importance of ecological factors is also strongly implied by physical change, and ecology may also play a part in the two biological explanations. The historical alternatives to dispersal, the ideas of vicariance and refuges, are presumably covered by combinations of Patterson's evolutionary and physical headings. Thus, if distinctly biogeographical processes exist at all, then we should be looking for them in the form of characteristic combinations of processes drawn from within the complex of ecology, evolutionary biology and earth science together. For instance, a biogeographical point that has often been neglected, especially in the ecological literature, is that ecological factors alone rarely account for how and where particular species originate.

Even so, biogeography is not characterized by a combination of processes alone. The 'geography' suffix indicates that we must also conceive that biogeographical processes must operate, or have their

Table 2.1. Summary of the main ideas that biogeographers have believed to be important parts of the biogeographical system, grouped into three kinds of explanations for biogeographical distributions. Few of the processes have actually been observed or adequately tested. Some are not testable as currently conceived, and some are mutually exclusive. Note that there are ecological processes listed in the left hand column that are not confined to 'ecological biogeography'. The terms 'contemporaneous' and 'antecedent' are introduced for palaeobiogeographical purposes

Concepts and Processes	*Ecological biogeography*	*Historical biogeography*
	Short-term (contemporaneous) processes	*Long-term (antecedent) processes*
Maintenance		
Dispersion	★	
Geoecology	★	
Equilibrium theory and island biogeography	★	
Barriers	★	
Geoecological assemblages	★	
Provinces and provinciality	★	★
Distributional change		
Range expansion	★	★
Jump dispersal	★	★
Equilibrium theory and island biogeography	★	★
Earth history (TECO events and changing palaeogeography)	★	★
Environmental tracking as a response to:		
ecological change	★	★
the effect of earth history on ecological change	★	★
Adaptation of a taxon to different conditions	★	★
Origination		
Empirical phylogenetic biogeography		★
Vicariance as a response to:		
earth history		★
ecological change or the effect of earth history on ecological change		★
Fringe isolation as a response to:		
ecological change		★
the effect of earth history on ecological change		★
Jump dispersal		★
Centres of origin (in part)		★
Geoecology through time		★

effect, on a geographical scale. This introduces an unavoidable arbitrary element into our concept of biogeographical processes, because there is no obvious fundamental break which separates geographical scale from other spatial scales (Section 2.6; Chapter 1).

It is convenient to make a conceptual separation between the three groups of biogeographical processes already mentioned: maintenance, distributional change, and origination (adapted from Rosen's earlier (1984, 1985) distinction of maintenance, accumulation and speciation, respectively). These can be understood in terms of Young's (1971) homeostatic concept of biological systems as follows.

(1) Maintenance represents the currently operating processes by which an organism maintains its presence in a particular area (or conversely, the currently operating processes that exclude it). Ambient conditions are assumed to be in a steady state and in phase with the organisms' ecophysiology.
(2) Distributional change represents the processes by which an organism survives in response to changing environmental conditions, or by which its distribution changes for whatever reason. Here the ambient conditions in an earlier area occupied by a taxon are presumably out of phase with its ecophysiology, or some event brings the organism into conditions and areas that are new and different. There is of course no prior reason to assume that a species has changed its distribution at all, but biogeographers have generally conceived that this does happen, and the idea must be included in this review.
(3) Origination represents the evolutionary survival of living matter via germ plasm in the form of new taxa when the original species itself did not or could not survive. Conversely, it also covers extinctions – failure of species or lineages to survive. There is in fact no fundamental reason to assume that originations should be related to geography at all, but biogeographers have evidently believed this for a long time. Only with cladistic methods do we seem to have found a method for exploring the idea effectively.

The idea of levels within a system has already been mentioned, and for each of the above three groups of processes there are two levels to consider: taxonomic and scale. Ideas about the origin and distribution of a species, for instance, may be very different from those about a family or phylum, but not necessarily incompatible. As it happens, geographical differentiation between lower level taxa is generally much more pronounced than that between higher level taxa, and in this respect the first is generally more interesting. Not all higher level grouping is formally taxonomic however, there being, for example, an extensive biogeographical literature devoted to deriving and explaining such features as diversity patterns (Chapters 3 and 4) and biogeographical provinces, and these represent statistical groupings of taxa into higher level assemblages.

The two scales to consider in biogeography are those of time and space. Scale can only be handled relatively: it is difficult in our current state of knowledge to give any absolute figures for envisaged processes. Moreover, by extrapolation from Southwood's (1977) discussion of habitats and ecology, absolute values will probably not be the same for all organisms, but are likely to be related to the size, life style and life history patterns of individual taxa. This should be borne in mind in the sections that follow.

2.5.3 Maintenance

Maintenance is basically ecological and the key underlying concept is that of the geographical range of a taxon, or area over which the taxon is routinely found, taking into account all its life history stages. For some organisms like birds, it is usual to distinguish habitat occurrence from areas of migrant visitation, but the overall area is what matters. Although the idea is simple, establishing the precise range of an organism may be difficult, and generally involves some level of approximation or generalization, especially when dealing with fossil distributions.

I use maintenance to refer to those processes by which a species maintains its presence within its range, together with those external processes which effectively determine its overall survival within its range area. It follows that maintenance processes apply to an organism within the time span of its own existence, and that they are therefore relatively short term processes. Ecophysiological responses represent the shortest term factors, followed by environmental events like tides, hurricanes, earthquakes and floods. Slightly longer are lunar, seasonal and annual cycles, human interference and short term climatic fluctuations. All these are readily observable within a human life span, or evidence of them is otherwise available from human historical records. We can therefore discover more about the processes involved than we can about longer term factors. Reproductive and dispersion biology are also an important part of biogeographical maintenance.

Ideas about maintenance processes are also relevant to understanding whether the absence of a species at a particular place is due to a temporary change of conditions (which may also be linked with geological processes), either because the conditions prior to a disturbance event have not yet been restored, or because the prior conditions have returned but the populations have not yet recovered.

Dispersion. This refers to the routine way in which individual members of a species distribute themselves within the current geographical range of its whole population, not to historical events. Distinguishing the two however has been a major problem for biogeographers.

Geoecology consists of (1) the behavioural and ecophysiological processes by which a taxon maintains its presence within its existing area, and (2) the environmental factors that favour its presence, when both are considered on a geographical scale. Obviously there is a feedback relationship between the two. The ecophysiological side of maintenance represents an organism's homeostatic response to the conditions around it, i.e. its means of surviving in the presence of other organisms and in relation to physical and chemical environmental factors (Chapters 6 and 9). Ecophysiology is not of itself biogeographical however, because an organism does not respond to geography *per se*. It acquires a geographical dimension only when related to the geographical limits of a taxon's environmental tolerances.

Environmental processes operate on every spatial scale from microhabitat to global, though it is only the larger scales that are directly relevant to biogeography. There is no obvious absolute threshold beyond which ecological factors can also be called geographical. Even so, the primary ecological interface is between an organism and its immediate surroundings. Spatially, therefore, this part of ecological biogeography is effectively an integration of these ecological processes on a geographical scale. This integration might be by direct areal summation of particular habitats, like rocky shores, forest canopy or reefs, whose separate occurrences are relatively local. Or the features themselves may actually be geographical in scale, i.e. the concept of biomes. On the largest scale are global features like continents and oceans, the tropics, or the poles.

Indeed, latitude itself is often treated biogeographically as if it were a mega-habitat, especially with respect to latitudinal diversity gradients. This also applies to other physical parameters like temperature, rainfall, nutrient levels, altitude, water depth and salinity, or biotic ones such as primary productivity (Chapter 3). It does not follow however that every environmental factor has a geographical dimension, and extension of ecological processes on a geographical scale may in itself be hypothetical. Vermeij (1978), for example, has argued in favour of geographical variations in processes like predation, grazing and competition.

It is useful to distinguish the geographical-scale extension of ecology from the ecological processes themselves, by referring to it as geoecology, especially in the applied context of using biogeography for palaeogeographic purposes (Chapter 14).

Equilibrium theory and island biogeography (MacArthur and Wilson 1967), has already been mentioned as a major influence on ecological, evolutionary and biogeographical thinking. In particular, it offers hypotheses about how organisms colonize and occupy territory, as discussed in Chapter 15, though it seems to be potentially applicable to two distinct biogeographical conditions. In the first, it might explain how one or more species colonize an area previously uninhabited by those particular species, though other species may be there in the first place. In the second, it provides hypotheses about how species repopulate an area recently devastated by events like a hurricane or forest fire. But whereas repopulation takes place within the existing geographical range of the species concerned, and is achieved, presumably, by dispersion processes (maintenance), true colonization takes place in an area not previously occupied by the species concerned, and therefore represents distributional change. In practice, there may be reasons why confident distinction between colonization and repopulation is difficult, but the conceptual distinction is clear.

There are two further biogeographical respects in which equilibrium theory is relevant to maintenance. Firstly, it offers a theoretical basis for the area effect on diversity, although this is not widely accepted (Chapter 4), and secondly, through the idea of saturation, it may explain why species do not always occupy areas that are otherwise apparently suitable for them (Chapter 9). Equilibrium theory has been applied to numerous biogeographical and ecological problems, not just to those relevant to maintenance, and I refer to it again below.

Barriers discussed in the introduction to Part III, are ostensibly an empirical concept denoting the observed geographical limits of a taxon's range. In practice, barriers are really hypotheses because they are mostly interpretations of distributional limits in terms of particular processes. Usually these processes are directly or indirectly ecophysiological, as when barriers are regarded as zones that are unsuitable for inhabitation by particular organisms, or that prevent migrant passage by the dispersal phases of particular organisms. An alternative to such ecophysiological ideas is provided by equilibrium theory through the idea of saturation preventing further colonization of an area by new species.

Barriers, though ecological in their immediate effect on organisms, might in turn be related to major geological features like a mountain range, a hypersaline sea, an isthmus or simply the land/sea interface. This geological link with ecology introduces a long-term element into maintenance processes that is an important part of ideas about distributional change. In this way the barrier concept is also used to reconstruct earth history from the distribution of fossil organisms (Chapter 14).

Geoecological assemblages. In the recognition of specific and repeated ecological assemblages, empirical connection is made between the presence of particular organisms and the habitat they occupy, whether on smaller spatial scales, in biomes, or larger global units. The related idea of provinces (see below) is often defined on the basis of assemblages, but here assemblages are interpreted strictly ecologically. The concept is of great importance in palaeogeography, where more than any other, it is used to infer ancient ecological environments (e.g. shelf facies, lagoonal environments, abyssal conditions), which in turn give clues about past geographical features.

Provinces and provinciality. The idea of biogeographical provinces is one of the oldest in biogeography (Nelson and Platnick, 1981), but it now incorporates several different concepts. Firstly, there are provinces that are based on presence or absence data of taxa alone, either through identifying regions of endemism (characterized by particular taxa that occur nowhere else: see Chapter 5), or statistically, as geographical assemblages. Either way, they can be viewed empirically, or as a response to particular processes. The processes invoked have either been maintenance-based (ecological) or evolutionary and long-term. In fact provinces are not in themselves a process, and do not belong strictly to the concept of maintenance alone. They are based on patterns obtained from various methods interpreted in various ways.

A different view of provinces, though one that is not completely incompatible with the ecological explanation, is to regard them as regions of endemism, whose characteristic taxa are confined to a particular region by the same biogeographical barriers (Chapter 5). Emphasis on endemism however has generally led to historical and evolutionary explanations of provinces, in which ideas about mechanisms of distributional change, geology and ecology of barriers, and allopatric speciation have dominated over pure ecology. A third possible source of confusion is the restricted use of 'provinces' (or 'realms') for latitude-based differences in distributions rather than pure areas of endemism.

Moreover, province concepts have not all been strictly based on presence or absence, there being a long tradition of combining purely distributional data with geographical or geological features to provide convenient lines of demarcation. This is probably due to the fact that the distribution of organisms is usually a continuum or overlapping mosaic, and there is often considerable debate about where to draw the lines when there are conflicting patterns between different biotas. Ways of overcoming this are reviewed in Chapter 14.

Nevertheless, the concept of provinciality (the number of provinces present at any one time) has had a very important theoretical role in explaining the regulation of biotic diversity (and conversely, extinction patterns) through geological time. For a review, see Jablonski *et al.* (1985).

2.5.4 Distributional change

Distributional change refers to all those processes, such as 'dispersal' or refuge theory, by which distributions of organisms are supposed to change through time on a geographical scale (Fig. 1.1), assuming that is, that they change at all. Generally, these amount to one or other, or combinations, of ecological or geological processes. Some origination theories, like those based on jump dispersal, also link distributional change to evolution, and these are treated under originations in the next section. I prefer 'distributional change' as a general term because (1) 'dispersal' has been used to mean so many different things, and (2) it avoids prior commitment to the special cases of expansion or contraction of range.

A methodological and conceptual problem posed by the idea is that of a suitable geographical frame of reference, because we have to bear continental drift in mind. For example, a taxon may have retained its exact geographical distribution within a particular continent, but the continent itself may have moved significantly with respect to other continents in the course of that taxon's existence.

Range expansion conceives that the area in which a taxon is distributed lies within a much broader, continuous area of ecologically favourable conditions. The taxon is conceived as having colonized the potential areas beyond its initial range over a period of time. Colonization might be through an exogenous process, like prevailing winds or marine currents, that transports members of the taxon into the potential areas. Such processes might affect the whole population or just one of its life stages (e.g. larva). Over time, a taxon's range becomes extended in the prevailing direction. Alternatively, colonization might result from endogenous processes, such as its normal means of locomotion. The resulting colonizations might be gradual or phased.

Range expansion is one of several processes invoked by 'dispersalist' biogeographers, especially in the specific case of centres of origin. This has inspired numerous biogeographical maps covered with arrows, especially before the geographical scale of geological change was fully realized. Indeed, on an earth whose continents are conceived as stationary, range expansion models are one of the only ways of explaining distributional change. More recently, it has been common to combine such ideas with palaeogeographical reconstructions, leading to ambiguity as to whether geological events, or range expansion, or both, are supposed to be the agents of distributional change. Sometimes, the palaeogeographical reconstructions used for plotting the 'dispersal' routes have themselves been based partly on these inferred routes, so weakening the conclusions with circularity. Nevertheless, there is a strong empirical relationship between geological longevity and geographical range of taxa (Jablonski *et al.*, 1985), even if range expansion, as usually conceived, is not the real cause.

For these and other reasons, centres of origin and related 'dispersalist' models of distributional change have received a great deal of criticism from vicariance biogeographers. This is discussed in Chapter 1 and Part IV and there is no need to repeat it here.

Jump dispersal is generally conceived as a species level process and is explicitly based on the concept of initial containment by a biogeographical barrier (Chapter 1). One or a few members of a species are envisaged as reaching another environmentally suitable area beyond the barrier through a rare chance event. In fact, some jump dispersal events are thought to be so rare that the colonizing individual(s) remain in isolation long enough to speciate (Section 2.5.5). The process is therefore supposed to operate on all time scales from short-term to historical, though at the shortest end it blurs with dispersion. Whether historical or not, the initial barrier is envisaged as remaining for at least a significant amount of time after the 'dispersal' event.

Jump dispersal has also been invoked to explain the process of biogeographical expansion in connection with the idea of centres of origin, the argument being that the longer a taxon exists in geological time, the greater the cumulative effect of its rare dispersal events.

Equilibrium theory and island biogeography has been mentioned already in the previous section, but is also relevant to distributional change because it provides a model for colonization of new terrain. It can be invoked (1) as an alternative to the other models of distributional change, (2) as an integral part, or (3) as an optional part of some of them.

Earth history includes geological factors in the broadest, earth science sense of tectonic, eustatic, climatic or oceanographic (TECO) events, or a combination of these (Rosen, 1984). However, since biogeography is concerned with distributions, the geological factors include not just the events themselves but also the physical geography of the earth in the past (palaeogeography). Hence palaeogeographical change is widely regarded as a cause of distributional change. For convenience, I use 'earth history' to refer to this combination of events and palaeogeography.

The simplest notion of earth history as a factor in distributional change is an empirical one, based on documenting distributional changes of taxa or assemblages of taxa (including provinces) through geological time. It is here that the question already raised about geographical frames of reference is most relevant. We can either assume, now unacceptably, that the earth has remained unchanged, and record distributional data at chosen stratigraphic horizons on the modern globe, or use palaeogeographic reconstructions for the appropriate stratigraphic horizon. Since there is always uncertainty about the accuracy of a palaeogeographic reconstruction, there is a beguiling objective merit in using the modern configuration of the earth as a biogeographical frame of reference, even for fossil distributions. In fact, the logical conclusion of this argument is initially to disregard the geographical frame of reference altogether and handle distributional data simply in terms of disconnected, contemporaneous sample points (Rosen and Smith 1988; Chapter 14).

It is common practice however to plot past distributions on palaeogeographic reconstructions and then look for empirical evidence of geological cause and distributional effect. From an applied point of view, this raises the difficulty that unless this is done in the particular context of a test or comparison, it prejudges (or obscures) any palaeogeographical conclusions that might be drawn from the distributions.

Two important models exist for distributional change in connection with geological factors in particular (though they might also result from other factors). (1) Divergence is the division of the range into two or more disjunct populations, in which case the new disjunct areas might lie entirely within the boundaries of the preceding one, or outside them. Divergence is an essential part of the vicariance model of originations (Section 2.5.5; Part IV). (2) Convergence (or hybridization or intermixing of previously separate populations or biotas) can take two forms. In the first (true convergence), the biotas of previously separate regions are brought into contact with each other while they are still living (contemporaneously), and in theory at least they can then also intermix. In the second, the biotas are brought together, long after they have become extinct in those regions, through geological events bringing together the crustal terranes that they once occupied. With respect to terrestrial biotas, McKenna (1973) has referred to these two models as 'Noah's Arks' and 'beached Viking funeral ships', respectively, and notwithstanding the choice of metaphors, the concepts also apply to marine biotas.

Environmental tracking as a response to ecological change. Environmental tracking refers to the way in which the overall distribution of a species changes passively in time because of changes in the geographical extent of the ecological conditions which favour it (Chapter 1). In contrast to range expansion, or equilibrium theories, environmental tracking envisages that a taxon is confined to its particular area at a given time by ecological barriers. At least a part of the new suitable area must overlap geographically with the previous one long enough for the species to 'follow' the environmental shift, otherwise, depending on the detailed circumstances, the species would either not change its distribution at all, or become extinct. The idea can be extended to explain how assemblages and provinces might also change their location through time. Time scale is hard to define in that ecological conditions change all the time, some more long-lastingly than others, and often cyclically.

Environmental tracking might result in range expansion (which implies that environmental tracking is yet another possible explanation of centres of origin), contraction of range (cf. the idea of refugia), or just a geographical displacement of a range (i.e. range shift). Boucot (1975) has illustrated some of these various models in a simple graphical fashion, reproduced here in Fig. 2.1. Environmental tracking might also result in divergence or convergence.

Note that with all these ideas of distributional change, there are subtle conceptual differences with respect to taxonomic rank. The distributions of the subcategories within a taxon will often differ from that of the whole taxon which represents a summation of the distributions of its lower categories. Above the species level, the distribution of a taxon may change more through the origination of new sub-taxa, than through a wholesale shift of any of its existing sub-taxa. Thus the disjunct distribution of a supraspecific taxon may consist of different species in different non-adjoining areas (a pattern of greater significance in phylogenetic methods of historical biogeography).

Environmental tracking as a response to the effect of earth history on ecological change. The processes involved in this model are simply a combination of the effects of earth history and ecological change (above) in which TECO events drive the ecological changes which in turn drive the distributional changes. There are four main ways in which this could happen. Firstly TECO events could move areas of the earth in and out of different geoecological zones. Secondly, they could be the direct cause of areal changes of ecological parameters like sea temperatures, deserts, ice sheets or shelf seas. Thirdly they could bring ecologically similar regions together, allowing organisms to swim, float, fly or walk to places that they could not previously reach by dispersion. Fourthly, they could divide or fragment a previously homogeneous region by creating hostile territory (hence barriers) within it – like sea between land masses or vice versa.

As a model, rather than as a method, the attraction of these geological-ecological ideas over the purely geological concept of distributional change (above) is that it offers a plausible causal link between earth

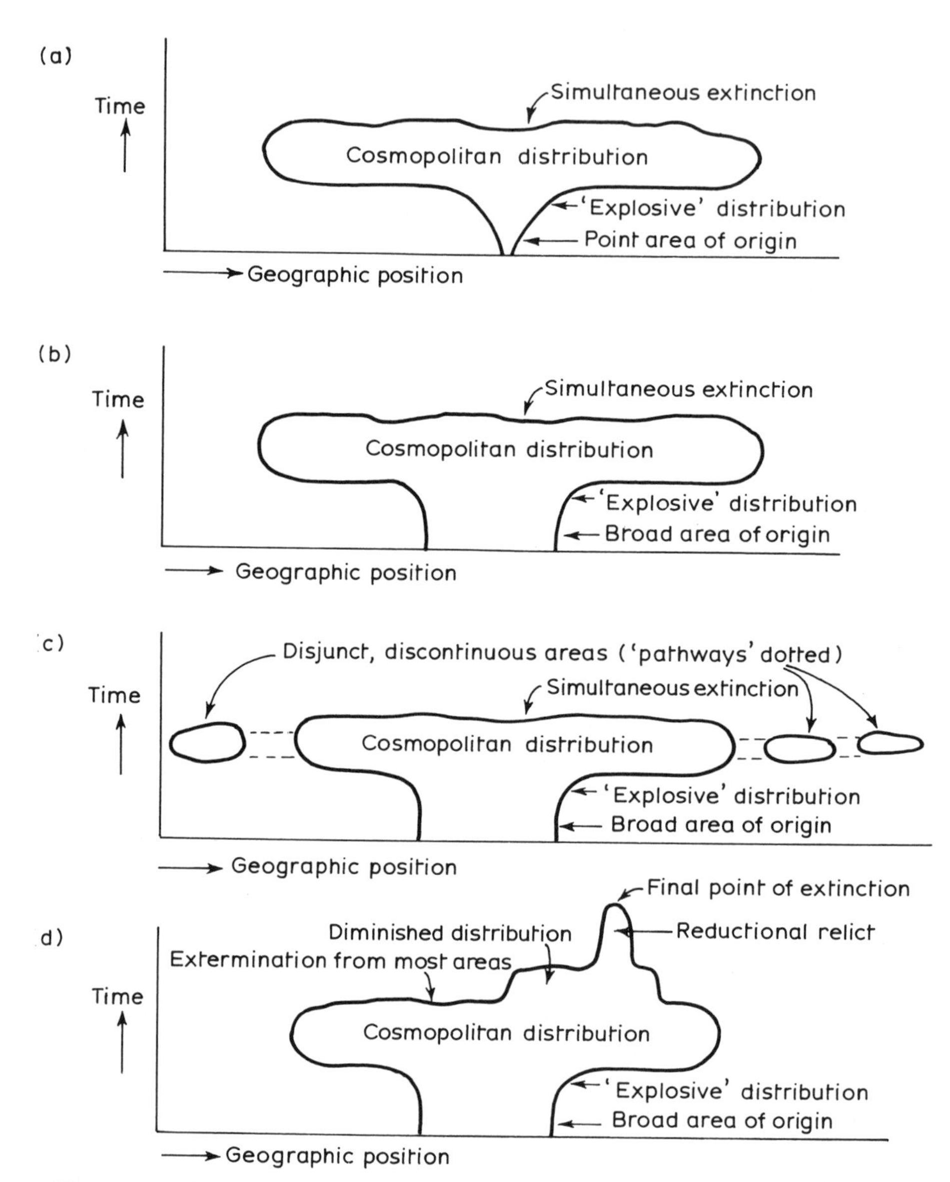

Figure 2.1 Some models of distributional change through time (from Boucot 1975, fig. 42). Note that diagram (c) is a dispersalist model for disjunct distributions, though they might equally arise from vicariance. All four examples of distributional change are models in which areas of occupation either increase, or increase before decreasing. Further possibilities include decrease from an initially broad area without any increase phase, e.g. (d) without the cosmopolitan phase, and wax-wane sequences.

history and organic distributions. The disadvantages are methodological rather than conceptual, because gain in plausibility is offset by loss of empirical simplicity, a view which is held by cladistic biogeographers in particular (Chapter 12).

Adaptation of a taxon to different conditions is another idea which is based on the barrier concept, but envisages that conditions beyond the existing area of occupation are overcome by adaptation of the taxon in question, like a shift in its temperature tolerance (Chapter 6). This is essentially an evolutionary process. Strictly therefore, under the heading of distributional change, only those adaptations that are infraspecific are relevant. In this model, a species is conceived as being able to change its environmental tolerances through time. This is different from the adaptation that can be conceived for supraspecific levels, for which the tolerances of member species need not change through time, but as new species appear, their tolerances may differ from earlier forms. This may result in a net change in the tolerances of the higher taxon. Such supraspecific adaptation obviously links distributional change with originations.

2.5.5 Origination

In a biogeographical context, origination refers to those processes, in which there is also a geographical dimension, that result in the appearance of new taxa. There is no prior reason for causally linking evolution to geography. However, biogeography has long embraced the idea that geographical factors, in the form of ecological or geological processes, or both, are evolutionarily important. In data terms, the geographical component of originations is the time and place of occurrence of new species (and, by extension, new higher-level taxa), a subject which is universally taken to be part of 'historical biogeography'. Extinctions, as the converse of originations, should also be included here, though the exact processes involved may well reflect 'failure' of maintenance factors rather than evolutionary processes *sensu stricto* (Chapter 8).

As with distributional change, the biogeographical concepts that link origination with geography are of two kinds. Firstly there are empirical models by which biogeographers explore their basic data for correlations, using either the geography of the present globe, palaeogeographical reconstructions, or geographically detached sample points. Secondly, there are process models mostly built around allopatric speciation (Chapter 7), and these generally also incorporate one or other of the various models of distributional change. Any other speciation model can only have a much more indirect or obscure relationship with geographical factors.

Empirical phylogenetic biogeography is based on the idea that phylogenetic relationships may be related to geography, either of the present earth, or in the past, or both. In its purest empirical form, this approach is based on cladistic methods, and no assumptions are made as to how or why these relationships might be connected with geographical factors (see Part IV). The emphasis is simply on the search for consistent patterns (c.f. transformed cladistics). This is a derivation from cladistic vicariance biogeography, without the initial proposition of vicariant allopatric speciation. In this respect, it is not a model at all, in any explicit sense, though as explained by Hull (see earlier discussion), there are hidden notions of process even in empirical cladistics.

Vicariance as a response to earth history. Vicariance is now an established concept in historical biogeography, though it is not in fact a recent idea. Originations (allopatric speciation) are envisaged as resulting from division of previously continuous distributional ranges of a species by the appearance of a new barrier, i.e. divergence (as covered above under distributional change). The kinds of event which cause divergences are taken to be geological in the widest sense (TECO events). These events, and the resulting divergences and speciations, are all taken to be temporally interlocked or, if there is a time lag, TECO events must precede divergence, which in turn precedes speciation. In distinction from other modes of speciation there must also be correspondence or direct correlation between the spatial scale of all these events.

There is a tendency to equate the idea of the process of vicariance with the methods of cladistic biogeography and panbiogeography, as well as with the ideals of hypothetico-deductive science. Although there are good historical reasons for this confusion, all three have to be seen independently from each other. Thus, vicariance is often incorporated into narratives and dispersal can be inferred from the methods of cladistic biogeography. As with the pure pattern approach (above), there are degrees of empirical purism in the application of vicariance concepts, but it is difficult to imagine a connection between earth history and vicariance that is not also related to the idea of an ecological barrier, or just possibly a 'barrier' created by equilibrium (saturation) conditions.

Vicariance as a response to ecological change, on its own, or as a result of earth history events. This differs from the preceding idea in explicitly incorporating ecological factors. These models are combinations of ideas about environmental tracking (Section 2.5.4), and vicariance. Following divergence, speciation is taken to occur in at least one of the disjunct populations. This might be due to geological events like continental fragmentation, or to purely ecological events like loss of habitat in an intermediate part of a species' range, or regionally differentiated changes in predation or grazing patterns. In many cases such ecological changes may themselves have resulted from geological events.

Fringe isolation as a response to ecological change is based on the idea that part of a species' population becomes isolated at the edge of its range because conditions are 'marginal' and the population in such areas is 'stressed'. Speciation is thought to result from such circumstances (see Chapters 6 and 7).

Fringe isolation as a response to the effect of earth history on ecological change is similar to the preceding idea except that geological factors are also introduced. Many 'dispersal' models belong here.

Jump dispersal. These models of origination are simply an extension of the idea of distributional change by jump dispersal. With sufficient infrequency, and on a sufficiently long time scale, the process is envisaged as giving founder individuals or populations plenty of time to speciate through isolation.

Centres of origin have been treated so fully in the biogeographical literature that there is little to add here except to make the point that there is not one model but many, as has already been implied by reference to this topic under several different headings. As already mentioned, centres are primarily patterns, not processes. Moreover, as various authors have pointed out, there seem to be several different kinds of pattern that qualify for centres of origin.

Geoecology through time takes ecological control of distributions as its starting point but considers the role of geological factors (TECO events) through time in changing global and regional ecological patterns. This combination of geology and ecology is envisaged as a stimulus for speciations and extinctions, and hence as a diversity regulator. Jablonski *et al.* (1985; see also Chapter 8) give numerous examples. Its time scale is historical, though this kind of biogeography is not usually referred to as 'historical biogeography', probably because of its ecological emphasis. The evolutionary component is general in the sense that it addresses extinctions, diversities and originations of major clades or major ecological assemblages, usually statistically. It is therefore concerned with higher taxonomic levels than vicariance or fringe isolation models and illustrates well the application of the concept of levels in biogeography.

In the first place, geoecology is empirical in its interest in correlations between temporal and spatial taxonomic data with geological conditions and events. In practice, it usually also incorporates a whole range of process assumptions drawn from evolutionary ecology (such as equilibrium, competition, niche theory, resource partitioning). It considers geology (via ecology) as an evolutionary driving force, not so much in the vicariance sense, but as a cause (through natural selection, presumably) of adaptations in organisms or in whole clades. For discussion of the application of equilibrium theory on a historical time scale see, for example, Mark and Flessa (1977, 1978), Cowen and Stockton (1978) and Carr and Kitchell (1980).

2.6 LEVELS, AND THEIR IMPLICATIONS FOR HISTORICAL PATTERNS

At this point, I draw together the discussions of previous sections in order to answer the question, 'what are historical patterns?' Earlier, I concluded that historical patterns were those that have been obtained in the search for historical biogeographical events, whether or not such patterns are actually historical in origin. Conversely, a historical pattern may be any pattern that has an inferred historical content, whether or not the aim of the original investigation concerned was historical.

This left the question, what is meant by 'historical'? In Section 2.4.3, I summarized the three main preoccupations of what is traditionally referred to as historical biogeography: understanding distributions (pure biogeography), and using patterns of distributional change to infer TECO events and palaegeography (part of applied biogeography). The problems of time-scale for this aspect of applied historical biogeography are discussed in Chapter 14. Here I concentrate on the biogeographical system itself.

There is no simple way of establishing an absolute time-scale for defining what is historical in biogeography. For the many biogeographers who have studied distributions in the context of originations, 'historical' implies an evolutionary time scale (whether or not one accepts that originations are actually linked to geographical factors). Some notion of an absolute figure for an evolutionary time-scale can be derived from estimates of species' longevities based on the fossil record. These vary from group to group and range from 1–2 Myr in insects, mammals, trilobites and ammonites to 20–27 Myr in Foraminifera, with an approximate mid-point for all organisms of 7.5 Myr. (Stanley, 1979).

On the other hand, many historical biogeographers are also concerned with distributional change, whether or not this actually occurs in any of the ways that biogeographers envisage, and whether or not there is actually any link with originations. The time-scale of distributional change can obviously be anything from short-term to long-term, where long-term would be comparable to an evolutionary time-scale. Much distributional change conceivably occurs within the time span of a taxon's existence, and some is sufficiently short-term to be observable.

Taxonomic level also has a bearing on what is historical. The biogeographical history of subspecific categories like demes and clines may represent the preliminary phases of allopatric speciation or evolution in action (Chapter 7), and would have to be regarded as short-term if we use the above species-level criterion.

It is therefore difficult to go beyond an intuitive concept of defining 'historical' as broadly long-term, qualifying this where necessary by reference to the three main groups of processes already discussed (Table

2.1), and taxonomic level. Maintenance processes in general however, can reasonably be excluded from historical biogeography, though many of these processes are also incorporated into ideas about long-term distributional change.

Historical biogeography has usually carried with it an implicit notion of large spatial scale, i.e. the geographical scale of regions, large scale geological units, global patterns, etc. Relevant processes are those that (1) act on this scale (e.g. continental drift), or (2) have a possible geographical component (e.g. vicariance), or (3) are the net outcome or integration of smaller scale processes on this geographical, rather than local scale (e.g. the geographical expression of an ecophysiological tolerance to drought or temperature).

A final relevant point arises from the neo-/palaeobiogeographical distinction made earlier. From a neocentric point of view, non-historical processes would largely be those that are happening now and observable, whilst historical processes would be those that happened in the past, or that are acting gradually on the historical time-scale above. Historical processes are mostly not directly observable. From a palaeocentric point of view, all processes that have a bearing on a past distribution have occurred in the past and none can now be observed (except in a general uniformitarian sense). It is useful nonetheless to maintain the neocentric distinction of scales in palaeobiogeographical studies. Thus short-term processes would be those that are contemporaneous with a particular distribution at a particular geological time, and long-term processes would be those that were, in effect, antecedent to that geological time (Table 2.1). It is also important to recognize however that, in a palaeobiogeographical context, the study of short-term factors (like palaeoecology) is broadly 'historical' because the biogeographical events supposedly resulting from them lie in the distant past.

2.7 SUMMARY AND CONCLUSIONS

(1) 'Pattern' is an ambiguous word, but here I regard patterns as representational. They are distinct from the real truth which in many cases might be 'unknowable'.

(2) Science is a formalization of everyday psychological processes by which we understand the world. This understanding consists of chains of perceptions. Perceptual processes affect all stages of an investigation from our observations, through the patterns we find in our results, to the interpretation of these results. Perceptions are broadly the same as hypotheses. At any point along the chain, consciously or otherwise, we develop a particular perception/hypothesis through the filter of other perceptions along the chain.

(3) Perceptions consist of interpretations that range in rigour from

informal narratives through models to rigorous and testable hypotheses. Following Gregory (1981), perceptions are inductive and probabilistic, and ultimately psychological and subjective. Deduction only works within this inductive framework. Gregory rejects Popper's view that this subjectivity invalidates the power of induction, arguing that whatever the demands of pure logic, the reality of learning and understanding in general, and of science in particular, is that it is inductive. Science works pragmatically on what is probably true, rather than by searching only for what is absolutely true.

(4) Thus demands for greater use of hypothetico-deductive methods in biogeography, while justifiable in their own right, emphasize only a part of scientific methodology and misleadingly diminish the value of the rest. Note that although the inductionist approach generally recognizes the importance of hypothetico-deduction, the reverse is not true because of the particular arguments used to underpin the supremacy of hypothetico-deduction.

(5) Nevertheless, whatever one's position on the induction/deduction controversy, there is no doubt that biogeography has suffered from not being explicit or rigorous enough about its assumptions and methods. To complement the efforts of hypothetico-deductivist biogeographers to overcome this defect, I have concentrated here on perceptual aspects of biogeographical methods, advocating in particular a more explicit and integrated approach to modelling the biogeographical system.

(6) The biogeographical system can be thought of as consisting of three groups of processes: maintenance, distributional change and origination, with the different constituent processes acting at different levels (spatial scale, temporal scale, and taxonomic rank).

(7) Adopting a perceptual view, historical patterns in biogeography are therefore those that result from methods used to investigate historical processes, as well as those obtained in other ways but used for this same historical purpose. For the purposes of this broad definition, it does not matter if the method or investigation is temporarily or ultimately unsuccessful in its historical aim.

(8) Thus, for any particular geological horizon, including the present, historical patterns are those that demonstrate (or appear to demonstrate) long-term changes in distributions of organisms, together with those that relate (or appear to relate) such changes to long-term causal mechanisms. Long term processes are primarily evolutionary, but include geological and ecological factors either in their own right or in conjunction with evolutionary processes. Because investigation of short-term processes in a palaeontological

context (e.g. palaeoecology) can also be regarded as 'historical' in a broad sense, there is a need to distinguish antecedent and contemporaneous influences on distributions. Broadly, long-term processes are antecedent with respect to a particular geological horizon, including the present day.

(9) In an applied context, historical patterns are those in which distributions of organisms are used to infer non-biogeographical processes that act on a long-term time scale, these being most commonly geological and geoecological processes, usually in the context of reconstructing geological history (TECO events and palaeogeography).

(10) What remains are patterns that arise from the investigation of short-term processes. For any particular geological horizon, including the present, these demonstrate (or appear to demonstrate) short-term changes in distributions of organisms, or relate (or appear to relate) such changes to short-term causal mechanisms, principally ecological factors. They also include geological factors either in their own right or in conjunction with ecological processes, and in this respect ecological and historical biogeography overlap. Broadly, short-term processes are contemporaneous with respect to a particular geological horizon, including the present day.

ACKNOWLEDGEMENTS

I have been fortunate in being able to take advantage of the wide range of interests and depth of expertise of my colleagues at the British Museum (Natural History), though they do not necessarily agree with the ideas in this chapter. In particular, I thank Peter Forey, Richard Fortey, Nick Goldman, Chris Humphries, Dick Jefferies, Noel Morris, Gordon Paterson, Colin Patterson, Andrew Smith (who also read the manuscript), and Dick Vane-Wright. Jill Darrell bore the brunt of the 'Sturm und Drang' generated by the task, and the editors advised helpfully on the manuscript, as well as tolerating my delays with good nature.

I should like to thank my parents for first stimulating my interest in learning and perception a long time ago, this being the main influence behind the framework for this chapter. I also gained a great deal from the experience and relevant discussions with numerous colleagues during the time I spent helping to develop ideas and prepare exhibition material for the Hall of Human Biology in the British Museum (Natural History).

CHAPTER 3

Species diversity

J. H. BROWN

3.1 INTRODUCTION

Ever since 19th century European naturalists first visited the New and Old World tropics, biogeographers have been intrigued by the enormous latitudinal variation in the richness of plant and animal species. A simple explanation eluded Wallace, Bates, Darwin, Humboldt and the other great naturalists of the last century. In the intervening hundred years, biogeographers, ecologists, and palaeontologists have identified many other spatial and temporal patterns of species diversity, but the latitudinal gradient from the teeming richness of the tropics to the stark barrenness of the poles remains one of the most striking features of the distribution of life on earth – and one of the least well understood.

The goal of the present chapter is to describe the major features of geographic variation in the diversity of species and to evaluate the various hypotheses that have been proposed to account for these patterns. To document the patterns is a relatively simple task. Many of the spatial gradients in species richness have long been apparent to naturalists, although only in the last few decades have they been quantified by rigorous studies. To explain these patterns is another matter. Numerous alternative, but often not mutually exclusive, hypotheses have been proposed. Few of them have been unequivocally supported or rejected, either by logical arguments or empirical data. The relative merits of these hypotheses have been debated by biogeographers for over a century, and they remain one of the greatest sources of controversy and challenge for current practitioners of the discipline.

The failure to explain adequately the variation in numbers of species from place to place over the earth's surface does not reflect the ineptness of biogeographers. After all, the problem stymied Darwin and Wallace, who made great progress in resolving other questions relating to the diversity of living things. Instead, it testifies to the magnitude and complexity of the problem. Biogeography is a discipline that does not stand in isolation, but is closely dependent on related fields, such as ecology, palaeontology, systematics and evolutionary biology. The essential contributions of all these disciplines is nowhere more apparent than in the effort to explain the patterns of species diversity.

3.2 DEFINITION AND MEASUREMENT

Ecologists, even more than biogeographers, have struggled to understand spatial variation in the abundance and distribution of species. In this endeavour ecologists have encountered the two problems of definition and quantification.

3.2.1 Diversity indices

The first problem is how to assess diversity. Not only does the number of species differ between regions, but the relative abundances of the species vary as well. In addition, patterns of diversity have changed over time as discussed by Marshall (Chapter 8). Motivated by the consideration that a biota composed of a few common and many rare species would in some important sense be less diverse than one composed of an equal number of species which were all equally abundant, ecologists have devoted considerable effort to developing various indices of diversity that combine both the number of species and their relative population densities in a single number. The most frequently used are the Shannon–Wiener Index,

$$H' = -\sum_{i=1}^{S} p_i \ln p_i$$

and the Simpson Index,

$$D = 1/\sum_{i=1}^{S} p_i^2$$

in both of which p_i represents the fractional abundance of the *i*th species. The derivation, properties, and uses of these indices are discussed thoroughly in the ecological literature (Peet, 1974; Pielou, 1975; Whittaker, 1977).

Although the utility of diversity indices in ecology can be debated, their value in biogeography is very limited for at least four reasons. Firstly, the accurate data on population densities needed to calculate the indices are not usually available for species on a geographic scale. Secondly, the distribution of abundances within different biotas tends to be quite similar. Although the same specific species–abundance relationship may not characterize all assemblages, virtually all floras and faunas are comprised of a few common species and many rare ones (e.g. Preston, 1962; Williams, 1964; May, 1975). The result is that most indices give values that are very closely correlated with each other and with the total number of species, *S*. Thirdly, although some ecologists are inclined to discount the importance of species of low abundance, rare species usually are as interesting to biogeographers as common ones. Many of the rare forms are highly restricted endemics that may give important clues to the biogeographic history of regions, as discussed in Part IV of this book. Finally, problems of differential preservation of fossil forms make it particularly hazardous to apply these indices to extinct biotas, and especially to compare them to extant faunas and floras. (See Chapter 8 for a discussion of problems in estimating changes of diversity over geological time scales.) For these reasons biogeographers usually are required to quantify diversity simply in terms of species richness, *S*, but it is easy to justify this practice on scientific grounds. Consequently, in the remainder of this chapter I shall make no distinction between species richness and species diversity.

3.2.2 Spatial scale of measurement

The second problem that ecologists have grappled with – the spatial scale on which diversity is measured and compared – must be of equal concern to biogeographers. Ecologists have defined a set of terms to classify this scale: 'alpha diversity' for diversity within a local site or ecological community; 'beta diversity' for spatial turnover in species composition between sites; and 'gamma diversity' for differences in biotic composition between widely separated sites, such as different continents. Despite efforts to make these classifications precise and quantitative (Whittaker, 1977; Wilson and Shmida, 1984), ecologists have not been able to overcome the problems inherent in dividing an apparently continuous range of spatial variation into discrete units. As discussed in Chapter 4, diversity increases with size of the area sampled, and the quantitative nature of this species-area effect is due largely to the nature of the underlying heterogeneity in the physical environment.

In this chapter I shall avoid this ecological terminology, but it must be remembered that all measures of species richness are highly dependent on the spatial scale of sampling. For example, owing largely to the area and distance effects discussed in Chapters 4 and 15, large regions will tend to contain more species than small ones, and nearby areas tend to be more similar in diversity than widely separated ones. Biogeographers must take the complexities of spatial scale into account if their comparisons and the inferences that they draw from them are to be rigorous.

Whenever possible I shall illustrate my points with examples from the work of investigators who have made concerted efforts to sample equal areas with comparable methods.

3.3 THE PATTERNS

More species live in some regions than others. The distribution of organisms over the earth is neither random nor uniform. As pointed out in Chapter 9, it has become fashionable to challenge the existence of patterns and to subject purported patterns to statistical tests against null or random hypotheses. Nevertheless, it is easy to document highly non-random geographic variation in species richness.

3.3.1 Geographic gradients

Four features characterize much of this variation. First, it is closely correlated with variation in physical characteristics of the earth's surface: with such variables as latitude, elevation, aridity, salinity, water depth, and so on. Effects of these physical factors in limiting distributions of organisms is also evidenced in the relationship between adaptations and climatic variation discussed in Chapter 6. Secondly, number of species varies gradually and incrementally with variation in these physical variables, so these patterns are justifiably referred to as gradients of species diversity. Thirdly, these trends are characteristic of nearly all kinds of organisms: plants, animals, and microbes. The few taxa that are exceptional are not sufficient to cast doubt on the generality of the patterns, but they may eventually contribute to an understanding of what causes the patterns. Fourthly, it is easy to show statistically that many of these gradients are highly non-random patterns. Simple regression techniques are sufficient to demonstrate highly significant relationships between number of species and the physical environmental variables. Similar relationships can be found independently in different taxa in the same gradient within the same region and in the same taxa in the same kind of gradient in different regions.

Below, I review some of the best examples of geographic gradients of species diversity.

Latitudinal gradients

The dramatic increase in the diversity of living things from the poles to the equator is perhaps the most striking pattern in biogeography. This biotic variation is closely correlated with an equally pronounced physical gradient in solar radiation, temperature, seasonality, and other factors

Table 3.1. Latitudinal gradients in richness in various taxonomic groups

Taxon	*Region*	*Latitudinal range*	*Range in species (or generic) richness*	*Source*
Land mammals	North America	8–66°N	160–20	Simpson, 1964; Wilson, 1974
Bats	North America	8–66°N	80–1	Simpson, 1964; Wilson, 1974
Breeding land birds	North America	8–66°N	600–50	Cook, 1969; MacArthur, 1972
Reptiles	United States	30–45°N	60–10	Kiester, 1971
Lizards	Northern Hemisphere	20–67°N	20–1	Arnold, 1972
Lizards	United States	30–45°N	15–1	Schall and Pianka, 1978
Snakes	Northern Hemisphere	20–67°N	25–2	Arnold, 1972
Amphibians	United States	30–45°N	40–10	Kiester, 1971
Anurans	Northern Hemisphere	20–67°N	15–1	Arnold, 1972
Marine fishes	California coast	32–42°N	229–119	Horn and Allen, 1978
Papilionid butterflies	New World	0–70°	80–3	Scriber, 1973
Papilionid butterflies	Old World	0–70°	50–4	Scriber, 1973
Sphingid moths	South America	0–50°S	100–2	Schreiber, 1978
Ants	South America	20–55°S	220–2	Darlington, 1965
Calanoid crustacea	Atlantic coast of North America	25–50°N	80–10	Fischer, 1960
Gastropod molluscs	Atlantic coast of North America	25–50°N	300–35	Fischer,1960
Bivalve molluscs	Atlantic coast of North America	25–50°N	200–30	Fischer, 1960
Permian brachiopods	Northern oceans	10–55°N	80–40	Stehli *et al.*, 1969
Hermatypic corals (genera)	Western Pacific	0–30°	45–0	Stehli and Wells, 1971
Hermatypic corals (genera)	Northwestern Atlantic	20–40°N	20–0	Stehli and Wells, 1971
Planktonic foraminifera	World oceans	0–70°	16–2	Stehli, 1968
Trees	Eurasia	45–70°N	12–2	Silvertown, 1985
Orchids	New World (South America)	0–66°S	2500–15	Dressler, 1981

(e.g. Schall and Pianka, 1978). Although the rate of increase in species richness with decreasing latitude varies greatly among different taxonomic groups, the qualitative trend is almost universal for the highest taxonomic categories (Table 3.1). In most groups, it is only at the level of orders and families that conspicuous exceptions begin to appear, and even these are in the great minority. Nevertheless, it is incontrovertible that some groups, such as penguins and sandpipers, are most diverse at very high latitudes, whereas others, such as coniferous trees, ichneumonid wasps (Janzen, 1981), salamanders, and voles (Rose and Birney, 1985) have most of their species in the temperate zones.

Elevational gradients

Just as the change of physical conditions with altitude resembles in many respects the variation with latitude, so the decreasing diversity of most organisms with increasing elevation mirrors in most respects the latitudinal gradient of species richness (Yoda, 1967; Kikkawa and

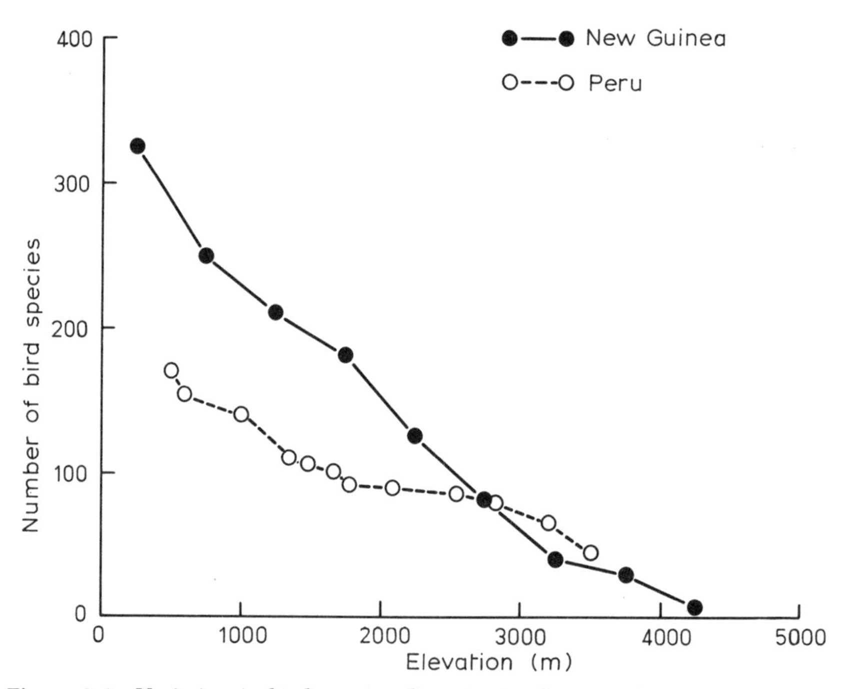

Figure 3.1 Variation in bird species diversity in elevational gradients in New Guinea (from Kikkawa and Williams, 1971) and on the Amazonian slope of the Andes in Peru (from Terborgh, 1977). The data were collected in different ways so they should not be compared quantitatively, but note that both studies show qualitatively similar decreases in diversity with increasing elevation.

Williams, 1971; Terborgh, 1977). Although the elevational diversity gradient is less well documented than the latitudinal one, it is probably just as general and there are some excellent examples (Fig. 3.1). Because relationships between organisms and climate appear to be evolutionarily constrained (Chapter 6), many of the same exceptional taxa (e.g. conifers, voles, and salamanders) that attain their greatest species richness at temperate or arctic latitudes also reach their greatest diversity at intermediate to high elevations on tropical and temperate mountains.

Aridity gradients

Although it is common knowledge that deserts support fewer species than more mesic regions at comparable latitudes and elevations, there

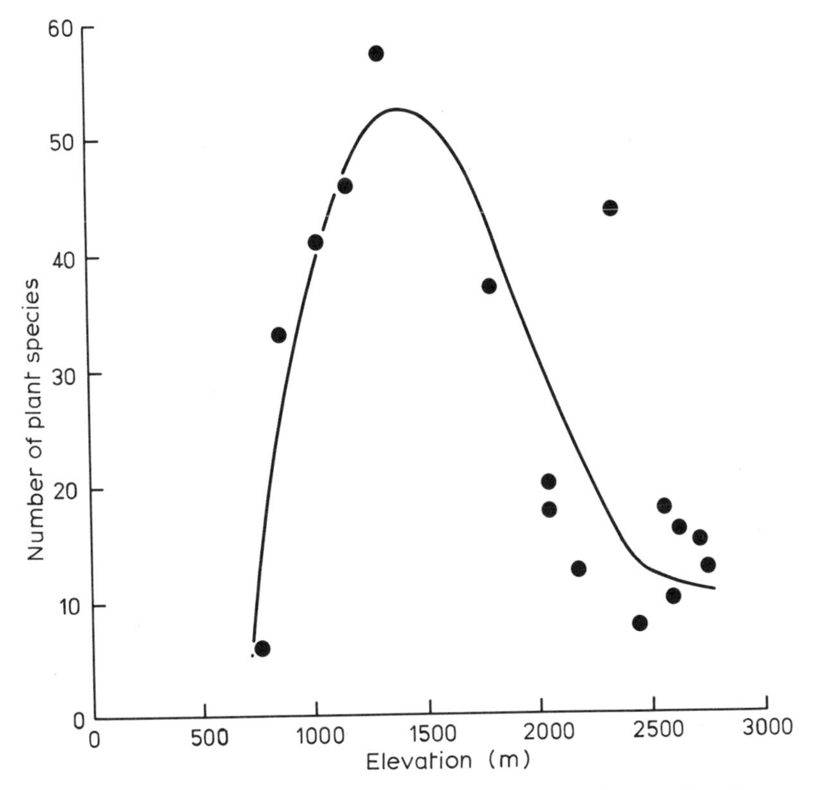

Figure 3.2 Variation in plant species richness as a function of elevation on a desert mountain, the Santa Catalina Mountains of Arizona (plotted from data in Whittaker and Niering, 1975). Note that the highest diversity occurs at intermediate elevations, because of the limiting effect of aridity at low elevations and of temperature at high elevations.

has been little careful quantification of variation in species richness along aridity gradients (but see Brown, 1973; Whittaker and Niering, 1975; Brown and Davidson, 1977). One problem is that variation in moisture availability is often correlated with other factors, such as elevation and proximity to seacoasts, that complicate the interpretation. For example, on the mountains in the south-western United States and other desert regions plant species attain their highest diversity at intermediate elevations (Fig. 3.2). This is commonly interpreted to reflect the antagonistic influences of aridity and elevation (primarily temperature), but it is difficult to isolate the separate effects of these variables.

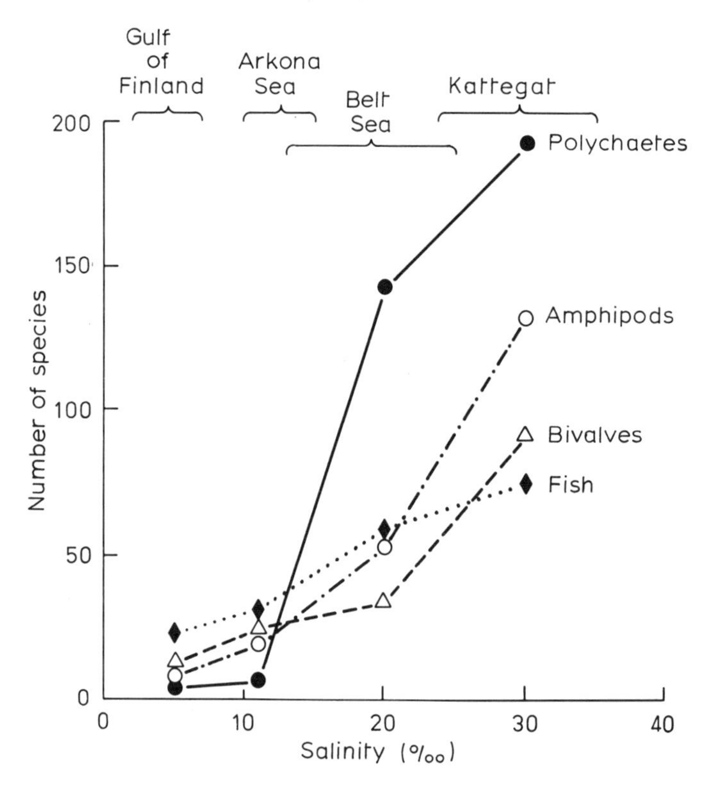

Figure 3.3 Variation in diversity of several kinds of marine organisms as a function of salinity in the Baltic Sea (from data in Segerstrale, 1957). Note that as salinity decreases from normal sea water (35 ‰) at the mouth of the Baltic to almost fresh water in the Gulf of Finland, all groups of marine organisms decrease in species richness, but at different rates.

Salinity gradients

In several coastal regions marine organisms encounter local gradients of decreasing or increasing salinity. Almost invariably, diversity declines as the concentration of dissolved solutes deviates from normal sea water, approximately 35‰ (Fig. 3.3; Segerstrale, 1957; Kinne, 1971). Diversity of freshwater species also declines once concentrations exceed about 2‰. One consequence of these relationships is that estuaries and other brackish habitats where sea and fresh waters meet typically have low species diversity, although these waters may be highly productive and support dense populations of some species (Kinne, 1971).

Depth gradients

In aquatic environments, both marine and fresh water, diversity generally decreases with increasing depth. Because this is also usually a gradient of decreasing temperature, it invites comparisons with elevational gradients on land. However, the comparisons are misleading, because light and seasonality also decrease and pressure increases with water depth. In most bodies of water depth patterns are further complicated by the fact that the greatest diversity occurs not in the shallowest waters, but some distance below the surface (Fig. 3.4; Vinogradova, 1959; Sanders, 1968; Rex, 1981).

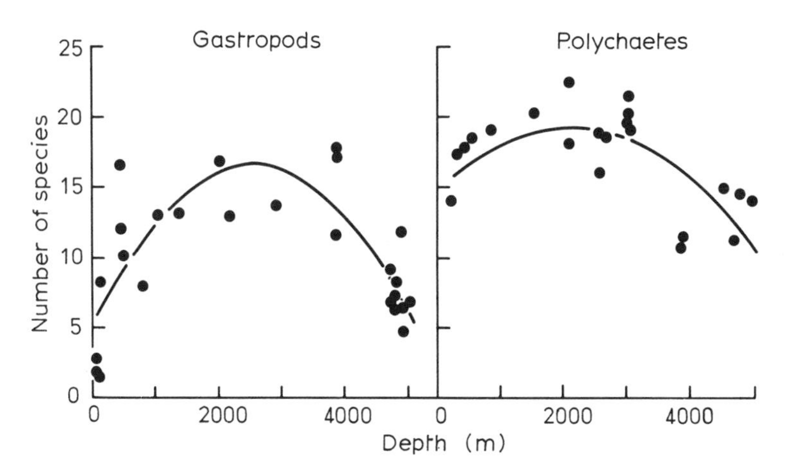

Figure 3.4 Variation in species richness with depth in two groups of benthic marine invertebrates, molluscs and polychaetes (after Rex, 1981). Note the curvilinear pattern in both groups, with low diversity on the continental shelf and abyssal depths and greatest richness at intermediate depths on the continental slope.

Island-mainland gradients

Small, isolated islands almost invariably support fewer species than local sites of comparable size and habitat on nearby mainlands. As many biogeographers and ecologists have noted, this generalization applies not only to true islands of land surrounded by water, but also to almost all small, isolated patches of habitat that are separated from other, larger patches by environments that constitute barriers to dispersal. This pattern is treated in much more detail in Chapters 4 and 15, but it is relevant to any general treatment of species diversity.

Other spatial gradients

There are many other gradients of species richness: with depth in the soil; light intensity in caves; temperature in thermal springs and hydrothermal vents; concentrations of both nutritious and toxic substances in soil, water, and air; frequency or intensity of physical disturbance by storms, fires, and other agents (for additional examples see Sousa, 1984; Begon *et al.*, 1986). Many of these patterns are expressed primarily on small spatial scales, so they have been studied more by ecologists than biogeographers. Nevertheless, since they, too, embody complex ecological and evolutionary relationships between the number of species and spatial variation in the physical environment, they are directly relevant to any attempt to evaluate alternative general explanations for patterns of species diversity.

3.3.2 Temporal patterns

Species diversity varies in time as well as in space. Like the spatial patterns, the temporal variation is expressed over a wide variety of scales. At the risk of oversimplifying, I shall claim that patterns of increase in species richness following a major disturbance tend to be qualitatively similar, independent of spatial and temporal scale. Thus

Figure 3.5 Increases in diversity in empty environments with time following disturbance. (a) Recolonization of small mangrove islands by arthropods following removal of the entire fauna (after Simberloff and Wilson, 1970). (b) Buildup of insect and bird species in successional habitats following clearing of forest and abandonment of cultivated fields (from data in Brown and Southwood, 1983 and Johnston and Odum, 1956, respectively). (c) Evolution of mammalian families during the Cenozoic following the mass extinction of reptile groups (after Lillegraven, 1972). Note that all of these curves have qualitatively similar trajectories, with diversity increasing rapidly and then stabilizing at a level close to or somewhat below the maximum.

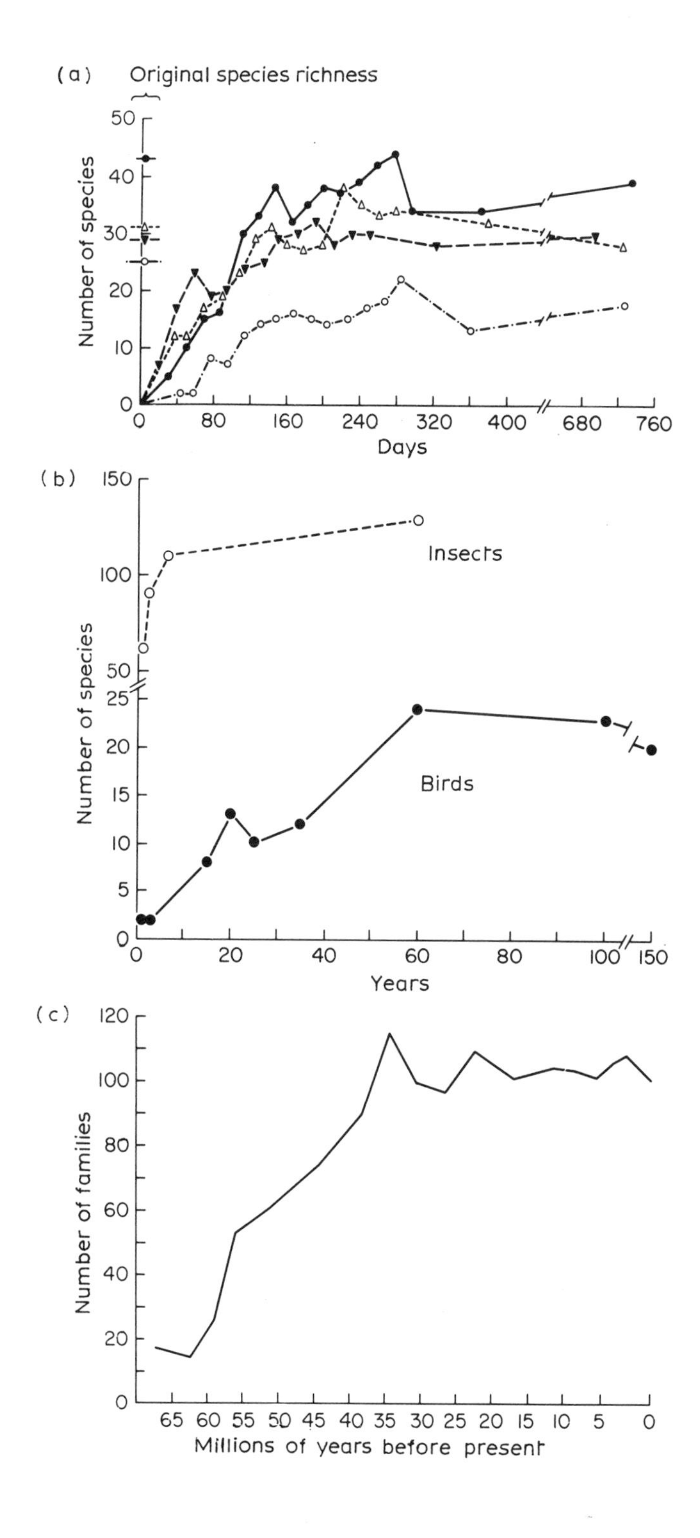

(a)
Original species richness
Number of species
Days
(b)
Insects
Birds
Number of species
Years
(c)
Number of families
Millions of years before present

recolonization of artificially emptied islands (Simberloff and Wilson, 1969, 1970; Chapter 15), ecological succession on sites disturbed by natural agents such as fire, flood, or volcanic eruption (Dammermann, 1948; Johnston and Odum, 1956; Odum, 1969; Bazzaz, 1975; Fridriksson, 1975; Simon and Dauer, 1977; Brown and Southwood, 1983), and adaptive radiations in the fossil record (Lillegraven, 1972; Knoll, 1986; Chapter 8) all suggest that number of species initially increases rapidly and then almost levels off or in some cases decreases (Fig. 3.5).

Whether diversity would continue to increase if the environment were to remain constant is an intriguing theoretical question. In reality, however, the natural world is never really constant; it is continually perturbed by disturbances of varying frequency and magnitude. Examples of catastrophic disruptions in the fossil record are given in Chapter 8, and the ecological literature is filled with examples of more local and transitory disturbances (e.g. Sousa, 1984). Since these perturbations often cause extinctions and hence temporary reductions in diversity, the diversity of any region reflects a balance between the effects of the most recent and severe perturbations and the rate of regeneration of diversity through colonization and speciation. Consequently, any attempt to explain spatial gradients in diversity must also consider the magnitudes of the most recent disturbances, the times since these events, and the dynamics of extinction (Chapter 8), colonization (Chapter 15), and speciation (Chapter 7).

3.4 HYPOTHESES

Numerous hypotheses or theories have been advanced to account for the geographic patterns of species diversity. Some of the most frequently cited ones are outlined briefly below:

3.4.1 Time since perturbation

Low diversity has often been attributed to historical changes on either ecological or geological time scales that have eliminated or at least prevented the establishment of species. Regions may still be 'undersaturated' if there has not been sufficient time for colonization and speciation to produce a rich biota adapted to present conditions. There are many variations on this theme. The latitudinal gradient has long been attributed to the effects of Pleistocene glaciation and climatic change causing wholesale extinctions in polar and temperate latitudes, whereas the tropics were thought to have offered relatively constant environments that permitted the survival of many taxa (Fischer, 1960; Fischer and Arthur, 1977; Stanley, 1979; Chapter 8). Other explanations for latitudinal and other gradients invoke more recent disturbances such as storms, fires, floods, and volcanic eruptions. There are abundant examples that such perturbations operate on ecological scales to reduce diversity by causing local extinctions and initiating or resetting successional processes. Sanders (1968, 1969), put forward the stability-time hypothesis to account for the purported effects of both seasonal and longer-term perturbations in preventing the build-up of as much diversity in temperate waters and on shallow continental shelves as in supposedly more stable tropical latitudes and deeper continental slopes farther offshore (but see the critique of Abele and Walters, 1979).

On the other hand, the exact opposite argument – that disturbance (either physical, such as storms and fires, or biological, such as predation or herbivory) of intermediate magnitude and frequency actually promotes diversity by not allowing sufficient time for competitive exclusion to occur – has been advanced by Connell (1978) and others (Darwin, 1859; Jones, 1933; Paine, 1966; Grubb, 1977; Lubchenco, 1986). For example, to explain the species richness of trees in tropical forests, Connell (1978) and Hubbell (1979, 1980; Hubbell and Foster, 1986) have suggested that fairly frequent natural disturbances, high rates of seed and seedling predation (Janzen, 1970), extremely similar resource requirements, and long lifespans act in concert to prolong resolution of competitive interactions and result in high diversity, non-equilibrium assemblages.

3.4.2 Productivity

Because the greatest species richness typically occurs in highly productive habitats (e.g. tropical rain forests and coral reefs), several investigators have suggested mechanisms by which increased energy availability tends to result in proliferation of different species rather than increased populations of existing species (Hutchinson, 1959; MacArthur, 1965, 1972; Brown, 1973; Brown and Gibson, 1983; Wright, 1983). In general, these explanations all suggest that more productive environments are able to support more specialized species. Because these specialists have smaller populations per unit of productivity, they are subject to high rates of extinction and unable to persist in unproductive environments. Such specialists are also able either to use resources that are not sufficiently abundant to support any species in unproductive habitats, or to use abundant resources more efficiently than generalized species so that many specialists can exclude a few generalists when productivity is sufficiently high. However, any general theory attributing diversity to productivity must account for the well-documented decrease in species richness that accompanies certain increases in productivity, such as those owing to nutrient pollution of marine or freshwater habitats, natural eutrophication of lakes, and fertilization of pastures (Whiteside and Harmsworth, 1967; Rosenzweig, 1971; Tilman, 1982; Abramsky and Rosenzweig, 1983).

3.4.3 Habitat heterogeneity

Correlations between species richness and various measures of physical habitat structure (MacArthur and MacArthur, 1961; MacArthur *et al.*, 1966; Pianka, 1967; Rosenzweig and Winakur, 1969; Cody, 1974, 1975) have stimulated formation of hypotheses to explain relationships between these two variables. In Chapter 4 it is suggested that spatial heterogeneity may largely account for the positive relationship between species richness and area sampled. These correlations imply that physical structure directly facilitates coexistence because the species are able to tolerate different physical conditions, take refuge from their enemies, and reduce competition by exploiting different resources. The ecological literature is filled with examples showing that when diverse species coexist in heterogeneous environments, they often differ from their competitors and avoid their predators by specializing on different components of habitat structure (Begon *et al.*, 1986; Chapter 9). On the other hand, lack of physical habitat structure may not limit the development or maintenance of high diversity on a geographic scale. Hutchinson (1961) coined the term 'paradox of the plankton' to focus attention on the many species of small aquatic organisms that coexist in water masses with seemingly little structural heterogeneity (but see p. 84).

3.4.4 Favourableness

Environments that support diverse species appear to have mild physical conditions that could be tolerated by many species that do not occur there. In contrast, habitats that are depauperate in species often have harsh and/or unpredictable physical conditions that require special adaptations and would be extremely stressful to most species that occur in other regions (Chapter 6). Ecologists and biogeographers have struggled to understand the basis of these relationships and to define 'favourableness' and 'harshness'. Most concepts of favourableness have been somewhat circular: harsh environments are those that cause physiological stress, and conditions that cause stress are those that an organism does not encounter in the environment where it normally occurs. Terborgh (1973) and Brown (1981) have attempted to define favourableness and harshness more rigorously by recognizing that these terms involve implicit interspecific comparisons; i.e. favourableness must be judged not by the response of the native, presumably well-adapted species, but by the ability of alien species to tolerate the physical conditions. Thus, Terborgh pointed out that favourable environments are common and geographically widespread. In a similar vein, I noted that harsh environments are not only physically extreme, but also small, isolated, and ephemeral; not only are rates of colonization low, but also rates of extinction, even for native, well-adapted species, tend to be high. These problems are explored in more detail later.

3.4.5 Niche breadths and interspecific interactions

It has often been suggested that, compared to organisms at higher latitudes, tropical species show the following characteristics: (1) are limited less by physical factors and more by biotic interactions; (2) are affected less by intraspecific and more by interspecific interactions; (3) overlap more in their requirements and compete more intensely; (4) are more specialized and compete less intensively; (5) have such similar requirements and competitive abilities that interactions are resolved only very slowly; (6) are kept in check by such intense predation that they compete less; (7) have fewer, but more specialized predators, parasites, and pathogens; (8) are involved in more obligate mutualistic interactions with other species (Dobzhansky, 1950; Paine, 1966; Janzen, 1970; Connell, 1975; Menge and Sutherland, 1976; Hubbell, 1980; Hubbell and Foster, 1986). These conjectures are extremely difficult to evaluate rigorously, because of problems in measuring niche variables (see discussion of niche concepts, p. 148) and comparing the intensity of interspecific interactions in widely separated regions where the potentially confounding effects of other variables cannot be held constant. For example, temperate species may indeed be able to tolerate a wider range of temperatures or fewer kinds of parasites than their tropical relatives, but the significance of these differences is difficult to evaluate when the temperate forms also encounter more temperature variation and fewer parasite species in their environment.

3.4.6 Origination and extinction rates

It has often been suggested that geographic gradients of diversity can be attributed to differences in colonization, speciation, or extinction rates (Rosenzweig, 1975). For example, many of the recent attempts to explain the high species richness in lowland wet tropical forest have invoked high speciation rates (Janzen, 1967; Haffer, 1969; Prance, 1978, 1982b). Despite the recent popularity of these ideas, there are few direct data to support them. Climatic change (Chapters 6 and 10), habitat fragmentation (Chapter 10), and other perturbations that might influence species dynamics are being increasingly well documented. However, as pointed out in Chapter 8, actual rates of origination and extinction are difficult to measure accurately.

3.5 EVALUATION OF HYPOTHESES

3.5.1 Levels of explanation

Any attempt to evaluate the various hypotheses must recognize that they are not necessarily independent or mutually exclusive of each other. On the contrary, those listed above differ fundamentally in the level at which they attempt to explain observed patterns. Some hypotheses offer primary explanations in the sense that they try to account directly for the effects of the physical variables that characterize the diversity gradients. Other hypotheses provide only secondary or tertiary explanations, because they are couched in terms of biological properties of organisms, such as the breadth of niches, the intensity of interspecific ecological interactions (Chapter 9), or the rates of speciation (Chapter 7). These are indirect responses of organisms to the physical factors that ultimately cause the diversity gradient. This distinction between levels of explanation was made by Pianka (1966; see also Hutchinson, 1959; Brown, 1981), but it has been ignored by many subsequent investigators and much confusion has resulted.

The ultimate cause of all deterministic patterns of species diversity must be variation in the physical environment. To understand why this must be so, ask why there are more species in the tropics than in the temperate zones. The answer must be either by chance alone or because the physical characteristics of the regions differ in the present or have differed in the past. As mentioned above, there are those who would explain the higher diversity in the tropics in terms of specialized species, more intense competition (or predation), or higher speciation rates (or lower extinction rates). These could all be true, but what is it about the tropics in particular that causes these attributes of organisms to be different? If the physical environments have always been identical, then the organisms would differ only by chance. Therefore, the highly non-random differences in the biotas must ultimately be attributed to differences in the past or present physical environment.

This is not to say that niche relationships, interspecific interactions, and speciation and extinction rates are not important parts of the explanation of diversity gradients. Indeed, our knowledge will remain incomplete until we understand the dynamic relationships between the characteristics of the extrinsic physical environment that determine its capacity to support organic diversity and the intrinsic biological processes that generate and maintain species richness. However, we must recognize that the biological processes play a secondary role, and they come into operation only in response to primary differences in the physical setting.

3.5.2 Equilibrium

To what extent are the diversity and composition of contemporary biotas responses to current conditions, and to what extent are they the legacy of historical events? The recent interchange of ideas between biogeography, ecology, evolution, palaeontology, and systematics has hardly resolved this difficult problem, but it has helped to put it in clearer perspective. Since the 1960s, most of the investigation of species diversity has been done by ecologists. The small-scale, open communities that most ecologists study were usually assumed to be close to an equilibrium between opposing rates of local immigration and extinction. MacArthur and Wilson (1963, 1967) extended these ideas to a larger scale in their theory of island biogeography, which is discussed in Chapter 15. Later, the concept of the species equilibrium was generalized (Wilson, 1969; MacArthur, 1972) and applied widely to biogeography at all scales (Flessa, 1975; Rosenzweig, 1975; Brown, 1981; Brown and Gibson, 1983; Hoffman, 1985a). Although they were criticized with increasing frequency (Sauer, 1969; Carlquist, 1974; Wiens, 1977; Gilbert, 1980), the elegantly simple steady-state models stimulated a great deal of productive research. Like many good scientific ideas, however, they often sowed the seeds of their own destruction. Because the equilibrial models made robust, qualitative predictions, these were soon tested and frequently falsified (Brown, 1971; Diamond, 1972, 1975).

Limitations of equilibrial models of species diversity have become apparent. Perhaps most importantly, they obscure the important relationship between the intrinsic processes that tend to move a system in a constant environment toward some steady state, and the extrinsic perturbations that tend to prevent the system from ever attaining such an equilibrium. It is becoming increasingly apparent that in many systems the frequency and magnitude of these disturbances is so great that species diversity and composition is constantly changing, even in local ecological communities that are open to relatively unrestrained immigration (Wiens, 1977; Brown, 1987). There is equally good evidence that patterns of diversity on a geographic scale reflect continuing adjustments to changes in the extrinsic environment. For example, Brown and Gibson (1983) give several examples of native species that have greatly expanded their ranges within the last two centuries in response to changes in habitats caused by human activities.

3.5.3 History

Present conditions and historical perturbations have often been advanced as alternative explanations for particular patterns of diversity. However, there is no logical reason why contemporary and historical explanations must be mutually exclusive. On the contrary, there is every reason to expect that they will be intimately interrelated, in part because biotas interact with the contemporary environment even while they are still responding to changes in the past, and in part because environments that are different today were probably also different in the past (Croizat, 1964; Brooks, 1986). There is irrefutable evidence that some of the variation in species richness must be attributed to historical events.

The most unambiguous cases are provided by isolated habitats that have been in existence only since the late Pleistocene, little more than 10 000 years ago. These include both continental islands that were once joined to the mainland by land bridges and distinctive patchy habitats that were formerly broadly interconnected. The number and composition of species on these islands reflects their historical connections as well as their present isolation. For example, Diamond (1972, 1975) has shown that islands that are currently separated from New Guinea by such shallow seas that they were connected by Pleistocene land bridges have more bird species than islands of comparable size, habitat, and distance from New Guinea that are separated by such deep straits that they were never connected. Furthermore, the Pleistocene land bridge islands are inhabited by certain species that are absent from the oceanic islands, presumably because they are poor over-water colonists. Lawlor (1983) has documented a very similar situation in the small mammals inhabiting the land bridge and oceanic islands in the Gulf of California. These Pleistocene land bridge islands provide good examples of how historical perturbations can promote diversity.

Insular biogeography also provides excellent examples of historical legacies of reduced diversity. On the Indonesian islands of the Krakatau group species richness of taxa that are poor over-water colonists is still depressed as a consequence of a volcanic eruption that caused the extinction of virtually all life in 1883 (Fig. 3.6; Whittaker *et al.*, 1984; Thornton 1986). Similar effects of the more recent eruption of Mount St. Helens are described in Chapter 15. I have shown (Brown, 1973) that historical events are necessary to explain why some isolated localities in the south-western United States have fewer rodent species than expected on the basis of geographic variation in current environmental

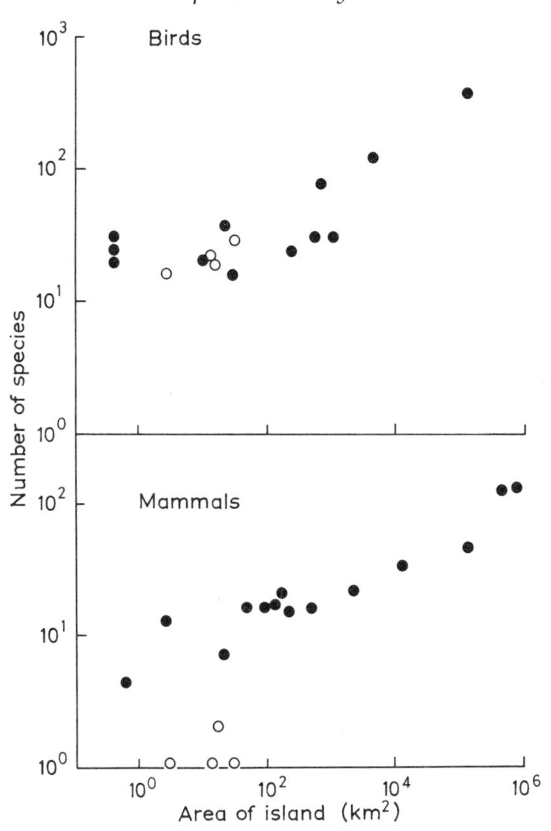

Figure 3.6 Effects of the eruption of Krakatau on the recolonization of islands by birds and mammals as shown by deviation from regional species–area relationships. The diversity on islands sterilized by the eruption of 1883 (unshaded circles) is compared with the richness on islands not disturbed by major volcanic eruption within recorded history (shaded circles). Note that birds, which are notoriously good over-water colonists, have built up comparable numbers of species on both disturbed and undisturbed islands. This is true even for Anak Krakatau, which was devastated by recent eruptions in 1952 and 1957. In contrast, mammals, which are poor over-water colonists, still show reduced diversity on all disturbed islands, three of which have had a century to be recolonized. From data in Thornton (1986) and other sources.

conditions, which quite adequately account for most of the variation in species richness in less isolated regions.

These and many other examples from island biogeography are sufficient to demonstrate unequivocally that historical perturbations profoundly affect contemporary local patterns of species diversity. It is much more difficult to attribute any of the major diversity gradients mentioned above simply to the legacy of history. Virtually all regions of the earth have experienced major, periodic climatic changes during the Pleistocene, but it is not easy to explain why the effects of these perturbations on diversity have been more severe in some regions than others. Certain climatic zones and habitats have been much contracted in the past compared to their present extent, but this is true of those, such as tropical rain forest, that support very high numbers of species (Chapter 10), as well as those, such as temperate coniferous and deciduous forest, that contain many fewer species. It is also apparent that long time periods may be required for the re-establishment of diversity following climatic and geological changes. For example, recolonization of northern North America by tree species following the retreat of glacial ice and warming of the climate has occurred at different rates in different species, has required most of the last 10 000 years in some species, and may still be continuing in others (Bernabo and Webb, 1977; Betancourt, 1986; Davis, 1986; see also Silvertown, 1985). When speciation, as well as colonization, is involved in the regeneration of diversity following a severe environmental perturbation, the imprint of history may potentially last for a very long time. For example, speciation seems to be necessary for large lakes to build up diverse fish faunas, because only a small number of species typically colonize lakes from rivers and streams. Consequently the Great Lakes of North America appear to be depauperate because they have had only a few thousand years to acquire species following Pleistocene glaciation, whereas in contrast the Great Lakes of central Africa have been in existence for millions of years and they have accumulated diverse fish faunas largely as a result of speciation (Fryer and Iles, 1972; Barbour and Brown, 1974).

Despite some well-documented examples, historical explanations for diversity patterns are usually very difficult to test. This is perhaps best illustrated by the fact that the two main historical hypotheses for high diversity in the tropical Amazon Basin use directly opposite logic. The older idea was that the lowland tropical forest habitats represented very stable environments that were relatively unaffected by the geological and climatic events of the Pleistocene that caused extinction, prevented recolonization, and thereby reduced diversity at higher latitudes. In contrast, the more recent Pleistocene refugium hypothesis (Haffer, 1969; Chapter 10) argues that the climatic perturbations of the Pleistocene were pronounced in the Amazon Basin and that these were an indirect cause of high diversity. Alternating cycles of dry and wet climate are now thought to have caused the periodic retraction of rain forest to form isolated patches separated by savanna vegetation. These temporary forest refugia have been hypothesized to have stimulated high rates of speciation, contributing to the build-up of diversity in such different taxa as birds, frogs, butterflies, and trees. Recently, however, Endler (1982a) and Lynch (Chapter 10) have pointed out serious problems with this hypothesis.

Note that a logical consequence of this refugium hypothesis is that the effect of forest fragmentation in enhancing speciation rates sufficiently outweighed the expected effect of reduced forest area in increasing extinction rates (Chapter 8) so that the climatic perturbations resulted in a net increase rather than an equally possible decrease in species diversity. Just the opposite effect of Pleistocene climatic change has classically been invoked to account for the low diversity in polar regions (Fischer, 1960; Fischer and Arthur, 1977). The effect of glaciers and cooler temperatures is hypothesized to have caused the restriction of high latitude forms to isolated refugia, where higher extinction than speciation rates resulted in the loss of many species.

Clearly, if it is theoretically possible for either stable or changing climates to promote high diversity, and if climatic perturbations could in theory cause either higher or lower diversity, it is difficult to frame and test unambiguous hypotheses about the effect of historical climatic changes. With hindsight, it is distressingly easy to concoct *ad hoc* scenarios that are consistent with almost any set of observations. As we shall see later, it is not sufficient to show that speciation or extinction has occurred as a result of isolation of populations in Pleistocene refuges; it is necessary to demonstrate that one of the processes has predominated to cause a net increase or decrease in diversity.

3.5.4 Favourableness

Most diversity gradients are correlated with variation in the contemporary physical environment, and those environments that support the greatest variety of species tend to have generally similar physical conditions. On land the most diverse habitats are tropical rain forests, which are warm but not too hot, wet but not waterlogged, and not subject to great seasonal variation. Coral reefs, the most diverse marine habitats, are also relatively warm and aseasonal. Both are extremely productive and support high biomass as well as high species diversity. In contrast, habitats with low species diversity typically have extreme physical conditions (swamps, hot springs, and caves), and often they are also subject to wide fluctuations in abiotic factors (estuaries, temporary pools, deserts, and tundra). As indicated in Chapter 6, organisms have only limited capacity to evolve fine-tuned adaptations to deal with such unpredictably varying environmental stresses. Many, but by no means all such areas are insular: small, spatially isolated patches of habitat that differ markedly from surrounding environments. Some of them are unproductive (caves, deserts, and tundra), but others are highly productive (hot springs, salt marshes, and estuaries).

This suggests a direct causal relationship between physical environmental variables, such as temperature, moisture, light, and salinity, and the capacity to maintain diverse species. That is, some environments appear to be inherently more 'favourable' than others. As pointed out earlier, biogeographers and ecologists have long struggled to develop a non-circular, rigorous, operational definition of favourableness so that it can be used to make predictions and test hypotheses.

History and favourableness

I shall depart somewhat from previous treatments by suggesting that history holds the key to developing a useful definition of favourableness. Living things have special properties and requirements because they share histories in common environments and constraints owing to common ancestry. Thus environmental conditions that are favourable for a particular taxon reflect both environmental history and evolutionary constraints. The history of the environment and the constraints on different lineages of organisms are not independent. Each taxonomic group has a history of restriction to a limited geographic region, and its structural and functional attributes reflect adaptations to conditions in that region (Chapter 6). These acquired constraints tend to limit a lineage to a restricted range of environments, and hence to inhibit expansion of the taxon into other regions with different environmental conditions.

Although the climate and geology of the earth has changed throughout the history of life, in many respects the present distribution of environmental conditions reflects historical conditions. For example, at present and for the last hundreds of millions of years the frequency distribution of aquatic environments has been bimodal with respect to salinity; there has been a lot of ocean, a moderate amount of freshwater lake and stream habitats, and relatively few estuaries and hypersaline lakes. This is reflected in the bimodal distribution of species richness of aquatic invertebrates as a function of solute concentration (Fig. 3.7; Kinne, 1971). A similar plot for fishes would show a qualitatively similar pattern (Nelson, 1984). The present distribution of species diversity reflects the current distribution of aquatic environments, because aquatic habitats have a similar distribution now as in the past. If all the waters in the world had been either as fresh as Lake Superior or as saline as the Great Salt Lake (approximately seven times the concentration of sea water) until ten or 10 000 or even 10 000 000 years ago, the pattern of aquatic species diversity would undoubtedly be very different from what it is today.

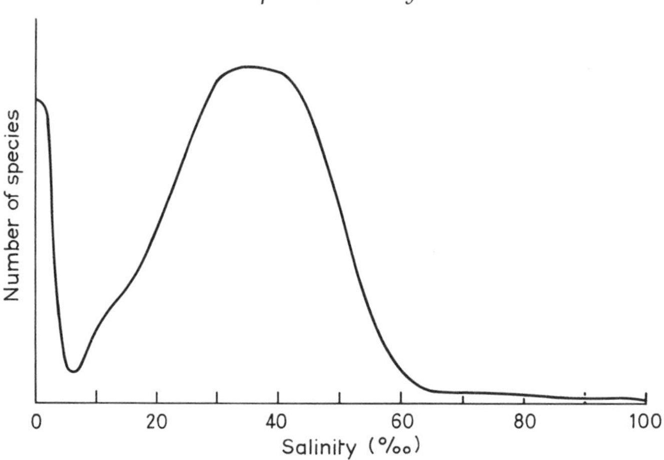

Figure 3.7 Estimated diversity of aquatic invertebrates throughout the world as a function of salinity (after Kinne, 1971). Note that the pattern is bimodal, with distinct peaks corresponding to fresh water (<2‰) and sea water (30–40‰).

The influence of evolutionary constraints on contemporary distribution and diversity is illustrated by those exceptional groups that attain their greatest diversity at higher latitudes rather than in the tropics. I shall choose the voles (the subfamily Microtinae of the large rodent family Muridae), but I could equally well choose pines or penguins, salamanders or seals. Voles are most diverse in temperate to arctic habitats (Rose and Birney, 1985), because they are adapted to these environments. They have acquired morphological, physiological, and behavioural specializations to tolerate the physical conditions, exploit the plant food resources, take advantage of the mutualistic microbial symbionts, withstand the competitors, and avoid the predators, parasites, and pathogens. Their attributes reflect a history of confinement to high latitudes and of adaptive response to the unique physical and biotic environments of these regions. They have migrated, been isolated, formed new species, and become extinct as a result of glaciation and climatic change during the Pleistocene, but they have remained in cold climates because they are adapted to cope with them; and because they have been restricted to these conditions, their adaptations to them have been reinforced.

This may sound circular, but it is not. It is simply an expression of the positive feedback loops that tend to make for conservatism in evolution, ecology, and biogeography. The constraints of evolutionary history are only relative, not absolute. Indeed, some species of temperate North American mammals (but not voles) have managed to evade their constraints sufficiently to colonize South America since the completion of the interamerican land bridge about three million years ago. Some of these have since begun to speciate and diversify, giving rise to new South American lineages adapted to the regional conditions (Marshall *et al.*, 1982; Marshall, 1985).

Insularity and favourableness

Such an historical interpretation of favourableness would seem to account for the low diversity in a wide variety of insular environments. The unusual physical conditions and spatial isolation of these habitats make them difficult to colonize. Their small size and often ephemeral existence cause frequent extinctions among the few species that manage to inhabit them. This combination of low colonization rates and high extinction rates tends to prevent the build-up of a diverse, well-adapted biota.

Under certain circumstances, however, insularity may tend to increase rather than decrease species richness. Sometimes the degree of isolation may be just sufficient to cause such high rates of speciation that they outweigh the combined effects of low colonization and high extinction. This appears to be the cause of some spectacular insular radiations, such as the genus *Drosophila* on Hawaii (Williamson, 1981; Carson and Yoon, 1982) and the cichlid fishes in the lakes of central Africa (Fryer and Iles, 1972). The Pleistocene refugium hypothesis to explain the diverse biota of the Amazon Basin (see p. 76, Chapter 10; Haffer, 1969), is another example in which a special combination of insular conditions is supposed to have resulted in increased species richness by causing high rates of speciation.

In general, however, these are exceptional examples and diversity on islands and insular habitats is low. For example, despite the large number of species of *Drosophila* and a few other taxa, the overall richness of the Hawaiian biota is much less than in a comparable area and climate on any continent. A delicate balance of present and past conditions seems to be required for sufficient speciation rates to generate high diversity in insular environments. Although these conditions may be met for a few taxa inhabiting certain islands, they are usually so restrictive that they do not cause high overall species richness. The Pleistocene refugia hypothesis may be an important exception to this generalization, but current evidence hardly provides unequivocal support for it (Chapter 10).

I note parenthetically that habitat fragmentation can normally be expected to decrease local or 'alpha' species diversity. If, however, it is maintained for sufficient time for speciation to occur in the fragmented areas, it will tend to increase large-scale or global diversity. Thus the fact that approximately 39% of the world's 21 500 fish species inhabit freshwater habitats even though these account for less that 1.5% of the earth's aquatic surface area (Nelson, 1984), can be attributed to the speciation of fishes in small, isolated areas of fresh water. Local diversity of fish species within lakes and streams is typically lower than in marine habitats of comparable temperature and productivity. Fusion of formerly long-isolated regions or habitat patches should generally have the opposite effect; so if Gondwanaland were reunited, a substantial number of species would become extinct, but local diversity would increase because the surviving species would overlap more in their geographic ranges.

Productivity and favourableness

The unfavourableness of other low diversity environments such as tundra, desert, and the deep sea can hardly be explained by their insularity. These environments are extensive, and I am not aware of any hard evidence that they have experienced any more turbulent or transitory histories than forests or shallow coastal waters in temperate or tropical latitudes. In trying to account for these patterns, Hutchinson (1959) asked why, if some taxa, such as polar bears and cacti and anglerfish, were able to inhabit these environments and adapt to their severe conditions, were they so limited in their ability to speciate and other groups so unable to colonize that diversity remained low? Hutchinson suggested that the answer lay in one common feature of these spatially and temporally extensive, but low diversity environments: they are unproductive. Primary productivity and standing crop biomass are much less than in equally extensive but more species-rich habitats. The low productivity can be attributed to 'harsh' physical conditions: lack of water in deserts; cold, and extreme seasonality of solar radiation in arctic tundra; and low temperature and absolute darkness in the abyssal depths.

To explain why these conditions limit productivity, we must again invoke history and evolutionary constraints. This time, however, the constraints are fundamental limitations on the structure and function of all living things: photosynthesis requires sunlight; cellular activity must occur in an aqueous medium; rates of biochemical reactions increase exponentially with increasing temperature, but the chemical bonds in proteins and other organic molecules become increasingly unstable as temperatures exceed 45–50°C. These constraints are not absolute. There are microbes (and sometimes other organisms) that have at least partially circumvented all of them: to fix energy from sulphur compounds by chemosynthesis; to survive complete desiccation; to grow under the ice in permanently frozen Antarctic lakes; and to live in hot springs where temperatures may exceed 70°C (Brock, 1985; Stetter, 1986). Occasionally, as in the case of hot springs and hydrothermal vents, these alternative systems are able to achieve high productivity that supports local ecological communities of moderate diversity (Grassle, 1985; Jannasch and Mottl, 1985). However, they require such special physical conditions that they do not form the basis of geographically widespread, highly productive ecosystems based on alternative modes of biophysical and biochemical function.

It is still necessary to explain why high productivity may contribute to the maintenance of high species richness. As outlined earlier, there is not necessarily a positive relationship between productivity and diversity. Under what circumstances is increasing productivity and biomass divided among increasing numbers of species rather than simply among more individuals of a few species? The general answer appears to be, whenever there is a 'cost of commonness' (F. Hopf, pers. comm.). More than one mechanism may cause negative effects as population size increases. If there is a trade-off between less efficient use of a wide range of resources and more efficient exploitation of a narrow range, then a few species of abundant generalists would tend to be replaced by many species of less common specialists as productivity increases. If common organisms are more susceptible than rare ones to predators, parasites, and pathogens, then as productivity increases enemies will tend to facilitate the addition of new species rather than the further increase of those already present. Because both of these mechanisms depend largely on the ability of relatively rare species to invade and persist, and they are only likely to do so if they can maintain sufficiently large population sizes to avoid extinction, total population size and areal extent of productivity are likely to be at least as important as local population density and productivity per unit area.

This is essentially the reasoning used by Wright (1983) to develop what he calls species-energy theory. Species richness is conceived as being promoted by total productivity (not productivity per unit area) on a spatial scale sufficient to maintain populations of the rarest species. Thus islands of similar size but different productivity (because of differences in latitude, for example) would be predicted to have different diversity and so would islands of different size with the same productivity per unit area, because in both cases the island with the greater total productivity would be able to support more species. Because of the trade-offs associated with the cost of commonness, more diverse communities should in general be characterized by species that are more specialized, both in their resource requirements and in their relationships to other species.

Conclusions about favourableness

Thus, those aspects of contemporary environments that make them 'favourable' and able to support large numbers of species cannot be understood without a knowledge of both the history of these environments and the evolutionary constraints that limit the tolerances and requirements of their inhabitants. Attributes of contemporary organisms, including the composition, distribution, and diversity of entire biotas, reflect their unique histories, including the histories of the environments where they have lived. In this most basic sense, the explanation of all patterns of diversity must be both historical and ecological.

3.5.5 Habitat heterogeneity

It is difficult to evaluate whether habitat diversity is a cause or merely a correlate of species richness. The latter is so well documented as to be incontrovertible, but the high productivity that is characteristic of most species-rich habitats also tends to result in complexity of physical structure. Tropical rain forests and coral reefs provide the most dramatic, but by no means the only, examples. It is also clear that in these structurally diverse habitats, coexisting species often appear to specialize and to avoid competitive exclusion by differential use of the physical structure (Chapter 9). For example, bird and epiphytic plant species are typically confined to particular vertical strata in tropical forests (Terborgh, 1980, 1985) and numerous fish and invertebrate species are absolutely dependent on holes, caves, or similar physical refuges in coral reefs (Molles, 1978; Sale and Dybdahl, 1978; Sale, 1979). Experimental studies in which habitat structure has been manipulated but other variables have been controlled have shown changes in the composition and diversity of species (Rosenzweig, 1973; Sale and Dybdahl, 1975; Molles, 1978). These responses can be attributed to the fact that individual species have evolved varying degrees of dependency on structural components of the environment, just as they may require other physical and biotic conditions for their existence.

It is not clear, however, to what extent structural complexity is necessary for the development and maintenance of large-scale geographic patterns of species diversity. If habitat diversity directly promotes species richness, then we should not only expect to find highly productive but structurally simple and species-poor habitats; we should also find evidence for alternative hypotheses, such as past perturbations or harsh physical conditions, insufficient to account for the low number of species in these environments.

In addition, we should expect a different division of biomass and numbers of individuals among species, since the number of species but not the numbers of individuals per species would be limited by requirements for particular structures. Despite the large literature on species–habitat and species–abundance relationships, I am not aware of any strong evidence of this sort. If anything, the high diversity of some open-water habitats (Hutchinson, 1961; Hayward and McGowan, 1979; McGowan and Walker, 1979) suggests that physical structural complexity may not be essential for the accumulation of many species. On the other hand, recent studies in both limnology and oceanography have documented surprising microspatial heterogeneity in what had once been assumed to be large, uniform water masses (Wiebe, 1982; Harris, 1987; Reynolds, 1987). Exactly how this structural diversity may solve 'the paradox of the plankton', however, remains to be determined.

I suspect that structural diversity can best be thought of as another component of favourableness. Most organisms are evolutionarily constrained to require physical structures to grow on, live in, or provide refuges from their enemies. Most productive environments have long provided a variety of such sites and shelters. In the sense that present requirements reflect past requirements for and availability of certain conditions, diverse structural elements are another component of favourableness that promotes species richness.

3.5.6 Interspecific interactions

As pointed out above, biological attributes of organisms cannot provide a primary explanation for patterns of diversity. Faunas and floras reflect long histories of interaction between organisms and their environment. Other organisms are important features of the environment, and the different ways that interspecific interactions can affect geographic distributions and species diversity are discussed in Chapter 9. But the presence or absence of other species and the outcome of interspecific interactions must ultimately be determined by physical factors. Two regions that have had identical physical conditions throughout their histories will not, except by chance, differ in diversity; therefore it follows that two regions that differ systematically in diversity must also differ in their present or past physical environments.

The build-up and maintenance of species richness obviously depends in large part on the structure and dynamics of interactions among the coexisting species. But despite an extensive literature (e.g. R. H. MacArthur, 1955; May, 1973; J. W. MacArthur, 1975), the relationship between species diversity and community stability remains poorly understood. For one thing, even simple pairwise interactions may have sufficiently complex dynamics so that in temporally and spatially variable environments they are not resolved either rapidly or predictably. For example, theoretical analyses suggest that extreme similarity between competing species can greatly prolong the expected time to exclusion (Caswell, 1982), and coexistence can further be prolonged by spatial heterogeneity (Horn and MacArthur, 1972) and temporal fluctuation (Huston, 1979; Chesson, 1986). Community ecologists are also coming to realize the inadequacy of direct, pairwise models of interspecific interaction. Both theoretical treatments and empirical studies indicate that the organization of diverse biotas must be viewed in terms of realistically complex networks of direct and indirect interactions that link the dynamics of many species (Brown *et al.*, 1986). Community ecology has much to contribute to understanding species diversity on all scales including biogeographic ones, but clear, satisfying answers cannot be expected immediately.

3.5.7 Species dynamics

In a very important sense the diversity of biotas is a consequence of the dynamics of the species units. Species richness can be increased only through colonization from another biota or by speciation, and it is decreased only through extinction. Thus the diversity of a biota is a direct consequence of the rates of species' origination and extinction. However, failure to understand the implications of these species dynamics has led to much confusion in the effort to explain patterns of species diversity.

Some investigators have failed to recognize that differences among regions in rates of origination and extinction are not simply intrinsic properties of biotas. Species dynamics are consequences of the size, spatial and temporal distribution, and genetic structure of populations, and these ultimately are caused by spatial and temporal variation in the physical environment. Any complete explanation of differences between regions in rates of origination and extinction must invoke the history of refuges, barriers, disturbances, and other physical characteristics. Some taxonomic groups (clades) may indeed exhibit different rates of speciation, colonization, and extinction. But these are dependent not solely on the intrinsic attributes of the taxa, but on the interaction between these biological traits and the physical environment in which they occur.

A more serious problem is that many investigators have implicitly assumed that the rates of species origination and extinction are independent, so that high diversity can be attributed to some factor that enhances speciation while the extinction rate remains unchanged, or vice versa. To see why this assumption may be incorrect, consider the

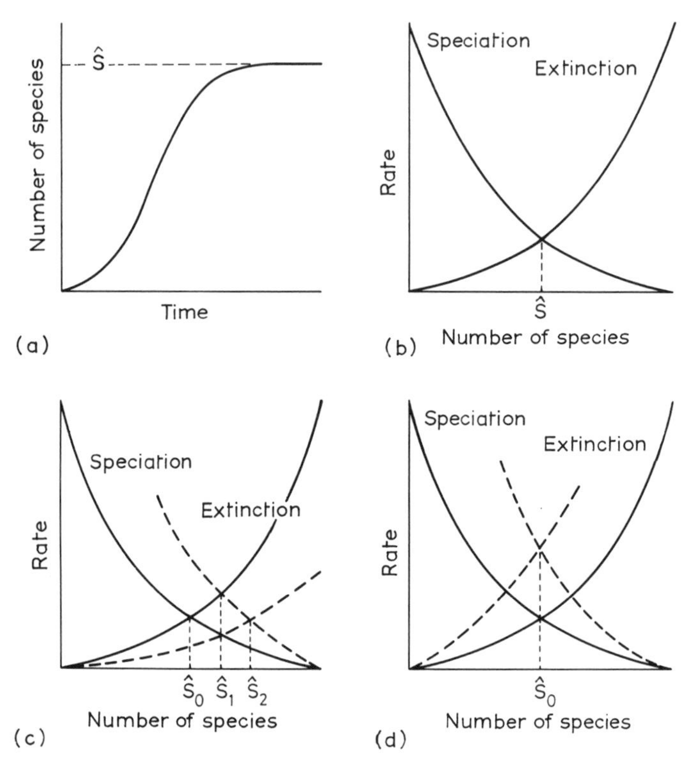

Figure 3.8 Graphical equilibrium models of species dynamics. (a) Increase in species richness to approach the species equilibrium, $\hat{S}$, in an initially empty environment as a function of time since disturbance. (b) Variation in the rates of speciation (or colonization) and extinction with the number of species present resulting in a stable species equilibrium, $\hat{S}$. (c) Increasing speciation rate or decreasing extinction rate (dashed lines) increases the equilibrium diversity from $\hat{S}_0$ to $\hat{S}_1$ if the other rate remains constant and to $\hat{S}_2$ if both rates change simultaneously. (d) If an increase in speciation rate is compensated by an increase in extinction rate the result can be little or no change in diversity from the equilibrium, $\hat{S}_0$, despite more rapid turnover of species. See text for further explanation, and MacArthur and Wilson (1963, 1967), Wilson (1969), and MacArthur (1972) for derivation of these or similar models.

close analogy between species dynamics and population dynamics. Speciation and birth are multiplicative processes, colonization of species is analogous to migration of individuals, and extinction and death are irreversible losses. Just as the population densities of different species may bear little or no relationship to their birth rates, so the species richness of different regions may bear little or no relationship to the speciation rates in those regions. Just as any kind of 'density–

dependence' in population dynamics will cause the death rate to vary positively within the birth rate, any kind of analogous 'diversity–dependence' will cause positive covariance of extinction and speciation rates.

This is easiest to visualize in an equilibrial model (Fig. 3.8; MacArthur and Wilson, 1967; Wilson, 1969; Chapter 15), in which habitats or regions have a 'carrying capacity for species' analogous to the carrying capacity for individuals in logistic and other models of density–dependent population growth. The fact that changes in diversity as a function of time since a major perturbation often resemble logistic population growth curves (Fig. 3.5) provides empirical support for such a model. But the carrying capacity need not remain constant and biotas need not be near equilibrium for species-dependence to operate. Some regions may differ in species richness largely because they have different histories of disturbance so they are in different positions on a species build-up curve (Fig. 3.8a) after having been temporarily perturbed from equilibrium (Fig. 3.8b). On the other hand, the rapid build-up of diversity to near original levels following many kinds of perturbations (Figs. 3.5 and 3.6a) suggests that many systems spend most of their time near the equilibrium species richness. When a region is near its carrying capacity for species, a change in either some of the organisms or in their abiotic environment that increases speciation rate without affecting extinction rate will result in increased diversity; so will a change that decreases extinction rate while leaving speciation rate unchanged (Fig. 3.8c).

I suspect, however, that many changes in organisms or their extrinsic environment that alter speciation (or extinction) rates are accompanied by a compensating change in extinction (or origination) rates, so that the equilibrium species richness is little changed. Thus Knoll (1986) presents evidence that suggests that when one plant lineage acquired adaptations that increased its speciation rate, the diversification of such a clade was often accompanied by increasing extinction rates in other clades, which Knoll attributes to competition. The result was that for long periods of geological time land plant diversity remained relatively constant (close to an implied equilibrium) despite major shifts in the composition of taxonomic groups. Similarly, one could imagine that habitat fragmentation owing to climatic change might increase the speciation rate of some kinds of organisms but increase the extinction rates of others, with the result that whether species diversity ultimately increased, decreased, or remained nearly constant would depend on whether and how the carrying capacity for species was affected. In some such cases diversity might remain very similar despite higher turnover of species owing to increases in both speciation and extinction rates (Fig. 3.8d).

3.6 CONCLUSIONS

This has been a brief, and necessarily simplified review of a very large and diffuse subject. In contrast to some authors, I have devoted only a little attention to problems of how to measure species diversity. These are important, but conceptually relatively straightforward. Refinement and standardization of methodology should make future studies more rigorous, but I do not believe that the most challenging questions hinge on how to define or measure species diversity.

The geographic patterns of diversity have been described fairly briefly. Some of these may still be controversial, but most of them have been recognized for decades and recent work has only served to document them more convincingly. Most of these patterns can be characterized as gradients of diversity, because the number of species varies relatively continuously with geographic variation in physical environmental variables, such as temperature, solar radiation, air or water pressure, soil water potential, and solute concentration of water. Some of the gradients, such as those with respect to latitude, elevation, and water depth, reflect simultaneous variation in several different and potentially important physical factors.

Most of the chapter has been devoted to evaluating the numerous hypotheses that have been proposed to explain the patterns. These include: time since perturbation, favourableness, productivity, habitat heterogeneity, interspecific interactions, and rates of speciation and extinction. These diverse explanations are difficult to treat systematically, because they are not alternative hypotheses in the sense that they are necessarily mutually exclusive. On the contrary, some offer different levels of explanation ranging from primary effects of differences in present or past physical environments to the higher-level, biological processes, such as interspecific interactions or the dynamics of speciation and extinction.

In evaluating these hypotheses, I have tried to raise and clarify the questions, rather than provide definitive answers. The patterns may be well documented, but complete mechanistic explanations must await a better understanding of ecological and evolutionary processes and their interacting effects on biogeographic spatial scales. In evaluating the various hypotheses, I have focused on the problems: logical inconsistencies, hidden assumptions, circular arguments, use of the same process to explain opposite patterns and vice versa, and lack of independence between different processes. In order to overcome these problems, biogeography, like all disciplines, could certainly benefit from increased analytical rigour. To a large extent, however, the confusing state of these hypotheses simply reflects the fact that they have been contributed by diverse kinds of investigators trying to extend their limited tools and experience to grapple with one of the most complex problems in science.

There has been real progress, as shown by the studies reviewed here and in the other chapters in this book. We are beginning to understand the causes, as well as the correlates of geographic variation in species richness. It is becoming increasingly clear that both historical events and ecological processes must be part of any complete explanation of any pattern of diversity. On the one hand, ecological conditions, both present and past, determine the capacity of regions to support species, the barriers to dispersal that inhibit colonization and promote speciation, the population sizes and structures that affect probabilities of speciation and rates of evolution, and the selective agents that promote adaptive evolutionary change. On the other hand, the attributes of each species reflect the unique history of its phylogenetic lineage, and the composition of each biota reflects the unique history of the region where it occurs and its relationships to other regions that have been the source of colonists. To account adequately for geographic patterns of diversity will require an understanding of geological history, past and present physical environments, phylogenetic history and evolutionary constraints of the taxa, the dynamics of species origination and extinction processes, and the ecological relationships of species with both their physical environment and other kinds of organisms.

The present resurgence of interest in biogeography has led to great expectations, but these are likely to be fulfilled only if the discipline becomes increasingly interdisciplinary. We cannot hope to understand completely the diversity of life and its distribution over the earth until we completely understand the history of the earth, the history of its inhabitants, and the relationships between the various kinds of plants, animals, and microbes and their biotic and abiotic environments. This is a tall order, but it makes for an exciting future.

ACKNOWLEDGEMENTS

I am grateful to R. Craw, N. Ferguson, K. Flessa, P. Giller, P. Hastings, B. Maurer, A. Myers, D. Thomson, and M. Williamson for valuable discussions or helpful suggestions. M. Pantastico called my attention to the pattern shown in Fig. 3.6. My research has been supported by the National Science Foundation and the Department of Energy.

CHAPTER 4

Relationship of species number to area, distance and other variables

M. WILLIAMSON

4.1 INTRODUCTION

It is common knowledge that different species have different distributions worldwide, and this remains true at all spatial scales. In general, both the centre of distribution and the spread around the centre are different for each species, as can readily be seen in any atlas of distribution. The consequence of this is that, again in general, larger areas will contain more species than smaller areas. However, the species–area relationship is not simply a reflection of different distribution patterns. As will be seen, it can also arise from sampling a set of species all with the same distributions but with different abundances.

In this chapter I shall first describe the phenomena, the shape of the species–area relationship and its variability, and then consider the main explanations that have been put forward to explain these results. This involves some discussion of the mathematics of sampling, and of the patterns of abundance shown in ecological communities, and of what is often, rather curiously, referred to as island biogeography theory, namely the equilibrium theory of MacArthur and Wilson (1963, 1967; Chapter 15). Other theories that explain the patterns of biogeographical distribution, both on islands and in general, involve a consideration of the nature of environmental heterogeneity and of the geometrical patterns called fractals. These theoretical considerations are influenced by three phenomena that affect almost all aspects of biogeography: sampling, dispersal and evolution.

Although data on area and species number are those most often reported and indeed the easiest to obtain, both can be regarded as secondary variates. The number of species reflects the population dynamics of the individual species, encompassing the four primary processes of birth, death, immigration and emigration (Kostitzin, 1939;

Williamson, 1972). Although MacArthur and Wilson have been described as using an area *per se* hypothesis (McGuinness, 1984), they used area in lieu of better information: "More often area allows a large enough sample of habitats, which in turn control species occurrence. However, in the absence of good information on diversity of habitats, we first turn to island areas" and "Neither area nor elevation exerts a direct effect on numbers of species; rather, both are related to other factors, such as habitat diversity, which in turn controls species diversity" (MacArthur and Wilson, 1967, pp. 8, 20). In Section 4.5, I consider what is known about the quantitative effect of variables other than area, linking back to the diversity of explanations of the species–area relationship.

Biogeography, in all its aspects, relates the distribution of individuals and species to the variability of habitats in space and time. It is arguable that the study of the species–area relationship provides one of the best approaches to studying habitat variability, of quantifying it, and so of indicating the effect of habitat heterogeneity on species distribution and the structure of ecological communities. I will consider this possibility in the final section of this chapter.

A somewhat comparable species–area relationship is found with insects on plants (Strong *et al.*, 1984a). However, while one variable here is the number of species, the other is not the area, but the area occupied by a given plant taxon. That is, the plot shows the number of insects on different plant taxa, the abundance of each plant taxon being measured by the area of its distribution. A clear indication that such plots are different from the standard biogeographic species–area plots is that, quite often, many of the plant taxa have no insect species at all in the group studied. For instance in the British Umbelliferae (Apiaceae) 44 species have no Agromyzid dipterous flies feeding on them, while only 18 species do have Agromyzids. This chapter is not the place for any further discussion of this interesting but distinct phenomenon, except to say that the explanations offered for it by Strong *et al.* (1984a), and the conclusions they draw, match the conclusions of this chapter quite well.

4.2 DESCRIPTION OF THE PHENOMENA

4.2.1 The shape of the relationship

It has been known to botanists at least since the 1920s, both that the number of species increases systematically in samples of larger area, and that the rate of increase of species number decreases with progressively larger area. That is, the curve of species (on the ordinate) against area (on the abscissa) flattens off (Goodall, 1952; Williamson, 1981). That is, the curve has the shape shown in Fig. 4.3, though the axes there are logarithmic, and in the curve being discussed here arithmetic. It is sometimes said, for instance by Connor and McCoy (1979), that species–area curves become asymptotic for large areas, but this is not so. Species numbers continue to increase at an ever slower rate as area increases. This would be perfectly possible even if the total area were infinite; as the area of the earth is finite there is no need to postulate an asymptote and it is not observed. In general, the number of species increases with increasing area.

It is well known that it is easier to comprehend and compare straight lines than curved ones, and the usual approach has been to transform either the number of species, or the area, or both, so that the relationship is linear. An example is shown in Fig. 4.1, for the number of bird species in different areas of the British Isles. This figure shows both breeding counts and wintering counts and, because of the pattern of bird migration in the British Isles, the set of species recorded at one place is different in the two seasons. Nevertheless both sets of counts give a satisfactory straight line on the log-log plot. Fig. 4.1 can be regarded as typical of the majority of species–area plots, which are linear on this transformation and have a slope between 0.2 and 0.35, as will be seen.

Collecting data for a plot such as Fig. 4.1 involves some problems that should be mentioned. The first is deciding what qualifies as an occurrence. For wintering birds any individual will count, but for breeding birds one pair is the minimum, and the ornithologists have defined rules to distinguish confirmed breeding, probable breeding, and possible breeding. Sharrock (1976) lists 18 criteria for arriving at these three main categories, with some variants from species to species. Similarly, in plants, it is usual to record any vegetatively growing species, which implies excluding those present only as seeds, or not restricting counts to those plants successfully reproducing. A second problem is deciding over what period to count occurrences. This may be a day, a short survey period, a whole breeding season, or many breeding seasons, and may not be the same for all points in a data set. A related problem is that the number of species recorded is a minimum, and some may have been missed. Conversely the area is a maximum, and some parts may not have been searched, and indeed for large areas, will not have been searched. Again it is not always clear what area should be measured. For islands this is usually the area above high tide (or maybe mid tide) mark, but if an island contains a large lake this area may be too large for interpreting the distribution of, say, angiosperms. Rafe *et al.* (1985) note that for birds the number of breeding species on cliffs is probably determined more by ledge area than by cartographic area. All this means there is little point in being pedantically rigorous in the analysis of species–area plots; the points are fuzzy.

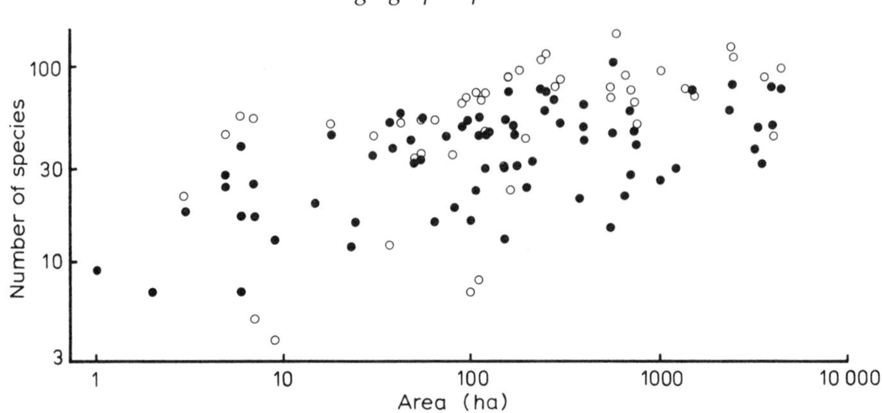

Figure 4.1 The species–area relationship for breeding species (●) and wintering species (○) from the British Trust for Ornithology's Register of sites throughout the British Isles and in various habitats. Note the logarithmic scales on both axes. Adapted from Rafe *et al.* (1985).

One further point deserves special mention. Imagine an archipelago consisting of a set of islands identical in size and set of habitats. The species found on one island will, under this hypothesis, be the same as those found on any other. In such a case the appropriate area to plot is that of a single island. Real archipelagos are variable, yet the point applies. Plotting the species found in the area of the archipelago as a whole does not appear to be appropriate. Each island, or on a mainland each distinct sample, is best treated separately.

What shapes do species–area plots have in practice? There have been a number of surveys studying the linearity of plots under different transformations, and three of the most important surveys are shown in Table 4.1. All three examined the log-log plot (first championed vigorously by Arrhenius, 1921) and the plot of species against the logarithm of area (Gleason, 1922). Two of the three surveys also examined other

Table 4.1. The shape of species–area relationships, showing the number of studies found to fit the Arrhenius, Gleason or neither relationship in three surveys. From Connor and McCoy (1979), Dony (1970) and Stenseth (1979)

	Arrhenius	*Gleason*	*Neither*
Connor and McCoy, best fit	43	27	49
Connor and McCoy, satisfactory	75	38	69
Dony	47	12	17
Stenseth	20	9	–

transformations, but the conclusion is quite clear: the Arrhenius plot is the commonest satisfactory and good fit, but there are plenty of data sets not made linear by it. So any explanation of species–area effects must explain both the commonness of the success of the Arrhenius plot and the reasons for exceptions to it.

There are some minor problems in deciding which is the best fit. Dony (1970) examined the plots by eye to see if they were linear, and Connor and McCoy (1979) examined the residuals for systematic deviation, again by eye. Either is generally sufficient, but there are standard tests that could be applied for trends in the residuals, and which might have been used in critical cases. In view of the variability of the sets, the subjective testing of fit is of little consequence.

The Arrhenius plot implies:

$$\log S = c + z \log A \tag{4.1}$$

where S is the number of species, A the area, z the slope of the line, and c another constant usually referred to as the intercept, though there is no true zero on a log-log plot. Vincent (1981) urged that non-linear least-squares should be used to fit the equivalent expression:

$$S = cA^z \tag{4.2}$$

but I agree with Harvey *et al.* (1983) that there is no *a priori* reason to prefer any equation with the same number of parameters to any other. The results from fitting these two equations will differ slightly, because in one the least squares being minimized are in S, in the other in $\log S$. Technically, the least squares fit of (4.1) is a biased fit of (4.2), but then the least square fit of (4.2) is a biased fit of (4.1). The emotive word bias refers here to the statistical concept, not the everyday concept of bias. Empirically, as can be seen in Fig. 4.1, the residuals from equation (4.1) are usually well behaved, that is homoscedastic (with the same variance at different values of the abscissa), and normally distributed. The commonest exceptions are a tendency to a few large negative residuals, particularly at small areas, which can be seen in Fig. 4.1 for wintering birds.

Wright (1981) claims better fits by non-linear least squares, on the grounds that the total residual sum of squares is reduced, and on the theoretical ground that the error is independent of area or, to put it another way that the error scales with S, not $\log S$. He seems not to have examined the residuals individually, and in my experience, and as shown for instance in Slud (1976), the residuals are wildly hetereoscedastic before the logarithmic transformation and well behaved after. That is, the variance of the number of species is very different at different areas, increasing rapidly with increasing area, while the variance of the logarithm of the number of species is more or less the same at all areas. Wright's empirical claim is also weak as his studies were on very small sets, his mode being at four points only, and half his studies were on fewer than ten points.

Fitting a regression of some function of the species number against the same or another function of the area, by whatever method, is clearly not ideal in view of the remarks made above about the errors in both variables. Connor and McCoy (1979) suggest that the reduced major axis could be used, and that will inevitably give a steeper slope. This suggestion has not been followed up, mainly no doubt because most earlier authors used a regression, and it is more helpful to produce results in a form that can be compared with previous results. The reduced major axis also implies an equality of errors in the two variables, another unjustified assumption.

It might have been more interesting to search for the best transformation by the methods pioneered by Box and Cox (1964). Essentially these involve using one of a number of criteria derived from the analysis of variance to determine the goodness of fit produced by various transformations. These procedures allow a systematic search for the best transformation for a data set (Atkinson, 1985). In practice it is usually found that a range of transformations all give satisfactory fits, and it is unlikely, though the work is still to be done, that a transformation better than the standard log-log fit would be found for the generality of species–area relationships.

4.2.2 The slope of the relationship

The earliest suggestion that the Arrhenius relationship is appropriate seems to be in Watson (1835). He says (p. 41–42) "On the average, a single county appears to contain nearly one half the whole number of species found in Britain; and it would, perhaps, not be a very erroneous guess to say that a single mile may contain half the species of a county". The area of Britain is a little less than 89 000 square miles, and, assuming that Watson meant a square mile in that quotation, this implies a z, in equation (4.1) or (4.2), of 0.12, but also a typical county area of a little less than 300 squares miles which is distinctly on the large side. Dony (1970) suggests a typical slope for the British flora of 0.1873, which might well be thought to be consistent with Watson's statement, allowing for the variable size of English counties, and his uncertainty about the actual number of species to be found in areas smaller than a county.

Along with the surveys of the shape of the species–area curve, there have been surveys of the slope. These have all shown a wide variation and the distribution is invariably skew, with the longer tail towards the

Table 4.2. Distribution of z, the slope of the Arrhenius plot, found in three surveys. The distributions are specified by Tukey's (1977) 'five figures': extremes (1), hinges (i.e. rounded quartiles) (H) and median (M) (cf. Fig. 4.4 for a plot of such figures from other data)

	Figure					
	1	*H*	*M*	*H*	*1*	$n =$
Connor and McCoy (1979)	−0.276	0.185	0.275	0.410	1.132	100
Dony (1970)	0.086	0.146	0.249	0.340	0.550	47
Flessa and Sepkoski (1978)	0.05	0.21	0.30	0.37	1.10	54

higher values of slope. So I report the variation of the slope in the manner suggested by Tukey (1977), namely five figures of the two extremes, with the median and the two hinges (or quartiles) in Table 4.2. The extremes are the smallest and largest observations, the median the middle one in rank order, while the hinges are midway in rank order between the extremes and the median. It can be seen that the median value is in the range 0.25 to 0.3, in conformity with the expectations from the log normal curve discussed below, but it can also be seen that any value from zero up to 0.5 or more is quite often recorded. Dony's survey is entirely of land plants while the other two surveys are mostly of animal studies. It has often been said that botanical surveys give flatter slopes, but it is doubtful if Table 4.2 can be held to support that. Similarly, it has often been claimed that species–area slopes are steeper on sets of islands than on samples from the mainland, but again these wide surveys do not support this generalization either (Williamson, 1981).

To be explicit, and taking these surveys together, ordinary islands give slopes between 0.05 and 1.132, habitat islands (lakes, mountain tops, coral patches) 0.09 to 0.957, while mainland samples range from −0.276 to 0.925. For insects on plants, Strong *et al.*(1984a) give 0.14 to 0.90. A slope of zero means that the same number of species is found in all samples, while a slope of one means that doubling the area doubles the number of species. Consequently, apart from effects caused by markedly different climates, species–area slopes will be between zero and one. These surveys show that more or less the whole of this range is found for islands, habitat islands, mainland samples and for insects on plants.

The variations between the results of the different surveys will reflect the range of habitats used, as much as the organisms studied, both aspects reflecting the results available in the literature. For British vegetation, Dony certainly regarded values ranging between 0.1 and 0.2

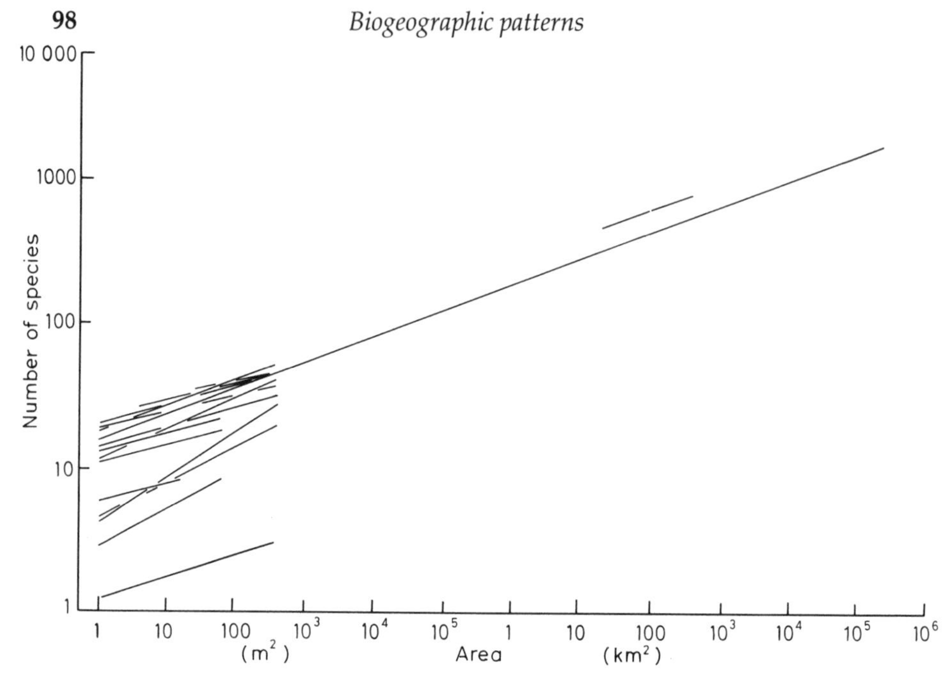

Figure 4.2 Fits of the Arrhenius relationship to various botanical surveys in the British Isles. Each line terminates at the largest and smallest area studied. The long line is Dony's "suggested normal species–area relationship". Note the logarithmic scale on both axes. Modified from Dony (1970).

as the norm, and others requiring special explanation. Figure 4.2 summarizes his compilation on British surveys. The end of the line at the top righthand corner is the total number of vascular plants recorded in a standard British plant list. There are two surveys, both in the floristically richer southern areas of England, centring on an area of 100 km². The other surveys are all in areas from 1 m² up to 1 000 m². Dony's suggested normal species–area relationship in the British Isles, with a slope of 0.1873, is also shown. This is based on a figure of 15 species per square metre, which is indeed higher than most of the surveys, but conversely these surveys were mostly done, for obvious practical reasons, in species-poor vegetation such as bogs, marshes and shingle banks. Small areas are generally more homogeneous than large and consequently there will be a greater variation of species found between surveys of small areas than between surveys of large ones. Fig. 4.2 brings out the point that the slope of the species–area curve, and indeed its shape too, will depend, within limits, on the range of habitats chosen for study.

Despite all this variation the important questions remain:

(1) Why is the Arrhenius relationship the commonest?
(2) Why do some surveys clearly not fit the Arrhenius relationship?
(3) Why is the slope of Arrhenius relationship in general between 0.15 and 0.40?
(4) Why is there so much variation in this slope between surveys?

4.3 EXPLANATION OF THE SPECIES–AREA EFFECT

There are several possible explanations of the species–area effect, and these explanations are not necessarily contradictory; some of them are complementary. There is a natural tendency to try to find an explanation which fits what is perceived as the commonest case, that is to say the Arrhenius plot or the Gleason plot depending on the author, but any satisfactory explanation, or set of explanations, must address all the four questions listed above. For convenience I have grouped these explanations under four headings, those related to random phenomena, those related to the MacArthur and Wilson theory, those related to environmental heterogeneity, and other miscellaneous factors.

4.3.1 Explanations based on randomness

Preston (1962) seems to have been the first to suggest that the species–area effect could be a sampling effect. If all the species in the area are distributed randomly across the area but have different abundances, then with increasing sample size increasing numbers of species will be found. Real species of course have different distributions from each other, and most species show a non-random distribution, but nevertheless this sampling effect will exist, and is well-known from day to day experience in the field. On entering a new area, the number of species recorded in the groups of interest will at first rise quite rapidly, and then will progressively slow down. Examples of such sampling curves for birds are shown by Slud (1976). A distribution that comes in here is that known as the log normal. This is a distribution with the shape of the normal (or Gaussian) curve, but with the difference that, while the ordinate is arithmetic, the abscissa is on a logarithmic scale. Preston (1962), using various approximations, claimed that if the distribution of species abundances followed the log normal curve, then the species–area plot would be of the Arrhenius type, with a slope of about 0.26. This agreement between a typical species–area curve and Preston's theory has led some authors to claim that the observed species–area effects lend support to the view, originating with Preston, that species abundances follow a log normal distribution (Sugihara, 1981). There are two major difficulties with this explanation. The first is that Preston's approximation is too crude, and that random sampling does not lead to a straight line on the Arrhenius plot. The second is that species abundance curves are only approximately log normal; empirical distributions are often skew, and there are theoretical reasons for believing that log normal could not be met exactly (Williamson, 1981).

As the log normal explanation is so well embedded in the literature, it is perhaps desirable to discuss it in slightly more detail. As Brown and Sanders (1981) point out, log normal distributions have been proposed in many diverse areas, and the underlying idea has always been some form of the law of proportion of effect. That is, if it is natural to take logarithms of the variable and the central limit theorem can then be applied, a log normal distribution will follow. The problem with applying this recipe to species abundance distributions, as Pielou (1975) points out, is that the argument is only valid if the counts of different species can be regarded as samples of the same thing. In practice, workers have studied relatively homogeneous groups, such as birds or butterflies or copepods (McGowan and Walker, 1985). There is clearly a possible argument that the count of each species of, say, copepod is an independent sample from the category copepod, and therefore a log normal distribution might be expected, rather approximately, following the arguments above. Certainly if one takes a larger taxonomic spread of plankton, then the distributions found are obviously not log normal (Williamson, 1972).

All this argument is rendered somewhat academic, at least for small areas, by Coleman's (1981) derivation of the exact sampling distribution under rather tightly defined conditions. He assumes a finite area in which we know exactly the abundance of every species being studied. If, as before, each species is randomly distributed in this area, Coleman showed that taking samples of successively larger size within the area produced a species–area curve which is concave on the Arrhenius plot (and convex on the Gleason plot). To be precise, at the smallest area, where only one species is sampled, the slope of the Arrhenius relationship will be 1, declining progressively to h(1) when the full area has been sampled (where h(1) is the proportion of species in the area represented by exactly one individual). Examples found where Coleman's random placement curves fit real data can be seen in Coleman *et al.* (1982) and in Fig. 4.3 derived from Hubbell and Foster (1983). In Fig. 4.3, the slope can be seen to be 1.00 at the lefthand side, and it flattens to 0.11 because there are 21 species out of 186 on the plot represented by only one individual. Extrapolating the curve at this slope predicts 274 species for the 15.6 km^2 island, against 270 known.

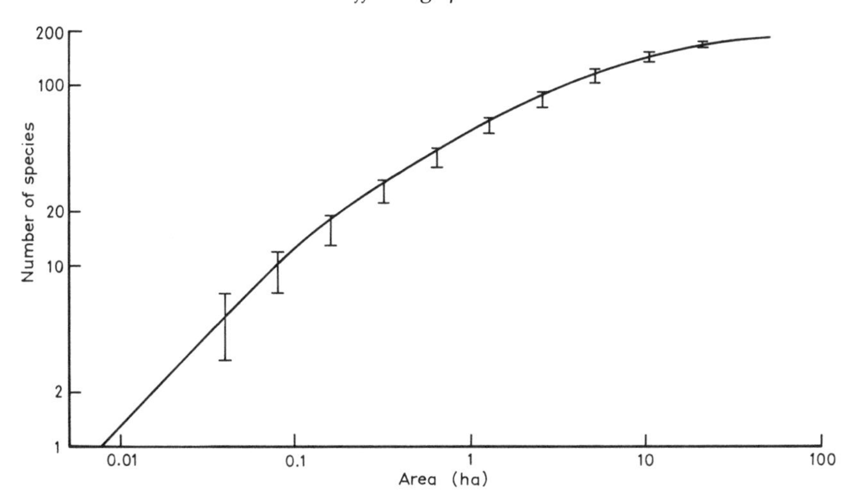

Figure 4.3 A species–area curve from random sampling. The continuous curve is a Coleman curve derived from the abundance of 186 tree species (greater than 20 cm dbh) in 50 hectares in Barro Colorado Island, Panama. The bars indicate the number of species observed in different sub areas. Modified from Hubbell and Foster (1983).

Note that this theory only predicts the shape of the curve up to the largest area in which the abundances of all the species are known. As Coleman also gives the variance of his relationship (and the mean and variance are also given be Hubbell and Foster) it is possible to test any observed distribution for goodness of fit to Coleman's theory. Clumping within the species, and the non-random distribution of species presence, are bound to make this theory break down above a certain area. It is a matter of empirical investigation to discover how large the area is in which the fit is satisfactory. Coleman's theory does at least explain the steepening of the species–area curve frequently seen at the left-hand side of an Arrhenius plot (Williamson, 1981).

Another very different argument suggesting that the species–area slope is an artefact of random sampling comes in Connor and McCoy (1979), and Connor *et al.* (1983), replying to a comment by Sugihara (1981). Their argument is that if three values are known, then the slope, z, of the Arrhenius plot is fixed. The values they take are the standard deviation of the area, the standard deviation of the number of species, and the correlation between these two variables. Their conclusion follows from their premise, though it is not clear why they think that the three variables named should behave in the way that they are observed to. More generally one can say that for any bivariate distribution, studied by least squares with two variables labelled x and y, there are five primary quantities to be calculated, namely Σx, Σy, Σx^2, Σy^2 and Σxy. These are generally combined to give estimates of the two variances and the covariance, giving three empirically observed variables. From these three it is possible to calculate the regression of y on x, the regression of x on y, the correlation coefficient, and the standard deviations of x and y. It seems to be no easier to explain the observed values of the variances and covariances than it is to explain the observed slope of the Arrhenius relationship. All these variables are measured empirically. While it could be argued that the variance of the area is a consequence of the sampling design, the other values, the variance of the species and the covariance, reflect what actually happens in nature, and are in no sense artefacts. The only curve in this field that could be called an artefact is Coleman's random placement curve, in that it will be found if the sampling design is restricted to samples of different size within an area, and if the species are randomly distributed. However as there is no necessity *a priori* for the species to be randomly distributed, and indeed as they are well known in general not to be so, even Coleman's curve cannot reasonably be regarded as an artefact, but rather as a test for deviation from randomness.

In small areas, the effects of random sampling will dominate the species–area curve. Coleman's theory is important, and displaces much that has been written earlier on the subject, but it will often be difficult to apply because it requires a complete enumeration of all species. The shape of the curves produced by this theory is demonstrated in Fig. 4.3, where it can be seen that, for tropical forest trees, 'small' is as much as 20 hectares. Fig. 4.2 shows that for herbaceous terrestrial plants in Britain, small is at least four orders of magnitude less, at 20 m^2 or less. In future surveys it would be desirable if Coleman's effect was always looked for.

4.3.2 The MacArthur and Wilson theory of island biogeography

Many recent authors ascribe to MacArthur and Wilson, the idea that the Arrhenius plot will be straight and claim that this relationship is predicted by their theory. Both claims are wrong. The history has been outlined above; as will be seen, the theory in simple form makes no precise prediction about the shape or the slope of a species–area relationship, though in more elaborate forms, predictions are possible which do not match the Arrhenius plot.

Of the basic features of the MacArthur and Wilson theory (Chapter 15), I need only mention the main constituents. They postulate that the immigration rate will be a declining function of the number of species on an island, reaching zero at a point P, the pool number of species. The extinction curve is postulated to be an increasing function of species number, and the number of species on an island will be in equilibrium where the immigration and extinction curves cross. The simplest elaborations of the theory suggest that immigration will be dependent on distance, and extinction on island size. No scaling is suggested for any of the variables, so they are defined implicitly merely as monotonic functions. According to MacArthur and Wilson (1967, p. 25) "Any transformation of the ordinate is permissible, though not all immigration and extinction curves can be simultaneously straightened", and it is clear that they envisage that either I (the immigration rate), E (the extinction rate), or S (the number of species on the island), could be subject to transformation to produce linear plots, though some of the subsequent algebra done by them on these plots is not valid after transformation. The essence of the theory is that the number of species on an island will be at an equilibrium between immigration and extinction, and that there will be fewer species on more distant islands and on smaller ones. As the theory was devised to produce an explanation of these two latter points, it cannot be tested independently by examining them. A different interpretation is given in Chapter 15.

There are in fact only three phenomena put forward by MacArthur and Wilson (1967) that can be tested as predictions from the simple theory in the present context. These are firstly, that the species–area relationship will be steeper on more distant islands, secondly, that the ratio of variance to mean of the number of species on an island over a period will be about 1.0 initially, falling to 0.5 later, and thirdly, that the colonization curve of an empty island will be asymptotic. Of these, the first is discussed below and appears to be wrong. The second is not true from the data known to me (Williamson, 1981, 1983b, 1987). The third is a very weak prediction, in that almost any theory of colonization would predict an asymptotic curve. Experimental testing of various aspects of the model is considered in Chapter 15. What the theory does not predict is the shape or slope of the species–area curve. McGuinness (1984) has surveyed tests that have been made of the theory, and notes that 60% of them have tested to see whether the Arrhenius plot is straight, while the other 40% have tested the colonization curve. It is sad to note that the former are irrelevant and the latter too weak to be of any value.

As I have noted before (Williamson, 1981), there have been two elaborations of the MacArthur and Wilson theory which do allow a prediction of the shape of the species–area relationship. These are by Schoener (1976b) and by Gilpin and Diamond (1976), and neither predicts either an Arrhenius or a Gleason relationship. Schoener's main conclusion is that:

$$z = 1 - (1/(2 - (S/P)))$$

namely, for a set of islands all with the same P, the pool of number of species, the slope will vary with the equilibrium number of species, S, in a systematic way and so will not be a straight line. Both this study and Gilpin and Diamond's assume that there is a number P that can be determined. As the species–area relationship, empirically, applies at all finite areas, so a particular size P must relate to a particular size area. Both theories indicate that the equilibrium number of species will approach the pool number as the area goes to infinity, or, to put it another way, imply that the pool species are found in an area of infinite extent.

Gilpin and Diamond's theory is very much more elaborate, and they attempt to fit a great variety of functional forms to the data on the number of bird species in lowland areas of the Solomons, taking into account the area and the isolation of the islands. Contrary to the usage in MacArthur and Wilson, their pool is the total number of species in the archipelago. For our purposes, their conclusion about the shape of the species–area relationship is the most important. Referring to Gleason and Arrhenius plots, they say, "for a sufficiently wide range of S and A values [the results are] linear on neither graph", and they conclude that of the two the Gleason plot is the better. At least they have avoided the usual fallacy of thinking that MacArthur and Wilson theory indicates that the Arrhenius plot is to be expected.

So, in relation to the four questions posed in Section 4.2.2, the MacArthur and Wilson theory offers no helpful answers.

4.3.3 Environmental heterogeneity

All naturalists know that environments are heterogeneous, and that different species have different distributions. So it is a natural and frequently made suggestion that environmental heterogeneity is the explanation of the species–area relationship. The nature of environmental heterogeneity is discussed in more detail in Section 4.4, as the complexities that are now appearing require treatment on their own.

In recent years the habitat heterogeneity explanation has fallen out of favour (McGuinness, 1984). Apart from the attractions of a more abstract theory like the MacArthur and Wilson one, there are difficulties in quantifying the way the environment varies, and relating this to the number of species. Further, it has not been clear that the environment varies in such a way as to answer the questions posed in Section 4.2.2 helpfully. Yet the importance of habitat heterogeneity is obvious in figures such as Fig. 4.4. This figure shows the variability in surface geology of a small area in north-west England. The striking thing about the figure is that it is linear on the Arrhenius plot, and that a variety of

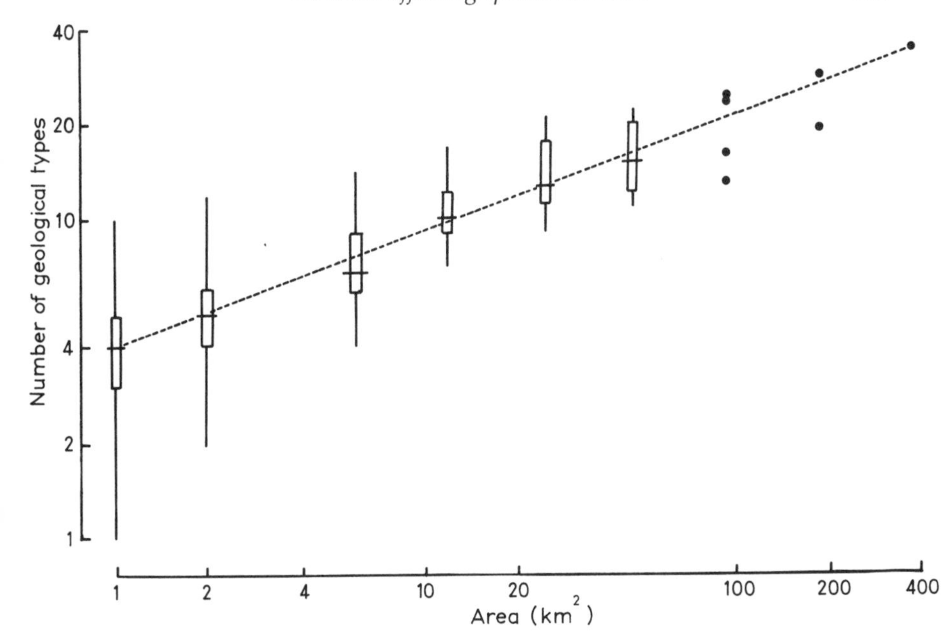

Figure 4.4 Number of geological types in areas of different sizes in Cumbria, England. Note the logarithmic scales. The spread at each area is shown by Tukey (1977) box and whisker plots. Each box spans the hinges (i.e. rounded quartiles), and has the median marked in it. The whiskers go from the hinges to the extreme values. From Williamson (1981). cf. Table 4.2, which gives hinges etc. for another data set.

slopes can be derived from it if samples had been taken in the ordinary botanical way of starting at a point and using increasingly large quadrats around that point. Indeed, given the sort of relation seen in Fig. 4.4, the variety of relations seen in Fig. 4.2 follow naturally. So, in principle, environmental heterogeneity may be involved in answering all four questions of Section 4.2.2. I will return to this possibility in Section 4.4.

4.3.4 Evolutionary and other effects

The relationship of species number to variables other than area is considered in Section 4.5. Here I mention some well-known effects on the number of species in a given area which would affect a species–area relationship, though not give rise to one.

The first is the effect of climate and other geographical variables on species richness. These are considered in detail in Chapter 3. Here we need only note that data sets that include samples from rather different climates may well give misleading species–area results. For instance, one reason why MacArthur and Wilson (1967) thought that species–area curves were steeper on more distant archipelagos was that the archipelagos that they chose to represent distant areas tended to have drier and otherwise less favourable climate for birds than those they chose for near areas.

Another effect, noted in more detail in Williamson (1981) is that on some islands there is a deficiency in species compared with that expected, apparently because dispersal to the islands has been very slow, and evolution has not yet had time to make up the deficiency. To repeat the aphorism put forward there, on oceanic islands evolution is faster than immigration. Examples discussed in detail in Williamson (1981) include the picture-wing *Drosophila* on the large island of Hawaii, and *Anolis* lizards on the lesser Antilles. This effect, like the climatic one, will tend to make some species–area plots more variable, and so go some way to answering questions (2) and (4) of Section 4.2.2, namely why the Arrhenius relationship does not always fit and why the relationship is variable between different studies. More systematic effects of evolution, leading to regularity in species–area curves, will be considered in Section 4.5.

4.3.5 Conclusions

The causes of species–area effects turn out to be rather simple. In small areas passive sampling effects will be important, possibly overwhelmingly so. In large areas, the effect of habitat heterogeneity will take over. There is some complication caused by systematic change in environmental variables such as climate and, in a few cases, by the effects of evolutionary history. MacArthur and Wilson's theory of island biogeography turns out to have rather little to say about the shape of the species–area relationship. It is now time to have a closer look at the complexities of environmental heterogeneity.

4.4 THE NATURE OF ENVIRONMENTAL HETEROGENEITY

4.4.1 The reddened spectrum

It has been observed independently by many authors that every point on the earth is slightly different in its characteristics from every other point. This holds true both in space and time. Further, closer points are more like each other than distant points, or, conversely and on average, the further off in space or time the more different. These well-known observations can be put into mathematical form, and can be related to the concept of fractals (Mandelbrot, 1983), as described in Section 4.4.2.

Time series are normally analysed by spectral analysis, and space variations may be studied in this way too. In the same way that Isaac Newton showed that white light can be split into a spectrum of different colours, so other phenomena including time and space series can be resolved into sets of waves of different frequencies or wavelengths. In the spectrum of light, each colour is specified by its wavelength, and the brightness of a colour is measured by the amplitude (the root mean square deviation) or the power (the mean square deviation) of its wave. The end result of a spectral analysis of light or of space variations is a plot of the power at a given frequency against the frequency. Frequency is the reciprocal of the wavelength, and the power can be thought of as the variance associated with points spanning a particular wavelength. In practice it is often found that plotting the logarithm of the power against the logarithm of the frequency gives an approximate straight line. This line shows the greatest power (or amplitude) at the longer wavelength, i.e. the lower frequencies, or, by analogy with light, at the red end of the spectrum. That is, there is a general rule that environmental heterogeneity has a reddened spectrum (Fougere, 1985; Steele, 1985; Williamson, 1981; 1987).

Colloquially, what this amounts to is that, for many environmental variables, the further apart measurements are made in space or time, the more different they will be. Points one metre away from a datum point, such as the point at which you are sitting, are in their environmental variables more like the datum point than points one kilometre away, and again the points one kilometre away are more like the datum point than the set of points one thousand kilometres away. What the reddened spectrum does is to put this very familiar phenomenon in to mathematical form, and to show that this familiar phenomenon is a systematic and continuous one.

The slope of the reddened spectrum varies in different studies from -1 to -3, plotting the logarithm of the power against the logarithm of the frequency, or equivalently from $+1$ to $+3$ for the logarithm of the power against the logarithm of the wavelength, as the logarithm of the wavelength is simply the negative of the logarithm of the frequency. Sayles and Thomas (1978) brought together twenty-three studies of surface roughness of fields and other natural surfaces, and of artificial surfaces such as ball-bearing races. They showed that the spectra grouped neatly around a value of $+2$. The position of the spectra, the intercepts, which Sayles and Thomas called the topothesy, varied enormously, over fourteen orders of magnitude, from a larva flow to a ground disc. The topothesy defines the statistical geometry, the apparent roughness at a standard scale, while the spectrum shows how the roughness varies with scale.

In the typical case, when the logarithm of the variance scales as twice the logarithm of the wavelength, note that, because the standard deviation is the square root of the variance, the logarithm of the standard deviation will scale directly as the logarithm of the wavelength. This standard deviation is a linear measurement of environmental heterogeneity, so I will give it the symbol H. If the wavelength is called L, then $\log H$ is proportional to $\log L$, i.e. $\log H = \log L +$ a constant. This is the basic mathematical representation of the observation, discussed above, that the further apart they are measured, the more different environmental measures are.

4.4.2 Fractals

At first sight, the invention by various mathematicians over the last hundred years of curves that are not differentiable at any point seems purely abstract, and such curves were given rude names such as pathological. An example of such a curve is the Koch snowflake curve, constructed as follows. Take an equilateral triangle, and erect a smaller equilateral triangle on the centre third of each side. That gives a six-pointed figure with twelve sides. Repeat using the twelve sides to get an eighteen-pointed figure, with thirty-six sides, each one ninth the length of a side of the original triangle. This process can be repeated indefinitely with smaller and smaller triangular projections. It can be shown that while the area inside the curve is limited, finite, the length of the perimeter, the length of the curve itself, becomes infinite. The Koch curve is self-similar: any small bit consists of a set of small equilateral triangular projections on a line and is similar to any larger bit, with larger equilateral triangular projections on a line. Eventually, in the limit, each part of the curve is either a point or an indentation, and so cannot be differentiated in the mathematical sense. That is, there is no definable tangent at any point. Stages in the construction of this curve are shown in Mandelbrot (1983), but are easily made with pencil and paper. The final figure is a fuzzy six-pointed star, hence the name snowflake curve. The limit curve is of infinite length, differentiable nowhere, but encloses a finite area and its general appearance is easily envisaged.

Order was brought to this part of mathematics by the realization that such curves could be described by a dimension between one and two. A curve of dimension one is an ordinary euclidian curve, while curves of dimension two fill a two-dimensional space exactly. In between, curves like the Koch curve have a fractional dimension between 1 and 2, 1.262 in this case. These curves have been named fractals by Mandelbrot (1983). In a series of books and papers he has shown that, far from being 'pathological', many fractal curves imitate nature rather beautifully. Many aspects of the real world may be fractal. For instance, Morse *et al.* (1985) showed that the leaves and stems of various plants have fractal edges, and that this has consequences for the arthropods living on them.

Fractals may be constructed in various ways, but in general will necessarily have a reddened spectrum, revealing more detail at smaller scales. Self-similarity is an aspect of this. The construction of fractals with a variety of different slopes to their spectra is possible using a method derived from a famous posthumous paper of Weierstrass (Berry and Lewis, 1980). However, another method of constructing fractal curves, much used by Mandelbrot, is to use a generating curve as in the construction of the Koch curve. In this case the displacement generated about a segment is directly proportional to the length of the segment. This displacement can be measured as a standard deviation: taking logarithms and using the argument in Section 4.4.1, such curves will have a reddened spectrum of slope two, the same slope as that found for environmental heterogeneity.

From the two statements that environmental heterogeneity frequently has a reddened spectrum from slope two, and that there is a class of fractal with a reddened spectrum of slope two, it does not necessarily follow that environmental heterogeneity is fractal. Logically, there is an undistributed middle. But the similarity of the statements, and the insight that they may give into the nature of environmental heterogeneity is clearly worth more study.

How does all this relate to species–area curves? Noting that area A involves the square of the length L, and ignoring the constants, the number of species in an area S can be related to the standard deviation of the heterogeneity H (defined in Section 4.4.1) as follows:

$$\log S = z \log A \quad \text{(the Arrhenius relationship)} \quad (4.3)$$

$$\log A = 2 \log L \quad \text{(by definition)}$$

$$\log H = \log L \quad \text{(the reddened spectrum equation)}$$

$$\text{therefore } \log S = 2z \log H \quad (4.4)$$

It might be expected that the number of species would scale exactly as the standard deviation of the environmental heterogeneity, but this is only so if the Arrhenius slope, z, is 0.5. In the usual cases, with z between 0.15 and 0.4, the increase in number of species is reduced compared with the increase in heterogeneity. The explanation of this may lie in the variation of the range of habitats used by each species, but that is only a suggestion.

The pattern of environmental variation is to some extent fractal. This fact, and the variation in fractal dimension, may, in time, lead to a deeper understanding of the causes of the variability of species–area curves, and to quantitative answers to the questions of Section 4.2.2.

The set of tentative arguments above at least allows an explanation of species–area curves based on environmental heterogeneity. Although Sayles and Thomas (1978) found a modal slope of 2.0 for log (H^2) against log L, the variation in individual slopes was from 1.2 to 2.8. This is the same proportional variation as is seen between the hinges of Table 4.2, suggesting that much of the variation in z might be found to be related to the rate of change of environmental variables in different ecosystems. If the Sayles and Thomas result is typical, there will be other explanations as well and some of these are considered below. But my major conclusion is that the observed pattern of environmental heterogeneity, as shown both in Fig. 4.4 and the reddened spectrum, is sufficient to answer all four questions of Section 4.2.2.

4.5 THE EFFECT OF OTHER VARIABLES ON THE SPECIES–AREA RELATIONSHIP

4.5.1 Distance

In island biogeography, the physical variable most frequently studied after area is undoubtedly distance (Chapter 15). Darwin and Wallace characterized oceanic islands as having, among other features, no terrestrial mammals and no amphibia. The observation that volcanic islands have no frogs or toads goes back to Bory de Saint-Vincent (1804). Modern studies involving multiple regression and related techniques, reviewed by Williamson (1981), put into more exact form the well-known fact that islands distant from continents have fewer species than near ones, and that the distance effect for a particular taxon is related to dispersal ability. To quote just one recent example, Opdam *et al.* (1984) found that the logarithm of the number of woodland birds in small lots in the Netherlands was related not only to the logarithm of the area, but also to the logarithm of three other variables, the area of forest within a 3 km radius, the distance to the species pool (i.e. the area of continuous forest) and the distance to the nearest woodlot.

However, in many of these cases, the effect of distance is negligible. Westman (1983), by taking standard size samples of plants on the California channel islands and on the Californian mainland, and by restricting his samples to one vegetation type, xeric shrubs, found that the difference between islands and different parts of the mainland related primarily to moisture availability, and not to the area or distance offshore of the islands. His work emphasizes in a different way the conclusion from all species–area studies: comparisons between different

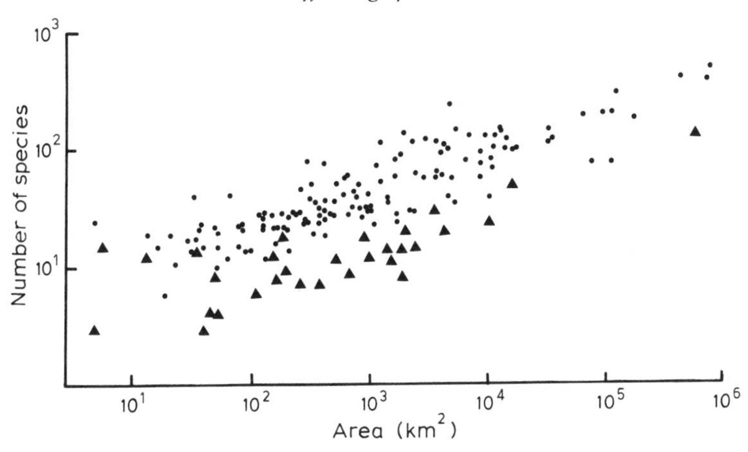

Figure 4.5 Species–area plot for the land birds of individual islands in the warmer seas. (▲) Islands more than 300 km from the next largest land mass, or in the Hawaiian or Galapagos archipelagos. Note the logarithmic scales. From Williamson (1981).

places should either be made on areas of the same size, or the effect of area should be controlled statistically. For islands, distance should be factored out. In comparisons between different biogeographic or climatic zones these statistical operations may be difficult.

The effect of distance, when it is appreciable, is to reduce the number of species at a given area, in other words to depress the species–area line, as can be seen in Fig. 4.5. MacArthur and Wilson (1967) predicted that the line would also become steeper. The example they gave is vitiated by using archipelagos rather than single islands, and by including archipelagos of different climate or deriving their avifaunas from different sources. Schoener (1976b), Connor and McCoy (1979) and Williamson (1985) all noted that the line tends to be flatter the more distant the archipelago. Williamson (1985) quoted, with some caveats, slopes of 0.28, 0.22, 0.18, 0.09 and 0.05 for birds in the East Indies, New Guinea, New Britain, Solomons and New Hebrides (Vanuatu) archipelagos, respectively, which are progressively more isolated.

4.5.2 Evolution

While the filter effect of different dispersal abilities is to lower and flatten the species–area line, so that in distant archipelagos all islands have the same biota of strongly dispersing species, the effect of evolution is to lower and steepen the species–area line, different islands now having different species. This effect has been noted by Williamson (1981, 1985)

Table 4.3. Numbers of endemic *Scaptomyza* spp. (Drosophilidae) in the Tristan da Cunha archipelago. From Williamson (1983a)

	Island			
	Nightingale	*Inaccessible*	*Gough*	*Tristan*
No. of species	7	5	2	2
Age (Myr)	18	6	6	1
Area (km^2)	4	12	57	86

and is best recorded for the West Indies. For West Indian butterflies, z is 0.08 for widespread species, 0.36 for restricted ones and 0.52 for endemics (Scott, 1972). For birds, using stages of the taxon cycle which again go from widespread to endemic species, the slopes are: stage I, 0.07; II, 0.15; III, 0.32 and IV, 0.42 (Ricklefs and Cox 1972). So, as dispersal flattens curves and evolution steepens them, the net effect may be close to the fairly even lowering seen in Fig. 4.4. For islands, these effects help answer questions (2) and (4) of Section 4.2.2, but it is doubtful if this is so for mainland samples.

In extreme cases, as was noted in Section 4.3.4, the effect of evolution can be to decouple the species richness from area. A simple example is given in Table 4.3, of the number of endemic, more or less flightless, Drosophilid flies on islands in the South Atlantic. These islands are formed by eruption, over a relatively short period of time, of volcanoes from the depths of the ocean. Once formed, the islands slowly erode until they disappear under the sea. Consequently, the older the island the smaller. The relationship of species number is clearly to island age, which in turn is inversely related to island area. The result is a strongly negative species–area relationship.

4.5.3 Other factors

As with distance, other factors have been studied by multiple regression, in particular elevation and habitat variation (Williamson, 1981). Shape may also have an effect. I will deal just with these three variables in this section, though others, such as latitude, have been studied.

High islands may have a series of climatic zones. In the Canary Islands, for instance, high islands have four zones, a sub-alpine zone, pine forest, laurel forest and a xerophytic zone, all simply related to the effect of the trade winds (Bramwell and Bramwell, 1974). The lower islands only have the xerophytic zone, and the result is seen in Table 4.4: the number of vascular plant species is related strongly to elevation, though there is still an area effect as well.

Table 4.4. Vascular plant species on the Canary Islands in relation to island height and area. From Carlquist (1974)

	Island						
	Tenerife	*La Palma*	*Gran Canaria*	*Hierro*	*Gomera*	*Fuerte-ventura*	*Lanza-rote*
No. of species	1079	575	763	391	539	348	366
Highest point (m)	3711	2423	1950	1520	1484	807	670
Area (km^2)	2060	728	1534	277	378	1725	873

Habitat information is more difficult to express in a quantitative form. Latitude has also been used occasionally, and may be a habitat indicator as much as anything. Indeed all the discussion on the nature of environmental heterogeneity in Section 4.4 confirms the widely held view that area is primarily an indicator of habitat. One difficulty is that while area is a reasonably objective measure and applies to all groups, the choice of habitat variable could well be subjective and vary from taxon to taxon. At this point, the argument merges into the general discussion of the effect of climate and other habitat variables on species richness, which is discussed in Chapter 3.

Still, habitat information can be used objectively, even when crude, and can help clarify the causes of variation seen in species–area relationships. A simple example comes from Dony (1970). The vascular plant studies he surveyed had in general been made on different areas within one major habitat type. From his compilation I find values for z from woodland surveys between 0.27 to 0.38, in more open habitats from 0.13 to 0.33, while the extreme flat open habitats such as bogs gave values from 0.11 to 0.15. Some of this variation can be seen in Fig. 4.2. As Dony notes, extrapolating the flat impoverished relationship found in the third set gives implausible figures, such as 81 species (rather than 2137) for the British Isles from the lowest line of Fig. 4.2, though as he says "one can well imagine that if the British Isles were one vast *Zostera* marsh it would yield as few as 81 species".

Finally, there is the question of the shape of the area sampled. One consequence of environmental heterogeneity having a reddened spectrum is that long thin samples can be expected to encompass a greater variety of habitats than square samples of the same area. This effect is well known on the small scale of quadrats in vegetation (Goodall, 1952), and Williamson (1987) shows that the effect holds for geological types in areas measured in square kilometres. However, biological records from various sets of islands have failed to show any consistent, or generally significant, effect of shape (Blouin and Connor, 1985).

4.6 CONSEQUENCES OF THE SPECIES–AREA EFFECT

In recent years, discussions of species–area effects have been dominated by the MacArthur–Wilson theory, and to a lesser extent by the log-normal distribution. It is perhaps time to bring the discussion back into the mainstream of biogeography, and consider the importance of the species–area effect for the study of speciation, adaptation, extinction, endemism, species richness and community structure. With the present primitive state of knowledge of species–area effects and their causes, I will only offer rather general comments.

Species–area plots, by their nature, emphasize the continuity of natural variation. Much of biogeography is concerned with discrete events. One of the difficulties of all environmental biology is that each real biological system "hovers in a tantalizing manner between the continuous and the discontinuous" (Webb, 1954). Viewing nature discontinuously, there are distinct, classifiable, ecological communities. Viewing nature as a continuum, these become arbitrary divisions in the range of number of species, areas and habitat heterogeneity. Both views convey aspects of the truth, and both are needed for comprehension. In multivariate analysis, it is often illuminating to study the results both of cluster analyses and of ordination, of the discontinuous and the continuous. My own view is that the continuous more nearly represents the truth, but the human mind finds it hard to grasp without the labels that accompany the discontinuous description.

There are some concepts taken over from the laboratory and everyday life that are difficult, in practice, to apply to ecosystems. Stability (Williamson, 1987) and equilibrium are two. Species–area studies have brought out the pervasive effect of the reddened spectrum of environmental heterogeneity. The effect is important throughout ecology and biogeography. Because all phenomena are perpetually changing at all scales of space and time, stability and equilibria are at best relative, and need to be studied in the context of particular, clearly stated, scales of space and time.

Species–area effects need to be allowed for in the discussion of all biogeographical phenomena. The importance of dispersal, evolution and environmental heterogeneity on the pattern of biogeographical distributions all come together in the species–area effect. By starting with simple data and simple graphical plots, the importance of theory, of empirical observation, and the interplay between them is brought out in a comprehensible way. Nevertheless, there is still much to be cleared up. The quantitative study of the effect of environmental heterogeneity is still in its infancy. Studies of species–area relationships have been restricted to narrow ranges of taxa. The effects of environmental heterogeneity on a wide range of taxa, at all trophic levels, has largely yet to be studied either empirically or in theory. The techniques needed are appreciably more difficult; the results should be correspondingly more interesting.

CHAPTER 5

Endemism: a botanical perspective

J. MAJOR

5.1 INTRODUCTION

A taxon is endemic if confined to a particular area through historical, ecological or physiological reasons. Endemics can be old (palaeoendemics) or new (neoendemics) and the endemic taxon can be of any rank, though it is usually at family level or below. The phenomenon of endemism is important to biogeographers, evolutionary biologists and ecologists in general. Whilst both plants and animals can be endemic, this chapter will use plants as examples although notable differences do exist (Harper, 1968). A historical discussion of plant endemism is given by Prentice (1976).

The area of endemism can be large (more than one continent) or small (a few square metres). This latter may make it possible to study the birth of a new species *in situ* (Section 5.5.4) which can, for example, occur through a change in ploidy.

Plants' demands on the environment are simple. They require a soil or even comminuted rock or an organic or water substrate for rooting; water itself; carbon dioxide; a limited number of mineral elements from the soil; radiation in the 400–700 nm band and temperatures above freezing during their growth period. Unlike animals, plants can be strictly limited in distribution by substrate composition (such as calcicoles, serpentine endemics and heavy metal plants), although many such restrictions may be related to competition rather than physicochemical requirements (Braun-Blanquet, 1964; Gankin and Major, 1964; Ellenberg, 1978).

Plants compete with each other for the above mentioned resources and lack the behavioural possibilities that enable animals to avoid biotic interactions and climatic extremes. However plants are more labile in form, growth, and reproduction than animals and therefore tend to have wider tolerances of climatic factors. They have especially wide tolerances when competition with other plants is minimized (Valentine, 1972) as expected from the discussions in Chapter 9.

The immobility of rooted plants makes them theoretically mappable and countable and usually only the seed stage is more or less mobile. The genetic individual and the phenotypic individual may not coincide and clonal growth can produce very large numbers of more or less adjacent, genetically identical 'individuals'. Harper and White (1974) emphasize these properties.

Since environments vary on a variety of time scales, plants must be plastic in their phenological responses. Life histories vary in duration from only a few weeks for some annuals to scores of years for herbs, a few centuries for shrubs, probably many centuries for some clonal herbs, up to 4000 years for some trees, and even longer for the clonal desert shrub *Larrea tridentate* (11 700 years) (Vasek, 1980). During such long life spans the environment changes, but the plants often persist. They also show remarkable adaptability to their environment. This adaptation can be genetic or phenotypic, and Bradshaw (1972) discusses many of the consequences of genetic adaptations. Gene flow can be extremely wide, extremely narrow, or even non-existent in apomicts, which reproduce without fertilization. Polyploidy is common among plants, and polyploids may or may not be classified as different species. The consequences of polyploidy for the formation and persistence of endemics are evident.

The purpose of this chapter is to define and illustrate the range of endemism in vascular plants. Endemism has a multifactorial dependency and it is important to evaluate the several axes of variation which may operate separately or in concert. These relate to size and age of areas, extermination and re-invasion of the biota of glaciated areas, island isolation, age and degree of splitting up of land masses, presence of land bridges, ages of floras and habitats, latitudinal, altitudinal and humidity variation, and differences in soil parent materials. Endemism is considered from a variety of viewpoints.

5.2 BIOGEOGRAPHICAL SIGNIFICANCE

Endemism is important when the history and origin of a flora are considered (Tolmachev, 1974a; Keener, 1983; Goloskokov, 1984). According to Braun-Blanquet (1923) who studied the origin and development of the flora of the Massif Central in France, "the study and precise interpretation of endemism of a territory constitute the supreme criterion, indispensable for consideration of the origin and age of its plant population. It enables us better to understand the past and the transformations that have taken place. It also provides us with a means for evaluating the extent and approximate timing of these transformations and the effects they produced on the development of the flora and vegetation". In general large areas have more endemics than do small areas. Taxa occupying a very large area are often assumed to vary ecotypically throughout their area, but the evidence is insufficient to make this an all-inclusive or even reliable generalization. Frequently, ecotypes have been considered as endemic species (Walter and Straka, 1970). Actually, some widely distributed species are composed of many ecotypes (Ernst, 1978). On the other hand some very widely distributed species have a remarkably similar ecological relationship to site throughout their ranges, even when widely disjunct. *Pinus sylvestris* from north-western Europe south to Spain and eastward to the Pacific is one example, and such taiga understory plants as *Pyrola secunda, P. minor, Vaccinium vitis-idaea, V. uliginosum, V. myrtillus,* or steppe species as *Artemisia frigida, A. rupestris, Ceratoides papposa* and *C. lanata,* or artic–alpines as *Kobresia myosuroides, Carex obtusata, C. limosa, C. microglochin, C. lachenalii,* are others.

Endemism is not just a species property. Plant and animal communities can also be endemic, but a community is an assemblage of individual species, and we must clarify species endemism before we take on community endemism. Endemism at the community level can be reflected in endemism at the species level (Tauber, 1984) as is becoming increasingly clear in the preservation of rare and endangered species through preservation of the plant communities in which they are found.

Taxonomic relationships are frequently illuminated by studies of endemics (Street 1978). Endemics may be relicts or recent. Widespread species indicate historical and geographical relationships while endemics characterize individual areas. Takhtajan (1978) employs endemism at various taxonomic levels from family to species to characterize floristic regions of the world, from his six kingdoms down through 35 regions to 153 provinces. Two of these provinces are the islands Norfolk and Lord Howe, only 36 and 28 km^2 respectively, situated between New Zealand and Australia. Probably the largest provinces are the Sahara and Central Siberia with areas of more than 5×10^6 km^2. The latter extends from the lower and middle Yenesei almost to the upper Lena and to its tributary Aldan River, north to the tree line and south to the Sayan Mountains. According to Takhtajan, floristic provinces of such different sizes are more or less equivalent in floristic uniqueness as shown by their degrees of endemism. The biogeographical history of the plants and animals of the area within one floristic unit must have produced some unifying properties and relationships. However, it is difficult to generalize about endemics which are known in detail to only a few local, specialist researchers. Even then details on life histories, demography, physiology and reactions to competition may be largely lacking (Keener, 1983). However, Braun-Blanquet's study (1923) is exceptional. Any generalizations must be built on thorough field studies, supplemented by laboratory, garden or greenhouse, and herbarium, investigations. We therefore have few generalizations about endemics or endemism, but a meaningful classification of endemics would at least provide a framework and suggestions are discussed below.

5.3 A MEASURE OF ENDEMISM

The degree of endemism increases as an area homogeneous in ecological conditions and floristic history is increased in size. The figures for California (and its constituent parts), the Alps and the Tien Shan (Table 5.2) illustrate this point. Bykov (1979, 1983) uses this relationship to define an index to endemism.

The logarithm of the number of plant species occurring in a series of areas of increasing size increases linearly with the logarithm of area (Chapter 4). Evidently a similar relationship holds for percentage endemism and area (Fig. 5.1). Since the totality of all vascular plants is endemic to the land area of the earth, at this area endemism is 100%. It is less for any smaller area. Bykov (1979, 1983) suggests it is 1% at about 625 km^2. According to Tolmachev (1974a) the area of a concrete flora (defined as the flora of an area of the minimal size required to contain all the plant species of a region – a larger area producing different species in the same kinds of habitats) changes from about 100 km^2 in the arctic to 500 km^2 in the southern taiga. At any rate the small Aksu–Dzhabarlinskii Reserve has 1.3% endemism on 700 km^2. This size should give a concrete flora and not one truncated in number by the sample area being too small. This last point and that for the land area of the earth and its total flora define the straight line of Fig. 5.1 on a log-log plot. Places falling above the line have less than normal endemicity; below the line more. Deserts are above the line; mountainous and tropical areas are below it. As Bykov notes, the precise size of the area at 1% endemism is really not critical. The slope of the line changes little over the area range of 300 km^2 to a few 1000 km^2.

A quantitative index is $I_e = E_f/E_n$ or if $E_f < E_n$ then $I_e = -E_n/E_f$ where I_e is the index of endemicity, E_f is the factual percentage endemism and E_n is the normal percentage endemism read off the line of Fig. 5.1. A value of 1 indicates the area has the normal expected degree of endemicity; a value of above 1 indicates greater than normal and below 1 less than normal endemicity. As pointed out earlier, the degree of endemism appears to be a function of a multitude of factors. Unfortunately, there is still disagreement about delimitation of species, and the floras of tropical regions in particular are poorly known (Street, 1978; Gentry, 1986). The distributions of species are not completely mapped, and the history of

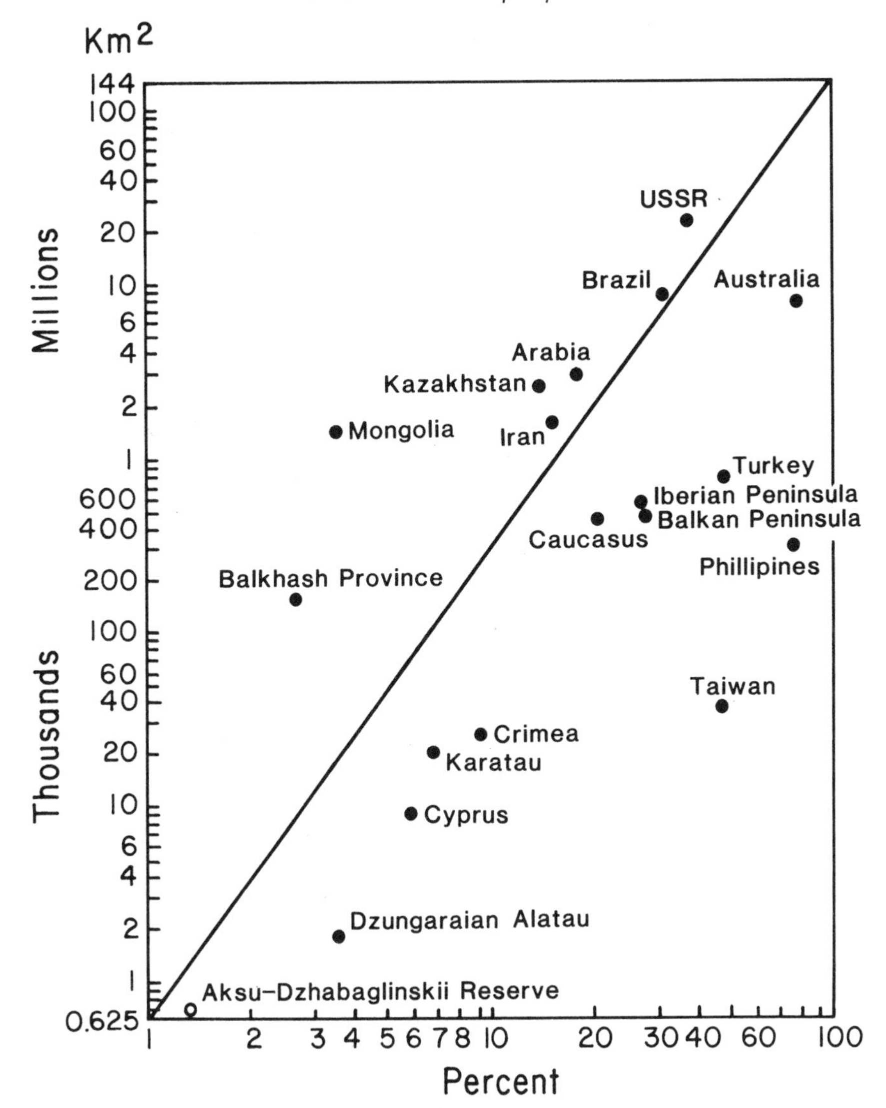

Figure 5.1 Nomogram for determining normal endemism. The ordinate is area, the abscissa % endemism. Points above the diagonal line have lower than normal endemism, points below have more. Data on the points labelled on the nomogram are in Table 5.1. From Bykov 1979, 1983.

development of species' areas is even less well known. Also the factors which we suppose determine endemism are themselves strongly hypothetical. Overall it seems premature to rely exclusively on any one index such as Bykov's to analyse endemism.

5.4 EXTENT OF AND ECOLOGICAL VARIATION IN ENDEMISM

A worldwide compilation of floras, their locations, areas and endemism is assembled in Table 5.2 and is supplemented by Table 5.1. The floras are listed geographically from the arctic through temperate climatic zones (maritime and continental), deserts, islands, tropics and the southern continents.

Data on endemism would be most useful if given by floristic provinces rather than by political subdivisions (see California in Table 5.2). However, Gentry's (1986) map of numbers of endemic vascular plants in the individual states of the United States shows several typical regularities. Many endemics occur in larger states with great contrasts in topography, on the borders of floristic provinces, and toward the south. There is tremendous variation between states, with California having 1517 endemics, Hawaii 883, Florida 385, Texas 379, Utah 169, Arizona 164 but all the mid-western and north-eastern states <6. The giant state of Alaska has only 80 endemic vascular plants although its interior, northern and southern coasts were unglaciated refugia, and it was

Table 5.1. Bykov's data on land areas with their numbers of vascular plant species and degrees of endemism (1979, 1983). (I_e is the index of endemicity)

Country	Area 10^3 km^2	*Flora*		I_e
		No. of spp.	*% endemism*	
Earth's land area	144000	230000	100.0	1.00
Soviet Union	22402	15000	37.5	−1.36
Brazil	8515	40000	30.0	−1.16
Australia	7631	12000	75.4	2.29
Arabia	3000	1808	17.5	−1.22
Kazakhstan	2756	5239	13.0	−1.69
Iran	1642	6150	14.6	−1.13
Mongolia	1565	1876	3.4	−5.30
Turkey	780	4650	48.0	3.44
Iberian Peninsula	582	5660	26.0	1.92
Balkan Peninsula	505	6530	26.9	2.24
Caucasus	440	5767	19.8	1.72
Phillipines	299	7620	72.6	7.41
Balkhash Province	160	755	2.6	−3.00
Taiwan	36	3265	48.0	10.66
Crimea	25.5	2200	9.0	2.25
Karatau	20	1472	6.6	1.83
Cyprus	9.3	1170	5.9	2.18
Dzungarian Alatau	1.9	2080	3.5	2.33
Aksu–Dzhabaglinskii Reserve	0.9	1306	1.3	1.06

Table 5.2. Degrees of vascular plant endemism in various regions of different sizes. % endemism is of total species numbers. Locations and areas have been taken from the references cited, Webster's dictionary, the Encyclopedia Brittanica, and scaled from available maps (e.g. Atlas Mira)

Region	*Latitude*	*Longitude*	*Area*	*Total flora*	*Endemics*		*References*
	°N,S	*°N,S*	*10^3 km^2*	*no.*	*no.*	*%*	
Far North							
Arctic	68–80°N	360°	4000	1000	±5	0.5	Yurtsev, Tolmachev and Rebristaya, 1978
Alaskan Arctic Slope	60–71°30′N	140–166°W	222	±500	1	0.2	Wiggins and Thomas, 1962
Bolshezemelsk tundra	68–70°N	60–65°E	38.0	475	1	0.2	Rebristaya, 1977
Greenland	60–83°N	20–53°W	2180	496	36	7.3	Böcher, Holmen and Jakobsen, 1968[1]
Iceland	63–67°N	14–25°W	103	540	17	3.15	Löve and Löve, 1963[2]
Alaska	55–71°N	141–168°W	1479	1366	80	5.9	Raven and Axelrod, 1978
Queen Charlotte Islands	52–54°N	131–133°W	10.30	593	12	2.02	Calder and Taylor, 1968[3]
Putorana mts.	66–70°N	85–102°E	31.3	569	5	0.9	Malyshev, 1976: 187–195
Yakutia	56–73°N	105–160°W	3065	1560	93	5.96	Tolmachev, 1974c, Karavaev, 1958
Uda River Valley	54°N	130–135°E	35	429	0	0	Doronina, 1973
Western pre-Okhotsk	54–56°N	130–135°E	40	480	2	0.42	Shlotgauer, 1978[4]
Suntar Kayata mts	62°30′N	140°10′–141°30′E	?	301	4?	1?	Yurtsev, 1968
Europe							
Europe	35–80°N	28°W–60°E	10000	10500	3500	33	Webb, 1978
Norden (Scandinavia +)	55–71°N	5–41°30′E	550	1938	55	2.84	Sjörs, 1956, Hulten, 1970
British Isles	50–55°N	2°E–10°W	308	1443	17	1.2	Raven and Axelrod, 1978
			226	1750	23	1.31	Street, 1978
Britain	50–59°N	2°E–8°W	?	?	21		Street, 1978
Ireland	51°26′–55°21′N	5°25′–10°30′W	82	1210	2	0.165	Street, 1978
Czech SSR	49–51°N	12–19°E	78.86	?	7		Hendrych, 1981
Slovak SSR	47°35′–49°30′N	17–22°45′E	49.01	?	25		Hendrych, 1981
Poland	49–53°40′N	14–24°E	310	2300	27	1.17	Szafer, 1966
Volyn Podolia (Ukraine)	48–51°N	28–31°E	130	1865	69	3.7	Zaverukha, 1980

Table 5.2. *(contd.)*

Region	*Latitude* °N,S	*Longitude* °N,S	*Area* 10^3 *km*^{2}	*Total flora* *no.*	*Endemics* *no.*	*Endemics* *%*	*References*
European high mountains							
Ural high mts	52°30′–68°N	58–66°E	50	521	44	8.44	Gorchakovsky, 1975
Carpathian mts	50–45°N	18–27°E	100	900	110	12	Valentine 1972[5]
Ukrainian Carpathians	48–49°N	22°10′–25°E	8	476	77	16	Chopik, 1976
Alps	43°30′–47°30′N	5°30′–16°E	110	1049	331	31	Valentine, 1972
Western Alps	46°30′–48°30′N	5°30′–8°E	30	805	105	13	Valentine, 1972
Eastern Alps	46°–47°30′N	7°30′–16°E	80	866	154	18	Valentine, 1972
Pyrenees	41°N	2°W–2°30′E	12	720	103	14	Valentine, 1972
Baetic Nevadan	37–38°N	2–4°W	4	349	125	36	Valentine, 1972
Central Appenines	42–43°N	13–14°E	9	268	36	13	Valentine, 1972
Corsica, mts	41°20′–43°40′N	9°E	3	142	54	38	Valentine, 1972
			<3	127	69	54	Contandriopoulos and Gamisans, 1974[6]
Balkan mts	40–44°N	15–25°E	110	1193	404	33.9	Valentine, 1972
				529	203	38.5	Valentine, 1972
Southern Greek mts	37–39°N	21–23°E	16	475	174	37	Valentine, 1972
Caucasus							
NW Caucasus	43°40′–44°N	40–40°30′E	1	819	287	35.1	Tolmachev, 1974b
Racha-Lechkhumi	42°30′–43°N	42°30′–43°30′E	3.8	1198	9	7.5	Gagnidze and Kemularia-Natadze, 1985[7]
Talysh high mts	38°25′–39°N	48–48°45′E	1.2	709	>72	10.1	Gadzhiev, Kuleva and Bagabov, 1979
Daghestan mts	41°15′–43°15′N	45–49°E	22		72		Elenevsky, 1966
Caucasus	41–44°N	40–49°E	65	6350	1600	25.2	Tolmachev, 1974b
United States and Japan							
Gray's Manual area (NE U.S.)	36°40′–51°N	53–96°W	3238	4425	599	13.5	Raven and Axelrod, 1978[8]
Carolinas	30–36°30′N	76–84°W	217	2995	23	0.8	Raven and Axelrod, 1978
Japan	30–45°N	128–145°E	380	4022	1371	34.1	Raven and Axelrod, 1978
Texas	26–37°N	94–106°W	693	4196	379	9.0	Raven and Axelrod, 1978[9]
South Dakota	43–46°N	96°30′–104°W	200	1585	5	0.3	Van Bruggen, 1976
Utah	37–42°N	109–114°W	220	2580	117	4.5	Welsh and Chatterley, 1985
Uinta Basin	39°15′–41°N	108–111°15′W	38	1600	30	2	Goodrich and Neese, 1986

Mediterranean							
Balkans	36°10′–45°40′N	13°30′28°30′E	470	?	1754	?	Horvat, Glavic and Ellenberg, 1974
Greece	36°10′–41°30′N	19°35′–26°30′E	129	±5500	±1100	20	Horvat, Glavic and Ellenberg, 1974
Albania	39°40′–42°30′N	19°20′–21°E	27.5	2450	157	6.41	Horvat, Glavic and Ellenberg, 1974
Corsica	41–43°N	8°30′–9°30′E	8.75	2148	279	13.0	Gamisans, Aboucaya and Antoine, 1985
Monte Albo, Sardinia	40°30′–40′N	9°20′–40′E	0.085	659	49	7.5	Camarda, 1984
Balearic Islands	39–40°N	1°30′–4°E	5.01	2230	156	7.0	Contandriopoulos and Cardona, 1984
Crete	34°50′–35°40′N	23°25′–26°10′E	6.29	1377	132	9	Valentine, 1972
Israel	29–33°N	34–36°E	20	2380	155	6.5	Shmida, 1984
Middle Asia							
Kuznetski Alatau	53°30′–55°N	88–90°E	8	332	16	4.8	Sedelnikov, 1979[10]
Altai mts	49–52°N	82–90°N	170	1840	107	5.8	Kuminova, 1960
Altai high mts	49°30′–50°10′N	86°15′–88°10′E	6	644	?	?	Krasnoborov, 1976
Western Sayan mts	51–54°N	88–97°E	100?	601	10	1.5	Krasnoborov, 1976[11]
Eastern Sayan mts	51–54°30′N	95–102°E	51.0	540	21	3.9	Malyshev, 1965
Stanovoi mts	55–59°N	112–120°E	120	602	15	2.5	Malyshev, 1972
Kazakhstan	40–55°N	47–87°E	2500	4750	550	11.6	Kamelin, 1973
Tarbagatai mts	46–48°N	80–84°E	40	1640	169	10.3	Stepanova, 1962
Tarbagatai high mts	47°N	82–84°E	4	479			Krasnoborov, 1976
Dzungarian Alatau	44–46°30′N	77°30′–82°30′E	12	2168	76	3.5	Goloskokov, 1984
Syr Darya Karatau	43–45°N	67–71°E	15			7.5	Tolmachev, 1974b
Tien Shan-Alai	41–43°N	70–82°E	180	6525	3370	51.6	Tolmachev, 1974b
Northern Tien Shan				2500	225	9.0	Tolmachev, 1974b
Western Tien Shan	41–43°N	69–74°30′E	45	2812	358	12.8	Tolmachev, 1974b
Talasski Alatau	42°19′N	70°25′–74°E	3.0	1491	25	16.8	Karmysheva, 1982
Nuratinski mts	40°20′–40°40′N	66–67°30′E	2	686	20	2.9	Zakirov, 1969, Kamelin, 1973
Nurat. mts + Kizyl Kum hills	40–41°30′N	65°30′–68°E	25	983	43	4.5	Zakirov, 1971
Zeravshan River Basin	40°N	66–69°E	5.0	2588	200	7.71	Zakirov, 1961
				2615	196	7.51	Tolmachev, 1974b, Kamelin, 1973
Kashkadarya Basin	39°40′N	67–67°40′E	2.4	1000	30	3.0	Kamelin, 1973
Hissar Range, alpine	39°N	67°40′–69°50′E	11	314	17	5.8	Tolmachev, 1957[12]
Fanski Range				1800	45	2.5	Kamelin, 1973
Barzob River Basin	38°30′–39°N	68°40′E	1.40	1535	28	1.75	Kamelin, 1973[13]
Zailiskii Alatau						2	Tolmachev, 1974b
Pamir	37°30′–39°30′N	73–75°E	25	700	32	4.6	Tolmachev, 1974b

Table 5.2. *(contd.)*

Region	*Latitude*	*Longitude*	*Area*	*Total flora*	*Endemics*		*References*
	°N,S	°N,S	10^3 *km*^{2}	*no.*	*no.*	*%*	
Karakorum	33–47°N	74–79°E	70			2.5	Tolmachev, 1974b
Tuva	50–53°30′N	89–99°E	170	1576	15	0.95	Krasnoborov, *et al.*, 1980, Sobolevskaya, 1953
Eastern Tannu- Ola Range	50°50′N	93°30′–95°20′E	3.8	973	0	0	Khanminchun, 1980
Mongolia	42–51°N	88–122°E	1565	2239			Grubov, 1982, Krasnoborov and Khanminchun, 1980
California							
California	32°30′–42°N	114–124°W	411	5046	1517	30.1	Raven and Axelrod, 1978
Calif. floristic Province	30°–43°15′N	115–124°W	324	4452	2125	47.7	Raven and Axelrod, 1978
Trinity Alps	40°06′–41°07′N	122°47′–123°07′W	0.437	571	7	1.23	Ferlatte, 1974
Marin County	37°50′–38°20′N	122°05′–123°W	1.37	1313	10	0.76	Howell, 1970
Mt. Diablo	37°45′N	121°55′W	0.146	525	6	1.14	Bowerman, 1944[4]
Santa Cruz mts	36°50′–37°50′N	121°35′–122°30′W	2.59	1799	27	1.50	Thomas, 1961
San Luis Obispo County	35°00′–35°45′N	119°30′–121°25′W	8.57	1287	21	1.63	Hoover, 1970[4]
Kern County	34°45′–35°45′N	117°40′–120°15′W	21.2	1463	17	1.16	Twisselman, 1967[4]
Hot Deserts							
Sonoran desert	27–35°N	109–117°W	310	2441	650	26.6	Raven and Axelrod, 1978, Shreve, 1951
Sahara (n, e & central)	17–31°N	17°W–35°E	9000	650	162	25.0	Ozenda, 1977
Saharan Africa	20–30°N	16°W–37°E	7000	1618	190	11.6	Quezel, 1978
Islands							
Canary Islands	28°N	14–18°W	7.27	1254	564	46.6	Kunkel, 1980[4]
Galapagos Islands	1°N–2°S	89–92°W	7.78	701	175	25.0	Raven and Axelrod, 1978
				529	211	41	Takhtajan, 1978
Hawaiian Islands	19–22°N	155–160°W	16.7	1897	1751	92.3	Raven and Axelrod, 1978
				2800	2700	>97	Takhtajan, 1978
Ascension Island	7°55′S	14°25′W	0.088	3	3	100	Takhtajan, 1978
St. Helena	15°55′S	5°42′W	0.122	39	38	97	Takhtajan, 1978

Juan Fernandez	33–34°S	80°W	>.135	195	136	70	Takhtajan, 1978
Desventurada	26°20′S	80°W		19	12	63	Takhtajan, 1978
New Caledonia	20–22°S	164–167°E	16.33	2700	2500	90+	Takhtajan, 1978, Schmid, 1982
Norfolk Island	29°04′S	167°56′E	0.036	174	52	29.9	Takhtajan, 1978
Lord Howe Island	28°S	159°E	<.028	226	70	31.0	Takhtajan, 1978
Kerguelen Islands	50°S	70°E	3.42	29	2	6.9	Takhtajan, 1978
Amsterdam & St. Paul	37°47′S	77°34′E	?	17	7	41.1	Takhtajan, 1978
Tropics							
Mt. Kenya, alpine	0°	37°20′E		248	225	81	Walter and Breckle, 1984
Guayana Highland	7°N–4°S	59–73°W	1000	8000+	6000+	75	Maguire, 1970
Southern Hemisphere							
Cape of South Africa	31°40′–34°30′S	17°40′–25°45′E	89	7000	6252	73.1	Goldblatt, 1978, Takhtajan, 1978
Cape Peninsula, S. Africa	34°S	18°25′E	0.47	2256	157	7.0	Goldblatt, 1978
Southwest Australia	15–26°S	116–124°E	900	3600	2450	68	Hopper, 1979
New Zealand	34–47°S	166–179°E	268	1996	1618	81.1	Raven and Axelrod, 1978

1. *Eliminating 27 endemics in Hieracium, Antennaria, Puccinellia, Calamagrostis and Taraxacum, % endemism is 9/479 = 2.1%. Probably < 1/4 of the area of Greenland is habitable by vascular plants.*
2. *The endemics are races of widely distributed species, including apomictic races. 'No sexually reproducing species can be considered as endemic'.*
3. *5 of these taxa are species, 7 subspecies.*
4. *Natives only.*
5. *Species figures are given as approximate by Favarger.*
6. *Arctic–alpine species only according to Contandriopoulos and Gamisans, 1974.*
7. *110 species of the Racha-Lechkumi flora are endemic to the Caucasus, 99 to the Bolshoi Caucasus, 54 to Georgia (p. 26).*
8. *This area is covered by Fernald (1950). The endemic species count includes 164 microspecies of Rubus and 67 of Crataegus, total endemic microspecies 231. Subtracting these from the total of 599 endemics leaves 368 for an endemic % of 8.77.*
9. *Raven and Axelrod give the area of Texas as 751000 km^2.*
10. *Endemics of the Altai-Sayan mountain region.*
11. *15 endemics for 2.5% endemicity for the entire Sayan Range.*
12. *10 of the counted endemics also occur in the nearby Zeravshanski Range, giving an endemic % of 7/314 = 2.2% for the high altitude area of the strictly Hissar Range (Tolmachev, 1957, 154).*
13. *47 additional species are endemic to the Hissar Range, making the endemicity of the whole Hissar Range equal to 4.9%.*

connected to largely unglaciated north-eastern Asia by the 1500 km wide Bering land bridge across which Asian plants now resident in Alaska and Alaskan plants now in Asia probably dispersed.

Geographically, properties of plant associations such as species richness, biomass, productivity, and endemism are interrelated.

5.4.1 Species richness

In general, vascular plant floras, as in most other organisms (Chapter 3), increase in species richness from the arctic to temperate latitudes, with highest values in mediterranean areas (California, Israel), mountain-bordered deserts (Texas, Utah, Kazakhstan, Zeravshan River Basin, Sonoran desert), and lands bordering on the humid tropics (Carolinas, Japan plus southern China and northern New Zealand). The comparatively rich flora of south-western Australia appears anomalous, but this area combines old land forms and old, nutrient-poor soils with sufficient but not excessive aridity (Stebbins, 1952; Stebbins and Major, 1965; Raven and Axelrod, 1978) and stability (Hopper, 1979). Then there is a precipitous decrease in numbers in both the cold (continental) and hot deserts (Sahara), followed by a rise to greatest species richness in the tropics (Gentry, 1986).

In general, the degree of endemism follows the same latitudinal trend as species richness (Table 5.2). Species richness is so low as to be almost non-existent in the arctic. The small arctic flora is certainly unique, but almost all its species occur somewhere in the mountains to the south (Hulten, 1968; Yurtsev *et al.*, 1978). The single arctic endemic genus, *Dupontia*, with one or two polyploid species is considered by Tsvelev (1976) to be a possible ancient hybrid between *Arctophila fulva* and a form of *Deschampsia caespitosa*. Both these latter species are arctic but with large extensions into the taiga and mountain ranges to the south. Two other examples are instructively arctic–alpine. *Phippsia algida* is exclusively arctic in Eurasia. In North America it occurs not only in the Brooks and Alaska ranges of Alaska but also in the Beartooth Mountains of north-western Wyoming and on Mt. Evans in the Colorado Rockies. *Pleuropogon sabinii* on the other hand is exclusively arctic in North America, but in Eurasia it occurs isolated to the south in the headwaters of the Kolyma River, on its eastern tributary, the Omolon, and on the Tom River in Siberia which is an eastern tributary of the Ob (Tolmachev, 1974a). The last site is about 51°N.

The Alaskan flora is well-known (Hulten, 1968; Murray, 1980) and provides an illuminating example when considering endemics and endemism. Alaska contains large parts of four of Takhtajan's floristic provinces. Although its low arctic north slope (part of the arctic floristic province, but a region according to Yurtsev *et al.*, 1978) has only 0.2% endemism in its limited flora (Table 5.2), the other Alaskan provinces have much more.

The arctic flora is best considered as an arctic–alpine flora. In general alpine floras have many endemics, but these are locally derived.

5.4.2 Islands

Islands form almost ideal natural areas for the study of endemism. They are finite and independent, easily delimited, naturally and not politically circumscribed, and their endemism has been much studied (Runemark, 1971; Valentine, 1972; Carlquist, 1974; Bramwell, 1979; Kunkel, 1980) and theorized upon (MacArthur and Wilson 1967). "Islands are closed communities [where] problems of isolation, dispersal and differentiation can be studied in a more purified form than is generally possible in mainland areas" (Valentine 1972). Island immigrants illustrate the founder principle, and later evolution demonstrates founder selection (Berry, 1983). Island populations are isolated from their former biotas, so for this reason they must evolve under new conditions.

In an ecological sense, islands can be land surrounded by water, mountains surrounded by lowlands, areas of unusual or azonal soils surrounded by zonal soils (Chapter 15; Walter and Straka, 1970; Walter and Breckle, 1984). True islands differ in their origins, subsequent histories, relative and absolute ages, possibilities of past migrations of disseminules to them, probabilities for plant establishment, plant competition, and extinction rates. As discussed in Chapter 15, islands in general have fewer species than continental areas of equivalent size. However, former continental connections (Canary Islands or New Caledonia) or nearby, now submerged, sea mounts may have contributed species and increased species numbers. Also, island floras are more or less 'unbalanced'. A measure of balance in a flora is the rank spectrum of species numbers in the ten vascular plant families with the most species in a flora (Tolmachev, 1974a). This balance is different for different floristic units at different scales in region, province and district. Island floras reflect difficulties of invasion (oceanic islands in particular) and evolution plus extinction in isolation. However, continental islands, such as the Canaries, may have rich, relictual floras preserved by their equable, maritime climate (Axelrod and Bailey, 1969) and on oceanic islands, plant evolution seems to have been speeded up (Carlquist, 1974; Carr and Kyhos, 1981).

Oceanic islands may have very high endemism, and their endemism is roughly proportional to their isolation. Continental islands have richer floras but lower endemism than oceanic islands. New Caledonia is quite exceptional, and Takhtajan (1978) considers it to be a floristic subkingdom with a level of endemism equivalent to the much larger African, Madagascan, Indo-Malaysian and Polynesian subkingdoms. He says (p. 250), "No other territory on Earth with a size comparable to New Caledonia possesses such a great number of endemic families and genera . . . The number of species in several endemic genera reaches into several tens. This shows that intensive processes of speciation have occurred here for a long time . . . and one encounters here 6 out of the 12 vessel-less woody genera of flowering plants". The island is old, continental, large, isolated, mountainous, tropical both now and when part of Godwanaland, and extremely diverse in soil parent materials, from peridotites to serpentine (Schmid, 1982).

5.4.3 Mountains

Mountains (e.g. Caucasus, Tien-Shan, Altai) are species-rich compared with lowlands, but the farther north the mountain ranges lie the poorer are their floras (Urals, Altai, Sayan). These latter were all once heavily glaciated and surrounded by tundra.

Mountains are typically rich in endemics. When located in deserts (Sahara, western US, middle and central Asia) they are mesic refugial islands. In the rich flora of middle Asia, more than three-quarters of the 7000 species are said to occur in the mountains (Tolmachev, 1974b).

Mountain ranges support a wide variety of climates, often from arid to wet and from warm to cold, within very short vertical distances. High mountains may have buffered the climatic changes of Pleistocene glaciations. The highly endemic forests of the Balkans (Horvat *et al.*, 1974), the high endemism of the Colchis (Takhtajan's (1978) Euxine province), and Talysh (Takhtajan's (1978) Hyrcanean province), areas of the Caucasus region, the endemic islands of *Acer* (maple) forests in the mountains of middle Asia (Kamelin, 1973) and in the Wasatch Mountains of Utah are all anomalies preserved from glacial and interglacial age extinction by their mountain habitats.

The degree of endemism increases with altitude (Table 5.2; Kamelin, 1973; Chopik, 1976; Agakhanyants, 1981) but only up to the subalpine. Alpine and nival vegetations are poor in species and poorer in endemics than regions just below. Chopik's data from the Ukrainian Carpathians show 1.7% of the total flora is endemic to the montane zone, 4.2% to the montane-subalpine, 9.9% to the subalpine, and only 2.1% to the alpine. Data from the western Tien-Shan (Table 5.3) show consistent maxima of species numbers, numbers of endemics, and percentage endemism at middle mountain elevations, above the adjacent valleys and below the subalpine and alpine belts. The percentage decrease from the middle

Table 5.3. Degrees of endemism in the altitudinal belts of the western Tien-Shan Mountains. From Pavlov in Tolmachev 1974b: 66, 68

Altitudinal belts (m)		*Species* *no.*	*Endemics* *no.*	*%*
Low mountain	500–600	1122	93	8.3
Middle mountain	800–2000	1729	211	12.2
Subalpine	2000–2800	739	98	13.2
Alpine	>2800	621	74	12.0

mountain belt to the subalpine is 43% for total species and 46% for number of endemics while the percentage decrease from subalpine to alpine is 8% for both total numbers of species and numbers of endemics.

Mountains form ideal refuges in times of climatic change; effects of a change in climate can be avoided by range shift up or down, tracking the original climate. Agakhanyants (1981) gives high figures for current rates of mountain uplift in the Pamir and Tien-Shan regions. At 10 mm/year, a few thousand years would move a given site up from one life zone to another, from subalpine to alpine or from alpine to nival.

5.4.4 Deserts

One might expect areas of low plant productivity to have a low degree of endemism. However many hot deserts have very high endemism in spite of their limited flora and vegetation. The Sahara (Ozenda, 1977; Quezel, 1978), transmontane southern California (Stebbins and Major, 1965; Raven and Axelrod, 1978) and south-western Australia (Hopper, 1979) are, or contain, hot deserts with concentrations of both relict and recent endemics. With the exception of Australia the endemism of these deserts is much increased by the mountains within them. South-west Australia is a special case, having little relief but with an adjacent well-watered coast.

Not all cold deserts have low levels of floristic endemism either. The valley and plains deserts of middle Asia have many endemics (Kamelin, 1973). Utah has remarkably abundant endemism in the cold deserts of its high plateaus and adjacent valleys, but in this case the endemism is clearly related to edaphic peculiarities inherited from specific geological formations (Graybosch and Buchanan 1983; Welsh and Chatterley, 1985; Goodrich and Neese, 1986).

5.4.5 Tropics

Tropical areas are highest of all in endemism. Even family endemism is prominent. We have little information on the extent of tropical species' endemism (Takhtajan, 1978), but Maguire's (1970) studies in Venezuela and Gentry's (1986) comparison of temperate and tropical endemics are illuminating. Tropical mountains (Mt. Kenya, Table 5.2) combine island and mountain properties of very high endemism. Tropical islands also have high endemism, but it is difficult to separate the effects of being both an island and in the tropics.

5.4.6 Substrate peculiarities

Edaphic endemism can be on a much finer spatial scale than that discussed above. Gabbro, serpentine, or heavy metal outcrops of only a fraction of a square kilometre often harbour several endemic plants.

In general, non-zonal habitats increase both species number and degree of endemism. Ferromagnesium rocks such as peridotite and serpentine probably increase overall species numbers in such different areas as the Urals, Swedish mountains (Rune, 1953), the Balkans (Horvat *et al.*, 1974), California (Kruckeberg, 1984) and Cuba (Iturralde, 1986). Many endemics growing on and often confined to serpentine, probably cannot compete with the normal, regional flora. Calcareous rocks, including dolomite, increase species number and preserve endemics at least in humid climates, but in many cases the increase may be due to persistent presence of open, dry, raw sites rather than to the chemistry of the substrate. Such sites are frequently called 'successional', but many really show no evidence of changing over time (Ehrendorfer, 1962; Galland, 1982). They are discussed in more detail below. In the 'sea of loess' of the southern Ukraine such unlikely and non-regional substrates as sand and granite as well as limestone outcrops have endemics (Walter and Straka, 1970). The conclusion is that any substrate different from the regional one so weakens competitive dominance by the regional vegetation that endemics find a place (Gankin and Major, 1964).

5.4.7 Productivity

Except on the unusual soil parent materials noted above, zonal vegetation productivity roughly follows the same trends as species richness. High species richness accompanies high endemism. Are there causal relationships here? What causes what? Or is high productivity simply associated with high endemism?

Phytomass actually seems to be independent of both species richness and productivity. For example, *Sequoia sempervirens* (redwood) stands in the southern part of Takhtajan's (1978) Vancouverian floristic province have phytomasses >346.1 kg/m^2 (Fujimori, 1977), and old growth

Pseudotsuga menziesii–Tsuga heterophylla (Douglas fir-western hemlock) and *Abies nobilis* (Noble fir) stands follow with 159.0 and 156.2 kg/m^2 (Fujimori *et al.*, 1976; Waring and Franklin, 1979). Non-zonal productivity seems to be highest in such disparate vegetation as the tall herb associations of oceanic Sakhalin and Kamchatka (Walter, 1981) and in the tugai (riparian) vegetation along such desert rivers of middle Asia as the Amu Darya (Oxus) and Syr Darya where maximum net sun radiation is combined with an unlimited supply of water and almost annual fertilization with sediment carried by the rivers from the Pamir and adjacent mountain ranges (Rodin and Bazilevich, 1967).

5.5 ENDEMISM FROM VARIOUS VIEWPOINTS

The natural phenomenon of endemism can be regarded from a variety of viewpoints: composition and structure, physiology, history, genesis, ecology and chorology. These headings are used below to shed light on some theoretical ideas or positions current in biogeographical analysis.

5.5.1 Composition and structure

Species richness varies over at least two orders of magnitude between floras; and it is unevenly divided among the plant families. Similarly, endemism is unevenly divided among the plant families of a given flora. Tolmachev (1974a) has used the ranking of numbers of species in the 10 richest families of a flora to characterize floras. In the flora of the Dzungarian Alatau (Goloskokov, 1984), 76(3.5%) of the 2168 species are endemic. When the families in the Dzungarian Alatau flora are ranked by number of species, the order of the first 14 is (1) Asteraceae, (2) Poaceae, (3) Fabaceae, (4) Brassicaceae, (5) Rosaceae, (6) Caryophyllaceae, (7) Lamiaceae, (8) Ranunculaceae, (9) Scrophulariaceae, (10) Cyperaceae, (11) Apiaceae, (12) Boraginaceae, (13) Chenopodiaceae, (14) Liliaceae. The order of endemism is different: (1) Fabaceae, (2) Asteraceae, (3) Boraginaceae, (4) Ranunculaceae, (5–7) Apiaceae, Lamiaceae and Scrophulariaceae, (8–10) Poaceae, Liliaceae and Rosaceae, (11) Brassicaceae, and (12–13) Cyperaceae and Rutaceae. Two genera of the Fabaceae alone, *Astragalus* and *Oxytropis*, have 18 and 8% of the total number of endemics, which ranks them just after Asteraceae and Boraginaceae among the families.

The processes of speciation, hybridization, immigration and extinction which form floras are taxonomically selective (Chapters 7 and 8). That is, these processes do not act upon all plant taxa in a uniform manner, but plants in different families react differently to the processes which form floras. The genera *Astragalus* and *Oxytropis* of the Fabaceae are known for their species richness and high endemism in both North America and extra-tropical Eurasia. The Dzungarian Alatau is unusual among northern hemisphere floras in its low proportion of endemism in the Caryophyllaceae and Brassicaceae. In western Europe, species of these two families are commonly endemic both on serpentine and on sites with concentrations of heavy metals such as nickel, lead, zinc and cadmium.

Further examples of variation in endemism over families of plants in some mountain floras of the southern part of the USSR are given in Table 5.4. Although species richness is comparable with Goloskokov's Dzungarian Alatau, endemism is two, three or more times as great in the nearby but more southern areas and 10 times greater in the Caucasus. The two parts of the Tien-Shan have rich floras, but overall endemism is comparatively low, although high in the larger, combined area. Again, the degree of endemism increases with size of area considered. The more northern and eastern (with better connections to the Siberian mountain ranges such as the Altai and Sayan) Tien-Shan has less endemism than the western part which extends into the deserts of middle Asia. The Pamir with its extremely high altitude, cold, and low precipitation has a comparatively poorer flora considering its large area, and endemism is low. Endemism is high in the partial Caucasus flora which overall is extremely rich in species (Table 5.2). Of the families which occur in all areas, the Poaceae has low endemism and the Fabaceae, Asteraceae, and Lamiaceae high, with Apiaceae, Brassicaceae, Caryophyllaceae, and Liliaceae, in between. Such northern families as the Ranunculaceae, Rosaceae and Scrophulariaceae have low endemism.

The distinct differences between Goloskokov's Dzungarian Alatau data and those of Table 5.4 in general fit the trends from the western to the northern Tien-Shan although absolute levels of endemism are lower in the more northern area. There seems to be an overall decrease in endemism with increasing latitude, and Apiaceae, Lamiaceae and Liliaceae are more abundant as endemics to the south and west toward the deserts although highest endemism throughout occurs in the Asteraceae and Fabaceae.

We turn from phylogenetic tu physiognomic considerations. Physiognomic plant classification has made little improvement over Raunkiaer's scheme (Braun-Blanquet, 1964) of life forms using position of perennating buds for the sole criterion. Phanerophytes (trees, Ph) have these buds above 2 m height above ground level, nanophanerophytes (shrubs, NPh) at 2–0.25 m height, chamaephytes (low shrubs, Ch) 0.25 m, hemicryptophyes (perennial herbs, H) at the ground surface, geophytes (with bulbs or rhizomes, G) underground, and therophytes

Table 5.4. Total numbers of endemics, species and % endemism in plant families of the Tien Shan, Pamir-Alai-Zeravshan and Caucasus Mountains. From Vykhodtsev, Pavlov, and Altukhov in Tolmachev 1974b: 14–18, 68 & 12

	Tien-Shan-Alai		*W. Tien-Shan*		*N. Tien-Shan*		*Pamir-Alai-Zeravshan*		*NW Caucasus*	
Family	*End/Total*	*% end*	*End/Total*	*% end*	*End/Total*	*% end*	*End/Total*	*% end*	*End/Total*	*% end*
Poaceae	159/456	34.8	11/257	4.3	10/211	4.7	7/254	2.8	15/73	20.5
Liliaceae	233/306	76.1	27/117	23.1	13/92	14.1	7/118	5.9	14/28	50.0
Caryophyllaceae	114/226	50.4	9/104	8.7	15/84	17.8	12/85	14.1	23/51	45.0
Ranunculaceae	82/189	43.3	–	6.4	14/99	14.1		3.8	22/37	59.4
Brassicaceae	134/350	38.2	21/177	11.9	15/140	10.7	13/154	8.4	16/34	47.0
Rosaceae	96/211	45.4	–	11.5	7/101	6.9		4.4	20/48	41.7
Fabaceae	645/837	77.1	75/284	26.4	56/232	24.2	40/253	15.8	17/44	38.6
Apiaceae	222/322	68.9	36/146	24.6	13/84	15.5	7/119	5.9	26/49	53.0
Plumbaginaceae	68/92	73.9	5/21	23.8	–	–	6/26	23.1	–	–
Boraginaceae	88/180	48.8	11/89	12.4	–	–	6/88	6.8	–	–
Lamiaceae	273/399	68.6	32/135	23.7	9/59	16.1	13/113	11.5	7/22	31.8
Scrophulariaceae	87/180	48.3	–	–	–	–	–	–	12/38	31.6
Rubiaceae	25/64	39.7	–	–	–	–	–	–	7/14	50.0
Asteraceae	620/1095	56.6	68/421	16.1	28/254	11.0	52/384	13.5	35/103	43.6
Total flora	3370/6525	51.7	358/2801	12.8	225/2500	9.0	196/2615	7.5	217/819	35.1

(annuals, Th) as seeds. The life form of hemicryptophyte and chamaephyte plants are abundantly represented among endemics of many areas (the Balearic Islands for example, Haeupler, 1983), specifically by chasmophytic (rock crevice) plants (Ellenberg, 1978; Horvat *et al.*, 1974; Runemark, 1971; Tauber, 1984; Iatrou and Georgiadis, 1985). Again and again such plants are mentioned as successful endemics. But there are many exceptions, such as the hemicryptophyte *Paraskevia cesatiana* (Boraginaceae) endemic in forests of southern Greece (Sauer and Sauer, 1980). Trees in the very rich Balkan forests are endemic (*Pinus peuce, Pinus heldreichii, Picea omorika, Abies cephalonica, Aesculus hippocastaneum*) (Turrill, 1958; Horvat *et al.*, 1974). The Balkans, having been outside the ice sheet that once covered Europe north of the Alps, having an extremely varied topography, with a variety of soil parent materials, and having a wet, fairly equable climate have been and are a species-rich refugium for trees as well as other life forms.

The actual distribution of endemics among life form classes changes with locality. It seems to be determined by climate, history of the flora and competition with the associated flora. For the Dzungarian Alatau, Goloskokov (1984) gives the following figures: H 51.3%, Ch 36.8% and NPh, Th and G 3.9% each. In the Ukrainian Carpathians (Chopik, 1976) the 74 high mountain endemics are 9% H and 4% NPh in the alpine belt; 54% H, 3% G, 4% Ch, 1% Th and 1% NPh in the subalpine; and 14% H, 4% G, 3% Th and 3% NPh in the montane–subalpine. In a quite different ecosystem, granite outcrops in the Piedmont of the south-eastern US from North Carolina to Alabama have shallow, gravel-filled depressions with a unique, highly endemic flora. Phillips (1982) shows that the life form spectrum of these species is shifted from the normal, regional high content of hemicryptophytes to a high content of therophytes. This makes the life form spectrum of the outcrop flora similar to a hot desert flora (Death Valley). Life forms are usually neglected when a flora is analysed. We need more data.

5.5.2 Physiology

Physiological plant ecology has burgeoned recently (Larcher, 1980; Chabot and Mooney, 1985), but endemic plants have not been the subject of much of the research. We need to know why specific endemic plants occur in particular, geographically delimited climatic zones. As yet systematic plant ecophysiological data are insufficient. Research has been on one plant at a time. However, plants exist in plant communities, competing with other plants.

Are more endemics C_3 or C_4 plants and in what ecosystems? The subscripts refer to the first products of two distinctively different photosynthetic pathways. Ecologically, C_4 plants have higher photosynthetic rates in general and at low CO_2 concentrations, at high temperatures, and under drought conditions. They transpire less per gram of production, become light-saturated only at very high light values if at all, have no photorespiratory loss of CO_2, and have higher growth and assimilation rates (Larcher, 1980). Do endemics actually have higher or lower rates of photosynthesis or of transpiration and in what ecosystems? Are endemics more or less successful than more widely distributed related species?

Endemics certainly grow better than non-endemics in some of their non-zonal habitats such as areas with high concentrations of heavy metals (see below). The serpentine endemic *Poa curtifolia* has a very high Mg content and a low Ca content for best yield (Main, 1981). Gottlieb (1979) established that the endemic *Stephanomeria malheurensis* differed from its parent *S. exigua* ssp. *coronaria* in such physiologically important morphological characters as leaf size, specific leaf weight, root/shoot ratio and, most notably, seed size and weight. It did not differ in the strictly physiological rates of photosynthesis or dark respiration. The two taxa grow in only slightly different sites, but associated vegetation which might distinguish the sites is as yet undocumented.

Extraordinary accumulations of heavy metals (lead, copper, zinc, cobalt, cadmium, nickel, even manganese and iron) by some plant species occur in soils with high concentrations of these elements, including serpentines and peridotites with their excessive Mg contents and usually low Ca (Horvat *et al.*, 1974; Wild and Bradshaw, 1977; Ellenberg, 1978; Kinzel, 1982; Papanicolaou *et al.*, 1983; Reeves *et al.*, 1983; Kruckeberg, 1984). There has been much interest in such plants and such sites worldwide. These specialized plants are frequently endemics. Kinzel (1982) has discussed the unique chemistry of species of Caryophyllaceae. Is this involved with many of its species' endemism on metal-rich sites?

Parts of the life cycles of several of the cedar glade endemics of Tennessee and Kentucky have been investigated. The glades occur on shallow, limestone soils with open stands of cedar (*Juniperus virginiana*) and differ from the regional, species-rich deciduous forest. The annual *Lobelia gattingeri* is highly variable in size and seed production, and it competes poorly with the annual grass *Sporobolus vaginiflorus* although the latter is more susceptible to spring drought than the *Lobelia* (Baskin and Baskin, 1979). Spring drought therefore favours the *Lobelia*. The winter annual endemics, *Leavenworthia exigua* var. *laciniata* and *L. stylosa*, develop large soil-stored seed banks which carry their populations through climatic oscillations which could otherwise eliminate them (Baskin and Baskin, 1978, 1981). Although the perennial *Astragalus tennesseensis* did not decrease its transpiration rate under drought stress, it did drop some of its leaves – thus behaving similarly to many purely desert species (Baskin *et al.*, 1981). Unfortunately we know too little of the physiology of desert annuals to carry this analogy to a wider conclusion.

Generalizations for this section are mostly questions. I have outlined some of the problems, but data for solutions are rare. The endemics best studied from a physiological standpoint are from unusual edaphic sites in the regional vegetation.

5.5.3 History

Engler (1882) originally differentiated two kinds of endemics, palaeoendemics and neoendemics. The former are relicts, isolated phylogenetically by extinction of any close relatives, and are often of high polyploidy. Neoendemics are newly evolved, and evidently closely related phylogenetically (Stebbins and Major, 1965; Prentice, 1976). Tolmachev (1974a) makes the point that young endemism lacks generic endemism.

In California palaeo- and neoendemics show clearly different patterns of distribution. Palaeoendemics are concentrated in the climatically equable, precipitation-rich north-western corner of the state (roughly the range of redwood, *Sequoia sempervirens*) plus the wet Klamath mountains just inland and also in the driest desert of the south-eastern part of the state (Stebbins and Major, 1965). The relict *Podocarpaceae* of Chile and New Zealand, *Araucaria* of New Caledonia, Chilean conifers such as *Fitzroya*, and *Microbiota decussata* of the Sikhote Alin mountains of coastal eastern Asia are palaeoendemics and must have an ancient history if phylogenetic uniqueness can be used as a criterion. Desert palaeoendemics include *Welwitschia mirabilis* (Walter and Breckle, 1984) of south-west Africa, the boojum tree (*Idria columnaris*) and ocotillo (*Fouquieria splendens*) of the Sonoran Desert (Humphrey, 1974).

On the other hand, recent endemism in California is concentrated in the border regions between humid and arid climates. The numerous species in such genera as *Clarkia, Downingia, Gilia, Linanthus, Plagiobothrys, Mimulus, Phacelia* and in the subtribe *Madiinae* are examples. Arid climates receive less precipitation than would be evaporated by the incoming heat regime; humid climates receive more. Budyko's index of aridity, $R_n/(L \times \text{precipitation in cm})$ where R_n is net incoming radiation in Jcm^{-2} and L is the amount of heat required to change the state of one gram of water from liquid to gas (or about 2500 Jg^{-1}; actually $L = 2501 - 0.24\ T$ with T in °C) is a reasonable climatic parameter. The index equals one on the border between humid and arid climates, and it is very sensitive at this borderline (Major, 1977) – as is neoendemic speciation (Stebbins and Major, 1965; Valentine 1972). This is also the border between pedocal and pedalfer soils, steppe and forest, etc. This is reminiscent of the effect of climatically induced physiological stress encouraging adaptation (Chapter 6).

Polyploidy can be used as a tool to trace the origin and history of plant taxa (Section 5.5.4). The relicts mentioned above mostly have high chromosome numbers and are assumed to be ancient polyploids. Younger meso-polyploids (with lower chromosome numbers) whose ancestors are probably known, can be migrants of exotic origin or native, depending on whether their immediate ancestors exist outside or within the floristic region where we now find them. Krogulevich (in Malyshev, 1976) determined and assembled chromosome numbers for 448 of the 569 species in the Putorana flora of north-central Siberia. Conclusions were that karyologically the youngest species are the arctic and boreal ones; the oldest are the montane and alpine.

Direction of range expansion can be read in some cases from a series of ploidy levels for one species. For example, Knaben (1982) has traced the postglacial history of several endemic Scandinavian species using polyploidy levels. Contandriopoulos and Gamisans (1974) find that the high mountain, endemic flora of Corsica is mostly derived from the Mediterranean element, and only 17% of the 137 taxa (Table 5.2) are arctic–alpine. Karyological data (Contandriopoulos, 1962) show these latter taxa are derived from alpine, not arctic species and therefore they must have immigrated to Corsica in the Tertiary when Corsica was physically connected to the continent, not during the Pleistocene when there were only lagoon connections.

The implicit assumption here is that all these plants have very limited powers of dispersal – as most plants do. However, Carlquist (1974) believes otherwise so far as island floras are concerned. The contradiction here may not be real since, as noted below, the Hawaiian flora is not primarily an immigrant one but overwhelmingly evolved *in situ*, and ancient sea mounts near the Hawaiian islands may have greatly expanded the time available for immigration (Rotondo *et al.*, 1981). Actually, systematic and quantitative data on propagule dispersal are very deficient (Harper, 1977; Kubitski, 1983). The test of such data is how they predict seed rain, and for specific sites this is almost totally unknown.

The geological history of endemism in the islands of the western Mediterranean is excellently reviewed in depth in Bramwell (1979) and by Haeupler (1983).

5.5.4 Genesis

Polyploidy is a genetic phenomenon useful in the study of plant endemism, as demonstrated above. In fact, Favarger and Contandriopoulos (1961) base their classification of endemics into palaeoendemics, patroendemics, apoendemics, and shizoendemics on polyploidy. Palaeoendemics are subdivided by Stebbins (1974) into those so old their derivation is unknown e.g. Austrobaileyaceae x = 22, Winteraceae x = 43 and those younger, meso-polyploids whose derivation probably is known e.g. *Antennaria* at n = 14, derived from *Gnaphalium* at n = 7. Patroendemics are diploids whose polyploid derivatives have wider ranges. Apoendemics are polyploids derived from a widespread diploid. Shizoendemics have retained the same ploidy level as their progenitors. Palaeo- and patroendemics are the old, relict endemics of Engler; apo- and schizoendemics are young. The relative representation of these classes in a flora is a guide to that flora's history and therefore genesis. Palaeoendemics indicate the ancient connections of a flora; neoendemics indicate the flora's current evolution. These categories have been effectively used to discuss endemism in the floras of Corsica and the Balearic islands (Contandriopoulos and Cardona, 1984), in the islands of the western Mediterranean (Bramwell, 1979), the mountain floras of Europe (Valentine, 1972) and the flora of California (Stebbins and Major, 1965; Valentine, 1972; Raven and Axelrod, 1978). Wider use of this fruitful classification of endemics is limited by lack of systematic cytological data.

The evolution of species would be better understood if a biologist were present when it originated. Gottlieb (1977a,b, 1978a,b, 1979) has come very close. Using enzyme electrophoresis he has shown that the derivative, diploid, schizoendemic species, *Stephanomeria malheurensis*, does in fact have only a partial set of the genetic variability of its parent. Similarly the origin of the endemic *Clarkia lingulara* from *C. biloba* by chromosomal reorganization in a very local, small, marginal population has been confirmed by Gottlieb (1974) using enzyme electrophoresis. *C. franciscana* was shown as likely to be an old species as a neoendemic (Gottlieb, 1973). The sites of origin of the contemporarily formed *Tragopogon mirus* and *T. miscellus*, in eastern Washington and adjacent Idaho are known, as well as the identities of their invasive, weedy, parent species. Roose and Gottlieb (1976) found that the allotetraploids do in fact have all the enzyme multiplicity present in the parent plus some additional, novel combinations. *Parthenium* species which are geographically and ecologically isolated on limestone or gypsum habitats do have depauperate flavonoid patterns (Mears, 1980). *Pinus torreyana*, a very narrowly endemic pine of California also has almost no gene diversity (Ledig and Conkle, 1983) – but the authors point out that the widespread *Pinus resinosa* and *Thuja plicata* are only slightly more variable. On the other hand *Silene diclinis*, a Spanish endemic, is variable, perhaps due to its occurrence in open habitats in formerly wooded areas (Prentice, 1976, 1984). According to Prentice (1984), "Low allozyme variability characterizes populations of most endemic species that have been investigated", and "Isolation and historical events may have been as important as population size or edaphic specialization in causing this low variability". A review of life history characteristics and electrophoretically detectable genetic variation comes to much the same conclusions (Hamrick *et al.*, 1979), although the relationship to endemism is not stressed.

Solbrig and Rollins (1977) investigated the differences in life history characteristics associated with self-compatibility and incompatibility in *Leavanworthia* species. In general the self-compatible species have the narrower range of variation, but the widespread self-compatible species *L. exigua* has more variation in fruit characteristics than the two, mostly self-incompatible endemic species, *L. alabamica* and *L. crassa*. Are species with narrow ranges of variability less successful? Success is difficult to define for a plant even if lack of success is proved by extinction.

The Madiinae (Asteraceae) of Hawaii have undergone extreme adaptive radiation. Many species of Madiinae also occur in California and there and throughout the western United States they are mostly annual herbs. In Hawaii they are all endemics and range in life form from various kinds of herbs through shrubs and lianas to trees. "The remarkable array of life forms . . . exploits almost every conceivable terrestrial habitat in Hawaii . . . from very recent lava flows to mature rain forest" (Carr and Kyhos, 1981). They have probably achieved this remarkable evolutionary differentiation from a single introduction. Carr and Kyhos unravel the complicated relationships within the group by chromosome analysis.

Genetic studies can thus illuminate the nature of endemics, but much more work needs to be done for valid generalizations to emerge. The ecological setting of endemics needs definition, but there is little systematic ecology at all, much less adequate attention to endemic plants. If the 1400–2000 angiosperm taxa on the Hawaiian Islands have really evolved from about 250 introductions (Carlquist, 1974), then the very high Hawaiian endemism (Table 5.2) is the result of speciation and not immigration. Has any extinction occurred? Among the Hawaiian ferns the situation is different – 135 immigrants which have produced 168 contemporary taxa (loc. cit.). Spore-bearing plants such as ferns should cross ocean barriers more readily than seed plants. As an ancient group do they now also evolve more slowly? On the other hand, on the small islands of the Aegean Sea evolution is said to be almost at a standstill (Valentine, 1972). These now isolated bits of land were fairly recently (mid-Tertiary) part of a continuous land mass which then had a richer flora than now occurs on the islands. Extinction has since reduced the numbers of species, and relicts now outnumber recently formed species. Runemark (1969, 1971) invokes extinction by reproductive drift to explain otherwise inexplicable extinctions and vagaries of species' distributions on these islands. All these interpretations may appear speculative but they are based on much field experience with the pertinent floras. Unfortunately, the kinds of pollen data which have been used to unravel the glacial and post-glacial history of the floras, and even vegetation, of northern Eurasia and North America are simply not available in this case.

5.5.5 Ecology

Any explanation of the occurrence of an endemic on a particular site or kinds of sites must be ecological as well as historical (Section 5.5.3). A plant occurs on a given site partly for historical reasons of dispersal, and longevity, but it grows on that site because its ecological tolerances fit the environmental conditions offered by the site, including modifications of the environment resulting from competition with other plants.

Prentice (1976) discusses the idea that all restricted endemics lack morphological plasticity and flexibility in responding to ecological conditions. However, the idea seems to be factually wrong (Gankin and Major, 1964; Stebbins, 1978a,b, 1980). Many apparently stenotopic endemics show extreme morphological variations when relieved of competition. Stebbins (1978a,b) asks whether the narrowly restricted occurrence of so many Californian plant species is due to genetic properties, lack of time for spread, or ecological factors. He excludes 'senility', for which he finds no evidence. "The answer . . . is, it depends upon the species". This is a helpful conclusion since it emphasizes the importance of studying the autecology of endemic species.

The flora of California is rich, and it has many endemics (Stebbins, 1978a,b). Habitats vary from temperate rain forest to hot deserts, from areas with abundant orographic precipitation to rain-shadow deserts, from mountain glaciers of high altitudes to Death Valley and other deserts at low elevations, from very maritime to very continental climates, from extremely fertile soils to very infertile, including ancient hardpans and podsols, from serpentine substrates to granite, limestone and sand dunes. Ecological 'islands' combining extremes of many ecological factors are common. Finally, the many combinations of species of great physiognomic variety, from gigantic forest trees such as *Sequoia sempervirens* and *Sequoiadendron gigantea* to the belly plants of the deserts and vernal pools, that form California's extremely different plant communities, add further diversity to habitats. These ecologically various habitats have interacted with the diverse flora which includes both old, relict species and young, very recent ones. The result is a richly diversified vegetation (Barbour and Major, 1977) in which a wide diversity of endemic plants find suitable sites.

California is not unique in its combination of rich ecological diversity and high endemism. Similar juxtapositions of a variety of habitats along with a diverse old flora occur in Spain, the southern Alps, the Balkans, the Caucasus, middle Asian mountain ranges, the Soviet Far East, Japan, New Caledonia, and a diversity of tropical mountains and mountain ranges. All produce high endemism. However, the extremely high endemism of the Cape of Good Hope, the richest of all (Table 5.2) is difficult to explain.

Endemics occupy the gamut of site stability. Some occur only on the most transient disturbed sites, raw cutbanks, frequently burned areas, talus, scree, river gravels, snow avalanche tracks, or heavily fertilized spots such as wild animal lookouts. On the other hand, the abundance of endemic chasmophytes (cliff crevice plants) associated with both arid and humid mountains and with cliffs of islands such as those in the Aegean Sea is striking. Open, rocky, cliff and even some rapidly eroding sites are often the most stable habitats we know (Ehrendorfer, 1962). The plant communities on these open, non-zonal sites resist changes in climate since their local climates deviate so widely from regional climates. No plant succession takes place, and these kinds of vegetation, which habour endemics, are far more stable than climax vegetation which is impacted by every shift in climate as well as by fire, flood, hurricane, and human activity. Examples of open sites on calcareous rock outcrops, on granite outcrops in an otherwise limestone region and on eroding shale scree are given by Ehrendorfer (1962), Sutter (1969), Horvat *et al.* (1974), Ellenberg (1978), Galland (1982), Graybosch and Buchanan (1983), Keener (1983), Papanicolaoi *et al.* (1983), and Iatrou and Georgiadis (1985).

In fact, any 'island' site may harbour endemics. A striking example is a mesic, wooded terrace along the southern Donets River within the regional steppe of the Ukraine (Ivashyn *et al.*, 1981).

Fire is a factor in the ecology of some endemics. *Cupressus forbesii* is endemic to low mountains of southern, mediterranean California. It requires fire to open its cones for seed dissemination. But it also requires more than ten years to reach cone-bearing age and over fifty years for maximum seed production. Fires occurring too frequently would eliminate the species (Zedler, 1977).

Where competitive relationships between species in plant communities change with a drastic change in soil parent material, presence of limestone for example, the plants' ecological tolerances seem to change

(Gankin and Major, 1964). The plants discussed (*Picea mariana*, for example), are not calcicoles but they occur only on limestone at the southern limit of their range where their normal associates do not occur. Morphological aspects of competition between *Stephanomeria exigua* ssp. *coronaria* and its recent derivative *S. malheurensis* were not apparent. Nor were large and small individuals of the former species different genotypically. The larger seed size of the derivative is believed to be responsible for its continued existence (Gottlieb, 1977a,b).

Few population studies have been carried out on endemic plants. Gorchakovski and Zueva (1982, 1984) delineated the age structure of four endemic, alpine, steppe species of *Astragalus* in the Urals (seedlings, juveniles, immature, young, mature, old generative, and senile) and followed it for up to eight years. They found wavelike trends in the numbers of plants in the various age states since reproduction was sporadic.

Model ecological studies that include not only some discussion of physical factors determining an endemic's success but also the plant's status within its community are by Sutter (1969), Horvat *et al*. (1974), Ellenberg (1978), Graybosch and Bucanan (1983) and Tauber (1984).

5.5.6 Chorology

Distributions of many endemics have been mapped, for example by Hulten (1968), Jalas and Souminen (1972), Malyshev (1976), and Welsh and Chatterley (1985).

Knowledge of disseminule dispersal and distributional data show little correlation (Kubitski, 1983). This is a surprising conclusion, but only intensive study of which seeds land where and use of all the chorological data available can change it. The fact that truly oceanic islands, which were never part of a continent, are populated by plants proves the existence of long-distance dispersal (Carlquist, 1974), but the high endemism on such islands, which is roughly proportional to their isolation, proves that long-distance dispersal is rare and more or less unpredictable. The endemics are the result of evolution *in situ* (Carr and Khyos, 1981). The flora of continental islands is the final result of altogether different processes (Valentine, 1972).

Clayton and Cope (1980) point out that areal distributions of floras are not accurately known. By calculating similarities (Jaccard index) in species occurrences between (political) areas, they produced maps showing areas of concentrations of particular species. The chorological units were illustrated by isolines of species abundance. Computer methods (Clayton and Hepper, 1974) have the advantage of objectivity but the disadvantage of not unscrambling which species are most responsible for the resultant classification.

The chorology of the 1200 Iberian plant endemics was analysed by Sainz Ollero and Hernandez Bermejo (1985) using matrices of both Jaccard coefficient and chi squared similarities in species composition between 25 units and 70 subunits. The matrices of similarity values were analysed using various statistical models. The authors chose an arrangement with six main units of endemics (such as Balearic, Cantabro-Pyrenees, South Atlantic) and 27 sectors.

Birks (1976) applied mathematical clustering techniques to the distribution of the pteridophyte flora of Europe (Jalas and Souminen, 1972). He produced a floristic classification of areas, and the endemics fit very well. Those that do not are aquatics, which are known to have very wide distributions, and known alloploids, which may have polytopic origins. Endemics are strikingly limited to areas south of the last, Weichselian, ice sheet.

5.6 ENDEMISM IN CONTEMPORARY BIOGEOGRAPHY

Island biogeography is extensively discussed in Chapter 15 of this book, but it is relevant to how endemics arise and how they persist (MacArthur and Wilson, 1967; Simberloff, 1974; Berry, 1983). Unfortunately, present theory depends on three unknowns, immigration and extinction rates and equilibrium. The shape and scale of the relationships between plant dispersal, extinction rate and equilibrium numbers of species are unknown, nor is there any good evidence that equilibria result. In fact, hypothetical equilibria in biology and geography are an unfortunate inheritance from physical chemistry where they work because many chemical reactions can be completed over a short time span. In our landscapes we have had a recovery time of only a little more than 10 000 years since ice wiped out plant and animal communities over much of the northern hemisphere north of 50°, loess accumulated over vast areas, tropical forests were probably fragmented, North America and Eurasia were joined by a 1500 km wide land bridge, the snow line in mountains was lowered by 1000–2000 m, sea level lowered by 200 m, and Antarctica was covered by an ice sheet from which it still has not recovered so far as vascular plants are concerned. During these 10 000 years plants have re-invaded many areas to form successive forests and other plant communities, a period which experienced higher temperatures than those of today intervened, and soils formed or resumed formation.

Are there comparisons of rates of continental drift with rates of adjustment of the biota to such drift and does plant succession itself reach a steady state in species composition, biomass, or productivity? The time frame of a succession must be equal to or less than the period during which the factors determining the properties of the ecosystem (other than time) are constant (Jenny, 1980). We have little evidence for such hypothesized steady states in vegetation (Major, 1974). Many forests become productive, then decline in productivity. This is well documented where permafrost develops under an accumulating moss and litter layer (Viereck, 1970). Progressive paludification is hypothesized for forests of coastal southern Alaska (Zach, 1950), and it is an actual, current problem in western Siberia (Walter, 1977). Podsolization leads to impoverishment of both soil and plant community (Jenny *et al.*, 1969), although it may also eventually produce plant speciation and endemism, as it has in coastal California (*Pinus contorta* ssp. *bolanderi*, *Cupressus pygmaea*, *Arctostaphylos nummularis*). In arid regions, with water income less than potential evapotranspiration, pedocal soils form a caliche layer which restricts root space. In mediterranean California, impenetrable iron-silica hardpans have formed in soils on the oldest terrain, and vernal pools rather than native grassland, result. In the California foothills an exhumed lateritic soil of Eocene age (Singer and Nkedi, 1980) supports a depauperate vegetation, including several endemic plants (Gankin and Major, 1964).

5.7 THE FUTURE

It is not clear whether the patterns of endemism are simply a collection of special, unrelated cases or whether there are a number of underlying unifying processes.

Autecological studies of plant endemics are obviously needed. They should combine field, greenhouse and laboratory work, probably in that order. It is necessary that we gather data on endemism by floristic provinces, not by political subdivisions. Synecological studies are however also needed since endemic plants exist within communities. They should be studied in relation to their climates (heat and moisture availability), soil parent materials (including residual soils), relief (which influences local climate), successional time (from a defined t_0 and without assuming a steady state), fire (intensity, frequency, phenological time of year), and, probably of most importance, the associated flora and fauna. Of all ecological parameters, this last is the one most frequently neglected in both autecological and synecological studies. Seed rain (Ryvarden, 1975a,b) and the seed bank are aspects of the floristic factor about which almost nothing is known (Harper, 1977).

ACKNOWLEDGEMENTS

I thank Mary Burke and the editors of this volume for their help.

PART III

Biological processes in biogeography

Introduction

The uneveness of species distributions over the globe forms a basic characteristic of living organisms. Biogeographers have frequently tried to explain these distributions by reference to preconceived biogeographic processes. This has often resulted in authoritarian narratives that really neither test the process invoked, nor propose possible alternatives. A more rigorous explanatory approach is to search for consistent patterns amongst the various species distributions and then use these patterns to discover the processes which underlie them. Indeed, the discovery of such processes is a prime function of biogeographic analysis.

Rosen (Chapter 2) conjectures as to whether there are any uniquely biogeographical processes, but whether there are or not, geographical (tectonic, eustatic, climatic, oceanographic), evolutionary (adaptation, speciation, extinction) and ecological (dispersal, biotic interactions) processes are all implicated in pattern forming. Intuitively, the assumption is that large scale abiotic processes exert their effect in conjunction with evolutionary processes, through vicariance or through merging of biotas, whilst ecological processes cause changes in pattern through biotic interactions and colonizing jumps and are in turn affected by small scale abiotic factors.

ADAPTATION AND ECOLOGY

At a very basic level, to understand why a particular organism or group of organisms is found in a specific place, we must know something about their general biology i.e. the nature of the adaptations which allow the species to live where it does. Different genetic strains or phenotypes of a species, often called ecotypes, can themselves be differentially adapted to their local environmental conditions and hence be restricted to certain habitats. These intraspecific habitat and regional differences tend to broaden the ecological tolerance ranges of many species. This intraspecific level of adaptation is explored in Chapter 6.

In broad biogeographic terms, adaptation is important on two different scales; a spatial one, influencing distribution patterns at a given time; and a temporal one, involving the influence of a change in conditions on adaptation and the indirect effect on distribution patterns over time. But how can we define the organism's world in order to understand the biogeographic role of the process of adaptation? This is the problem of defining the ecological niche. We can conceptualize two states of the niche following Hutchinson (1957). The potential, or fundamental niche represents the sum of a species' physiological, morphological and developmental adaptations (or tolerances), and hence the range of environments within which the species can potentially survive and replace itself indefinitely. This potential range is usually reduced to an actual or realized niche by the activity of sympatric species (including competitors, predators and parasites) or by some other factors. In the biogeographic context, we are mainly concerned with the spatial aspect of the niche, the habitat and range dimensions. The fundamental niche can describe the maximum range, encompassing all suitable habitats, over the globe. The realized niche describes the actual range the species occupies, controlled partly by activities of other species at a local level and at a global level by geographical barriers, dispersal ability and chance historical events, such as vicariance. All of these determine the presence or absence of a particular species in otherwise suitable habitats.

Because most species occupy only a limited part of their potential global range and because adaptation leads towards accommodation of organisms to their environment, we often see that distantly related organisms have independently evolved similar form and function to fill similar ecological niches in similar habitats around the world, i.e. ecological convergence. The important feature of such convergence in the present context is in the power of natural selection to shape a species population to fit into the prevailing environment (within the constraints imposed by the very nature of the organism). These ideas can be extended to convergence at the community level, although the evidence is equivocal (Chapter 9).

What happens when environments change either in abiotic parameters (e.g. climate) or biotic parameters (e.g. intensification of interactions or changing resource levels)? This question was considered in the short term by Fisher (1930) using a simple theoretical model of adaptation (Chapter 6). It is based on the premise that no organisms will be perfectly adapted to the present conditions, since adaptations (due to a time lag in response) reflect conformity to past environmental pressures. Over longer time scales, Van Valen (1973) views the environment as constantly decaying with respect to adaptations of existing species, so natural selection operates only to enable the species to maintain its current state of adaptation rather than improve it (Red Queen hypothesis; see also Chapter 8). Sufficient plasticity in behaviour or physiology can allow short-term tracking of, or acclimation to, small scale changes in environments over ecological time (a slight, short-term or temporary shift in the realized niche). Niche shifts, leading to a reduction in habitat ranges in the face of competition, provide a pertinent example and are considered in Chapter 9. Longer term changes in the species, in response to larger or longer term environmental change, require permanent (i.e. genetic) alterations to their biology and result in a permanent shift in the fundamental niche.

What effect can long-term changes in the environment outlined in many of the chapters have on the biogeography of species? If species are able to re-adapt, or track the environment, as outlined above, the species population can maintain its spatial occupation in the face of changing conditions, by moving into a new fundamental niche. But what if the species population cannot re-adapt sufficiently in the face of environmental change? Changing conditions can be countered, to some extent, if the species is able to shift its spatial occupation in relation to a shift in its required conditions. The species may thus retain its niche, but alter its geographical distribution. This kind of forced range shift has been documented as an effect of glaciations, forcing temperate organisms to shift latitudinally as the temperate zones moved (e.g. trees (Davis, 1986), insects (Coope, 1987)). Continual changes in conditions might also act to reduce the size of the realized niche of a species population, causing gradual restriction of its range to those places where conditions have not changed, and eventually create refugia for species or even whole biota (Chapter 10).

EXTINCTION

Adaptation may be viewed as a very slow environmentally driven rearrangement of niches accompanied by gradually changing species, which are always incompletely adapted and change their spatial occupations in time with shifting adaptive zones. Ultimately, if conditions continue to change and a species has nowhere left to go (i.e. its niche disappears), then extinction will ensue. As pointed out in Chapter 8, the speed of extinction will depend on the intensity and rate of environmental change, and different species will respond to this differently, so extinction episodes will not affect all species equally. Extinction, as a process, can thus be seen to act as a filter, affecting different species populations differentially. One obvious effect therefore is to change the species pool of an area, which will undoubtedly have a concomitant effect on distributions of other species (Fig. 1.1).

Three levels of extinction can be determined, largely based on differences in taxonomic, spatial and temporal scales. Local extinction of one species may broaden the ecological opportunities open to similar species, and may also enable a previously excluded species to establish itself (expansion of a realized niche towards a fundamental niche – known as competitive release). This process is discussed in Chapter 9. On a larger scale, local and sometimes global extinctions of species may be viewed as resulting from competition with ecologically similar species. Knoll (1984), for example, has suggested that vascular plant radiations actually caused extinctions of previously dominant groups through competition over time. Extinctions at this level would again lead to ecological readjustments amongst the remaining species, which in turn would influence distribution patterns. Ultimately, extinctions of lineages would occur, e.g. dinosaurs and ammonites. The explanations of these so-called mass extinctions are varied and subject to considerable debate (Chapter 8). Such large scale extinctions might then allow great adaptive radiation amongst the surviving species and groups.

SPECIATION

As demonstrated in the fossil record, extinction is the ultimate fate of all lineages and species. However, whilst the process of extinction has acted to reduce the available species on the earth, and greatly influenced their distribution, its action has been countered by enrichment of biotas through the process of speciation. This process is important in biogeography in the provision of raw material for distribution and in the alteration of the biotic environment and hence distribution patterns of species already resident in an area (Fig. 1.1, species pool A–A3).

In the biogeographic context a workable definition of a species is 'a complex of populations which may be spatially vicariated, and which together form a potentially interactive gene pool'. In this light, speciation requires divergent evolutionary change in two or more independently evolving population complexes, to a point where genetic differences between them are so great that they cannot normally interbreed. New species may arise through changes in and eventual replacement of the parent species (pseudospeciation) or alternatively, speciation may lead to phyletic evolution – a long-term product of reproductive isolation which leads to creation of new lineages which colonize new regions.

How does speciation occur? Knowledge of the type of speciation is fundamental for an understanding of biogeographic patterns and processes. Speciation is usually classified as allopatric, parapatric or sympatric, largely on the basis of distribution patterns. Fig. III.1 outlines the principal pathways and the various mechanisms are discussed in detail in Chapter 7.

Speciation is assumed to occur through geographical isolation in many biogeographic hypotheses, but, as pointed out in Chapter 7, there

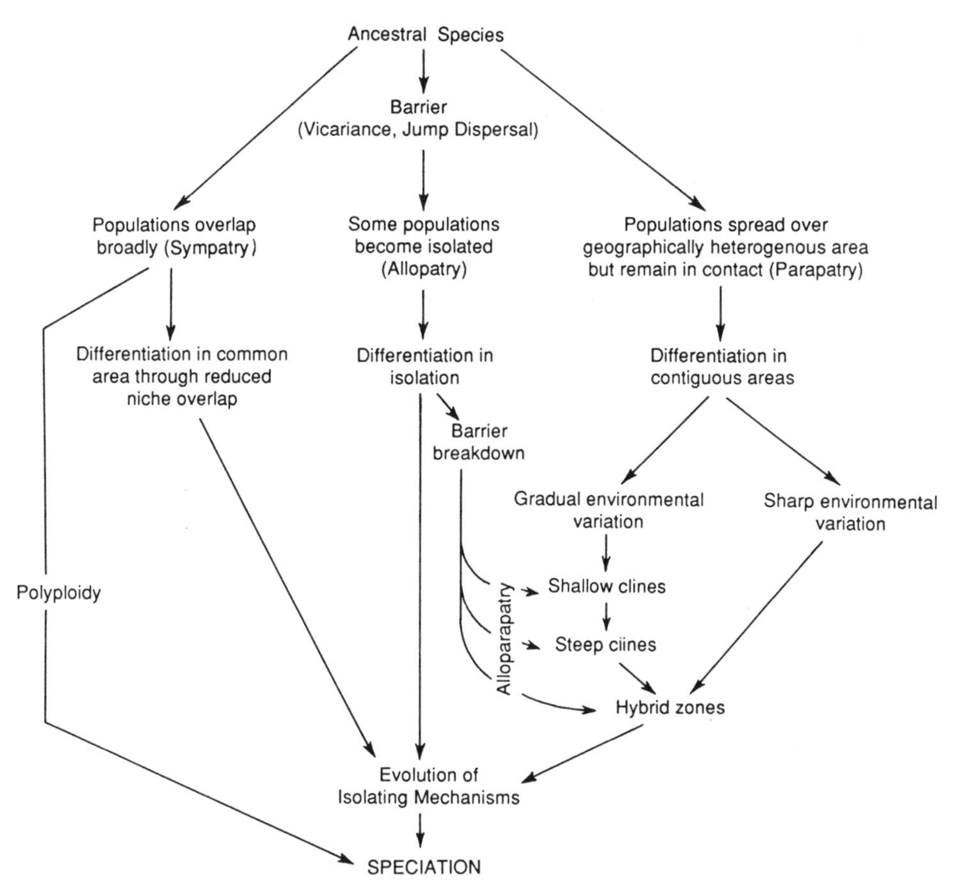

Figure III.1 Principal pathways of speciation. Sympatric, allopatric, parapatric, alloparapatric. (Modified from Endler, 1977).

are no empirical or theoretical grounds for supposing that rapid evolutionary divergence usually takes place in extremely small populations via the founder effect (as suggested by Mayr, 1982), nor is complete geographical isolation necessary for speciation. In theory therefore, divergence between species should be possible within continuous populations as well as within isolated ones (Fig. III.1).

Speciation, like colonization, does not, however, guarantee a future for the species in question. In fact, all forms of change usually reduce evolutionary fitness, as indicated in Fisher's (1930) model mentioned earlier. If the new characters are sufficiently adaptive, the new species can overcome its small population size, increase and rapidly expand into new habitats until the adaptive characters are countered by environmental barriers. This expansion will lead to invasion of ranges of related species, especially parent species, such that interspecific competition (Chapter 9) may well play a role in species selection, leading to failure in establishment of the new species or exclusion of the parent on establishment. Indeed, sibling species are almost always allopatric or parapatric. This kind of species replacement is often interpreted in terms of the parent species being ousted by daughter species from their original range, producing a pattern characterized by older forms at the periphery of a taxon range and younger forms at its centre. However, the reverse is also possible, with apomorphic specialists being restricted to the outer parts of a species range (see discussion of 'progression rule', Chapter 11). Such conflicting interpretations are common in the biogeographical literature, and since such hypotheses are untestable, their value to biogeographic analysis must remain questionable.

Of particular biogeographical interest is the question of whether there are certain processes, conditions or circumstances that promote speciation? Any processes which enhance extinction presumably also provide the circumstances that encourage speciation. Small species populations may create genetic bottlenecks, reducing the total genetic variability available and hence reducing evolutionary flexibility of the population which will inevitably lead to extinction (as seen in taxon cycles (see p. 155) and postulated in the Red Queen hypothesis). On the other hand, species populations which have become divided into small evolutionary units by vicariance events would be expected to exhibit more genetic variability than the same number of individuals in larger units. New mutations and gene combinations are incorporated more rapidly into small populations which increases the evolutionary rate. Fragmentation of phylogenetic lines thus has important implications for the rate of evolution. The large scale abiotic processes, such as climatic cycles (e.g. Pleistocene glaciations), plate tectonics, eustasy and other similarly disruptive geological conditions can serve to promote isolation and speciation by causing alternate contraction and expansion as well as fragmentation of populations and biomes. Vermeij (1978) discusses this in terms of the biogeography of tropical marine systems and in Chapter 10 this phenomenon is examined when discussing tropical rain forest refugia. The aforementioned abiotic processes can also lead to removal of barriers and fusion of previously isolated biotas. This can lead to formation of hybrid zones (merging of population characteristics in sympatry) and intuitively to extinction or coexistence of different species depending on degree of reproductive isolation and level of ecological similarity. This again highlights the importance of ecological interactions.

There are many cases of taxa found in small, insular areas which show extreme evolutionary diversification. For example, the famous adaptive radiations in the Galapagos finches, Hawaiian island honeycreepers and *Drosophila* and amongst cichlid fish of the great African lakes, where repeated episodes of speciation produce numerous kinds of descendants which remain largely sympatric within small geographical areas. This can occur largely because of the empty niche space within habitats, made available by the absence of taxa from colonizing biotas, or extinction of taxa which are normally associated with specific ecological roles from already assembled vicariant biotas. The filling of these roles by one taxon through adaptive radiation (often leading to morphological and ecological convergence with organisms from different taxa) can theoretically arise by two different mechanisms:

(1) Based on refugial isolation, speciation is promoted by cycles of isolation, invasion and reinvasion of areas, in the so-called taxon cycle (Ricklefs and Cox, 1972, 1978; Brown and Gibson, 1983). Such cycles may depend not only on a chance colonization following dispersal but also alternating connection and separation between islands due to eustatic and tectonic change (Part IV). Coexistence between sympatric sibling species on islands is then explained on the basis of the evolution of fortuitous ecological differences on different islands or parts of islands. The assemblage of such species has been hypothesized as being governed by rules termed 'assembly rules' (Section 9.3.2) based on the assumption that species which are too similar to residents will either be prevented from establishing on an island, or successful colonization will lead to extinction of the island resident through interspecific competition for resources.
(2) Based on the hypothesis of competitive release (niche expansion) of a species population, in terms of both resource use and morphology, in which it enters a relatively competition-free environment such as an island, and expands its population size and habitat range (Chapter 9). Intraspecific competition would then promote divergence in resource use by sympatric or parapatric populations within the enlarged and more variable parent population on the island. This would lead to radiation in form and function from the original parent stock, further driven by interspecific competition leading to resource partitioning (Giller, 1984). This will involve the coevolution of species.

There is no logical reason why a combination of the above two mechanisms could not occur. Previously isolated populations may meet and relatively small differences which have already evolved between them in isolation could become further elaborated by competition. This would lead to further resource partitioning and radiation. This idea involves alternate periods of allopatric and parapatric speciation (Fig. III.1 – alloparapatry). Given the obvious difficulties in testing long-term phenomena that occurred in the past, it is almost impossible to discern rationally which is the more correct hypothesis. This avenue of biogeographic analysis is thus likely to remain hypothetical.

The emergence of new adaptive zones and occurrence of mass extinction events which open up previously occupied adaptive zones (Chapter 8) provide similar opportunities for spectacular adaptive radiations. Since the beginning of the Palaeozoic era, certain animal and plant groups have periodically entered new adaptive zones not previously inhabitated by other forms of life. These groups then diverged from the typical mode acquiring major new adaptive innovations, e.g. Echinoderms in the Cambrian (Brown and Gibson, 1983) and plants during the Phanerozoic (Collinson and Scott, 1987).

The tempo and mode of speciation and evolution holds another important key to understanding the distribution of biota and this will be influenced greatly by the tempo and occurrence of extinction events (Chapter 8). The conflicting views on the temporal pattern of evolution of phyletic lines, i.e. gradual versus punctuated equilibria, are well known. Both views contain an element of truth; some periods or some taxa show slow rates with the accumulation of gradual changes, whilst other periods or taxa show fast rates with dramatic, geologically rapid speciation episodes, perhaps associated with periods of intense environmental change. The controversy may simply be a scale phenomenon, due to the fact that different rates of speciation can be inferred from analysis at different taxonomic levels (Chapter 8).

The link between extinction and speciation has led to the extrapolation of the theory of island biogeography (Chapter 15) to the concept of evolutionary equilibrium, where the species number is brought about by a balance between rates of speciation and extinctions over an evolutionary time scale (Rosenzweig, 1975; but see also Hoffman, 1985a). This approximate constancy in taxon diversity over extended periods of time has long been advocated by biologists. The similar trajectories of curves showing increasing diversity following a disturbance event (Fig. 3.5) is also noteworthy. In animals, great adaptive radiations only occurred after mass extinction episodes as shown amongst the tetrapods for example (Webb, 1987). In contrast, successive radiations of trimerophyte progymnosperm, pteridosperm and angiosperm plants were well underway prior to the respective extinctions of rhyneophyles, trimerophytes, progymnosperms and previously important gymnosperms (Niklas *et al.*, 1983; Knoll, 1984). In other words, animal extinctions permitted radiations, but plant radiations caused extinctions. These episodes of speciation, radiation and extinction have occurred many times during the history of the earth, and are driven by the background adaptive changes in species populations in response to environmental change. The concept of equilibrium and taxonomic turnover (differences in absolute or relative rates of extinction and origination) is covered in Chapter 8.

BARRIERS

Whilst ecological processes may be responsible for maintaining or increasing high local diversity (Chapter 9), they depend on a large existing pool of available species. The question of how species accumulate is different from the question of how local diversity is maintained (Vermeij, 1978). Speciation increases diversity, extinction reduces it and migration from one region to another may act in either way. Speciation and extinction do not affect all species equally and this makes the analysis of the evolutionary basis of biogeography so important in the understanding of past and present distribution patterns. In a similar way, barriers are often viewed as filters of variable mesh size, which allow some, but not all taxa through. Even those taxa that do pass through will do so to differing degrees, so affecting the potential gene pool and establishment success in the newly colonized area. Barriers may be abiotic (geological or climatic) or biotic (ecological) in origin but their operation is at the ecological level. Physical barriers vary from large scale, i.e. tectonic, eustatic, climatic or oceanographic (TECO barriers: Rosen, 1984) to small scale topographic. A narrow strait of water is, for example, perceived by some tropical forest birds as an impenetrable barrier, even though such birds may fly long distances over land to obtain their food (MacArthur *et al.*, 1972; Diamond, 1973, 1975). Biotic barriers may relate to whether a species from a donor source can fit into an assemblage in the recipient area or tolerate the physical conditions there (Fig. 1.1, species pool B–B1 or B2). According to MacArthur (1972) a community or biota in which all the resources are utilized at the rate at which they are produced (saturated community) is highly resistant to invasion by outside species through the potency of species interactions. The MacArthur and Wilson (1963) model predicts that a species can only gain a foothold in a recipient biota if it can utilize resources which are inefficiently exploited, or not exploited at all, by the resident biota. The importance of establishment is considered in more detail later.

Barriers are probably always polarized as a result of oriented physical parameters such as winds and currents, and by differing ecological conditions (species richness and composition, degree of saturation) on either side of the barrier. A good example of one way migration is provided by that between the Red Sea and the Mediterranean, via the Suez Canal (Por, 1978; Vermeij, 1978). Since its opening to navigation in 1869, increasing migration of Red Sea taxa into the Mediterranean Sea

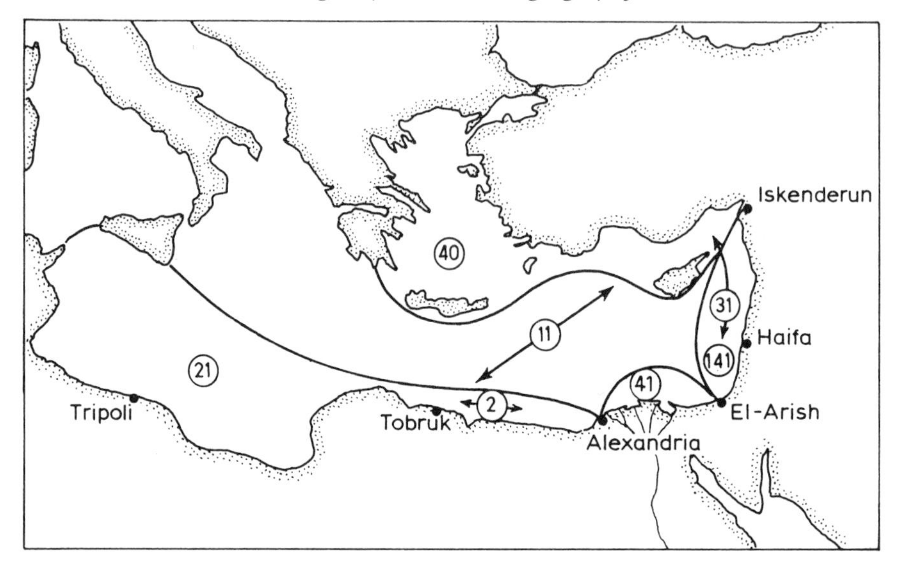

Figure III. 2 Numbers and distribution of animal and plant species thought to have migrated from the Red Sea to the eastern Mediterranean, through the Suez Canal. (Modified from Por, 1978).

has been documented (Fig. III.2.), as the salinity of the Great Bitter Lake has been reduced through leaching. On the other hand, there has been scarcely any evidence of migration in the opposite direction.

Barriers arise from time to time between once continuous areas, partitioning the biota, while others are destroyed or diminish in effectiveness, allowing the merging or partial merging of previously isolated biota (Fig. 1.1, p. 6). As discussed earlier, populations of organisms undergo periods of expansion and contraction in their geographic ranges over time in response to changing climatic, physical and biotic conditions. Even during periods of relative stasis in geologic and climatic conditions, interspecific and intraspecific interactions can result in expansion and shrinkage in the distributional range of taxa and in range shift (Chapter 9). However, movements away from parents are a normal part of the life cycle of all animals and plants and, as discussed in Chapter 1, range expansion is considered by most writers to be distinct from jump dispersal.

Jump dispersal models require dispersal to occur after the development of an isolating barrier, whilst vicariance models are independent of dispersal, although range expansion may have occurred prior to the vicariant event. How then might one determine whether dispersal pre-dated or post-dated a vicariant event? The discovery of a fossil of the same taxon dated older than barrier formation in an area beyond the barrier would falsify a dispersal hypothesis. Failure to find a fossil in the same circumstances would not, however, substantiate dispersal (see Chapter 14 for further discussion on use of fossils in biogeography). To overcome this problem, attempts have been made to relate time of isolation to the degree of differentiation in the taxon studied. When differentiation is less than would be expected given the timing of the vicariant event, a dispersal episode after the vicariant event is indicated. Unfortunately the assumption that evolutionary rates are uniform is probably unjustified, although Carson (1976) has suggested that amino-acid substitutions in the DNA may occur at a constant rate per locus. If substantiated, this would provide a useful clock for determining the timing of isolation in the taxa studied.

Proponents of dispersal biogeography have sometimes argued that barrier crossing need only occur very rarely since, over long periods of time, the highly improbable become probable and given long enough, the probable becomes highly probable (Simpson, 1952; Koopman, 1981). There are however, three inherent defects in this argument. Firstly, this 'maximum degree of probability' criterion is suspect (Ball, 1975) because expansion of the time scale to the exclusion of all other contributory and perhaps contradictory factors leads to a biased probability value. Secondly, theories with a high degree of corroboration are preferable to those with a high degree of probability (Popper, 1959) and thirdly, barrier crossing is only part of the process of colonization. Establishment is the key and is much more important than mere arrival. It is here that ecology plays an important role in biogeography, a fact that is too often ignored in dispersal hypotheses. The establishment process cannot be reduced to probabilities no matter how long the time scale. It is determined by the biological characteristics of the species concerned and the ecological characteristics of the recipient area, i.e. a suitable vacant or less efficiently exploited niche must be available on the other side of the barrier to allow establishment. Life characteristics include physiological tolerances, reproductive strategies, behavioural and genetic factors controlling, for example, minimum viable population size and ecological fitness in the recipient environment with its existing complement of organisms. Species diversity and ecosystem stability will almost always vary between donor and recipient areas. This may result in polarization, with the high stability ecosystem supplying biota to the low stability ecosystem more often than vice versa.

Oceanic islands have always played an important part in dispersal models. According to Peake (1981) they "provide unique opportunities to investigate complete biotas derived unequivocally by dispersal". However here again we return to the semantic problems inherent in the term dispersal and the distinction between range expansion and jump dispersal discussed in Chapter 1. New islands, including those which have been formed by volcanism in historical times, are presumably colonized by taxa from the species pool present in the wider area within which the island is formed. Jump dispersal need not necessarily be invoked. Where islands are formed in mid-ocean, long distance jump

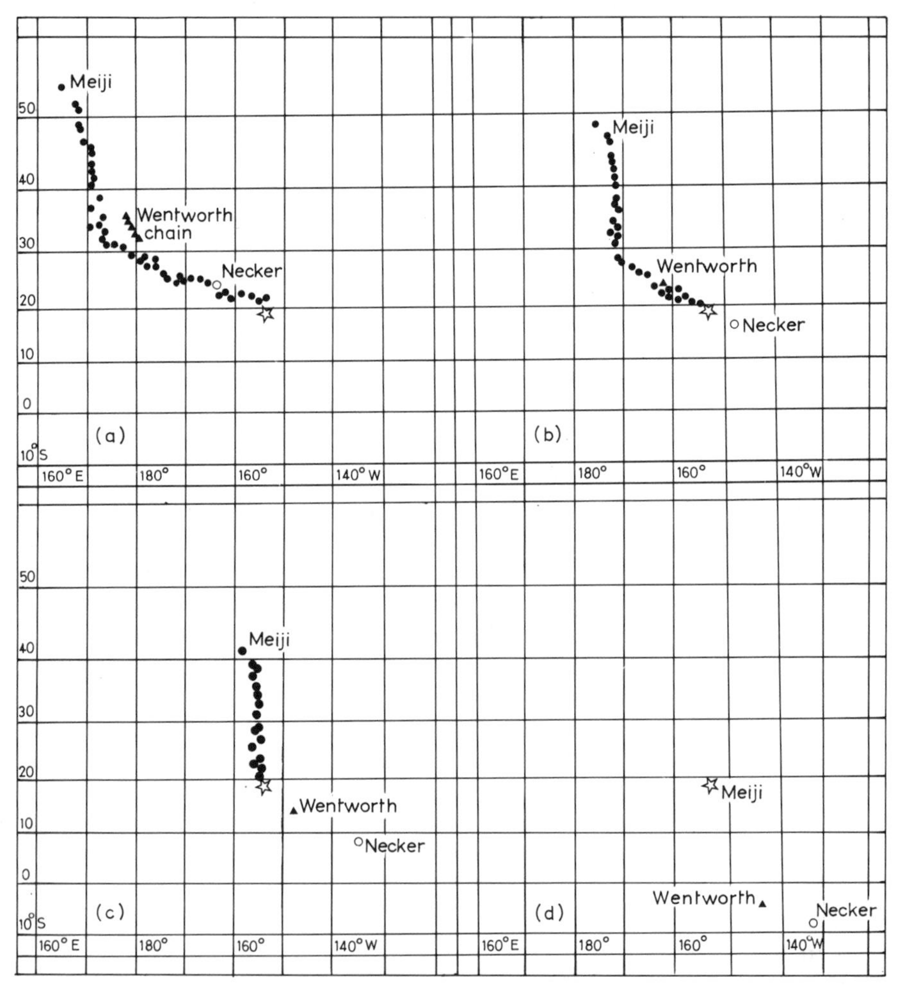

Figure III. 3 Hypothesized integration of the Wentworth (▲) and Necker (○) seamounts with the Hawaiian–Emperor chain of islands/seamounts (●) during the Cenozoic ((a) Present, (b) −20 Myr, (c) −40 Myr, (d) −70 Myr). ☆ = Hawaiian melting anomaly. Meiji is the oldest dated seamount of the Hawaiian-Emperor Chain. (Modified from Rotondo *et al.*, 1981).

dispersal may appear to be the only method of colonization and Hawaii, an archipelago formed over a melting anomaly or 'hot spot' on the Pacific plate, is often quoted as such an example. Rotondo *et al.* (1981), however, have shown that some islands in the Hawaiian-Emperor chain may be allochthonous to the chain (Fig. III.3). If substantiated, this implies that the Hawaiian biota may, at least in part, be a result of island integration. The biota of New Zealand may be derived from similar processes (Chapter 13). Shields (1979) has put forward evidence for a Jurassic opening of the Pacific Ocean, based on the theory of an expanding earth first proposed by Carey (1975). If correct, colonization of oceanic Pacific islands would not need to be explained solely by jump dispersal. Many extant distributional patterns may of course be a result of anthropochore dispersal which provides enormous opportunity for extensive barrier crossing. Examples such as the establishment of the New Zealand barnacle *Elminius modestus*, along the Atlantic coast of Europe are well documented (Crisp, 1958, 1960). Amphi-Atlantic distributions on the other hand may be due either to anthropochore dispersal or result from the opening up to the Atlantic seaway by seafloor spreading in the Tertiary. How important jump dispersal is in forming biogeographic patterns is a matter of considerable controversy among biogeographers and we will return to the problem in Part IV.

THE FUTURE

Biogeographic studies at the small-scale ecological level (incorporating studies of adaptation, the niche and ecological interactions) attempt to identify the processes that limit extant local distributions of species populations and maintain species diversity. An understanding of small-scale local distribution patterns is far from complete and more information leading to a greater understanding of the processes is clearly necessary.

But what is the significance of the small-scale processes to historical biogeographers attempting to reconstruct earth history? Ecological biogeographers alone have the option of investigating many of the processes experimentally, thus allowing a more rigorous evaluation of the value of biogeographic hypotheses based at this scale of analysis within a broader biogeographic context. Any resolution of questions about the significance of ecological biogeography must be framed at two levels; what is the effect of ecological processes acting now, and what effect did past ecological processes have on present distributions? The fundamentals of the role of adaptation over evolutionary time are generally accepted and further effort in this direction is unlikely to improve our understanding. However, the significance of ecological interactions, as mentioned above, is far from fully established. Extant ecological processes can perhaps be considered as 'background noise' to the control of species distribution and we must understand the extent of this noise before we can use modern distributions of taxa to infer historical processes.

The dispersal of a taxon from A to B is irrelevant if it does not survive at the new site. The importance of adaptation and ecological interactions

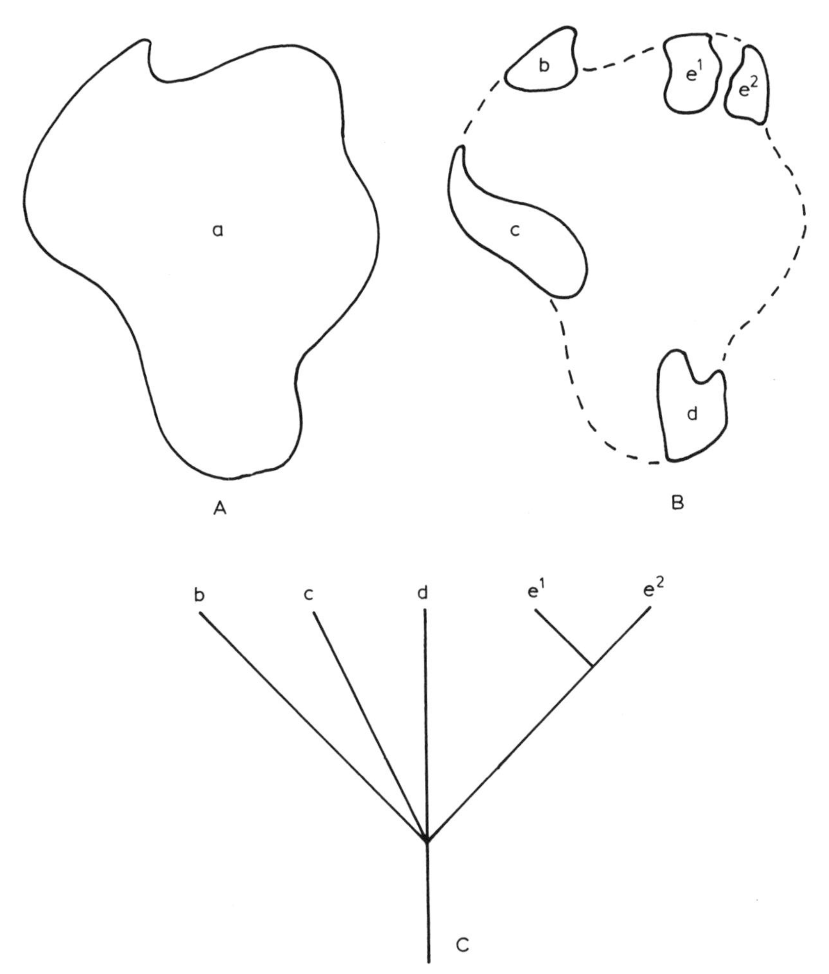

Figure III. 4 Polychotomous speciation during refugial episodes. A. Distributional range of taxon *a* before refugiation; B. Ranges of disjunct taxa *b-d*, e^1, e^2 after refugiation; C. Cladogram of relationships of taxa in disjunct areas shown in B. Refugiation of area A resulted in four contemporaneous vicariant areas in which taxon *a* initially evolved simultaneously into four taxa, one in each area. Later a further vicariant event resulted in taxon *e* evolving into two sub-taxa.

is apparent, as it is these processes that affect the survival and establishment of both immigrants and newly evolved species. These processes maintain an 'order' amongst distributional patterns by preventing most *ad hoc* establishment from random dispersal and mutations. A clear understanding of the potential effects of these ecologically based processes in recent time should allow the construction of null distribution models against which one can evaluate documented modern distributions. Only then could one be confident in using extant distributions to reconstruct historical processes such as vicariance (see Chapter 14 for a discussion of the role of geoecological data in historical reconstruction). The extent to which past ecological processes exert an effect on modern distributions is difficult to assess, although the effects of biotic interactions leading to local extinctions of palaeoendemics must have been profound, and the role of ecological processes is clearly relevant to discussion of post Pleistocene distribution patterns.

Information on the rates of speciation and extinction are essential for creating a time frame for the various reconstructed events of evolutionary history. Coarse data on rates of extinction are available from the fossil record, but this is likely always to remain incomplete. Rates of speciation are not a component of cladistic analyses, which attempt only to provide a sequence of evolutionary events. Cladistics must then rely on data outside its sphere of influence to date the sequential changes. If it became possible to incorporate this temporal aspect into the area cladograms constructed on the basis of differences in character state (i.e. identify the rate of change in character states), then such a methodology could provide an additional avenue to the dating of historical events, presently largely the domain of geologists.

Knowledge of the rates and modes of speciation are also important to the future consideration of refugia within biogeographical hypotheses and methods of reconstruction. In such cases, polychotomous speciation events resulting from the simultaneous speciation of multiple vicariant fragments of a species population might be expected. Such events are not decipherable through Hennigian nesting methods of phylogenetic analysis which assume dichotomous speciation (Fig. III.4 and see p. 304 for details of this analytical method). An ability to measure rates of speciation and extinction may therefore allow the construction of more robust hypotheses of historical events in distributional change through time, and perhaps permit projection to possible changes in the future.

CHAPTER 6

Adaptation

P. A. PARSONS

6.1 WHAT IS ADAPTATION?

"An organism is regarded as adapted to a particular situation, or to the totality of situations which constitute its environment, only in so far as we can imagine an assemblage of slightly different situations, or environments, to which the animal would on the whole be less adapted; and equally in so far as we can imagine an assemblage of slightly different organic forms, which would be less well adapted to that environment" (Fisher, 1930).

"Organisms are doomed to extinction unless they change continuously in order to keep step with the constantly changing physical and biotic environment. Such changes are ubiquitous, since climates change, competitors invade the area, predators become extinct, food sources fluctuate; indeed, hardly any component of the environment remains constant. When this was finally realized, adaptation became a scientific problem" (Mayr, 1982).

Following the above quote from Fisher (1930), one approach to an understanding of adaptation is to measure the degree of conformity between organism and environment. Because the environments of the living world have unpredictable components over time and space, perfect adaptation is not possible. Fisher (1930) developed a model of adaptation of existing organisms in relation to environmental and organismic change by comparison with a hypothetical perfectly adapted organism. Very small undirected changes in either organism or environment approach a 50% chance of being advantageous (Fig. 6.1) while very great changes are always maladaptive; even if in a favoured direction, organisms would end up in less than optimal environmental situations by overshooting points of increased adaptation. Thus as the magnitude of change increases and irrespective of the number of 'dimensions' used to assess the environment, the probability of improvement in adaptation will decrease towards zero. By parallel arguments, at the genetic level, new mutations will be expected to be

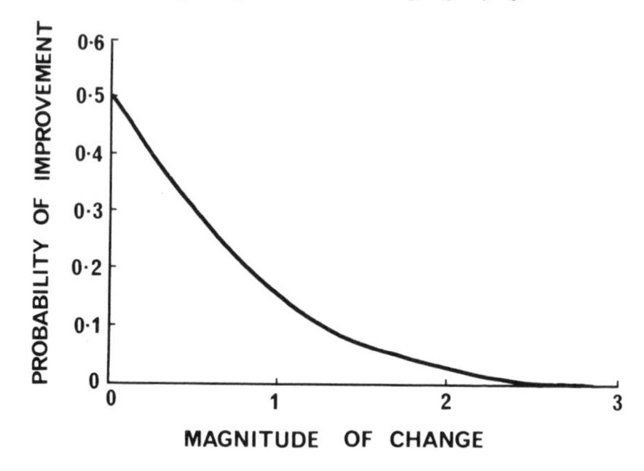

Figure 6.1 The relation between the magnitude of an undirected change and the probability of improving adaptation when the number of dimensions to the environment is large (after Fisher, 1930).

disadvantageous. At the extreme are deleterious mutants having large phenotypic effects. Genetic changes underlying adaptation more typically involve genes of lesser effects which are segregating in natural populations as polymorphisms. This means that genetic changes will follow the favouring of phenotypic extremes by the reorganization of segregating genes. Even so, the very existence of genetic variation means that perfect adaptation is an unachievable aim. In summary, at the level of gene, organism and environment, perfect adaptation must be regarded as a theoretically unattainable goal.

Following the above quote from Mayr (1982), adaptation has many facets including responses to the physical environment such as temperature and humidity conditions, protection against enemies, the utilization of resources and the adjustment of internal metabolic and physiological processes. For relatively small environmental changes, immediate responses of a physiological (e.g. acclimatization) or behavioural (e.g. learning) kind may be involved. However, beyond the limits set by these processes, genetic changes as a consequence of selection imposed by the environment at the phenotypic level may be a prerequisite for ultimate adaptation. It may be initially difficult to separate physiological and behavioural responses from genetic ones because an environmental change occurring in a particular direction over several generations could be the trigger for the genetic assimilation of physiological and behavioural responses consistent with the environmental change.

Over many generations therefore, a tendency towards adaptation to trends in the physical and biotic environments should develop, which will be menaced by more transient environmental changes of a largely unpredictable nature. In any consideration of adaptation, the time-scale is important and ranges over largely unpredictable short-term events to discernible trends over long periods of time which may be assessable at both the genotypic and phenotypic levels. The emphasis of Fisher (1930) has traditionally been on the shorter, and that of Mayr (1982) on the longer, time scale which is the main emphasis in any discussion at the biogeographic level (Chapter 1).

Biogeography is concerned with the differential patterns of distribution of organisms over a habitat template; a major aim is to show how the physical environment of a species, its biology, and its evolutionary history interact to bring about its distribution pattern. In attempting to interpret biogeographic change, adaptive explanations are normally put forward *a posteriori* following the collection of observational data within and among species at different times or in biota occupying different habitats. In particular, this means that any correlations cannot automatically prove cause-effect relationships, simply because of unknown important variables which may be revealed upon more intensive study. In spite of these difficulties, the theme of the following discussion concerns changes at organizational levels ranging from biogeographic to genetic in an attempt to seek unifying principles.

The occurrence of environmental perturbations of an unpredictable nature, usually involving climatic extremes, emphasizes the primary importance of physical factors in the determination of the distribution and abundance of organisms (Parsons, 1983, 1987). Even so, resistance to change is likely, for example the phenomenon of vegetational inertia whereby significant lag times separate climatic from consequential vegetational change (Cole, 1985). It will be argued in this chapter that adaptation over a long time scale can be expressed ultimately in terms of evolutionary responses to changing physical environments. For the majority of organisms, the physical environment may be becoming warmer or cooler, moister or drier, so providing circumstances under which there will be selection for increased adaptation. Initially, following selection at the phenotypic level, genetic changes within species will be expected, but where environmental changes are more extreme, extinctions, colonizations, and hence alterations in the total biota will occur so that the ultimate consequences of sufficiently extreme environmental changes will be obvious biogeographic change (Fig. 1.1).

6.2 SPECIES' DISTRIBUTIONS

In considering adaptation at the biogeographic level, it is appropriate to select a region of the world with a diversity of accessible habitats encompassing a wide climatic range. Australia, which occurs from

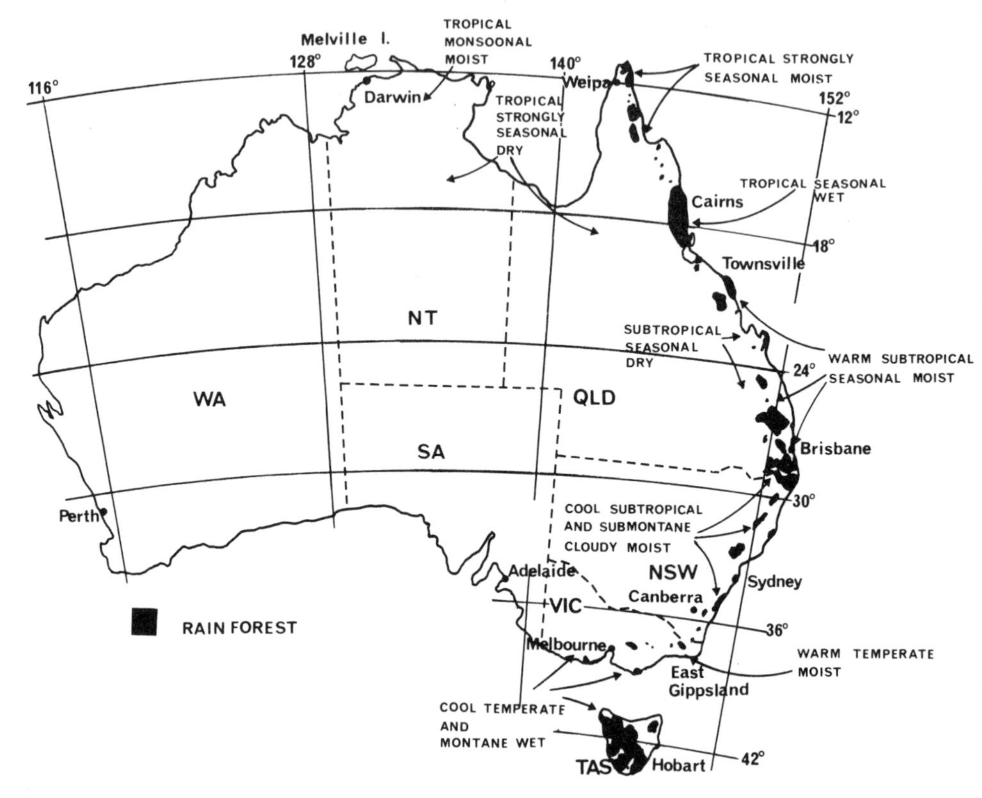

Figure 6.2 Distribution of Australian rain forests. The arrows indicate the approximate centre of distribution of each climatic type. The complex mesophyll vine forests that contain the greatest abundance of *Drosophila* species occur in the tropical seasonal wet forests inland from Cairns on the northeast coast of the continent. Lower species diversity occurs elsewhere especially to the south (after Webb and Tracey, 1981).

latitude 10° 30′S in the north to 43° 30′S at the south of Tasmania (Fig. 6.2) fulfils this criterion. Nix (1981) took the major climatic elements of solar radiation, temperature and precipitation, and identified the parameters of greatest significance for biology and ecology with reference to the thermal adaptation of the Australian flora. Three broad organismic groups were established:

(1) Megatherms with thermal optimum around 28°C and lower threshold temperatures of 10°C.
(2) Mesotherms with thermal optimum in the 18–22°C range and lower threshold temperature around 5°C.

(3) Microtherms with thermal optimum in the 10–12°C range and lower thresholds at or below 0°C.

Although mean temperatures provide useful insights, Nix's analyses of the adaptations of the Australian flora indicated that extremes of temperature are of the greatest biogeographic significance, for example the mean maximum temperature of the hottest week and the mean minimum temperature of the coldest week, since they correlate best with floral type.

Such associations necessarily apply to those insects intimately associated with the flora such as the dipteran genus, *Drosophila*. These flies are mainly associated with the micro-organisms involved in the initial stages of plant decay, encompassing a wide diversity of resources including fruits, flowers, fungi, bracken fern, and slime fluxes. The enormous radiation in the genus follows from this diversity, especially in the tropics. The dependence upon organisms causing fermentation and decay, being a rather aquatic mode of existence, leads to a rather narrow temperature range for resource exploitation compared with many animals. Such narrow ranges occur in undisturbed rain forests which form the centres of greatest abundance of *Drosophila* species (Fig. 6.2). These forests are mainly in the mesotherm category with maximum species diversity in the complex mesophyll vine forests of the humid tropics of north Queensland (Table 6.1). Complex interactions between climate, topographic and edaphic factors, as well as historical influences, determine the Australian rain forest vegetation type and its stability (Webb and Tracey, 1981). Climatic and topographic factors limit the distribution of a given rain forest formation, sub-formation, or structural type. The scenario is one of gradients of rainfall and temperature associated with gradual changes in floristics and abrupt replacements of one structural

Table 6.1. Species numbers of *Drosophila* in the Australian humid tropics and the temperate zone of Australia (summarized from Parsons, 1982)

	Rain forest[a]	*Urban*[b]
Humid tropics	53	14
Temperate zone	19	9

[a] *A complex mesophyll vine forest inland from Cairns (Fig. 6.2) was the site of the tropical, and a combination of cool and warm temperate forests in Victoria the site of the temperate zone collections. Only six species have been found in cool temperate* Nothofagus *rain forest habitats.*

[b] *Townsville, north Queensland was the site of the tropical and Melbourne, Victoria the site of the temperate zone collections.*

floristic type with another at certain climatic thresholds. *Drosophila* species' distributions necessarily follow these changes so that there is a broad tracking by *Drosophila* species of rain forest types and their resources (Parsons, 1982).

Drosophila species' diversity falls as forests become floristically simpler. In Australia the limits are (1) the heat-stressed fauna of the lowland tropics of northern Australia which contains a few widespread desiccation-resistant species, and (2) the cool temperate *Nothofagus* (southern beech) forest of the south, especially the island of Tasmania (Table 6.1). These limits grade into the megatherm and microtherm habitat categories respectively. Understanding the faunal distributions therefore depends upon identifying the most important variables – abiotic and biotic – underlying short and long-term gradients. From the long-term point of view, climate is of overriding importance. However within broad constraints exerted by climate, other variables may assume importance at a more local level. For example, in north Queensland rain forests, soil type is strongly correlated with floristic complexity whereby faunal species diversity is far higher amongst plants on soil of basaltic origin compared with diversity on granites and schists. Such faunal changes track soil type alterations over distances of just a few metres. In one survey, 19 *Drosophila* species, many rare, were collected in a complex mesophyll vine forest on basaltic soil while eight relatively common species were found in an adjacent complex notophyll vine forest on granitic soil (Bock and Parsons, 1977). This indicates that the rare species are mainly restricted to the floristically most diverse forests which presumably have a corresponding diversity of *Drosophila* resources. In addition, distance from permanent water and proximity to swampy regions are important; in any case, many *Drosophila* species are extremely sensitive to physical environmental stresses. While detailed research will increase our understanding at the microhabitat level, adaptation to climatic extremes, usually expressed as a function of temperature, is of paramount importance when the time-scale of concern extends to the historical/evolutionary level.

6.3 COMPARISONS AMONG SPECIES

Predictions can be made concerning the resistance of various *Drosophila* species to extremes of temperature and desiccation, based upon the wide array of tropical and temperate habitats in which *Drosophila* has been found. The seasonal pattern of cool wet winters and warm dry summers of the temperate zone means that flies from temperate zones should be more tolerant of climatic stresses than those from the tropics. For example, the *melanogaster* subgroup contains two widespread

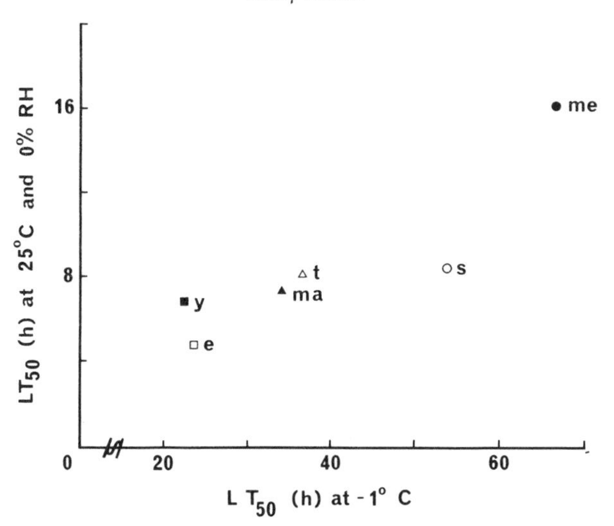

Figure 6.3 Relationship between desiccation resistance at 25°C and cold-temperature resistance at −1°C assessed by LT_{50} values (number of hours at which 50% of flies died from stress) for six species of the *melanogaster* species subgroup, comprising two widespread species (me, *melanogaster*; s, *simulans*) and four species restricted to the tropics (e, *erecta*; ma, *mauritiana*; t, *teissieri*; y, *yakuba*) (after Stanley *et al.*, 1980).

species present in both the tropics and the temperate zone, namely *D. melanogaster* and *D. simulans*, and a number of species restricted to the tropics. Comparative studies (Fig. 6.3) show that adults of the two widespread species are more resistant to the physical stresses of −1°C cold and desiccation at 25°C (and temperatures up to 34°C) than are the tropical species (Stanley *et al.*, 1980). The type of analysis in Fig. 6.3 has been subsequently extended to a wider array of 20 species made up of nine restricted to the tropics and 11 temperate zone species (Parsons, 1983). For cold tolerance there was no overlap across zones, while for high temperature/desiccation the overlap involved one species (*D. simulans*). Taken together with the importance of climatic extremes in the determination of *Drosophila* habitats in rain forests, these results suggest that stressful temperature and humidity regimes play an important role in the distribution and abundance of organisms.

Superimposed upon regular periods of climate-induced ecological stress on a diurnal and seasonal basis, there are likely to be occasional rare periods of climatic stress of such severity that biological extinctions may occur, as found by Ehrlich *et al.* (1980) in their studies of the effects of a major Californian drought on the dynamics of species of the checkerspot butterfly genus, *Euphydryas*. Several populations of different species became extinct, some were dramatically reduced, others remained stable, and at least one increased; the different responses relate to the fine details of the associations between insects and their host plants. Such 'ecological crises' may not be uncommon but because of their unpredictability and short duration, detection is difficult unless records are kept for a long period of time. Studies over shorter periods of time may therefore lead to ambiguous conclusions concerning distribution and abundance. At an extreme, it follows that all biological groups will be subject to crises, so that steady states in evolutionary and hence historic time cannot be expected even though there may be long periods of apparent stability, which in the long term are necessarily transient.

Populations that are periodically subjected to sufficiently extreme and unpredictable environmental stresses will not be in a position to evolve initimate adaptations to resources. The situation for insects such as *Drosophila* is, in any case, one where highly dissected rapidly changing resource patches are exploited, so that an equilibrium with resources is unlikely. The variability of stress levels imposed by sufficiently frequent and unpredictable periods of climatic extremes will tend to maintain populations at densities that would not severely deplete resources. This is consistent with a recent study of rain forest *Drosophila* where no evidence for interspecific competition was found (Shorrocks *et al.*, 1984). While this conclusion may appear contrary to much recent evidence, it is important to emphasize that much of orthodox competition theory refers to birds, lizards, and other vertebrates, which are only a small component of biotic diversity (Strong, 1983). In an analysis of over 150 field studies Schoener (1983a) concluded that there is stronger evidence for interspecific competition among primary producers (being plants competing for space, light and nutrients) and carnivores compared with herbivores especially among terrestrial and freshwater systems, while marine systems showed virtually no trend. In Chapter 9, the role of species interactions in causing species distributions is explored further.

The time-scale of most reported studies is, however, insufficiently long to incorporate the possibility of ecological crisis when there could be a tendency for phenomena such as competitive relationships to disappear due to the major perturbations imposed by the crisis, a scenario envisaged by Wiens (1977) from work on the North American shrub–steppe. Under this situation, it follows that except over very short time periods, a simplicity of analysis can be assumed such that, as a first approximation, adaptation at the interspecific level may be reduced to climate and climate-related variables although the extent to which this assumption is valid will depend upon the species being considered over adequate time intervals. It is important to note that convincing associations with climate come from experiments where the measure used is death resulting from the inability of organisms to withstand extreme temperatures and desiccation stresses. These are stresses where fertile survivors occur, yet where lethality is close. Therefore, the emphasis upon climatic extremes proposed for an understanding of species distributions has parallels with comparisons among species in terms of direct selection exerted upon organisms at the phenotypic level by thermal stress. It suggests a need to study biota under conditions of climatic marginality as developed in an agricultural context by Parry (1978) and discussed more generally by Parsons (1986).

6.4 MOLE RATS – A TRANSITION TO THE GENETIC LEVEL

The physical geography of Israel is distinguished by substantial climatic diversity within a small area, whereby the country is divided into two distinct intergrading parts – the northern mesic (cool and humid) Mediterranean, and southern xeric (dry) regions. The combined diverse climatic and geological background causes significant environmental variation over remarkably short distances. Correspondingly, there is extreme diversity biogeographically, including Mediterranean, European, Asiatic and African plant and animal communities. The varied physical and biotic background, combined with historical factors, results in rich biological diversity for analysing adaptation to habitats (Nevo, 1985, 1986).

Subterranean mole rats of the *Spalax ehrenbergi* complex form a vertebrate example where relationships with habitat have been established. In Israel there are four morphologically indistinguishable incipient chromosomal species for which the diploid chromosome numbers are 2n = 52, 54, 58 and 60. Nevo (1985, 1986) argues that these chromosomal species are dynamically evolving, undergoing ecological separation relating to different climatic regions in Israel: cool and humid upper Galilee Mountains (2n = 52), cool and drier Golan Heights (2n = 54), warm and humid central Mediterranean Israel (2n = 58), and warm and dry Samaria, Judea and northern Negev (2n = 60). The species of the complex appear to be morphologically adapted to the subterranean ecotype which is essentially closed, and hence forms a relatively stable microclimatic system. The animals are essentially cylindrical with short limbs, having no external tail, ears and eyes so that they are effectively blind. There is, however, a quantitative decrease in size in response to geographic variation in heat load from large individuals in the Golan to smaller ones in the northern Negev, presumably as a mechanism to minimize overheating. Likewise pelage colour varies from dark to light from the northern heavier black and red soils to the lighter soils in the south. The smaller body size and paler pelage colour, characterizing primarily 2n = 60, contribute to alleviate the heavy heat loads in the hot steppe regions approaching the Negev desert (Nevo, 1986).

These animals show several adaptations at the physiological level: basal metabolic rates (BMR) decrease progressively towards the desert so minimizing water expenditure and overheating; more generally, the combined physiological variation in basal metabolic rates, non-shivering thermogenesis, thermoregulation, and heart and respiratory rates, appears to be adaptive at both the macro- and microclimatic levels, and both between and within species, all contributing to energy optimization. Ecologically, territory size correlates negatively and population numbers correlate positively with productivity and resource availability. Behaviourally, activity patterns and habitat selection appear to optimize energy balance, and differential swimming ability appears to overcome winter flooding, all paralleling the climatic origins of the different species. In addition, differential viabilities of the four chromosomal species in a standardized laboratory environment substantiate other inferences that co-adapted gene complexes underly advantageous adaptations in the various habitats occupied by the four species.

The nature and extent of hybridization between the four karyotypes is consistent with the hypothesis that they represent young, closely related, chromosomal sibling species at early stages of evolutionary divergence. The hybrid zones between karyotypes (Chapter 7) occur primarily along climatic rather than geographic boundaries. While hybrids are at least partially fertile, their overall fitness is lower than that of the parental types. In addition, the dispersal of hybrids into parental territories is restricted by a combination of genetic, cytogenetic, ethological and ecological incompatibilities.

In summary, the incipient species are reproductively isolated to varying degrees representing different adaptive systems which can be viewed genetically, physiologically, ecologically and behaviourally. All relate to climate, defined most accurately in terms of humidity and temperature regimes whereby ecological speciation is correlated with geographical variation of increasing aridity stress southwards. Mole rats provide a particularly good example of a likely adaptive response to climatic change from a mesic northern regime to a warm xeric southern regime during the upper Pleistocene in Israel. The precision of the model derives from the life-history of mole rats as habitat specialists, whereby different species achieve greatest fitness in the precise microclimate to which they are uniquely adapted. Mole rats form an important transition to the discussion of variation within species which follows in Section 6.5. The considerations in this and Section 6.3 provide models of the types of phenotypes for which genetic analyses should be attempted at the intraspecific level.

6.5 VARIATION WITHIN SPECIES

While biological species are unique genetical and ecological systems, the development of geographic races within species shows that the environment across the distribution of species is not uniform. Indeed the number of species of animals and plants in which even a rather elementary analysis has not established the occurrence of at least some geographic variation is very small. The importance of climatic variation at the interspecific level leads to the prediction that geographic races depend, as a first approximation, directly or indirectly, upon climate. Traits of direct importance in assessing such variation, in particular the ability of organisms to withstand thermal stress, have been called ecological phenotypes (Parsons, 1983). In the Queensland fruit fly, *Dacus tryoni*, which is a pest of ripening fruit from northern Queensland to south-eastern Australia (Fig. 6.2) over a climatic range from tropical to temperate, tolerance of adults to cold and heat stress increases towards the south as temperature fluctuations increase in importance (Fig. 6.4). Analogous results have been obtained from the sibling species *Drosophila melanogaster* and *D. simulans* showing parallel climatic races in all

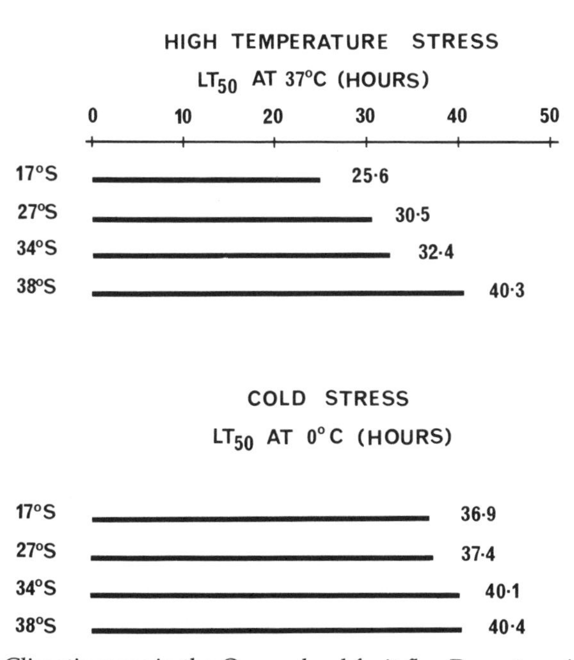

Figure 6.4 Climatic races in the Queensland fruit fly, *Dacus tryoni*, assessed by LT_{50} values (number of hours at which 50% of flies died from stress) for four populations from Cairns (latitude 17°S), Brisbane (27°S), Sydney (34°S), and East Gippsland (38°S) (drawn from data in Lewontin and Birch, 1966).

three species. The biological importance of thermal stress is therefore extended to the level of phenotypic variation within species.

The moss, *Tetraphis pellucida*, is a wide-ranging species usually growing on decaying wood in cool, moist coniferous forests of the north temperate zone. Using cabinets with controlled environments, it was possible to predict the likely distribution of the moss in the wild based upon variation in temperature, relative humidity, light intensity and pH. On a global scale, climatic data expressed as mean monthly maximum and minimum temperature dominate (Forman, 1964) implying that thermal stress is a major determinant of distribution.

Many other examples of physiological races related in some way to temperature in natural habitats are presented in Mayr (1963). Some additional examples of variation associated with climate include:

(1) In the nineteenth century, zoologists arrived at several ecogeographical rules. Races of many species from colder climates tend to be larger than races of the same species from warmer climates which is referred to as Bergmann's rule. The mole rats in Israel illustrate this point, since body size is smallest in animals from warm and dry habitats. Among warm-blooded animals, protruding body parts (ears, tails and legs) tend to be shorter relative to body size in races that live in cold climates than in those living in warm climates which is referred to as Allen's rule. Both rules are presumably connected with heat conservation or heat dissipation. Given the nature of the surface area to volume ratio, other things being equal it follows that large bodies having relatively short protruding parts are favoured in the cold whereas smaller and more slender linearly built bodies are advantageous in the heat.
(2) Clausen, Keck and Heisey (1940) studied races of a member of the rose family, the semi-perennial plant, *Potentilla glandulosa*, by taking individuals from sea-level, mid altitude and alpine zones of the Sierra Nevada region of California, and then transplanting across climatic habitats in all possible directions. In many instances, the transplants died or grew poorly in foreign habitats. In summary, altitudinal races were found to be different in appearance and best adapted to their native environments, so directly indicating the existance of climatic races.
(3) In recent years, a large literature has been accumulated on molecular polymorphisms detected by techniques such as electrophoresis. This has led to debate as to whether the molecular variation is adaptive or irrelevant to natural selection. However, discrimination between the two possibilities is usually difficult and frequently ambiguous. This unsatisfactory situation largely follows from *a posteriori* attempts at

providing causal interpretations of data from natural populations. Even so, gene frequencies of polymorphisms detected by electrophoresis often tend to be correlated with temperature based upon geographic data. There are many examples (discussed in Parsons, 1983), including the alcohol dehydrogenase (ADH) polymorphism in *D. melanogaster*, and the majority of enzyme or protein polymorphisms in man. In a survey of genetic variation in the slender wild oat, *Avena barbata*, which is a colonist of California from the Mediterranean, microgeographic variation measured from an assessment of electrophoretic variants was only interpretable in terms of adaptation to differences in temperature and aridity regimes among habitats; such selection played a major role in moulding the sample of genes from the ancestral gene pool (Clegg and Allard, 1972). Nevo, Beiles and Ben Shlomo (1984) studied the diversity of enzymes and proteins among populations, species, and higher taxa in relation to ecological parameters, demographic parameters, and a series of life-history characteristics. Figure 6.5 shows that ecological factors explain by far the highest proportion of genetic diversity. Even though this is an *a posteriori* approach, it is effectively based upon published data readily available to these authors. Given that the ecological parameters incorporate climate and climate-related variables, it appears that temperature is a factor of importance in change at the level of the genome itself.

The three examples above are consistent with the importance of temperature and associated variables in all biological processes. Such

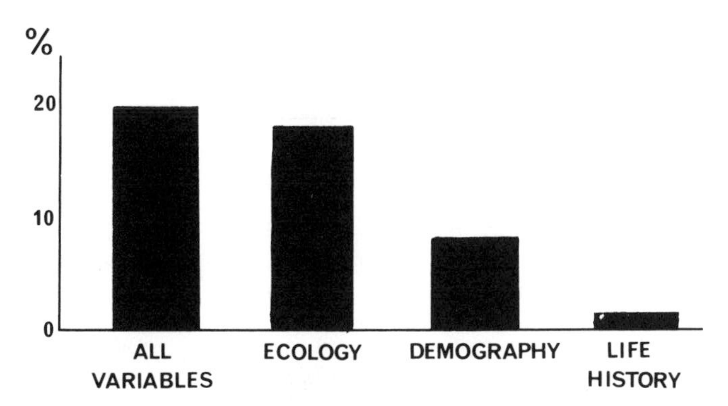

Figure 6.5 Coefficients of multiple regression (%) with dependent value heterozygosity, and as independent variables (a) 15 biotic (all) variables, and (b) a subset of 6 ecological, (c) 5 demographic and (d) 4 life history variables. The coefficients represent the proportion of genetic variance that can be explained by these variables (plotted from data in Nevo *et al.*, 1984).

examples are quite usual if adequate data are collected. Even so, the effects on gene frequencies are more realistically regarded as 'second order' responses to primary selection at the phenotypic level involving mortality after thermal stress. Temperature is of course not the only physical component of the environment, and examples of other stressful components of the environment have been studied in the form of clines (sharp phenotypic character gradients over short distances). Since clines normally reflect direct intraspecific adaptive differences at a local level, they are of unusual evolutionary interest (Endler, 1977, 1986). Among many examples that have been studied, three are listed:

(1) Environmental stress in grasses caused by heavy metals such as copper in the vicinity of old mines can result in extremely localized clines for resistance to the metals; copper tolerance has been shown to evolve in just one or two generations in the grass, *Agrostis tenuis* (Wu, Bradshaw and Thurman, 1975). The severe stress imposed by high concentrations of copper is shown by the observation that only a few species can colonize metal-contaminated habitats (Table 6.2), despite the fact that a wide range of species have had the opportunity to do so (Bradshaw, 1984). In the most polluted sites of all, only *A. stolonifera*, sometimes with *A. tenuis*, is to be found.
(2) In *Drosophila melanogaster*, ethanol in the environment underlies localized clines for tolerance to ethanol in the vicinity of wineries (Table 6.3).
(3) In the marine intertidal bivalve, *Mytilus edulis*, in Long Island

Table 6.2. Species to be found in mown grassland in copper-contaminated and uncontaminated areas at Prescot, Lancs. (from Bradshaw, 1984)

Copper in soil (ppm)	*Species found*	
< 2000 Adjacent to refinery	*Agrostis stolonifera* *A. tenuis*	*Festuca rubra* *Agropyron repens* *Holcus lanatus*
< 500 Away from refinery	*Ranunculus repens* *R. bulbosus* *Cerastium vulgatum* *Trifolium repens* *T. pratense* *Taraxacum officinale* *Rumex obtusifolius* *Prunella vulgaris* *Plantago lanceolata* *Bellis perennis*	*Achillea millefolium* *Hypochaeris radicata* *Leontodon autumnale* *Luzula campestris* *Lolium perenne* *Poa annua* *P. pratensis* *P. trivialis* *Dactylis glomerata* *Cynosurus cristatus* *Hordeum murinum*

Table 6.3. Percentage of adults emerging from 9% ethanol-supplemented media for strains of *D. melanogaster* inside and outside the wine cellars at Chateau Tahbilk winery and from Melbourne, 110 km to the south (summarized from McKenzie and Parsons, 1974). Note that the difference between the inside and the outside of the cellar, a distance of a few metres, leads to a greater difference in emergence than for the outside cellar vs. Melbourne contrast

	May 1972	*February 1973*	*March 1973*	*April 1973*	*Mean*
Inside cellar	44.1	48.8	36.7	38.9	42.1
Outside cellar	32.3	34.7	19.7	27.4	28.5
Melbourne	20.6	10.9	14.7	16.8	15.8

Sound, genetic variation at the leucine aminopeptidase locus responds to levels of environmental salinity, whereby alleles detected electrophoretically with high enzyme activity are favoured in high salinity environments (Koehn, 1978).

These examples of clines demonstrate microgeographic variation at the habitat level. More generally, variation among habitats, in particular that relating to biotic factors, may also lead to microgeographic variation. These local effects contrast with climatic factors which result in variability expressed in more regular gradients, and which dominate as the time-scale extends to the historic/evolutionary level.

Adaptation can therefore be clearly demonstrated at the phenotypic level in terms of ecological phenotypes having obvious relationships with habitat. The relationship with the genotype tends to be more obscure, since as a first approximation selection does not act directly at the genetic level as shown by the vast majority of data on variation detected electrophoretically. Indeed, underlying the phenotype is a harmonious integration of the physiological effects of many genes incorporating interactions within and between loci; this means that the effects of selection may be difficult to predict and will vary among populations differing in their genetic backgrounds (Mayr, 1963). However, there are situations where ecological phenotypes are controlled by one or two genes of major effect in which convergence between the phenotypic and genotypic approaches occurs, as in the *Mytilus edulis* example. A major field of investigation concerns the study of the connection between adaptation as expressed at the phenotypic level and the underlying genotypes. The prerequisite is to use traits that are demonstrably significant in determining distribution and abundance (Andrewartha and Birch, 1954) as the starting material. For this reason, it is important to consider adaptation at the biogeographic and interspecific levels, since in this way traits for detailed study at the genetic level can be realistically selected.

6.6 ADAPTATION AND STRESSFUL ENVIRONMENTS

At the outset, no precise definition of adaptation was given because it is a term variously used to describe an intimate association of organism with habitat, a general state, and sometimes a particular phenotypic condition. These are but components of a more generalized definition, namely traits that increase the chances of survival and reproduction of those individuals carrying them. In this chapter, the developing emphasis is upon adaptation in terms of the differential survival of organisms brought about by extreme stresses. This is 'hard selection' in the sense of Wallace (1981) whereby death occurs after various periods of stress irrespective of the density and frequency of neighbouring organisms. Adaptation is therefore being assessed following selection pressures on organisms whereby few survive, but those that do may carry genes promoting survival against a potentially lethal stress. Both experimental and distributional data within and among species emphasize the validity of this approach and it can be extrapolated *a posteriori* to a better understanding of the distribution of the species making up faunas and floras.

An understanding of adaptation in the biogeographic context therefore should be enhanced by seeking out and studying examples of change in natural conditions during periods of stress, which is not easy because of the normally short duration of stress periods in a given period of total time. A current study of the medium ground finch of Daphne Major Island in the Galapagos, *Geospiza fortis*, is informative since populations are subjected to droughts of such intensity that in one recent episode during 1977, population size decreased by 85%. During this stress period survival of finches was non-random, and large birds, especially males with large beaks, survived best because they were able to crack the large and hard seeds that predominated in the drought (Boag and Grant, 1981; Boag, 1983). The main cause of mortality was starvation, and differential survival with respect to morphology was at least partly a reflection of differential handling ability of the birds for foods remaining in the environment. Seven morphological characters were measured and were found to be highly heritable, but because of high correlations between them it was difficult to specify the precise targets of selection; however, weight and bill dimensions are the most important. Figure 6.6 shows an association between temporal change in finch numbers and principal component scores for a measure of overall body size. Therefore, occasional strong selection of highly heritable

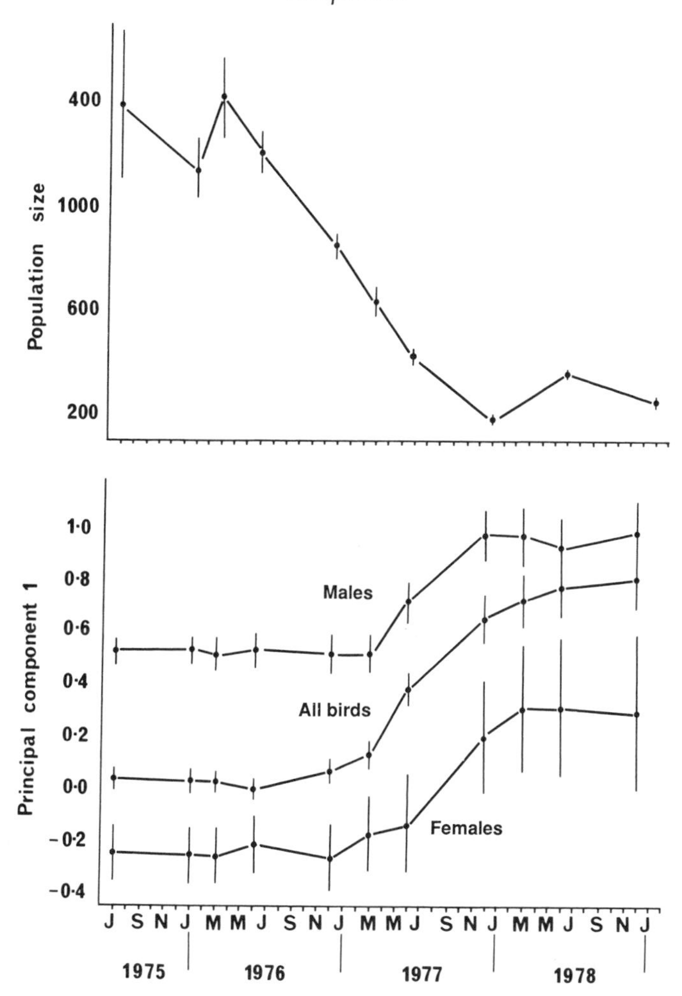

Figure 6.6 Temporal changes in finch numbers and morphology expressed in terms of a principal component which explains 67% of the total variance involved in body size; it is composed of seven measures: weight, wing chord, tarsus length, bill length, bill depth, bill width and bill length at depth of 4mm (after Boag and Grant, 1981). The major shift in morphology assessed by the principal component occurs immediately following the fall to low population size due to an intense drought.

characters in a variable environment may be one of the factors explaining the rapid evolution of the Geospizinae in the Galapagos. Another example without genetic studies comes from moles (*Talpa europaea*) during a hard long European winter in 1946–7 when the ground was frozen so access to normal food was prevented; a consequence was a high selective premium on small body size (Mayr, 1963). Hence, the evolutionary trajectory of even well-buffered species may be largely determined by occasional short periods of intense selection during a small part of their history brought about by the kinds of environmental changes described in Chapter 8. Therefore, rapid morphological evolution is most likely under ecological stress conditions when population sizes are reduced to transiently very low levels.

These results suggest that more emphasis should be placed upon the study of phenotypes of obvious relevance in the determination of distribution and abundance under stressful environments. Plants can produce useful material for such studies. For example, a study of $\log_{10}$ fresh weights of 43 homozygous races of the crucifer, *Arabidopsis thaliana*, gave genotypic variances of 0.0128, 0.0115, and 0.0592 at 25°C, 30°C, and the extreme temperature of 31.5°C respectively (Langridge and Griffing, 1959). The variance at 31.5°C was more than five times that of the lower temperatures which indicates that genotypic differences in growth are most evident at high and stressful temperatures. Indeed at 31.5°C, five of the races showed high temperature lesions consisting of pronounced morphological symptoms, suggesting that under extreme environments, specific metabolic reactions involving thermolabile enzymes are likely to be affected by stress. This means that interpretations in discrete molecular (and hence genetic) terms should ultimately become possible. It also indicates that genes of quite minor effect under optimal conditions may, under stress conditions, increase in effect sufficiently to convert a quantitative trait to one of high variability which can be studied by conventional Mendelian methods (Parsons, 1986).

Combined with much published data, the general conclusion is that the variability of such quantitative phenotypes and their underlying genotypes tends to be high under conditions of severe environmental stress, in particular involving the direct or indirect effects of temperature (Parsons, 1983). This conclusion, which applies directly to quantitative traits of importance in determining survival, also applies less directly at the level of genes controlling protein variation. This is a reflection of the direct effects of selection at the phenotypic level being translated into indirect selection at the protein level. Such experiments in the field and the laboratory are therefore consistent with the observation of rapid evolutionary change under stress observed in *Geospiza fortis*. There is also some evidence accumulating in the recent literature that *de novo* variation due to recombination and mutation in the broadest sense increases under stress (Little and Mount, 1982). Hence at stressful moments in evolutionary history when there is a premium on major adaptive shifts, variability of all types may be higher, which could trigger genomic reorganizations in rapidly changing environments with results that would be ultimately detectable in a biogeographic context.

6.7 CONCLUSION

For the biogeographer, a valid approach to an understanding of adaptation is to commence at the phenotypic level and concentrate upon traits contributing to survival under stress conditions. This emphasizes the interface of quantitative genetics and ecology. It is not surprising that the genetic basis of most adaptive change is poorly understood, because the analysis of the variation of quantitative traits in natural populations is a problem of considerable difficulty. This is exacerbated by the need to concentrate upon the critical and elusive interface involving a level of environmental stress whereby fertile survivors occur, yet where lethality is close. The problem of designing experiments to fall into this narrow environmental stress zone undoubtedly contributes to a literature where obtaining meaningful interpretations is not easy (Parsons, 1986), especially as the consequence of quite minor perturbations could be extinction.

The ideal situation is *a priori* to make predictions on correlations between variations at the ecological level of habitats, and of postulated important phenotypic traits. For example, it was predicted that resistance to extreme physical stresses would be greater in flies from temperate than in those from tropical zones (Figs 6.3 and 6.4). Hence for the *a priori* approach, it is important to select traits that are a direct reflection of ecological variations (ecological phenotypes). In other words, by considering physical extremes as ecological variants of primary concern, predictions concerning tolerances to extreme stresses under laboratory conditions can be made. To emphasize this approach further, it was predicted that the genetic variability of mortality following desiccation stress should be greater in populations of *Drosophila melanogaster* from the humid tropics, by comparison with populations from the heat-stressed wet-dry tropics and the temperate zone simply because there is strong selection in nature during certain times of the year in the latter two populations for desiccation resistance. In confirmation of this prediction, approximate intraclass correlations (which are measures of levels of genetic variability) from two humid tropical populations were 0.61 and 0.71, and from a wet-dry tropical and a temperate zone population, 0.07 and −0.03, respectively (Parsons, 1980).

When the time-scale is extended as in the historic context, the emphasis is almost totally on *a posteriori* assessments whereby adaptive explanations are formulated following the collection of observations. Even so, the study of adaptation does reveal the underlying importance of environmental stress, which then can be incorporated into *a priori* predictions of significance for biogeographic analysis.

6.8 SUMMARY

The approach taken in this chapter is reductionist, whereby the emphasis is on adaptation in terms of traits that are a direct reflection of ecological variations, or ecological phenotypes. By considering climate-related stress as the ecological variant of primary concern, it is possible to provide broad interpretations of distributions at the biogeographic and species levels, and of climatic races with species, although the translation from direct selection at the phenotypic level to more indirect selection at the genetic level presents interpretative difficulties.

Variability of ecological phenotypes and their underlying genotypes tends to be high under severe environmental conditions so providing circumstances conductive to rapid evolutionary change. Hence environmental stress is an underlying determinant of adaptation across organizational levels ranging from the biogeographic to the genetic, and may simultaneously be the environmental circumstance when evolutionary change is the most rapid. Even so, at the level of the genome, organism and environment, perfect adaptation is ultimately unattainable because of unpredictable variation at all levels, and because of phenomena involving complex interactions including competition.

CHAPTER 7

Speciation

N. H. BARTON

7.1 INTRODUCTION

Biogeography is concerned with the distribution of living organisms across space and through time. As discussed in Chapter 1, the subject can be studied over a wide range of scales. Biogeography usually concentrates on whole species or higher taxa, but it may be fruitful to extend the range of biogeographic scales to include genetic patterns of populations. First, the processes involved in the spread of genes may be analogous to those involved in the spread of species: comparison of the two may illuminate both problems. Second, biogeography is linked with population genetics through the process of speciation: inferences in biogeography may therefore depend on how new species arise. Finally, the wealth of molecular data now becoming available gives accurate information about phylogenies over the very long time-scales with which much of biogeography is concerned.

In this chapter, I will describe the various modes of speciation. That is, I will describe the various ways in which genes causing reproductive isolation may become established and then spread through a population. My main concern will be with the inferences about the process of speciation that can (and cannot) be made from current evidence. We will see that even given detailed genetic observations, it is hard to tell how present patterns evolved. However, some biogeographic inferences can be made about the way species spread, and these may be independent of the unknown processes by which the species first arose.

7.2 THE NATURE OF SPECIES

7.2.1 Definitions

Before considering how new species might evolve, we must be clear about what exactly is meant by a 'species'. Throughout, I will be referring to biological species, which are defined as 'groups of actually or potentially interbreeding natural populations which are reproductively isolated from other such groups' (Mayr, 1942, p. 120). In other words, a biological species is characterized by genetic differences which prevent it exchanging genes with other species under natural conditions.

This definition has become widely accepted because it is objective and unambiguous, and because it gives units which are genetically independent: genes which arise in one species cannot spread to another. However, it is hard to apply in practice. The simplest case is where two putative species live together in sympatry. Here, the absence of intermediate morphs would suggest that the two are good biological species. However, since distinct morphs may coexist within a single interbreeding population (e.g. polymorphic Batesian mimics (Sheppard, 1958), and some cichlid fishes (Kornfeld *et al*, 1982)), genetic information is needed to be absolutely sure of their status. When two forms abut in a parapatric distribution, details of the narrow zone of contact are needed to show whether or not they can exchange genes. Where the two forms are entirely allopatric, it may be essentially impossible to show whether they could exchange genes under natural conditions. Of course, assignment of fossils or asexual forms of species must necessarily be by arbitrary morphological criteria.

In practice, therefore, species are usually recognized by their morphology. Morphological species do not necessarily correspond to biological species. On the one hand, sibling species may be almost indistinguishable, and yet may be unable to exchange genes. On the other, very different phenotypes may be able to exchange at least some genes (e.g. the mice *Mus musculus*/*M. domesticus* (Gyllenstein and Wilson, 1987) and the toads *Bombina bombina*/*B. variegata* (Szymura and Barton, 1986)).

Despite the practical difficulties in its application, and despite the fact that the majority of species are in fact defined by the quite distinct criterion of morphology, the biological species concept gives the clearest basis for investigating speciation. The morphological characters which distinguish species may arise in a wide variety of ways and, indeed, the evolution of these features involves essentially the same processes that are involved in intraspecific adaptation (Chapter 6). In contrast, there are specific obstacles to the evolution of genetic differences which inhibit gene flow, and thereby cause some degree of reproductive isolation.

7.2.2 Obstacles to speciation

The essential difficulty is that genes which promote reproductive isolation will be selected against when rare; it is therefore hard to see how they can be established. The simplest case is where heterozygotes at some locus are less viable or less fertile: for example, individuals heterozygous for chromosome rearrangements are usually partially sterile as a result of improper disjunction at meiosis (White, 1978). The rarer allele or karyotype will be found mainly in heterozygotes, and so will be eliminated. Post-mating isolation may also be produced by the breakup of co-adapted combinations of genes: for example, hybrids between populations of butterflies which mimic different distasteful models will be eaten more often if different elements of their warning patterns recombine (as happens in the burnet moth *Zygaena ephialtes* (Turner, 1971)). Then, the rarer combination of genes will be found more often in heterozygous form, and so will be broken up by recombination and segregation more frequently. (In the case of Müllerian mimicry, the rarer form suffers an additional disadvantage through frequency dependent selection: predators are less likely to recognize the rarer form as being distasteful (Mallet, 1986)).

Similar difficulties impede the evolution of premating isolation. Paterson (1982) argues that male secondary sexual characters, and female preferences for those characters, must evolve together, and that a change in either will be selected against. Certainly, new male characters which are less attractive to females will be unable to establish themselves. Female preferences for novel male characters may also be disadvantageous, either because the preferred males are hard to find, or because if the females do find and mate with the rare males, their sons will inherit their father's characteristics, and so will be unattractive to the majority of females. However, there is considerable controversy over the evolution of female preferences, and it may be that pre-mating isolation can evolve rather more readily than post-mating isolation (Lande, 1981; Kirkpatrick, 1982; p. 198).

A useful metaphor for the difficulty of evolving reproductive isolation is provided by Wright's 'adaptive landscape' (Provine, 1986). This is a graph of the mean fitness of a population, plotted against the state of the population (represented by allele frequencies, means of quantitative characters, and so on) (Fig. 7.1a). Since, at least in simple cases, selection causes mean fitness to increase, populations will move uphill, and will come to rest on local 'adaptive peaks'. In general, the 'adaptive landscape' will have a complicated topography, and so a population may come to rest on one of many alternative peaks. When populations at different peaks hybridize, the progeny population is likely to be less fit, and so gene exchange will be impeded. Where the two populations occupy different parts of a continuous geographic range, they will be kept separate by a stable hybrid zone, in which selection against hybrids balances dispersal, and in which mean fitness is somewhat reduced (Fig. 7.1b). Thus, species can be thought of as resting at different equilibria, or 'adaptive peaks': these equilibria are stable both to new mutations, and to introgression from other populations (i.e. to incorporation of alien genes). From this perspective, the problem of

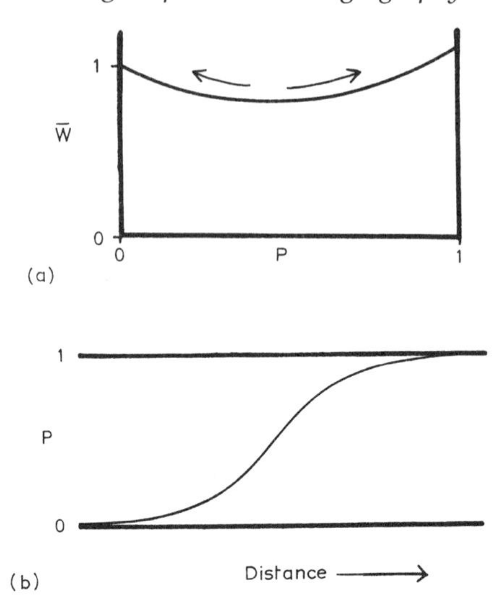

Figure 7.1 (a) The 'adaptive landscape': the mean fitness ($\overline{W}$) is plotted as a function of allele frequency (P); natural selection causes the population to move uphill, towards a local 'adaptive peak'. In this example, P is the frequency of a chromosome rearrangement which reduces the fitness of heterozygotes by 50%, but which increases fitness by 10% when homozygous. (b) When two populations at different 'adaptive peaks' meet, a stable hybrid zone will form.

speciation is reduced to the problem of how a single population comes to split into two populations resting on different adaptive peaks.

7.2.3 Species differences are polygenic

Though I have argued that selection will generally oppose the divergence of populations onto separate, reproductively isolated, adaptive peaks, this difficulty should not be exaggerated. Certainly it is hard to imagine how a new, fully isolated species could arise in a single step, as suggested by Goldschmidt (1940): the definition of a biological species implies that the "hopeful monster" could not mate successfully with any of its neighbours (Charlesworth, 1982). The only exception is where, as with some polyploid plants and parthenogenetic interspecific hybrids, the aberrant individual can self-fertilize, or can reproduce asexually (see Chapter 5 in relation to the origin of endemic plants). However, related species usually differ at many genetic loci, suggesting that reproductive isolation evolves in a series of small steps.

Evidence for this comes from both experimental crosses, and from natural hybridization. Laboratory crosses have shown that, in most cases, morphological and behavioural differences, hybrid inviability, and hybrid sterility, depend on many genetic differences (Muller, 1939; Stebbins, 1950; Barton and Charlesworth, 1984). Gottlieb (1984) has argued that, in plants, species differences may often be based on a few major genes (see also Templeton, 1980). However, though this may be true for particular taxonomic characters (Hilu, 1983), the overall number of genetic differences still seems large (Tanksley *et al.*, 1982; Coyne and Lande, 1985). Furthermore, laboratory crosses can greatly underestimate the numbers of genes responsible for species differences and for reproductive isolation: a few generations of crossing may not separate linked blocks of genes; genes with minor effects may cause substantial isolation in aggregate, and yet will be missed by Mendelian analyses; and finally, estimates can only be made for those characters chosen for study – genes affecting other characters will be missed.

Studies of hybridization in nature also show that even subspecific differences are, in the majority of cases, polygenic. Out of 21 zones of hybridization for which detailed surveys have been made, an average of 20% of enzyme loci are fixed for different alleles, or show substantial frequency differences. In general, hybrid zones detected by one character (such as a chromosomal difference) usually involve differences in a large number of other characters (Barton and Hewitt, 1985). Of course, not all these differences need contribute to reproductive isolation. However, in the few cases where the number of genes causing isolation can be estimated, this number is large: in the grasshopper *Caledia captiva*, $\geqslant$ 18 (Shaw and Wilkinson, 1980; Shaw *et al.*, 1982); in the grasshopper *Podisma pedestris*, ~ 150 (Barton and Hewitt, 1981); in the toads *Bombina bombina*/*B. variegata*, ~ 300 (Szymura and Barton, 1986).

7.2.4 Speciation and biogeography

The polygenic basis of species differences makes it easier to see how species could evolve through a series of shifts across shallow adaptive valleys. However, it also complicates biogeographic arguments. Speciation is usually seen as an abrupt bifurcation of one line of descent into two (Fig. 7.2a). But in fact, a population consists of a tangled collection of lineages (Fig. 7.2b), each consisting of the passage of a particular gene down through the generations. For example, the mitochondrial DNA is inherited from the mother; all the mitochondrial genomes in the human population can therefore be traced back through the maternal pedigree, and necessarily derive from a single ancestral genome. In this example, the sequence diverge between different mitochondrial DNAs allows this ancestral genome to be traced back by one method of analysis to a woman who lived ~ 100,000 years ago (Cann *et al.*, 1987). Similarly, all human Y chromosomes trace back through the male pedigree to a single common ancestor, as indeed do all discrete genetic loci. However, sexual reproduction ensures that different genes will trace back to different ancestors: the whole collection of lines of descent which make up a sexually reproducing population will continually be recombining with each other to produce different combinations of genes. Speciation

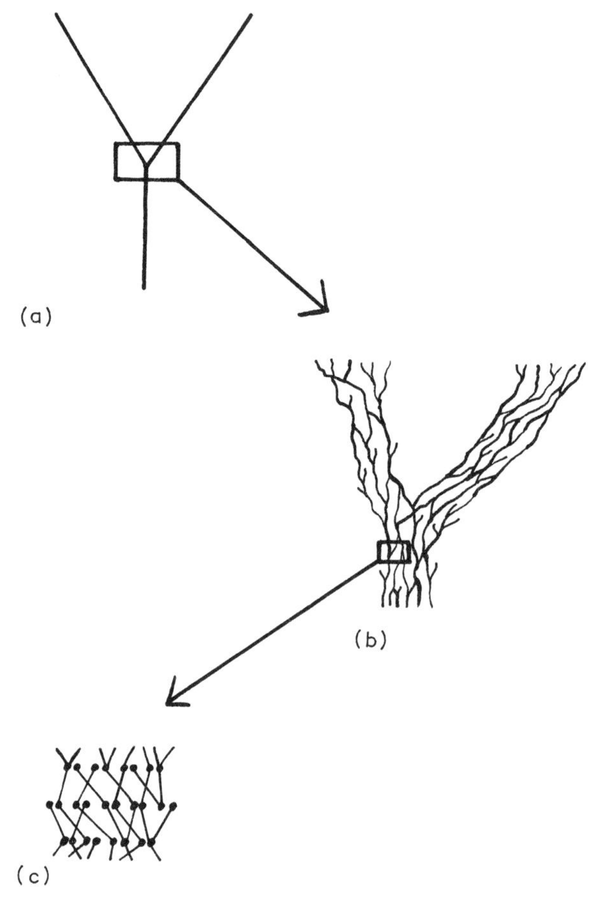

Figure 7.2 The multiple lineages within sexual populations. (a) Over very long time-scales, a population may appear to split cleanly into two separate lines of descent: speciation appears instantaneous. (b) In fact, speciation involves the gradual disentanglement of many intertwined lines of descent. The lines in this figure may represent either discrete geographic subpopulations, or on a finer scale, an approximation to the pedigrees of sexually reproducing individuals. (c) Seen in more detail, each diploid individual carries two copies of each gene (shown as a pair of dots); these may or may not be passed on to the next generation. This diagram only shows the descent of genes at a single locus; genes at other loci will descend in different patterns.

therefore consists not of an abrupt bifurcation, but of a complicated separation of a set of lineages into two separate streams (Fig. 7.2b). Just how many lines of descent lead to each species, and just how quickly full separation is achieved, are matters of controversy.

To what extent are biogeographic inferences affected by the tangled origin of species? We will return to consider this question in more detail later; for the moment, though, it is worth noting several reasons why biogeographic arguments may be largely independent of the mode of speciation. First, speciation may often be fast enough that, over the longer time-scales of historical biogeographers, it appears to give a clean bifurcation. This is certainly so when we are concerned with the evolution of higher taxa. However, the prevalence of hybridization at the level of species and subspecies, particularly in plants, shows that transitional stages in speciation cannot be neglected over such intermediate time-scales. The distribution of many hybridizing populations shows that they have been in contact since the last glaciation (~5–10 000 years ago). For example, the boundaries between three chromosomal races of the flightless grasshopper *Warramaba viatica* can be extended from the mainland of South Australia to Kangaroo Island, showing that their distribution has remained more or less fixed since the island was cut off by the rising sea level (White, 1978). Similar examples can be found in *Podisma pedestris* (Hewitt, 1975) and *Bombina* (Arntzen, 1978), (see Barton and Hewitt, 1985). Hybridization may of course have persisted for much longer: molecular evidence suggests that both *Mus musculus/M. domesticus*, and *Bombina bombina/B. variegata* have been diverging for several million years (Szymura *et al.*, 1985; Wilson *et al.*, 1985). It may therefore be that over periods of less than a few million years, we cannot safely assume that species form abruptly.

Fortunately, hybridization is not fatal to simple biogeographic arguments. We have seen that where two distinct populations meet and hybridize, they generally differ at a large number of genes. It is remarkable that patterns at these different loci are almost always closely coincident. We will consider why this should be later: the point here is that, at least as far as the divergent loci are concerned, we can treat the hybridizing taxa as coherent units. Genes may well be exchanged at some loci, so that the taxa may not be good biological species; however, this need not affect most arguments about their spread and distribution.

In some cases, however, we may be forced to follow the spread of each gene independently, and must abandon the simple view of speciation as a clean split of one set of lineages into two. For example, the mitochondrial sequence which characterizes *Mus musculus* has penetrated north into Sweden; in contrast, nuclear genes responsible for enzyme and morphological differences all shift from the *M. musculus* to the *M. domesticus* form in Jutland (Gyllenstein and Wilson, 1987). Similarly, the geographic patterns of chromosome arrangements in *Warramaba viatica* imply that at some stage, clines for different chromosome arrangements have passed through each other (Key, 1982; Hewitt, 1979). Such cases should not be seen merely as unwelcome violations of species boundaries: it may be that we can learn much about the processes involved in the spread of higher taxa from the behaviour of these much simpler genetic differences.

7.3 MODES OF SPECIATION

7.3.1 Mechanisms of speciation

How do populations come to rest on distinct adaptive peaks, despite the fact that shifts between peaks will be opposed by selection? If species accumulate reproductive isolation through a long sequence of peak shifts, then the selection opposing each shift may be weak. However, we must still explain how it might be overcome: in particular, we will see that the mechanism by which reproductive isolation evolves may influence the subsequent geographic relations between the genes involved.

Random drift

The most straightforward possibility, and the one which has received most attention, is that random sampling drift in small populations can knock a population from one peak to another (Fig. 7.3a). Random drift necessarily occurs in all natural populations: out of many examples, perhaps the best is the fluctuating frequency of inherited diseases in small human populations (Vogel and Motulsky, 1986). However, there has been long and heated argument over its significance relative to other evolutionary processes, and in particular, over its importance in speciation (Provine, 1986). Essentially, the difficulty in resolving this argument is that peak shifts are likely to occur rapidly, in small populations, and may have no immediate phenotypic effect. Evidence bearing on the relative importance of different processes must therefore be indirect and circumstantial.

Firstly, consider the relatively well understood example of chromosomal evolution. A rearrangement may cause non-disjunction in the meiosis of heterozygotes. The chance that a rearrangement which reduces heterozygote fertility by a factor s will be established in a local population (or 'deme') of N individuals decreases exponentially with Ns, and is negligible when this product is greater than 20 or so (Wright,

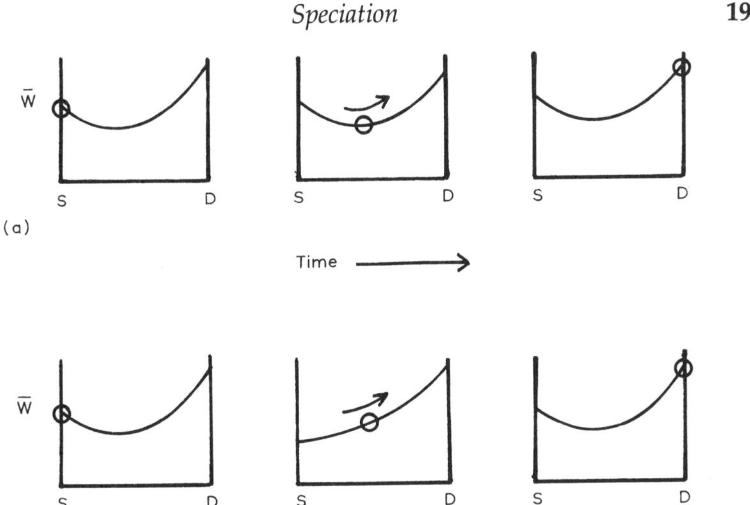

Figure 7.3 Alternative mechanisms of speciation, illustrated by the evolution of dextral coiling in *Partula suturalis*. Since matings between sinistral (S) and dextral snails (D) produce fewer offspring, a population will tend to be fixed for one or other morph. In the presence of *Partula mooreana*, the dextral allele gains an additional advantage: the corresponding peak is therefore higher. $\overline{W}$ is the mean fitness. (a) In a small population, random drift may fix the dextral allele. (b) Changes in selection may also push the population to a new peak. If *P. mooreana* is sufficiently abundant for some time, or in some place, the dextral allele will be advantageous even when rare.

1941; Bengtsson and Bodmer, 1976). On the assumption that new karyotypes spread by the random extinction and recolonization of isolated demes, the rate of evolution of the whole species is equal to the rate of evolution of a single deme (Lande, 1979; p. 212). Observed rates of chromosomal evolution then fit with plausible deme sizes and selection pressures. For example, in mammals centric fusions (i.e. fusions between two chromosomes with terminal centromeres) cause a fertility loss of $s = 0.05$–0.20. Such rearrangements arise spontaneously at a mutation rate of $\mu = 10^{-3}$–10^{-4}, and accumulate in species at a rate of 10^{-6}–10^{-7} per year. If random drift, and random extinction and recolonization, are responsible for their establishment and spread, then the effective size of local demes must be 30–200. The rate of accumulation of centric fusions in insects is much lower, implying a correspondingly larger deme size of 200–800 (Lande, 1979). These calculations show that simple drift is a plausible explanation of chromosomal evolution. However, since we have only a rough idea of actual deme sizes, mutation rates, and selection pressures in nature, and since we cannot be sure of how rearrangements actually spread through populations, this cannot be taken as strong support for sampling drift.

One reason why sampling drift has been accepted as the cause of chromosomal evolution is that it is hard to imagine other processes which could overcome the effects of meiotic non-disjunction. This argument is reasonable for rearrangements such as translocations and pericentric inversions (i.e. inversions which include the centromere), which cause $s \sim 50\%$ sterility; it is supported by the general restriction of chromosomal polymorphism to narrow zones of hybridization between homogeneous populations (White, 1978). However, the meiosis of heterozygotes involving those rearrangements found in nature is often much more regular than would be expected from the behaviour of spontaneous rearrangements in the laboratory (Capanna *et al.*, 1977; Barton, 1980; Shaw and Wilkinson, 1980; Shaw *et al.*, 1982). The weak effects of non-disjunction may often be less important than other effects of chromosome differences on (for example) spermatogenesis, DNA content, meiotic drive, and linkage relations (White, 1978; Charlesworth and Charlesworth, 1980, Moran, 1981; Westermann *et al.*, 1987). New karyotypes may therefore be established by straightforward selection, without any need to invoke random drift.

Although the great majority of species do differ in karyotype, these differences only contribute a small fraction of the net reproductive isolation. What can we say about the relative importance of random drift in the accumulation of overall isolation? Here, the evidence is even less direct. The main argument that random drift is a major cause of speciation is the frequent diversity found between small isolates (Mayr, 1942). The classic example is the dramatic radiation of Drosophilidae on the Hawaiian Archipelago. Here, hundreds of species of Drosophilidae have evolved striking morphological and behavioural differences in a relatively short time ($\leq$700 000 years on the main island of Hawaii; Carson and Kaneshiro, 1976; Williamson, 1981). Severe bottlenecks have undoubtedly occurred, both during the colonization of new islands, and during the colonization of new patches of forest ('kipukas') within islands. Carson (1975) and Templeton (1980) have argued that these 'founder events' have caused the striking diversification seen in this and other cases. It is important to note that founder events involve many processes other than random sampling drift (p. 202) and that, conversely, drift may be important within broadly continuous continental distributions (as, for example, in Wright's (1932) 'shifting balance' theory). The argument over the role of founder events in speciation is distinct from that over the role of drift. But in both cases, the evidence is ambiguous (Williamson, 1981; Barton and Charlesworth, 1984). The Hawaiian *Drosophila* have allozymes no less heterozygous than those of their mainland relatives (Craddock and Johnson, 1979; Sene and Carson, 1977), suggesting that any bottlenecks which have occurred have had little genetic effect. There are striking radiations on other islands, in which there is no evidence for any role of drift or founder events: for example, the land snail *Partula* on Moorea (Johnson *et al.*, 1986), cichlid fishes in Lake Victoria (Fryer *et al.*, 1983), and Darwin's finches *Geospiza* on the Galapagos (Grant, 1986). Even on Hawaii, it is clear that many of the *Drosophila* species have evolved bizarre morphologies as adaptations either to the unusual ecological community, or as a consequence of intraspecific sexual selection (Williamson, 1981). Some laboratory experiments show that population bottlenecks can promote partial reproductive isolation (Powell, 1978; Ahearn, 1980); however, isolation may also evolve under selection in large populations (Kilias and Alahiotis, 1982).

The best evidence that drift is an important cause of both chromosomal evolution and speciation comes from the strong correlation between the two, both within mammals and across vertebrates (Wilson *et al.*, 1974; Avise and Ayala, 1976; Bush *et al.*, 1977; but see Gold, 1980). The most obvious explanation of this correlation is that small, subdivided populations accumulate both chromosomal changes and reproductive isolation more rapidly; this interpretation fits with the inverse correlation between chromosomal evolution and enzyme heterozygosity (Coyne, 1984). On balance, however, though drift and founder events might be important causes of speciation, there is no decisive evidence that this is so.

Changes in selection

Populations may move to new adaptive peaks or equilibria as a result of changes in selection, as well as through random drift. If selection changes for a limited time or within some limited region, a population may move uphill into a new genetic state. When selection pressures change back, the adaptive valley separating the old state from the new may reappear, and the population will be partially isolated from its ancestors (Fig. 7.3b). Although, as with sampling drift, it is hard to see this process happening, and it may be hard to identify the selective agents involved, a few clear cases are known. For example, the snail *Partula suturalis* is divided into two races, which differ in the direction of shell coiling. The sinistral form meets the dextral form in a narrow (~1 km) band of polymorphism. In this region, fertilizations of one morph by the other are less frequent than expected by random combination; furthermore, the net reproductive output is reduced when different morphs mate (Murray and Clarke, 1980; Johnson, 1982). This combination of assortative fertilization and reduced fitness of cross-matings is caused simply by mechanical incompatibility, and leads to partial reproductive isolation. It is hard to see how the dextral form could have arisen from the sinistral (or vice versa), since the rarer morph would be at a selective disadvantage. However, the range of the dextral morphs parallels the range of another species, *Partula mooreana*, which freely mates with *P. suturalis*, but gives inviable offspring. One can show that when more than one third of *P. suturalis* matings are with *P. mooreana*, the dextral morph is at an advantage even when rare: the disadvantage of partial incompatibility with its own species is outweighed by the advantage of partial isolation from fruitless matings with *P. mooreana* (Johnson, 1982). Thus, a change in selective conditions (i.e. the presence of *P. mooreana*) allows the population to shift to a new state (Fig. 7.3b), which gives partial isolation in places where *P. mooreana* is absent.

Other examples where a change in selective conditions has caused a shift to a new 'adaptive peak' are found amongst Müllerian mimics. Here, shifts away from the current warning pattern are usually selected against: if the new pattern is not carried by any other distasteful species, then predators will not recognize it, whilst even if it does resemble other warning patterns, the resemblance may be too crude to be effective. However, even a crude resemblance may be selectively advantageous if that pattern is sufficiently common, and if the models are sufficiently distasteful. This seems to explain the evolution of new morphs in *Zygaena ephialtes* (Turner, 1971), and *Heliconius hermathena* (Brown and Benson, 1977).

I have contrasted random drift and changes in selection as alternative ways of reaching different equilibria. However, the distinction is not always clear-cut: for example, the deterministic selection pressures on any one species may change as a result of the random extinction of, and colonization by, other species. Turner (1971, 1976, 1982) has suggested that such 'faunal drift' in forest refugia may cause changes in the frequency of different models, and so may explain the divergence of mimetic patterns. More generally, adaptive radiations on oceanic islands may be due to the different assemblages of species on those islands, rather than to the random effects of genetic drift within species (Williamson, 1981).

Quasi-neutral models

The assumption so far has been that in order for strong reproductive isolation to evolve, a deep adaptive valley must be crossed – either by drift in a small population, or by substantial changes in selection. This most likely entails evolution of reproductive isolation in a series of

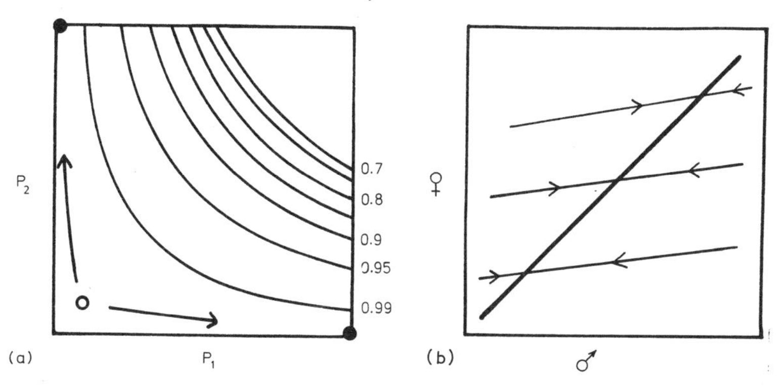

Figure 7.4 Quasi-neutral models. (a) Suppose that the double homozygote ($P_1P_1P_2P_2$) is inviable, whereas all other genotypes have equal fitness. The graph shows contours of mean fitness as a function of allele frequencies. When one or both recessive alleles are rare, the surface is almost flat. Weak mutation pressure will push the population towards one of two stable equilibria (•), in which one of the recessive alleles is fixed (Bengtsson and Christiansen, 1983). The equilibria for mutation rate $\mu = 10^{-4}$ are shown. (b) A line of neutral equilibria arises in the coevolution of a male trait, and a female preference for that trait (Lande, 1981). The horizontal axis shows the mean of the male trait and the vertical axis shows the value of the trait preferred by females.

weakly selected steps (Walsh, 1982: Barton and Charlesworth, 1984). Several models have been proposed which aim to avoid this constraint, by allowing strong isolation to evolve without any substantial barrier being overcome. In these models, the population avoids a deep valley by moving around it on a flat ridge (Fig. 7.4a). Though populations may drift freely into alternative states, a cross between two populations in different states may give substantial isolation (Wills, 1970; Lande, 1981; Bengtsson and Christiansen, 1983; Nei *et al.*, 1983). A rough geographic analogue of this scheme is provided by ring species, in which two taxa are fully isolated, and live in sympatry in one region, but are linked by a continuous gradation of intermediates in other places, for example in *Partula* (Murray and Clarke, 1980), or in salamanders of the genus *Ensatina* (Mayr, 1963, Wake *et al.*, 1986). 'Quasi-neutral' divergence of this sort may be possible whenever different, incompatible genotypes can fulfil the same function, and are linked by a more or less continuous series of intermediate genotypes.

Two general classes of examples should be mentioned. Firstly, under stabilizing selection, the optimal phenotype can be produced by a variety of combinations of polygenes ($+++---$, $-+-+-+$, etc.); thus, populations fixed for any of these combinations will be at stable adaptive peaks (Barton, 1986). Transitions between neighbouring peaks will only be opposed by weak selection, but substantial incompatibility may be revealed when very different combinations recombine (Cohan, 1984).

Secondly, there has been much argument recently about whether similar phenomena result from sexual selection, and can explain both the bizarre elaboration of male secondary sexual characters, and the diversity of mating behaviour and genital morphology between species (Templeton, 1980; Lande, 1981; Carson, 1986; Eberhard, 1986). These ideas stem from Fisher's (1930) suggestion that male characters and female preferences for those characters might coevolve in an unstable 'runaway process'. Explicit models of this suggestion show that male characters and female preferences may evolve to anywhere on a line of possible equilibria (Fig. 7.4b). The equilibrium point is arbitrary because, though female preferences for the male character must balance other selective forces against the character, the strength of these opposing forces is indeterminate. Populations may therefore drift along the line of possible equilibria, and substantial pre-mating isolation will be produced between populations which happen to reach different equilibria.

A concrete biological example of such 'quasi-neutrality' is the evolution of different chromosomal races of the shrew, *Sorex araneus*. These evolved from an ancestral acrocentric karyotype, and have accumulated different metacentric combinations of chromosome arms by centric fusions (Searle, 1986). For example, the Oxford race carries fusions between chromosome arms *k* and *q*, *n* and *o*, and *p* and *r*: (*kq*, *no*, *pr*). The Hermitage race carries (*ko*, *q*, *n*, *p*, *r*). There need have been little obstacle to the evolution of these races, because any metacentric combination will pair and disjoin from the constituent acrocentric chromosomes almost normally (e.g. (*pr*/*p*, *r*)). However, in crosses between races carrying different metacentric combinations, several chromosomes will attempt to pair at meiosis (e.g. ((*kq*, *no*)/(*ko*, *n*, *q*))); severe meiotic problems, and substantial sterility, ensue (Bickham and Baker, 1986).

These examples seem to show that substantial barriers to gene exchange can be produced without strong selective barriers being overcome. However, the situation is not quite so straightforward. There may be considerable hybrid breakdown when the divergent populations first meet. However, provided that some gene exchange is possible, and provided that the necessary alleles can be recovered, the missing links through which the populations evolved will eventually be restored (Barton and Charlesworth, 1984). This has happened in *Sorex*; in the hybrid zone between the Oxford and Hermitage races, the ancestral acrocentric chromosomes are common. This greatly reduces the level of non-disjunction in the field (Searle, 1986). Similar breakdown of isolation is to be expected in the other examples of 'quasi-neutral' speciation discussed above: the level of gene flow in nature may be much greater than would be expected from the first few generations of crosses. Thus, although these models do allow differences to accumulate under weak selection, the isolation which results is correspondingly weak.

Accumulation of incompatible mutations

Although the 'quasi-neutral' models discussed above avoid the crossing of deep adaptive valleys, they would not be expected to give much isolation in nature. This argument is based on the simple picture given by Wright's 'adaptive topography', in which a surface of mean fitness is plotted against two gene frequencies. In fact, many thousands of genes are polymorphic, and insofar as this genetic variation affects fitness at all, it is likely to do so in complicated, highly interactive, ways. The 'adaptive topography' should therefore be thought of as a complex surface which depends on a very large number of dimensions: a population may evolve in any of a large number of directions.

The simplicity of the usual picture of 'adaptive topography' may be misleading. Although two populations may have reached distinct states without crossing any strongly selected valleys, and may therefore be linked by a continuous series of intermediates, this intermediate path may not be easily recovered when the populations meet: it may simply make up too small a fraction of the enormous number of possible evolutionary paths to be found again. In the simplest case, suppose that the ancestral genotype is A/A, and that in one population, a new allele B is fixed, whilst in another, allele C is fixed. B and C might each be selectively advantageous relative to A ($BB>AB>AA$; $CC>AC>AA$), or they might be selectively equivalent, and simply have drifted to fixation. In either case if B/C heterozygotes are unfit, reproductive isolation will result. The argument of the previous section – that isolation will break down when allele A enters the population – only applies if A is still present, or can be regenerated by mutation. This may be likely in this simple case, where only one allele is needed (cf. the *Sorex* example), but becomes increasingly unlikely as more intermediate genotypes are required. In a sense, divergence is caused by sampling drift, since in an infinite population, all possible mutations would be available at all times. However, this model will operate even in an extremely large population, and so is distinct from the models discussed on p. 192.

The idea that reproductive isolation might evolve through the accumulation of different, incompatible alleles is in some ways the simplest model of speciation (Wright, 1940; Muller, 1942). In general, we would not expect independently evolving lineages to acquire the same combination of genes, even if environmental conditions were identical. This view is supported by the molecular basis of mutation. A typical coding sequence of (say) 1000 base pairs can mutate by a single substitution to one of 3000 possible new alleles. Though this is a large number, it is far smaller than the total number of possibilities ($\sim 4^{1000}$); the range of evolutionary possibilities is strongly constrained by the current state (Gillespie, 1984), and the reversal or reconstruction of any particular evolutionary path becomes essentially impossible.

7.3.2 The geography of speciation: initial establishment

I have contrasted two mechanisms of speciation. First, isolation may accumulate as populations cross adaptive valleys to reach new stable equilibria. This process may be driven either by random drift, or by changes in selection pressure. Alternatively, reproductive isolation may develop as different, incompatible mutations accumulate in different populations. The new alleles may be more or less neutral, or they may be selectively advantageous, relative to their predecessors; in the latter case, speciation occurs as a direct consequence of adaptation. The discussion so far has only been of the divergence of isolated, panmictic populations. This section deals with the way genes that cause reproductive isolation first become established within spatially subdivided populations. The central question is whether strict isolation is necessary for speciation. The way genes spread out over large areas once they have become established in some local region will be dealt with in the next section.

Classic allopatric speciation

In the classic model of allopatric speciation, some external barrier divides the species' range into two fully isolated parts, each geographically extensive. The isolated regions then diverge genetically; when they meet, they are either already fully isolated, or are at least sufficiently isolated that natural selection can reinforce post-mating isolation with ethological differences which complete the process. This 'dumb-bell' model is the basis for the methods of vicariance biogeography, which require that each speciation event corresponds to some extrinsic disjunction of the species' range. We will see that this simple view of speciation raises several problems; whether these affect biogeographic inferences is another matter, which I return to later.

The key difficulties with the 'dumb-bell' model (or at least, with its simplest interpretation) are that firstly, divergence is assumed to occur only with the aid of a physical barrier to gene flow, and secondly, changes within each isolate must occur smoothly and evenly. This would only be plausible if selection pressures were uniform across each isolate, and different between them, and if the response to selection were also the same everywhere. If, as is likely, selection pressures and genetic variation differed from place to place, then it would seem that divergence could occur anywhere in the species' range, and would not require the presence of an external barrier. Consider the mode of speciation which is most favourable to the 'dumb-bell' model: the accumulation of different, mutually incompatible, advantageous alleles. Each new allele should spread rapidly through the population; if each allele spread through the whole isolate before an alternative arose somewhere else, then each isolate would evolve as a homogeneous unit. From what we know of natural populations, such rapid spread of adaptations seems unlikely. For example, consider the spread of warfarin resistance in British rats after the introduction of this anticoagulant poison in 1953. One allele arose in the Scottish Lowlands five years later, in 1958. However, a different resistance allele at the same locus arose in the Welsh Borders only two years later, and spread outwards at a steady rate of 4 km per year. Since then, several more alleles have arisen in Britain (Bishop, 1981). Many similar examples can be found (Georghiou, 1972; Bishop and Cook, 1981; Templeton, 1981). Multiple and diverse responses to a common selective agent suggest that a broadly distributed species is unlikely to evolve as a homogeneous unit.

Two reactions can be made to the argument that if a widely distributed population can respond to selection or drift, it is likely to do so in a heterogeneous way, and so may diversify without the need for any external barrier. Mayr (1942, 1963) argues that gene flow across widely distributed species prevents them evolving significant reproductive isolation: only 'insignificant' clinal variation is possible. This leads Mayr to emphasize 'peripatric' speciation, in which speciation occurs in very small isolates (either peripheral or central). The alternative is that divergence is possible despite gene flow, and that speciation can therefore occur in parapatry, with no need for extrinsic barriers.

Peripatric speciation

If a small population becomes isolated from the bulk of the species' range, a 'genetic revolution' may occur, in which the population moves from "one well-integrated and stable condition through a highly unstable period to another period of balanced integration" (Mayr, 1963, p. 538). Several arguments led Mayr to propose this model of 'peripatric speciation'. The importance of divergence in small isolates was suggested by the frequent pattern of broad uniformity across continuous populations, contrasting with striking divergence in peripheral isolates (which often share apparently similar environments). Island radiations are the most extreme example of this pattern. Mayr argued that in addition to the homogenizing effects of gene flow, fitness interactions ('co-adaptation') combine with developmental homeostasis to prevent divergence. He argued that complete geographic isolation, and drastic founder events, are needed to overcome these conservative forces. Carson (1975, 1986) and Templeton (1980, 1981) have put forward similar theories of founder effect speciation, though the postulated genetic details differ somewhat.

The argument over whether species evolve most readily or most often in small geographic isolates overlaps with the argument over whether drift in small populations is an important cause of reproductive isolation. Indeed, one of the main problems in testing different models of speciation is to separate out all the different processes they involve. Speciation in small isolates may differ in many ways from speciation within broad continua; these differences might be reflected in the contrasting patterns seen on oceanic islands and on continents, but the exact cause of any difference is hard to establish. Peripatric speciation may involve a sharp bottleneck in population size, producing sampling drift within the species. There may be a severe ecological bottleneck, giving a limited and unrepresentative sample of species ('faunal drift'). The physical environment in a small region may differ from the average found over a larger area ('habitat drift'). In the long term, small population size may give a restricted range of new mutations, and cause loss of genetic variability. Populations diverging on islands will be completely isolated from related populations, whereas hybridization may allow some gene exchange for long periods where divergence is parapatric. Finally, the limited range of competing species in an isolate may make the species that evolve there less likely to succeed in competition with mainland species (Darwin, 1859, Robbins *et al.*, 1983). If this were so, peripatric speciation would not give such a significant supply of species to the main part of a taxon's range.

Certainly, the number of species within some taxonomic groups may be much higher on islands or archipelagos than in continental regions of similar area; the rate of speciation in these radiating groups may be high, though it is not clear that the rate is necessarily higher than on continental areas (Williamson, 1981). But which of the many possible factors is responsible for such patterns? Population bottlenecks must occur during colonization; this is reflected in reduced enzyme heterozygosity on some small island populations (Berry, 1986; Vogel and Motulsky, 1986). However, theoretical arguments suggest that, since adaptive peak shifts are opposed by selection, such bottlenecks should not produce much reproductive isolation (Barton and Charlesworth, 1984; Rouhani and Barton, 1987a). In any case, on many islands heterozygosity is not in fact reduced (Sene and Carson, 1977; Craddock and Johnson, 1979; Nevo *et al.*, 1984).

The effects of ecological bottlenecks are more striking than the effects of genetic bottlenecks: though it is often hard to identify the particular selective agents involved, groups have often made major niche shifts in response to changes in species composition (Carson and Kaneshiro, 1976; Williamson, 1981; Grant, 1986; Chapter 9). However, it should be borne in mind that changes in species abundance may occur on continents as well as on islands, and may have similar effects there; for example, Turner's (1971, 1982) argument that *Heliconius* butterflies have shifted their pattern because of shifts in model abundance in Pleistocene refugia only requires that species densities be patchy, not that there be strict geographic isolation. Similar arguments hold for the physical environment: on an isolate, this may well differ from the average over a continent, but then, so may the environment in any small region. 'Habitat drift' only has greater effects on islands than continents if gene flow in fact averages selection pressures over large distributions.

Perhaps the hardest arguments to assess are the proposals that loss of genetic variability, and lack of competition with other species, reduce the 'adaptedness' of island species. Releases of laboratory stocks into nature often (but not always) fail (Jones *et al.*, 1981; Williamson, 1981; Endler, 1986); however, this might well be due to loss of adaptations to nature, rather than to inbreeding and isolation *per se*. Mainland species have often had catastrophic effects on island faunas; however, this might be explained simply by the fact that there are more species on the mainland than on any one island, and so one of them is likely to succeed (Heaney, 1986): it is not obvious that island populations are at an inherent disadvantage. In any case, introductions between continents (either naturally, as when the Isthmus of Panama allowed northern mammals to cross into South America, or by man, as into Australia) have often had equally catastrophic effects (Chapter 8). One striking counterexample is the apparent origin of the widely distributed genus *Scaptomyza* from a Hawaiian Drosophilid species (Carson, 1986).

In summary, speciation on small isolates may differ in many ways from speciation in larger areas. However, it is not clear which of these differences are significant in explaining the net speciation rate, or whether either peripheral or central isolates are an important source of mainland species.

Parapatric speciation

Can new species evolve parapatrically – that is, within a continuous distribution – despite the homogenizing effects of gene flow? Both theoretical arguments and direct evidence show that gene flow does not prevent divergence. This is true for all the various mechanisms of parapatric speciation. Before considering each of these in turn, it is worth making the general point that the distinction between parapatric and peripatric speciation is not clear. Though most discussions of peripatric speciation involved peripheral isolates, Mayr also argues that species may evolve in small populations within the main range, but isolated from it. Indeed, it is hard to see why there should be any difference between peripheral and central isolates. But, in any finite population, small regions may fail to exchange migrants with neighbouring regions, just by chance. (This is true even if the population is statistically homogeneous.) Since divergence may occur quite rapidly, chance isolation for a few generations may allow new genotypes to be established at high frequency. In reality, no natural population is homogeneous. Even if it is not subdivided into distinct local demes, density and dispersal may vary somewhat from place to place. Since there is a continuous range of population structures from complete statistical homogeneity to strict subdivision into independent demes, the distinction between peripatric and parapatric divergence becomes arbitrary. Furthermore, as Wright has argued for many years, partial and impermanent isolation may be more favourable to both adaptation and speciation than strict subdivision, since a wider range of genetic variability is available, and since new combinations of genes can be tested against each other (Wright, 1982; Provine, 1986).

Consider now the particular case where a population shifts from one peak to another as a result of a change in selection pressure (Fig. 7.3b; p. 195). Such a shift can occur in parapatry, provided that the region in which it is favoured is sufficiently large. The critical area is typically no more than a few dozen dispersal ranges wide ($\sigma/\sqrt{2s}$, where σ is the dispersal range, and s is the selection pressure; Nagylaki, 1975; Barton, 1987). There are many cases of adaptation to quite localized selection: for example, the grass *Agrostis tenuis* has evolved resistance to heavy metals on mine tips a few tens of metres across (MacNair, 1981; Chapter 6). Such local adaptations cause some degree of reproductive isolation, both as a direct result of the reduced fitness of individuals which move to the the wrong habitat, as an indirect pleiotropic side-effect, and (possibly) because the mating system may evolve so as to reduce maladaptive gene flow between the habitats (Antonovics, 1968; MacNair, 1981). In *Partula suturalis*, where dextral coiling has evolved as an adaptation to prevent cross-mating with *P. mooreana* (Johnson *et al.*, 1977), the cline between the two coiling morphs is only about 1 km wide. Dextral coiling could therefore have evolved even if *P. mooreana* were only common in a region 1 km wide. In general, the maintenance of narrow clines despite free gene flow shows that localized selection pressures can cause divergence in parapatry (Endler, 1977; Barton and Hewitt, 1985).

The models of parapatric speciation which have received most attention involve rather more than simple adaptation to different selective regimes. Fisher (1930), Clarke (1966) and Endler (1977) propose that once a cline has been established in direct response to environmental differences, it may be modified so as to produce further isolation. One must distinguish two types of modification here. Firstly, different alleles may evolve in the different genetic backgrounds on either side of the cline. This process of co-adaptation might eventually lead to substantial reproductive isolation, largely unrelated to the original environmental difference. Since the different alleles would be favoured over large areas on either side of the cline, gene flow would not impede their establishment. The second possibility is more problematic. Changes in the mating system which reduced the rate of gene flow between the differently adapted populations would be favoured (Fisher, 1930; Dobzhansky, 1940). This seems to have happened in *Agrostis*, where differences in flowering time across the cline are exaggerated, and selfing is increased. Here however, modification is only favoured within a very narrow region, of the order $\sigma/\sqrt{2s}$, and so may be impeded both by gene flow, and by a lack of suitable variation (Barton and Hewitt, 1981). Indeed, reinforcement of mating isolation across clines or hybrid zones seems remarkably rare: only a few good examples are known (Butlin, 1987).

Reproductive isolation may also arise through random sampling drift. Here, the arguments are slightly more complicated; nevertheless, stochastic divergence is possible in parapatry as well as in allopatry. If the population is subdivided into partially isolated demes, peak shifts can occur provided that the number of individuals exchanged between demes (Nm) is small (<1 per generation; Lande, 1979, 1985). Most species have much higher rates of gene flow than this (Slatkin, 1985). However, a low level of gene flow is only needed temporarily: a new adaptive peak could be established if a small population were partially isolated for a few generations. In a truly continuous population, the rate of stochastic divergence is determined primarily by the neighbourhood size ($Nb = 4\pi\rho\sigma^2$, where ρ = density, and σ = dispersal rate). If this is $\leqslant 20$, a new adaptive peak stands a reasonable chance of being established in an area large enough for it to survive and spread (Rouhani and Barton, 1987b). It is unfortunately hard to get direct evidence that drift can cause the evolution of reproductive isolation within continuous distributions. The frequent division of continuous ranges into a mosaic of regions fixed for different chromosome rearrangments suggests that this is possible (White, 1978), but since one can always claim that drift caused divergence in a temporary peripatric isolate (Key, 1982; Hewitt, 1979), this argument may be largely semantic.

Speciation may occur through the accumulation of incompatible mutations, each of which is initially advantageous. Such alleles can clearly spread through continuous distributions; gene flow would only impede such divergence if it allowed advantageous alleles to spread through the whole species range before alternative alleles could arise. As discussed above, this seems unlikely (p. 201).

Sympatric speciation

The argument of the previous section was that since isolation by distance can be as effective as a simple physical barrier, parapatric divergence is no less plausible than allopatric. If gene flow does not prevent a broad continuum breaking up into parapatric races, we may try to take the argument further, and ask whether the free gene flow within a single, randomly mating, population will necessarily prevent speciation. Such sympatric speciation requires two things; firstly, the maintenance of a stable polymorphism by disruptive selection, in which intermediate genotypes have reduced fitness; and secondly, a response of the population to this disruptive selection in which the frequency of cross-matings is reduced and eventually eliminated (Maynard Smith, 1966; Felsenstein, 1982). There are difficulties with both components. Selection against intermediate genotypes (such as heterozygotes or recombinants) generally leads to instability, and hence fixation of one or other form: the best example is the general lack of chromosomal variation. This instability must be countered by strong frequency-dependent selection; for example, through the presence of a range of independently limiting resources. There has been much argument over the extent to which diversifying selection does in fact maintain polymorphism (Hedrick *et al.*, 1976; Hedrick, 1986). However, it is clear that in some cases at least, a single sexual population can support a variety of morphs which exploit different resources, for example, some cichlid fishes (Kornfeld *et al.*, 1982) and Batesian mimics such as *Papilio dardanus* (Sheppard, 1958). This could, theoretically, lead to the evolution of resource partitioning between sympatric species, where the disruptive selection may be due to competition. Sympatric speciation is also seen amongst plants with chromosomal changes such as polploidy, hybridization and self-fertilization, leading to apparent genetic isolation within one generation (Chapter 5).

The main obstacle to sympatric speciation may be in the response of a polymorphism to disruptive selection. In the example of Batesian mimicry in *Papilio dardanus*, the alleles responsible for different pattern elements are dominant, and are tightly linked in a 'supergene'; intermediate phenotypes are therefore rarely produced, and there is negligible selection for assortative mating. Thus, a polymorphism maintained by disruptive selection will not lead to assortative mating unless a high proportion of maladapted types are produced; if they are, the polymorphism will be hard to maintain.

There are also strong theoretical obstacles to the evolution of assortment: the genes involved in the polymorphism must be strongly selected, and must be tightly linked to the genes affecting mating preferences (Felsenstein, 1982). Species pairs are known in which a few genes are responsible for habitat preferences and for mating preferences (Bush, 1975; Tauber and Tauber, 1977a,b). However, it is not clear that these pairs did in fact evolve sympatrically: other interpretations are possible (Futuyma and Mayer, 1980). Laboratory selection experiments show that reproductive isolation can sometimes evolve as a response to disruptive selection (Thoday, 1972). In nature, there are many cases where species which are already reproductively well isolated, and coexist over a wide area, have evolved further interspecific ecological and ethological differences (Zouros and d'Entremont, 1980; Roughgarden, 1983; Diamond, 1986b; Grant, 1986). However there are remarkably few cases where mating isolation has been reinforced in hybrid zones (Barton and Hewitt, 1985; Butlin, 1987).

On balance, then, it seems that though sympatric speciation is possible in principle, there is rather little evidence that it occurs. However, as with all arguments about the way species originate, it is very hard to find clear evidence one way or the other. As far as simple biogeographic arguments go, it may not matter much whether divergence is sympatric or allopatric. In both cases, differences are likely to first become established in some limited region, and to spread out from there. The net pattern would then be independent of the original processes.

7.3.3 The spread of new alleles

Once an allele has become established in some limited region, how does it then spread out to occupy a large area? The various possible patterns of spread correspond to the various forms of selection that may act (Section 7.3.1). However, much of this section applies not only to the spread of genes through populations, but also to the spread of species through their habitat.

Diffusion

The simplest case is where individuals move independently, in a random walk. Then, genes will spread by diffusion, in the same way that one gas will diffuse through another. The most important feature of this process is that it is very slow. The average distance moved by a gene in T generations only increases in proportion to the product of the average distance moved in one generation (σ) and the square root of T ($\sim\sigma/\sqrt{T}$; Endler, 1977). For most genes, in most populations, diffusion is slow both because the dispersal rate is low, and because the effects of random displacements accumulate very slowly.

This is shown both directly, by mark-release studies, and indirectly, through the existence of narrow clines and localized fluctuations in gene frequencies (Endler, 1977; Slatkin, 1985). Though indirect genetic measurements give rather larger (and probably more accurate) estimates of dispersal than direct mark-release, both give low values (Barton and Hewitt, 1985; Slatkin, 1985). For example, some salamanders move only a few metres during their adult life (Yanev and Wake, 1981). Though the extent of juvenile dispersal is uncertain, electrophoresis shows highly localized, and highly variable, fluctuations (Larson and Highton, 1978; Yanev and Wake, 1981; Slatkin, 1985). Even in the butterfly, *Euphydras editha*, which might be expected to disperse much more freely, there is very little exchange between adjacent demes a few hundred metres apart (Ehrlich and Raven, 1969). Such random, localized movements would, by themselves, only allow genes to spread extremely slowly. This can be illustrated in two ways: by the pattern of relationship between recessive lethals in *Drosophila*, and by introgression after secondary contact.

A substantial proportion of chromosomes in natural populations carry recessive lethals at some locus. On average, these have slight deleterious effects even when heterozygous (~2%) (Wright *et al.*, 1942). Since the frequency of recessive lethals at any particular locus is very low, homozygotes are formed very rarely, and so this weak selection against heterozygotes is the main force keeping them at low frequency. Recessive lethals will persist for roughly $1/(2\%) = 50$ generations before being eliminated by selection. Their spread over this time is reflected by the probability that two recessive lethals taken from different places descend from a common ancestor (that is, are 'allelic'): one expects this probability to decrease over about $\sqrt{2 \times 50} = 10$ dispersal ranges. In fact, it decreases over only a few hundred metres (Wright *et al.*, 1942 (*D. pseudoobscura*); Paik and Sung, 1969 (*D. melanogaster*); Fig. 7.5a). This shows that on average, *Drosophila* do not move very far, and that the cumulative effects of this movement are small.

The slow effects of diffusive dispersal can also be illustrated by the limited extent of introgression after secondary contact. The fire-bellied toads *Bombina bombina* and *B. variegata* have been hybridizing in Eastern and Central Europe since the last glaciation (i.e. for at least 8000 years). The rate of range expansion in *Bombina* spp. is estimated to be about 500 m

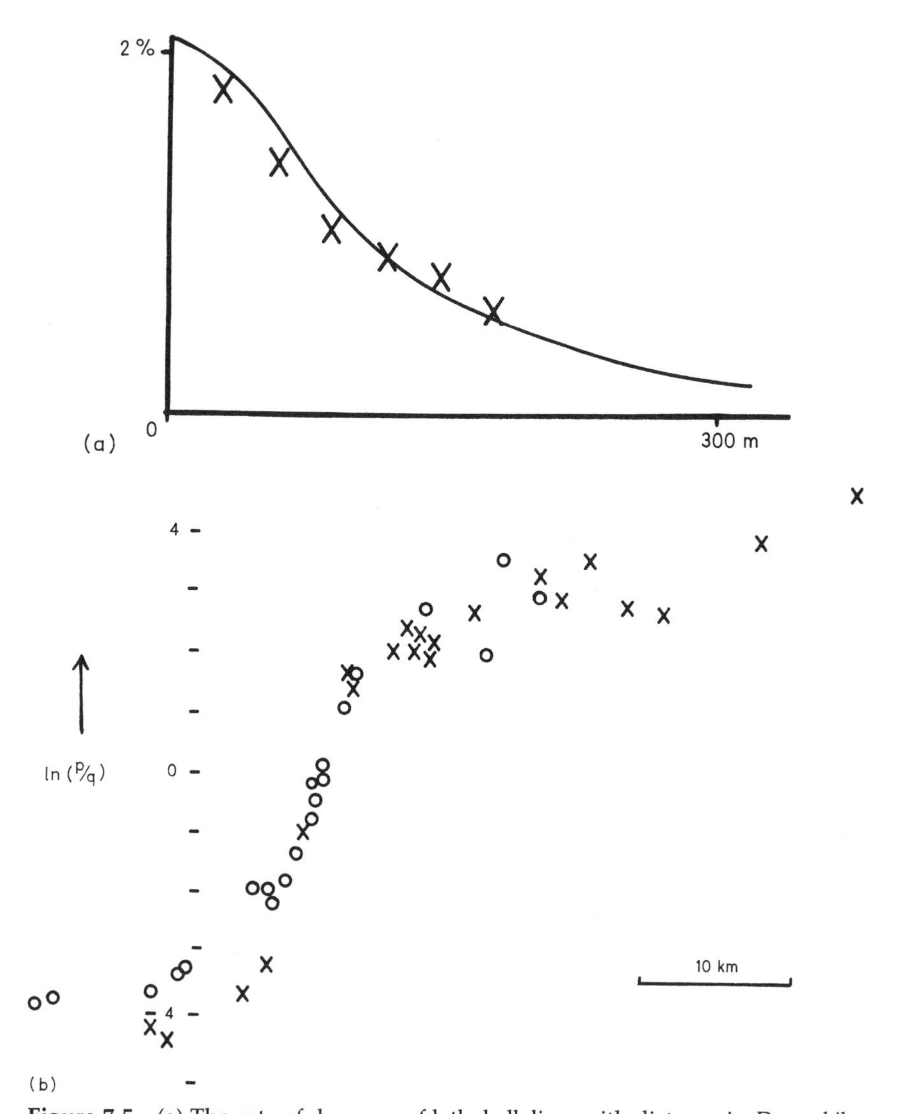

Figure 7.5 (a) The rate of decrease of lethal allelism with distance in *Drosophila melanogaster* (Paik and Sung, 1969). The solid curve shows the expected relationship if the dispersal rate is σ = 35 m, and selection against heterozygotes is s = 2% (from Malecot, 1948). (b) The hybrid zone between *Bombina bombina* (on the left), and *B. variegata*. The graph shows the average clines for five diagnostic loci near Cracow (Circles; Gpi, Ldh1, Ck, Ak, Mdh1; Szymura and Barton, 1986), and for six diagnostic loci near Przemysl (Crosses; Gpi, Ldh1, Ck, Ak, Mdh1, Np; Szymura, unpubl.). The allele frequency has been transformed to a logit scale ($\ln(p/q)$) to emphasize the long tails of introgression on either side; a simple sigmoid cline would be expected to follow a straight line on this scale. Points at ± 4 correspond to 1.8% and 98.2%.

per year; therefore one would expect that in the absence of selection, 8000 years of hybridization would have allowed foreign alleles to penetrate about 0.5 km $\times \sqrt{8000} = 50$ km on either side. Figure 7.5b shows the pattern of introgression of diagnosis electrophoretic alleles across two transects in Poland (Szymura and Barton, 1986; and unpubl.). The scale has been transformed to emphasize the tails of the cline; a straight line corresponds to an exponential decrease in allele frequency. In the centre, a sharp transition in frequency, over about 6 km, is caused by strong selection against hybrids, which acts as a barrier to gene flow between the two taxa. However, the gradients outside the hybrid zone are much shallower, and extend roughly the distance expected with neutral diffusion. This pattern, in which a sharp cline is flanked by shallow tails of introgression, is found across many hybrid zones (Barton and Hewitt, 1985); the limited extent of introgression shows that even several thousand years of gene flow may have a negligible effect on the bulk of the species.

It has been argued that, because the effects of random dispersal are so weak, gene flow is a negligible evolutionary force (Ehrlich and Raven, 1969; Endler, 1977). Certainly, it is true that gene flow will not prevent a response to local selection pressures (p. 204). However, though slow random diffusion may explain distributions over short periods, it fails over longer time-scales, and it does not take into account the distinctive behaviour of those genes which cause reproductive isolation. To understand the long-term evolution of species differences, we must take into account the effects of extinction and recolonization, and the way adaptive peaks spread through populations.

Extinction and recolonization

Gene flow often appears to have effects over distances far larger than would be expected by random diffusion. This is reflected in the frequent finding that the genetic distance between populations (as measured by electrophoretic differences) increases with geographic distance over very large areas (Endler, 1977; Jones *et al.*, 1981; Halliday *et al.*, 1983; Wijsman and Cavalli Sforza, 1984; Slatkin, 1985). This contrasts with the distribution of recessive lethals, which are typically much younger than enzyme alleles; a comparable case is the similarly localized distribution of (presumably deleterious transposable elements in *Drosophila* (Langley *et al.*, 1982). Populations separated by physical barriers are often much more different, genetically and morphologically, than populations separated by large areas of continuous habitat, even though gene flow across those large areas is expected to be negligible (Pounds and Jackson, 1981). Patterns of electrophoretic variation are often more even than would be expected from observed dispersal rates (Slatkin, 1985). This evidence of large-scale homogeneity is not easily explained by occasional long-distance dispersers: though there is often an excess of long-range dispersers ('leptokurtosis') (Jones *et al.*, 1981; Endler, 1977; Slatkin, 1985), a low rate of long-range dispersal will not, by itself, have much effect on the distribution of neutral genes. The most plausible resolution of this apparent paradox is that, over long periods, extinction and recolonization are the main cause of gene flow (Grant, 1980; Slatkin, 1985). Changes in the species' distribution will cause coherent movements, in which either many individuals move in a particular direction, or a few individuals leave many descendants in a new area. Such collective movements are a much more effective cause of gene flow than the independent, random dispersal seen within stable populations: I will refer to all such processes as 'extinction and recolonization', even though less extreme population restructuring may be more important than actual extinction and recolonization.

The importance of extinction and recolonization is seen most clearly when we consider the movements of whole species. Expansions in range after the last glaciation were relatively rapid, and tracked climatic changes quite closely (Coope, 1979). These movements were far too rapid to be explained by random diffusion (Skellam, 1973). For example, contrast the post-glacial spread of *Bombina bombina* several thousand kilometres westward into Poland and the Danube basin (Arntzen, 1978), with the subsequent allozymic introgression of only a few tens of kilometres. The fact that local dispersal may be negligible, relative to bulk movements of genes, allows the past history of a population to be reconstructed from present genetic patterns. For example, the detailed biochemical data available for the human population of Europe has allowed Ammermann and Cavalli Sforza (1981) to reconstruct the movements of neolithic farmers. Such reconstruction would be impractical if random diffusion had blurred patterns since their formation. However, before we can safely make such extrapolations, the complicating effects of selection must be considered more carefully.

I have argued that the observed homogeneity of species over large areas is caused by occasional expansion and contraction of whole populations, rather than by the continual dispersal of individuals. This argument applies in a straightforward way to neutral alleles, which will be carried passively from place to place; it may therefore explain the pattern of molecular variation. However, the genes we are most interested in are likely to be under selection: unfortunately, this substantially complicates the argument.

Rare, deleterious, alleles cause no difficulty, since they will be eliminated before being affected by long-term population restructuring:

such alleles will therefore be restricted to small areas, as is found for recessive lethals in *Drosophila*. Alleles which are advantageous everywhere will simply spread through the whole species. They will advance quite rapidly even if gene flow is entirely diffusive (Fisher, 1937); a few long distance dispersers will greatly speed the process (Skellam, 1973). The difficulty comes when we consider alleles which are favoured in some places, but not in others. These should quickly spread throughout the region in which they are favoured, and establish narrow clines at the boundaries of these regions. There is strong evidence that populations can indeed respond to local selection pressures (p. 204; Chapter 6). Therefore, any historical effects should soon be erased. Long-range correlations between genetic distance and either geographic distance or physical barriers could therefore only be explained if the characters involved were neutral, of if environmental differences were correlated with distance or barriers. This often seems unlikely (though see Section 7.4): for example, in the case described by Pounds and Jackson (1981), which involved morphological divergence across rivers, or the case of 'area effects' in *Cepaea* (Jones *et al.*, 1977). The same argument may apply to species as well as to genes, in the case of correlations between faunal similarity and distance (Flessa *et al.*, 1979; Fallow and Dromgoogle, 1980). Such correlations imply that faunas are not the direct consequence of their local environments.

In general, the pattern stressed by Mayr (1963), where divergence is confined to peripheral isolates, is hard to reconcile with the ease with which populations can respond to local environmental differences. This is true even if genes can move by extinction and recolonization, as well as by local diffusion. To resolve this problem, we must consider how the genes responsible for speciation itself can spread.

Tension zones

If reproductive isolation is to be firmly established, the two diverging populations must be at different stable equilibria under selection: that is, they must be at different 'adaptive peaks' (Section 7.2.2). If the two populations meet and hybridize, they will form a stable cline, in which selection against introgression is balanced by dispersal. Such clines behave very differently from the clines which form when advantageous alleles spread, or when populations respond to different environments; Key (1982) has proposed that they be called 'tension zones' (Fig. 7.1b). Their distinctive feature is that, because they are maintained by intrinsic genetic incompatibilities, rather than by external selection differentials, they can be formed anywhere, and can move from place to place. Their position is therefore somewhat arbitrary, and can be perturbed by a variety of factors.

One 'adaptive peak' may tend to expand at the expense of the other. The most obvious cause of consistent movement is an asymmetry in selection, such that individuals are fitter on one side than the other. However, other factors can move tension zones: dominant alleles will tend to expand at the expense of recessives (Mallet, 1986), and adaptive peaks which can maintain a higher population density or dispersal rate on one side will advance by weight of numbers – a form of group selection (Bazykin, 1969). These forces will cause a steady movement, until either one type is eliminated, or the tension zone reaches a point on an environmental gradient where the different selective forces balance. One therefore expects to find tension zones running parallel with ecological differences. This is often the case (Barton and Hewitt, 1985): for example, *Bombina bombina* and *B. variegata* meet at the boundary between lowland and highland habitats (Szymura and Barton, 1986); the coiling morphs of *Partula suturalis* meet at the edge of the range of *P. mooreana*; and many of the clines between the different mimetic races of *Heliconius* lie on ecological boundaries (Benson, 1982).

Tension zones are strongly influenced by population structure, as well as by selection (Barton, 1979). They will move so as to reduce production of maladapted hybrids, and so will tend to minimize their length (hence the term 'tension zone'). If the population density or dispersal rate is greater on one side than on the other, the zone will be pushed forward. Thus, tension zones will be trapped by local barriers to dispersal, and will be moved most effectively by large-scale population restructuring. This can be illustrated by the tension zone between two chromosomal races of the grasshopper *Podisma pedestris*. The zone follows the main ridge of the Alpes Maritimes, along the line where the races would first have met as the glaciers retreated: on a broad scale, its position is determined by expansion from alternative refugia. On a fine scale, the tension zone shows no correlation with any obvious ecological variables such as vegetation or altitude; instead, its position is determined by variations in population density and by local barriers (Nichols, 1984; Nichols and Hewitt, 1986).

The distinctive behaviour of tension zones has two important consequences. First, the distribution of populations which are trapped at different adaptive peaks will reflect their previous movements, as well as current selection pressures. For example, though the distribution of *Bombina* spp. is clearly strongly affected by adaptations to different environments (lowland *B. bombina* and highland *B. variegata*), it is also strongly influenced by the pattern of expansion after the last glaciation (Arntzen, 1978). Conversely, if the distribution of the two forms were determined solely by differential adaptations, then no inferences could be made about their previous history (Endler, 1982b,c). In general, the patterns discussed in the previous section are to be expected if differentiation involves divergence into alternative equilibrium states ('adaptive peaks'), but are puzzling if distributions are determined solely by the local environment.

The second major consequence is that because tension zones involving different sets of genes will all respond to changes in population structure in the same way, they will tend to clump together. In the most extreme case, if a species is reduced down to small refugia, which then expand into secondary contact, all their differences will be brought together. Hybrid zones do indeed usually involve large numbers of coincident differences (Section 7.2.4). One can argue that this coincidence is due to a convergent response to a single ecotone (Endler, 1977; Moore, 1977). However, it seems more likely that if each gene were responding independently to the external environment, these responses would differ. The aggregation of many independent differences into a single tension zone complicates attempts to find out how these differences first evolved; however, it simplifies biogeography by reducing the diverse genetic variation into a few coherent units.

7.4 BIOGEOGRAPHY AND SPECIATION

Species may evolve by a variety of mechanisms: by random drift, by changing selection, or by the accumulation of incompatible mutations. All of these can operate in parapatry as well as in allopatry. One might therefore expect species to break up into a mosaic of locally adapted races. Yet most species do form coherent and recognizable units. I have argued that this apparent paradox can be resolved if extinction and recolonization (or at least, concerted expansion and contraction of populations) move genes over large distances; and if populations evolve into alternative equilibrium states, so that they become bounded by a cluster of 'tension zones'. How does this view of speciation affect biogeographic arguments, which often assume simple allopatric divergence?

7.4.1 Ecological determinism, dispersal, and vicariance

Endler (1982c) contrasts three explanations of geographic distributions. In ecological determinism, present distributions are determined by direct adaptations to environmental differences, and evolved *in situ*. Alternatively, new forms arose in certain places, and then dispersed to fill their present ranges. This dispersal may have been across major physical barriers (jump dispersal), or it may be through a continuous habitat, occupied by related species or individuals (range expansion). Finally, taxa may have evolved in large, disjunct areas; this disjunction, or vicariance may have allowed or encouraged speciation. With vicariance, the present distribution is determined by a passive response to external forces, such as plate tectonics or Pleistocene glaciations.

There are clear examples of each process. Species may adapt to local environments (p. 204; Chapter 6). Dispersal has occurred during the colonization of oceanic islands, during infiltration across land bridges (as in the crossing of North American mammals over the Isthmus of Panama), and during the spread of individual advantageous genes. Vicariant events are clearly associated with divergence, for example, fishes across the Isthmus of Panama (Somero, 1986) or, more recently, in the secondary contacts following expansion from Pleistocene refugia in Europe and Australia. So, as with many biological controversies, the argument is over the relative importance of the different factors, rather than with whether they can or cannot occur. This is true even when we consider the evolution of any one species: since species differences are polygenic, and are likely to have evolved in many steps over a long period of time, the different steps may have occurred in different ways.

A more fundamental difficulty is that each of the three paradigms involves several components. Ecological determinism usually requires present distributions to reflect present ecology; the origin of the various differences to occur within a continuous distribution; and the distribution to have remained approximately in its original position. Dispersal, in its purest form, requires a species to have evolved all its components together, in one 'centre of origin', and that dispersal then occurred across pre-existing barriers: either definite barriers, or through isolation by distance (Chapter 1). Vicariance normally requires present distributions to be determined by external forces (either actual transport on drifting continents, or dispersal of individuals as a necessary response to changes in external climate); it may also require speciation to be encouraged or allowed by the presence of a barrier.

Much of the argument between these alternative views has been confused because separate components have been confounded. Species differences may well have evolved in parapatry, and yet their current distributions may be due to a long history of population expansion and contraction. Conversely, present distributions might have moved so as to parallel environmental gradients; yet they might have formed through secondary contact. I have argued above that the distribution of hybrid zones suggests that both possibilities are important. The argument between Endler (1982b) and Mayr and O'Hara (1986) has largely concerned the question of whether ecology, population structure, or history are more important in determining the current positions of parapatric contacts. But this argument is independent of whether the taxa first evolved in parapatry or in allopatry.

Many of the proponents of vicariance biogeography, and of allopatric speciation, have assumed that a barrier to gene flow will increase the rate of speciation. Theoretically, it is hard to see why this should be (p. 205). The factors which encourage speciation are largely independent of the factors which determine species distributions. For example, Pleistocene refugia have been seen as providing an explanation of the extraordinary diversity of the Tropics in general (Prance, 1982a; but see Chapter 10), and of the diversity of mimetic patterns in *Heliconius* in particular. However, Turner's (1971, 1976, 1982) suggestion that changes in the abundance of distasteful models cause divergence does not depend on any reduction in gene flow, and is not directly tested by the observations on present distributions on which most attention has been concentrated.

7.4.2 Does vicariance biogeography require allopatry?

The method of vicariance biogeography relies on the assumption that the separation of a population into a pair of lineages corresponds to the separation of the habitat by some external barrier. If this is so, then the phylogeny can give information about the sequence of disjunctions (Chapter 1). At first sight, this method requires allopatric speciation: certainly, vicariance biogeography is usually described in terms of the classic 'dumb-bell' model (p. 200). Can the method still be applied if species form in other ways?

This question is closely tied to the argument over whether species are bound together into coherent units by gene flow or by co-adaptation (p. 202). We have seen that, although genetic differences may evolve in a variety of ways, species or geographic races will still appear as coherent units if population restructuring is frequent enough to obscure the original pattern of divergence, and if the characters we observe are either neutral, or involve divergence into different 'adaptive peaks'. This seems a fair approximation: geographic races are usually separated by sharp boundaries, at which many genetic differences are clustered. Even where these boundaries now follow ecological gradients, they may well have been set up by secondary contact. Vicariance biogeography can therefore be applied without restrictive assumptions about the way the characters evolved.

7.4.3 Phylogeny and geography?

Where taxon boundaries become blurred, the separate distributions of the various genetic components must be followed. There are interesting parallels between vicariance biogeography and methods used in population genetics, which also depend on the relation between phylogeny and geography. Until recently, the most reliable phylogenies came from those Diptera with polytene chromosomes: for example, Carson and Kaneshiro (1976) have deduced the relationships of over 100 species in the picture-winged group of Hawaiian *Drosophila*. Here, most branches in the phylogeny do correspond to movements between islands. The sequence of movements is in rough accord with the history of the islands (i.e. with the order in which the barriers between islands were established), but the correspondence is not exact: lineages do sometimes move from younger islands to older. The argument here is of course rather different from the usual vicariance approach: barriers are established when new islands arise, and flies colonize them, rather than being imposed from outside.

Accurate phylogenies can be estimated for some other cases of chromosomal evolution (e.g. the morabine grasshoppers; Hewitt, 1979). However, the recent development of restriction mapping and DNA sequencing allows phylogenies to be estimated for any sufficiently long stretch of DNA; moreover, branch points can be dated. The clearest data come from mitochondrial DNA, which is inherited intact through the maternal lineage. Mitochondrial phylogenies have often revealed major disjunctions (e.g. Avise *et al.*, 1979a,b; Wilson *et al.*, 1985; Cann *et al.*, 1987). These seem most likely to correspond to disjunctions of the whole species, and not just the mitochondrial lineages; however, this is not always the case (Gyllenstein and Wilson, 1987). Individual molecular phylogenies do not, by themselves, show whether a split between different sets of lineages is due to dispersal of individuals, to movement of the whole population across the habitat, or to separation by vicariance. Comparison of phylogenies across different genes, or different species is needed, as in biogeographic arguments based on 'generalized tracks' (Chapter 13).

Construction of phylogenies for segments of nuclear DNA is complicated by recombination: the lineages of different segments will be different (Section 7.2.4; Aquadro *et al.*, 1986). However, the relationship between two populations can be estimated by averaging over a large number of genes, each of which will have descended through the population along a separate lineage: this is the basis of methods based on genetic distance. One may be interested in finding the relationship between populations which have remained more or less isolated (Thompson, 1975); or one may study continuously distributed populations, in which genetic distance increases with geographic distance (Section 7.3.3). Even in the latter case, there is a strong parallel with vicariance biogeography: the difference is that divergence indicates isolation by distance, rather than a sharp physical barrier.

7.5 CONCLUSIONS

Much of our knowledge of evolutionary processes comes from geographic patterns. Though I have argued that it is hard to infer the first causes of divergence, geographic patterns do contain a wealth of information on the history of population movements and disjunctions, and on the current processes of selection, drift and gene flow. However, much caution is needed in making inferences about even these relatively accessible aspects of evolution. Much of the evidence discussed here is disturbingly loose: one of the main needs is for consistent and statistically sound studies, designed to address particular questions. The necessity, and the difficulties, involved in moving beyond anecdotal evidence is most apparent in the argument over whether competition shapes communities (Chapter 9). Statistical treatment is badly needed in analysing the locations of parapatric distributions, both in asking whether they are significantly clustered, and in asking whether they parallel physical barriers or environmental gradients. It is remarkable that though the main evidence for Pleistocene refugia in South America has come from apparent parallels between different taxa (Chapter 10), these parallels may not be statistically significant (Beven *et al.*, 1984). Finally, estimates of phylogenies (whether derived by cladistic inference from morphology, or from molecular data) are often inaccurate and, moreover, it is hard to get any measure of just how accurate they are (Sokal, 1983a,b; Felsenstein, 1985). This poses a serious obstacle to attempts to infer the detailed mode of speciation (Cracraft, 1982; Thorpe, 1984); nevertheless, the combination of geography with phylogeny is a most promising approach to the study of speciation.

CHAPTER 8

Extinction

L. G. MARSHALL

8.1 INTRODUCTION

Extinction is to species what death is to individuals, an inevitable and irreversible end of existence (Gould, 1983). Of the millions of biological species that have existed on Earth most are now extinct, and about 250 000 extinctions are known from the fossil record (Raup and Sepkoski, 1984; Raup, 1986). Among fossilizable marine organisms about 72% of the families (2 400 of 3 300) and 78% of the genera (9 250 of 11 800) are known to be extinct (Raup and Sepkoski, 1982, 1986). It is estimated that 63 species and 52 subspecies of mammals, and 88 species and 83 subspecies of birds have become extinct since AD 1600 (Diamond, 1984c).

Extinction is only one part of the evolutionary process and serves the primary role of making room for the origination of new species (Gould, 1983). Origination is generally recognized as the 'obverse' of extinction (Van Valen, 1985b), yet extinctions need not be followed by originations and originations may occur without concurrent or preceding extinctions. Extinctions are thus not a prerequisite for originations and vice versa.

Traditionally, as dictated by Darwinism, extinction is viewed as the fate of species which lose in the struggle for survival; it is a constructive process for eliminating obsolete species (Gould, 1983). Yet evolution provides only for unspecified change and it is not a ladder in which each extinction-origination rung progresses a lineage one step closer to immortality.

It is now recognized that some extinctions are non-constructive and may result from 'unpredictable challenges' or 'Acts of God' (Gould, 1983); thus, a species' ultimate demise is no reflection of its 'goodness' as a biological organism. There is simply no way that a species can anticipate and accordingly pre-adapt to the environmental consequences of a meteorite impact of a nearby volcanic eruption.

Another view of biotic evolution, exemplified by the Red Queen hypothesis (Van Valen, 1973), maintains that extinction is neither constructive nor non-constructive, only inevitable. Species are in a continual race to maintain their existence by adaptation, but sooner or later they fail to adapt when the pace exceeds their capabilities (i.e. inherent genetic variability). Thus, "it takes all the running you can do to keep in the same place" (Red Queen in Lewis Carroll's *Through the Looking Glass*).

Literature divides extinctions into three basic time periods: (1) Recent or contemporary (Ehrlich and Ehrlich, 1981; Ehrlich *et al.*, 1984; Diamond, 1984c), (2) terrestrial Late Pleistocene/Holocene (Martin and Klein, 1984), and (3) pre-Pleistocene (Allaby and Lovelock, 1983; Raup and Jablonski, 1986). The first focuses on extinctions in ecological (short-term) time and on dynamics of populations and species; the second on extinctions in ecological and geological (long-term) time and on changes in species and genera; and the third primarily on extinctions in geological time and on genera and families (i.e. clades not species). These focuses are somewhat arbitrary but they serve to illustrate some basic differences between the three time periods which result from the quality of time resolution and taxonomic units being studied. Each has made significant contributions to understanding extinction patterns and processes, and attempts are now being made to integrate these contributions across 'time period' boundaries.

Extinction causes and processes were traditionally based on knowledge gleaned from the study of Recent biotas, and through inductive reasoning were extrapolated to the fossil record. This approach stemmed from the fact that causes and processes can be observed in Recent biotas as outlined in Chapters 9 and 15, while only patterns are preserved in fossil biotas. It is now realized that some extinctions in the fossil record may have resulted from causes or processes that have no observable or documented contemporary counterpart. The present can thus no longer be viewed as the key to the past (Graham and Lundelius, 1984). For example, extinctions were long explained solely by earth-bound processes, and biologists, geologists, and paleontologists were given the task of identifying pattern–process relationships. With the development and at least partial acceptance of the impact theory (i.e. that meteorites striking the earth may have caused mass extinctions on one or more occasions), some scientists are now entertaining extra-terrestrial causes. As a result, astronomers and physicists may now be making significant contributions to our understanding of extinctions in the fossil record.

In recent years, theories about extinctions, particularly mass extinctions, always far outstrip the facts. Yet, "heuristic models work best when they are a bit overstated – those are the ones that generate the testable predictions" (Flessa and Jablonski, 1983).

As shown below, case studies fall along a continuum with respect to cause, duration, and intensity (Diamond, 1984c). "Our failure to be explicit about particular instances of extinction is not because such an event is esoteric and inexplicable in detail but, quite on the contrary, because there are so many possible detailed explanations that we cannot choose among them" (Simpson, 1953).

I begin with a discussion of factors which effect changes in diversity because extinction results in a decrease in diversity. Next I discuss aspects of taxonomic turnover and identify the interrelationship of extinction and origination as these processes effect changes (decreases and increases) and maintenance of diversity. Then I discuss the biases effecting extinction patterns (i.e. the data sets used for documenting extinctions of populations, species, clades and/or biotas) and the patterns themselves. Consideration is then given to susceptibility of species and clades to extinction, followed by a discussion of the causes and processes of extinctions (the factors or explanations which account for the patterns). In the conclusions section I discuss the role of extinctions in biogeography and aspects of commonality in divergent extinction paradigms.

8.2 DIVERSITY

Standing diversity comprises all taxa (such as species, genera, families) that exist in any clade or biota (local, regional, global) at any moment in time. Many factors influence standing diversity as discussed in Chapters 3, 4 and 15, and these factors must be taken into consideration when assessing diversity.

In Recent biotas standing diversity of biological species can be securely determined by simple species counts. In fossil biotas standing diversity is based on counts of morphological species which may include more than one biological species. Many biological species are distinguished only on the basis of soft part anatomy, and these features are seldom preserved or their former existence rarely recorded in fossil taxa. A few exceptions exist, e.g. Lagerstätten faunas such as the Burgess and Mazon Creek shales (Raup, 1979a), and the recorded diversity in these biotas may approximate that of a previous biological community. In fossil floras, diversity estimates may be inflated when parts of the same plant (e.g. trunks, leaves, stems, flowers, pollen, seeds) are attributed to different species, genera, or even families.

The overall trend of taxonomic diversity in land plants, marine animals and terrestrial tetrapods has seen a continued increase through Phanerozoic time (the last 600 Myr) (Fig. 8.1). Although overall diversity for species, genera, and families of marine organisms (and presumably other groups as well) are similar (Sepkoski *et al.*, 1981), the diversity increase within each rank is not necessarily proportionate. The post-Paleozoic record of

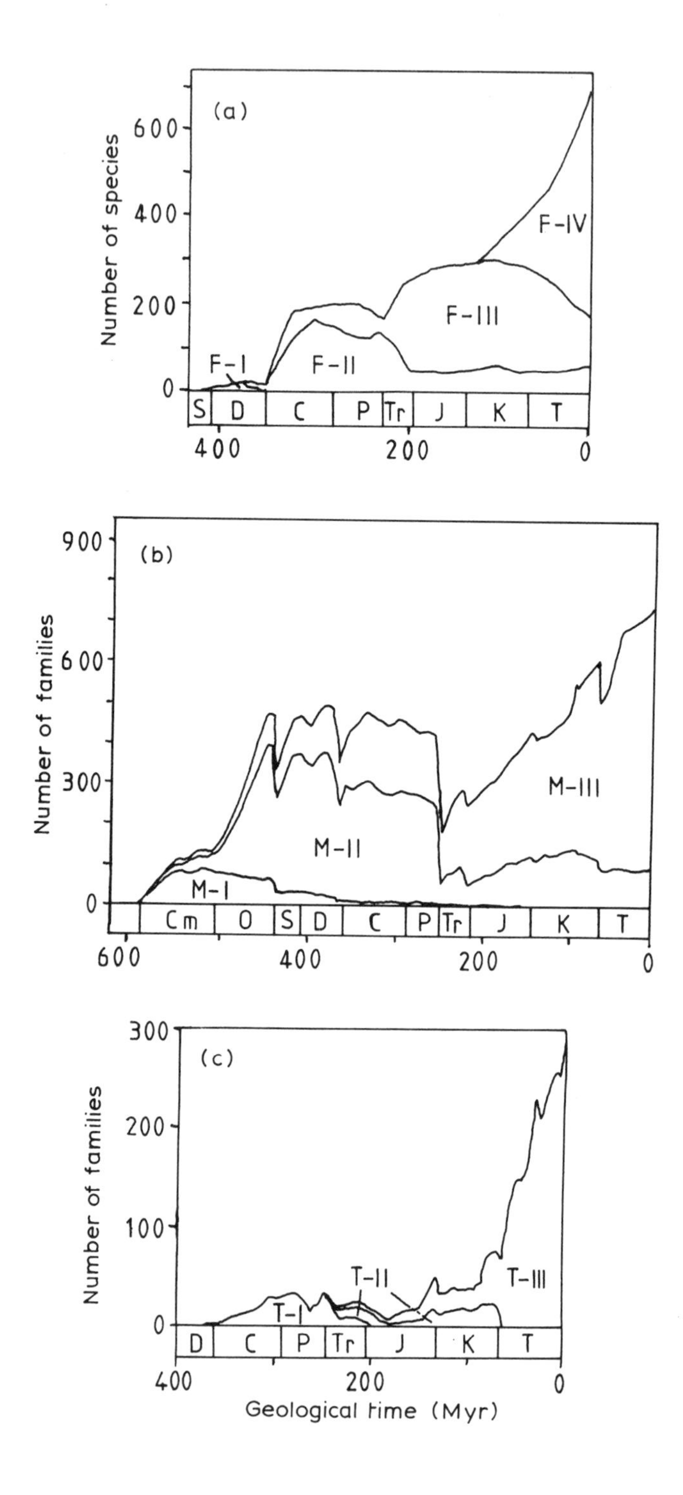
(a)
Number of species
600
400
200
0
F-IV
F-III
F-II
F-I
S D C P Tr J K T
400
200
0
(b)
Number of families
900
600
300
0
M-III
M-II
M-I
Cm O S D C P Tr J K T
600
400
200
0
(c)
Number of families
300
200
100
0
T-III
T-II
T-I
D C P Tr J K T
400
200
0
Geological time (Myr)

skeletonized marine organisms shows a three to one increase in species richness relative to families, and this ratio is five species to one family in the Late Phanerozoic (Fig. 8.2); thus, there is apparently an increase in species per family through time.

The quality of the fossil record increases with time because the probabilities of preservation are greater in progressively younger rocks which generally have broader geographic occurrences and less attrition from erosional and diagenic processes. This feature may account for the overall increase in taxonomic diversity during the Phanerozoic. Part of this increase in quality of the record through time has been attributed to the 'Pull of the Recent' (Raup, 1979a) which is primarily due to the tendency to assign fossils to living taxonomic groups. Thus, the fossil taxa with Recent representatives may have long temporal durations simply because the Recent represents a complete sampling point not prone to vagaries of the fossil record. The damping effect of this pull is normalized by some workers (Raup and Sepkoski, 1984) by deleting taxa with living members from data sets prior to analysis. This practice may, however, obscure real patterns in the history of life due to the deletion of potentially significant information (Hoffman, 1985b).

Adaptive radiations are patterns of exponential diversification (Stanley, 1979), and "the multiplication of species from the same ancestral stock" (Diamond, 1984b). These radiations can result in either an overall increase in diversity or in recovery of diversity levels following mass extinctions. They supposedly result from originations that follow (1) the opening of previously occupied adaptive zones by extinction, (2) the appearance of a new adaptive zone either by creation of new ecologies or by achievement of innovations by organisms, (3) immigration of taxa into new areas (e.g. colonization of islands), or (4) a combination of the above (Stanley, 1979; Van Valen, 1985a,b,c). Adaptive zones "are the

---

Figure 8.1 (a) land plant species diversity during Phanerozoic time and the successive radiation of four 'evolutionary floras' (F–I, early vascular plants; F–II, pteridophytes; F–III, gymnosperms; F–IV, angiosperms) (after Niklas *et al.*, 1983). (b) family diversity of heavily skeletonized marine animals during Phanerozoic time and the successive radiation of three 'evolutionary faunas' (M–I, Cambrian fauna; M–II, Paleozoic fauna; M–III, Mesozoic–Cenozoic or Modern fauna) (after Sepkoski, 1984). (c) family diversity of terrestrial tetrapods during Phanerozoic time and the successive radiation of three 'family assemblages' (T–I, labyrinthodont amphibians, anapsids, mammal-like reptiles; T–II, early diapsids, dinosaurs, pterosaurs; T–III, the Modern groups, including frogs, salamanders, lizards, snakes, turtles, crocodiles, birds, mammals) (after Benton, 1985). These figures demonstrate diversity patterns through time and evolutionary relays. Cm, Cambrian; O, Ordovician; S, Silurian; D, Devonian; C, Carboniferous; P, Permian; Tr, Triassic; J, Jurassic; K, Cretaceous; T, Tertiary.

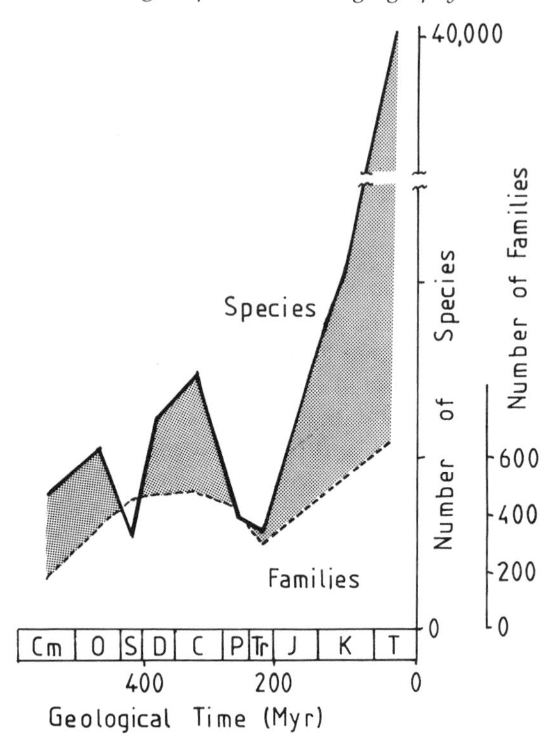

Figure 8.2 Diversity of average number of families (dashed line) and approximate number of species (solid line) of skeletonized marine organisms during Phanerozoic time. Abbreviations as in Fig. 8.1. After Flessa and Jablonski (1985).

niches of higher taxa. They exist independent of any occupants; there are empty adaptive zones and others where two or more higher taxa are in long-term competition. They need not have sharp boundaries, and indeed their boundaries in the resource space . . . may depend on their occupants" (Van Valen, 1985c).

8.3 TURNOVER

Fluctuations in biotic diversity through time result from differential rates of taxonomic turnover (i.e. differences in absolute or relative rates of the opposing processes of extinction and origination). In Recent biotas, diversity changes result from extinctions and immigrations (= originations, see below) of local populations (Diamond, 1984a,b,c). In fossil biotas, turnover is documented by the stratigraphic range of taxa in which origination is defined by the first appearance, and extinction by the last appearance, of a taxon. Origination implies evolution, and the appearance of new taxa is inferred to result from speciation. Yet, the immigration of taxa into new areas can result in their 'first' appearance (origination) in both fossil and Recent biotas. Thus, origination can result from two distinct processes (1) evolution (speciation) and (2) immigration (including colonization, recolonization, and introductions). The relative contributions of these two processes vary considerably, with taxa and with island and continental location (Diamond, 1984a,b).

As discussed in Chapter 15, the Equilibrium Theory of MacArthur and Wilson (1967) predicts that with ecological time a region will become saturated with taxa and reach a level where turnover is stochastically constant. The resulting diversity at equilibrium is determined in part by area (large areas will have greater diversity, smaller areas less) which sets the extinction rate (large areas have lower rates) and isolation which sets the origination (colonization/immigration) rate (see Chapter 4; Section 8.7.7; and Rosenzweig, 1975 for an application of this theory to the relation between area and probability of speciation). Equilibrium will tend to persist until it is disrupted by (1) the appearance of a new biotic group that can result in competitive interactions of taxa, (2) a change in physical environment that can alter ecologies, or (3) a combination of the above. Disruption can result in either a temporary or permanent change in diversity.

Because extinction and origination may be interactive processes a decline in diversity may result from a high extinction rate, a low origination rate, or a combination of the two (Benton, 1985). If origination rates increase standing diversity to a point above the equilibrium or saturation level, as occurred following the immigration of families of North American mammals into South America in the Late Cenozoic (Marshall, 1981) then a period of extinction will predictably follow.

8.3.1 Extinction with replacement

This kind of turnover results when extinction is followed by origination, so that diversity either remains constant or increases. Replacement may be viewed in either ecologic or geologic time, and may occur on a taxon-to-taxon level or among clades which occupy one or more of the same adaptive zones. The examples of replacement turnover discussed below are part of an interrelated continuum; as a result their distinctions are somewhat arbitrary. Also, many of the scenarios, models, and theories used to explain turnover in the fossil record are speculative and do not necessarily represent observable, demonstrated, or accepted processes.

The appearance of a new taxon by either speciation or immigration may result in active competition with a previously established taxon: if this is prolonged or intense the new taxon may replace the old. Competitive replacement occurs between ecological vicars (this term must not be confused with the biogeographic principle of vicariance; see p. 7) – "groups which have a similar role in nature or occupy the same trophic level within an adaptive zone" (Van Valen and Sloan, 1966). The vicars need not be closely related taxonomically and may be in different families, orders, or classes (Marshall, 1981). If the new taxon is more efficient in consuming or pre-empting available limited resources (such as food and living space) then it may replace the previously established taxon (see Chapter 9 for discussion of competition in ecological time). Competitive replacement may be inferred if the appearance of a new taxon and disappearance of a vicar are more or less synchronous. Immigration-induced biotic turnover episodes (Webb, 1984a) result from replacement of native taxa by immigrant vicars. Examples include: extinction of the dog-like marsupial *Thylacinus cynocephalus* following introduction of the dingo *Canis familiaris* into Australia (Diamond, 1984c); and extinction of native terrestrial herbivores such as large ground birds, lemurs, and tortoises following introduction of cattle, suids, and caprids into Madagascar (Marshall, 1984).

Passive replacement occurs when environmental changes or unpredicted phenomena cause extinction of a taxon and its role is subsequently filled by a native taxon or a timely immigrant. The successor taxon simply fills a vacated niche or adaptive zone, but it is not necessarily competitively superior to the taxon it replaces (Marshall, 1981). In such instances "it is not physical or behavioural limitations which guide a . . . [taxon's] . . . evolutionary potential or success, but merely the opportunity to exploit an available adaptive zone, which, because of the nature of the fauna, was open" (Hecht, 1975). Passive replacement results from Burger's Axiom – what survives may succeed (Marshall, 1981). Examples of passive replacement or environmentally-induced biotic turnover episodes (Webb, 1984a) are: the replacement of some native South American land mammals by immigrant land mammals from North America following the appearance of the Panamanian land bridge in the Late Cenozoic (Marshall, 1981; Marshall *et al.*, 1982); and mammalian radiations following dinosaur extinctions at the end of the Cretaceous (Simpson, 1953).

Competitive and passive replacements result in evolutionary relays (Simpson, 1953; Newell, 1967; Flessa and Imbrie, 1973), eco-replacement (Van Valen, 1971), or iterative evolution (Cifelli, 1969) – i.e. the successive abrupt or gradual replacement of taxa occupying the same adaptive zone (Fig. 8.1). Since extinction and origination may be interacting counterparts of the same evolutionary process, it may be difficult to determine a cause–effect relationship for evolutionary relays. In other words, are sequential originations 'permitted' by prior or concurrent extinctions or are they the cause of those extinctions? Evolutionary relays result from adaptive radiations of sequential clades which are characterized by high turnover rates in the early part of the radiation and lower turnover rates in the later parts of their history (Simpson, 1953; Sepkoski *et al.*, 1981; Niklas *et al.*, 1983; Van Valen, 1985c). Thus, clade radiations approach "a levelling-off in the increase in the number of species and ultimately decline in number as a new group radiates" (Niklas *et al.*, 1983). Examples of evolutionary relays include: replacement of some native carnivorous marsupials and predatory ground birds by immigrant placental carnivores in the Late Cenozoic of South America (Marshall, 1977); and successive replacement during the Phanerozoic of 'evolutionary floras and faunas' of marine organisms and families of terrestrial tetrapods (Fig. 8.1).

Taxotely is a statistical measure designed for comparing turnover rates of clades and is identified by significantly high or low turnover in a taxon when that taxon is compared with others of the same rank (Raup and Marshall, 1980). It can be used to identify relays among ecologic vicariant taxa. For example, a synchronous taxotelic relationship of high extinction rates in a native clade and high origination rates in an immigrant clade may signal replacement of the native by the immigrant (Marshall, 1981).

Origination rebounds permit diversity to be re-established to levels which existed before a mass extinction. The premise for this type of turnover is that extinctions leave room for successful originations, or, put another way, extinctions allow rebounds in origination rates (Van Valen, 1985b). In contemporary biotas, rebounds result from immigration or recolonization by species from populations outside an area effected by extinction (Chapter 15). In fossil biotas, rebounds can result from a combination of speciations (adaptive radiation) and immigrations, although the former are typically credited as the principal contributors following mass extinctions. As an example, Kitchell and Pena (1984) note that the severity of the Late Permian and Late Cretaceous mass extinctions "allowed the rate of successful originations to substantially increase in the lagged recovery periods." These rebounds apparently resulted from a combination of adaptive radiations of new clades and re-radiation of clades affected by the extinctions. Both processes permitted the recycling of adaptive zones made available by extinctions (Van Valen, 1985c).

Van Valen (1984b, 1985c) suggested that recycling of adaptive zones by mass extinctions or creation of new adaptive zones by innovation (e.g. the appearance of terrestrial adaptations) can "reset the turnover clock". However, Kitchell and Pena (1984) show that mass extinctions in the Late Permian did not reset the clock as suggested by Van Valen, but that the extinction rates of higher taxa decreased in a constant and permanent pattern throughout the Phanerozoic.

8.3.2 Extinction without replacement

Extinction without replacement and at least a temporary decrease in diversity can result when extinctions are not balanced by originations. This kind of turnover is most apparent in species and genera in ecologic time. Given sufficient time, some sort of replacement will predictably occur, unless extinctions result from the disappearance of niches or adaptive zones which could cause extinction of a taxon in geologic time.

Turnover without replacement may be related to predators, environmental changes, or a combination of the two. The introduction of the predatory mongoose into Jamaica resulted in extinction of many bird species (Diamond, 1984c); rats introduced as 'passengers' on canoes and ships onto islands resulted in the disappearance of many native species of molluscs, large insects, birds, lizards, snakes and frogs (Diamond, 1984c, 1985); and humans contributed to, or caused, extinction of many taxa on islands and continents as a result of over exploitation of food animals (Martin, 1984a,b; see man-related extinctions, Section 8.7.12). The disappearance of the mammal genera *Tapirus* and *Glyptotherium* from North America at the end of the Pleistocene is attributed largely to environmental change (Anderson, 1984), and Horton (1984) believes that climatic change was primarily responsible for biotic extinctions in the Late Pleistocene of Australia. A combination of man-the-hunter and climatic change is believed responsible for extinction of such large mammals as *Mammut* and *Mammuthus* in the Late Pleistocene of North America (Anderson, 1984).

It has been argued that Late Pleistocene/Holocene extinctions in North America represent non-replacement (i.e. the niches are still there, but the animals are gone; Martin, 1984a). Other workers (Guthrie, 1984; Graham and Lundelius, 1984) argue that the niches once occupied by these now extinct taxa are gone, precluding opportunities for replacement. However, the two alternative positions may not be comparable as they are based on different definitions of the niche, namely whether it is organism centred (autecological) or community centred (synecological).

8.3.3 Turnover rates

A prediction of the equilibrium theory (MacArthur and Wilson, 1967; Chapter 15) is that the time required to attain optimal diversity and equilibrium for any biota is largely a function of area, although distance from the colonization source(s) is also important: large areas (continents) will have a higher species diversity and lower per-species turnover rate and will require a longer time span to attain equilibrium than smaller areas (islands). Under equilibrium conditions turnover may be slow, but not necessarily on a one taxon-to-one taxon basis where one extinction is followed by one origination. This is a frequently stated misapprehension that in no way follows from MacArthur and Wilson's theory. Under that theory, extinctions and originations (immigrations) could occur independently of each other but yield an equilibrium species diversity; or species packing could be so tight that immigrations led to extinctions and vice versa. The MacArthur and Wilson theory is neutral as to where along the axis between these extremes truth lies (Diamond, pers. comm.). It is also recognized that extinctions and originations are a function of diversity (i.e. number of species) which implies some level of interaction between these two processes.

On the other hand, a perturbation that causes disruption of equilibrium may result in multiple extinctions which may be followed by a wave of originations (i.e. rapid turnover; Diamond, 1984a). The MacArthur and Wilson theory was formulated to explain diversity and turnover patterns on true (oceanic) islands, and was subsequently applied to ecologic (habitat) islands on continents (Diamond, 1984a; Patterson, 1984; Chapter 15), and continental (Webb, 1969; Marshall *et al.*, 1982) and global (Sepkoski, 1984) systems where the predictions generally hold true. Species–area effects (see Section 8.7.7) are easy to study in living biotas (Chapter 4); but difficult to assess in the fossil record (Stanley, 1984).

Rate is simply a ratio of some measure of change with respect to time. Rate calculations address the question, how fast or at what intensity do extinctions occur? (Dingus, 1984). Several popular methods are used (see Chapter 15 for discussion of turnover rates in ecological time).

A simple, but not very informative, method of comparing total extinction intensities is made by computing absolute number of extinctions in two or more time intervals (e.g. ages or stages). For example, if five land mammal genera become extinct in the Pliocene and fifteen in the Pleistocene, then the 'relative' rate of Pleistocene extinctions is three times greater than Pliocene extinctions. This rate, however, normalizes neither for the fact that the Pliocene and Pleistocene are of unequal duration, nor for differences in standing diversity (i.e. taxa available for extinction) during these time intervals.

Extinctions in a given time interval are cumulative indices (Lasker, 1978) that can be normalized for time intervals of unequal duration by dividing the total number of extinctions (E_i) of a particular taxonomic rank (species, genus, family, etc.) that occur in that interval by the duration (d, usually expressed in millions of years of that interval). Extinction rate (E_r) is expressed in terms of extinctions per millions of years; thus, $E_r = E_i/d$. Origination rates (O_r), based on first appearances, are calculated in the same manner (Webb, 1969). Turnover rates (T) are the average of extinction rates and origination rates for a given interval of time; thus, $T = (E_r + O_r)/2$.

To normalize for differences in standing diversity between time intervals, the number of extinctions (E_i) is divided by the number of taxa present (S_i), giving an extinction percentage (E_p); thus, $E_p = E_i/S_i$. To further normalize for time intervals of unequal duration, the value of E_p is divided by the duration (d) of that interval to give an extinction percentage rate (E_{pr}); thus, $E_{pr} = (E_i/S_i)/d$. This percentage rate measures relative intensities of extinction, rates per taxon, or the number of "taxa at risk" (Raup and Sepkoski, 1984). It has been translated into what Van Valen (1984b, 1985b) calls "probability of extinction per unit time" which expresses extinction as a proportion of the total number of extinctions per unit time (percentage of extinctions per million years).

The overall rate (Raup and Sepkoski, 1982, 1984) and probability (Kitchell and Pena, 1984) of extinction of marine families shows a uniform decline over Phanerozoic time. Background extinction rates (see below) decrease from about 4.6 families per million years in the Cambrian to about 2.0 families per million years today. The conventional explanation for this trend is that more recent taxa are simply more extinction resistant due to an increase in Darwinian fitness through time (Raup and Sepkoski, 1982). An objection to this explanation is that the decline may only be apparent, not real, and the trend is simply an artifact of the fossil record and/or disparate taxonomic treatment by different authors. The higher apparent extinction rates resulting from shorter stratigraphic ranges, hence durations, of older families may be due to 'lack of record' of the early and/or later parts of their true temporal ranges (i.e. stratigraphic ranges are only minimal records of true durations; Gould, 1983). Younger families have lower apparent extinction rates due to attributes of the Pull of the Recent. An alternative explanation for this trend is provided by Flessa and Jablonski (1985). They show that species/family ratios have, on the average, increased through time, so that families living today have more species than families in the Triassic (Fig. 8.2) and that species-rich families are generally more extinction resistant than species-poor families.

The increase in marine diversity through time may be related more to a decline in real or apparent extinction rates than to an increase in origination rates. Raup and Sepkoski (1982) observe that if the Cambrian rate of 4.6 families per million years had been sustained, about 710 additional families would have become extinct. The 710 families 'spared' by declining extinction rates are nearly the same number (680) that originated over the same time period. This observation is intriguing, although its relevance is unclear (Gould, 1983). Also, if family origination had remained high, either family-level diversity would have increased or many more families would have become extinct (J. Brown, pers. comm.).

While marine organisms show a declining trend in total and per-taxon rates of extinction, non-marine tetrapods show an overall increase in total extinction rates and only a minimal decline in probability of extinction through time (Benton, 1985). This difference suggests that generalities about overall rates of extinction through time may be valid only within clades, or that taxonomic discrimination may be higher among vertebrates than invertebrates.

8.4 BIASES AFFECTING EXTINCTION PATTERNS

8.4.1 Tracking

Some diversity studies are based on species (Niklas *et al.*, 1983; Gingerich, 1984), estimates of species (Flessa and Jablonski, 1985), genera (Webb, 1984b), or families (Marshall, 1981; Marshall *et al.*, 1982; Raup and Sepkoski, 1982, 1984, 1986; Benton, 1985). When genera or families are used it is assumed that the diversity patterns obtained will track or mirror those of the species "without any significant bias" (Flessa and Jablonski, 1985). Yet the basic unit of evolution is the biological species and use of families results in data sets that are three distinct taxonomic steps removed – biological species, morphological species, genus, family. The sort of disparity that can result from comparing data sets based on morphological species and families is illustrated by the following extreme example. Extinction of 10 monotypic families in a stratigraphic stage translates into 10 species extinctions in that stage, while extinction of two families each with 30 species in another stage translates into 60 species extinctions in that stage. The diversity information provided by the species and families is thus notably different and the patterns do not track. There are a few published diversity studies which demonstrate the invalidity of the tracking assumption: e.g. families of Late Cenozoic land mammals in South America suggest replacement, while genera suggest enrichment (Fig. 8.3); species of marine invertebrates show a sharp decrease in the Toarcian (early-middle Jurassic) while families do not and vice versa in the Late Eocene (Hoffman, 1985b); and analysis of marine families shows eight extinction episodes over the last 250 Myr while genera show these same episodes

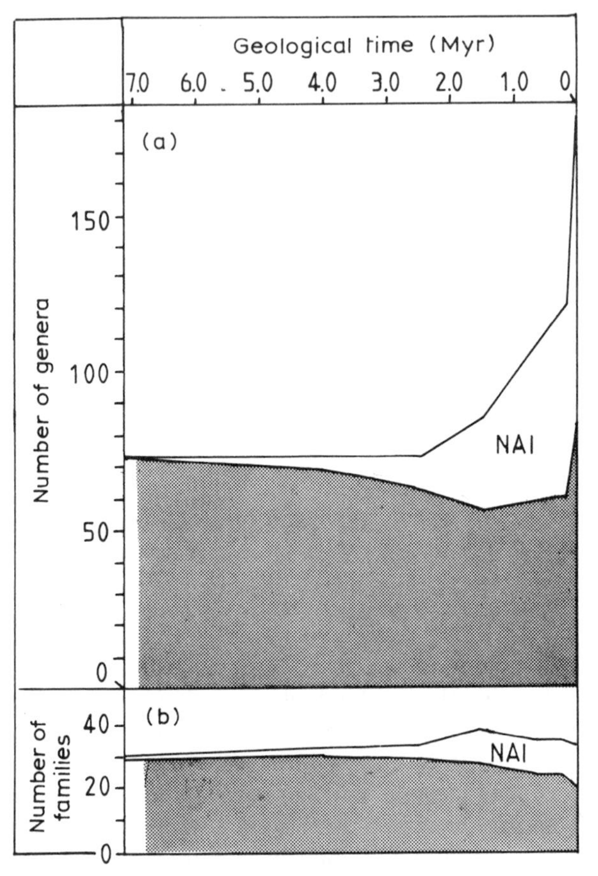

Figure 8.3 Number of known genera (a) and families (b) of land mammals known in South America during last 7.0 Myr. Native taxa (stippling) and North American immigrants (unshaded–NAI). Note that following appearance of the Panamanian land bridge and immigration of North American taxa to South America about 3.0 Myr that the generic diversity suggests enrichment while the family diversity suggests replacement; thus, the two data sets do not track. After Marshall (1981).

to be more pronounced and indicate the existence of two others (Raup and Sepkoski, 1986). Thus, contrary to conventional wisdom, there is no guarantee that diversity patterns based on genera or families represent real events in the history of life. The tracking assumption which is employed by some paleontologists underlies an important difference that exists for assessing extinction patterns in living and fossil biotas.

8.4.2 Hiatuses

Gaps or missing time intervals in the fossil record may produce 'apparent' truncations of stratigraphic ranges of taxa and give the appearance of a major extinction or origination event (Flessa and Jablonski, 1983). In reality the ranges of some or all of these truncated taxa may have extended into younger or older age rocks but are not recorded due to erosion and/or non-deposition. Benton (1985), for example, concluded that "apparent mass extinctions" of terrestrial tetrapods in the early and late Jurassic are simply artifacts of poor fossil records in immediately succeeding time periods.

8.4.3 Lazarus effect

The fossil record of a lineage (species, genus, family) is never 100% complete. There are always time intervals when a taxon is unkown but was surely present somewhere because it is known from preceding and succeeding time periods. Such 'apparent' absences produce a 'Lazarus effect' (Jablonski, 1986b) and may result from hiatuses, non-preservation due possibly to small population sizes, greatly restricted geographic ranges, and/or unknown records (i.e. the fossils exist somewhere but have not yet been found). These Lazarus taxa are useful in estimating probabilities of non-preservation (Raup, 1986). If the total known record of these taxa is not taken into consideration, then study of the time interval(s) that represent the 'apparent' absence will give a false impression of 'extinction' and 'origination' in preceding and succeeding biotas respectively.

8.4.4 Signor-Lipps effect

In a similar manner, the last surviving individual of a lineage will seldom if ever be documented in the fossil record (Signor and Lipps, 1982). L. Alvarez (1983) argues that the highest occurrence of the last known dinosaur a few metres below the Cretaceous–Tertiary boundary in Montana does not (*contra* Archibald and Clemens, 1982) represent the last dinosaur either in this or other areas. Vagaries of preservation will predictably result in premature truncations of true temporal ranges, particularly for taxa with small populations. The Signor-Lipps effect has been used to justify, in part, the extension of ranges of extinct taxa in specific time intervals to the upper boundary of those intervals (see also discussion on smearing, p. 237).

8.5 EXTINCTION PATTERNS

Patterns reflect the data sets used to document extinctions of populations, species, clades or entire biotas. Although patterns are data set specific, there are some pattern categories which are recognizable.

The permanent disappearance of local populations of a broadly distributed species will result in a range reduction, local or regional extinction of that species (e.g. the North American Condor was widely distributed in the Late Pleistocene but survives today in a restricted area of California). The same pattern can occur in clades by local extinction of species and genera; the order Proboscidea (elephants) was widely distributed in Africa, Europe, Asia, and the Americas in the Late Pleistocene and survives today only in Africa and India. These patterns can be observed in ecological and geological time.

Terminal, biological, blanket, global, real, total, or true extinction patterns represent what Martin (1984a) regards as extinction forever, the total global disappearance of a species or higher taxon. These sorts of extinctions apply to species (e.g. Dodo, Great Auk, Passenger Pigeon) and clades (e.g. ammonites, trilobites), and represent the disappearance of lineages.

Background and mass extinction patterns represent conceptual antipodes of an extinction continuum. 'Background' patterns represent the typical, normal, baseline, or low extinction intensities of all taxa during most time intervals. 'Mass' patterns represent excessively high extinction intensities of multiple clades on a global scale during relatively short time intervals (Gould, 1983; Sepkoski, 1984). A third category, 'Events', is represented by abnormally, but not usually significantly, high extinction intensities during short time intervals; they are typically clade or region specific. The most important variables for distinguishing these three categories are duration and intensity (magnitude/duration). A fourth category, 'Major patterns' or 'Episodes', includes all patterns above background (i.e. both events and mass).

It is the ill-defined event pattern that results in the intergrading nature of background and mass categories to form an extinction continuum. Events which involve one or only a few clades will conceptually plot in the upper level of the background category, while events involving several or many clades will plot in the lower level of the mass category. Conceptually, a large event can be regarded as a mass pattern as exemplified by Stanley's (1986) analysis of a 'regional mass extinction'. Definitions of background, event, and mass are thus relative and data set specific, just as the terms small, medium, and large are meaningful only among items in a given sample. To illustrate the intergrading nature of these categories one may equate extinction of species with death of individuals. Background deaths result from accidents, old age, and disease; death events from epidemics and regional wars; and mass deaths from global wars. Historical demographers, like palaeontologists, have difficulty in isolating these categories for any given time interval.

Raup and Sepkoski (1982) identified background intensities from about one to eight family extinctions per million years by plotting a linear regression for the 76 extinction points in their data set as a function of time (Fig. 8.4). Benton (1985), in a study of non-marine tetrapod families, found no evidence of mass extinctions above background (normal) levels. Van Valen (1985c) estimated that the background extinction rate of all species is about one per year. The major causes of background extinctions are apparently phyletic change (i.e. pseudoextinctions, see Section 8.7.1) and species interactions.

Mass extinctions are conceived as resulting from disruption of community structure by a catastrophic event. Catastrophe has been defined as "biospheric perturbations that appear instantaneous when viewed at the level of resolution provided by the geological record" (Knoll, 1984), and "a single event that set in motion a chain of other events, thereby causing major biological changes and extinctions within at most a few

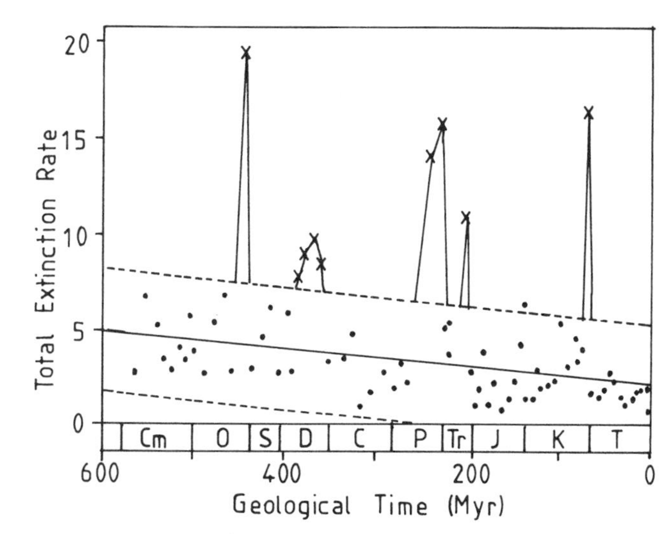

Figure 8.4 Extinction rate (families/Myr) of marine animals during Phanerozoic time. The 'big five' mass extinctions (peaks with crosses) occur in the Late Ordovician, Late Devonian, Late Permian, Late Triassic, and Late Cretaceous. Background rates (dots) occur within dashed lines; solid line is a regression. Abbreviations as in Fig. 8.1. Simplified after Raup and Sepkoski (1982).

thousand years" (Archibald and Clemens, 1982). Mass extinctions are sometimes identified by abnormally large decreases in standing diversity. Raup and Sepkoski (1982, 1984) attribute decreases in marine families solely to significant increases in extinction rates, while Benton's (1985) study of non-marine tetrapods revealed that none of the 'mass extinctions' were associated with statistically high extinction rates but "were the result of a slightly elevated extinction rate combined with a depressed origination rate." These studies identify two discrete mechanisms for decreasing diversity and hence for defining mass extinctions (Section 8.3).

To qualify as a catastrophic mass extinction (as opposed to a gradualistic mass extinction, see below) a large decrease in diversity must occur in a short period of time – ranging from within one generation of the longest lived species to 10^3 years (Clemens, 1982; W. Alvarez *et al.*, 1984; Dingus, 1984; Van Valen, 1984a). Some mass extinctions may occur over longer time frames and these may be labelled gradualistic – those occurring somewhere between 10^2 and 10^7 years (Clemens, 1982; W. Alvarez *et al.*, 1984; Stanley, 1984; Van Valen, 1984a).

Data amassed by W. Alvarez *et al.* (1984) show that the Late Cretaceous mass extinction included both gradualistic and catastrophic components: (1) a gradual decline of some taxa and overall diversity between 1 and 10 Myr before the Cretaceous/Tertiary boundary, and (2) sudden truncation of ammonites, some cheilostomate bryozoans, some brachiopods, and some bivalves at or just below the boundary. A similar two-fold pattern is reported for dinosaurs (Sloan *et al.*, 1986). Consequently, acceptance of one component (gradualistic or catastrophic) does not necessitate rejection of the other since both may be operating either synchronously or sequentially during any given interval of time. Moreover, these components are largely conceptual, not operational, because "it seems unlikely that we can distinguish episodes of extinction lasting 100 years or less from episodes lasting as long as 100 000 years" in rocks of this age in the fossil record (Dingus, 1984).

Unfortunately, not all workers attempt to distinguish catastrophic and gradualistic components within their Late Cretaceous (or other) data sets: these components are different from background and mass intensities discussed above, and are part of the mass intensities only. Rather, they take all last appearances and extend them (Section 8.4.4) to the Cretaceous–Tertiary boundary (Clemens, Archibald and Hickey, 1981). Thus, no attempted distinction is made "between last occurrences that clearly occurred before the upper boundary from those that occurred at the boundary" (Dingus, 1984). By ignoring the existence of these components it is implied that "extinctions in those lineages were synchronous at some level of temporal resolution", yet they are diachronous to some degree (Dingus, 1984). Grouping extinction patterns that are in part or in whole diachronous result in smearing of the data set (W. Alvarez *et al.*, 1984; Section 8.4). The resulting extinction intensity purported to occur at the end of a time interval is potentially a gross misrepresentation of the real pattern of events in the history of life. Given the reality that time resolution for determining durations of events generally decreases with increasing age, it is probable that the effects of smearing will be more prevalent in older sediments. Thus, older mass extinctions will have potentially greater probabilities of including unrelated and diachronous events than will those of younger rocks.

It is not fortuitous that the best known mass extinctions occur at or near the ends of geologic time intervals. This occurs because "stratigraphers who established the geologic time scale in the first half of the 19th century chose major faunal breaks as boundaries for the principal subdivisions" (Raup and Sepkoski, 1982). The boundaries between Paleozoic, Mesozoic, and Cenozoic are thus marked by two of the largest mass extinction events in the history of life. Likewise, recognition of epoch subdivisions of Cenozoic time was originally based on consideration of percentage of extinct taxa in each interval (Berry, 1968; see also discussion of Lyellian Curves in Stanley, 1979).

In Phanerozoic marine biotas, Newell (1967) recognized mass extinctions, based on percentage of families that became extinct, at or near the ends of the Cambrian, Devonian, Permian, Triassic, and Cretaceous. Raup and Sepkoski (1982) identified mass extinctions at the ends of the Ordovician, Devonian, Permian, Triassic, and Cretaceous (Fig. 8.5); they later (1984) observed 12 extinction peaks during the last 250 Myr with the four highest peaks in the Late Permian, Late Triassic, Late Cretaceous, and Middle Miocene. Stanley (1984) observed mass extinctions in the Cambrian, Late Ordovician, Late Devonian, Late Permian, Late Cretaceous, Eocene–Oligocene, and Pliocene. The mass extinctions at the ends of the Ordovician, Devonian, Permian, Triassic, and Cretaceous are now known as the 'big five' (Fig. 8.5). The one in the Late Cretaceous is best documented and shows extinction of about 13% of families, 50% of genera, and 75% of species of marine taxa during Maestrichtian time. Raup (1979b) notes that the Late Permian episode shows disappearance of about 50% of shallow water marine families and an estimated 96% of species.

A study of Phanerozoic land plant diversity shows sharp drops at the ends of the Devonian, Permian, Triassic, and Cretaceous (Niklas *et al.*, 1983). In the Permo–Triassic, diversity dropped by about 20%. The episodes of land plant extinctions in the Late Devonian, Permo–Triassic, and Late Cretaceous correspond roughly to those of mass extinctions in

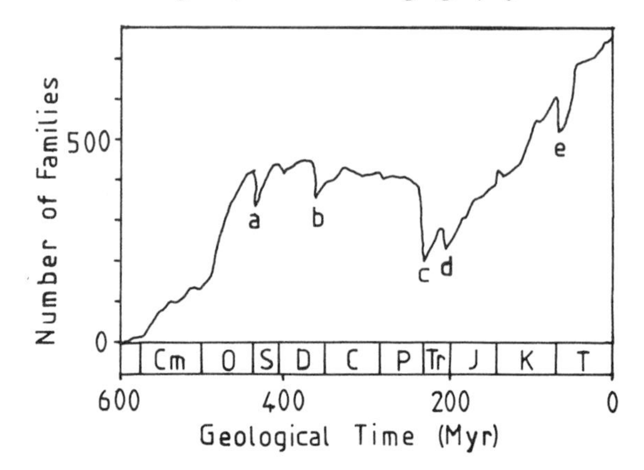

Figure 8.5 Standing diversity of families of marine animals during Phanerozoic time. The 'big five' mass extinctions are recognized by abrupt decreases in diversity as compared with immediately preceding and succeeding stages and include (% drop in diversity): a, Late Ordovician (−12%); b, Late Devonian (−14%); c, Late Permian (−52%); d, Late Triassic (−12%); e, Late Cretaceous (−11%). Abbreviations as in Fig. 8.1. After Raup and Sepkoski (1982).

marine animals, although "there is no evidence for globally synchronous mass extinctions in the fossil record of land plants" (Knoll, 1984).

In a study of non-marine tetrapod families, Benton (1985) identified six mass extinctions during the Phanerozoic (percentages are of families that become extinct): Early Permian (58%), Late Permian–Early Triassic (49%), Late Triassic (28%), Late Cretaceous (12%), Early Oligocene (8%), and Late Miocene (2%).

Major extinction events of land mammals in the Cenozoic of North America are reported in: (1) Middle Oligocene with loss of at least six families (Prothero, 1985): (2) Late Miocene with loss of about 60 genera (Webb, 1969, 1984b), and (3) Late Pleistocene (Wisconsin age) with loss of 7 families, 33 (69%) genera, and 78 (72%) species before about 11 000 years ago (Martin, 1984b).

The extinction episodes noted above are only representative and each differs in magnitude from the others. The intensities of each are relative since each is defined solely on consideration of known taxa immediately prior to and after each episode.

A periodicity in timing of extinction intensities above background level in marine taxa over the past 250 Myr is reported to occur at cyclic intervals of about 32 Myr (Fischer and Arthur, 1977), 26 Myr (Raup and Sepkoski, 1986), 30 Myr (Rampino and Strothers, 1984), and 31 Myr (Kitchell and Pena, 1984). If this periodicity is real, then it may be triggered by a common recurrent cause; both earthbound (Fischer and

Arthur, 1977) and extra-terrestrial (L. Alvarez, 1983; Rampino and Strothers, 1984) mechanisms have been suggested. An extra-terrestrial mechanism (Section 8.7.10) has been linked with the ±26 Myr cyclic passing of a proposed dark companion star (Nemesis) through the existing Oort cloud of comets around the sun (deflecting some of the comets toward the Earth; Whitmire and Jackson, 1984; Davis *et al.*, 1985), and to an increase in the influx of comets due to a ±33 Myr vertical oscillation of our solar system about the galactic plane (i.e. solar z-motion; Rampino and Strothers, 1984; Schwartz and James, 1984) and Planet X (Whitmire and Matese, 1985). The Nemesis and Planet X scenarios are reputably testable (W. Alvarez, 1986). Study of impact craters larger than 10 km in diameter (there are less than 20) on the Earth's surface shows peak impact intensities about 28.4 Myr (W. Alvarez and Muller, 1984) or 31 Myr (Rampino and Strothers, 1984). These data are consistent with impact theory (Section 8.7.10); yet Grieve *et al.* (1985) provide data on crater ages which show no statistically significant evidence of periodicity.

The periodicity of major extinction episodes as reported by Raup and Sepkoski (1984) is questioned by Hoffman's (1985b) analysis of three features of their data set and statistical methodology. Firstly, culling of their data set distorted the true picture of extinction intensity and pattern. Beginning with stratigraphic range data of about 3 500 families, Raup and Sepkoski deleted those with questionable ages, uncertain taxonomic affinity, and living representatives (to normalize for pull of the Recent), leaving only 567 (16.2%) of the families upon which they based their analysis. Secondly, several widely accepted geologic time scales are available, and if Raup and Sepkoski had used another their findings would have been considerably different (e.g. boundaries for Mesozoic stages differ by 10–14 Myr between the time scale of Harland *et al.* (1982) used by Raup and Sepkoski, and that of Odin (1982)). Thirdly, the nature of their time units (stages) gives cause for concern – "there is a 1 and 4 probability that any stage will stand out as a major extinction, given a random distribution, and with stages averaging 6.2 million years long, a 26-million year signal (4 × 6.2) is statistically inevitable" (Lewin, 1985b). The sources of error pointed out by Hoffman may simply be potential 'noise' in the Raup and Sepkoski data set, and such sources should serve to degrade a signal of periodicity, not enhance or produce it (Gould, 1985).

Analysis of extensive data sets of land plants (Niklas *et al.*, 1983; Knoll, 1984) and non-marine tetrapods (Benton, 1985) during Phanerozoic time shows no clear evidence of periodicity as reported in marine organisms, although some extinction episodes in the three data sets do coincide.

In summary, the existence of periodicity in both mass extinctions and crater ages is debated and unresolved. However, the indications available in support of such periodicity appear to justify further consideration of its existence.

8.6 EXTINCTION SUSCEPTIBILITY

Extinction episodes are filtering processes in the history of life. The "victims and survivors are not random samples of the pre-extinction biotas" (Raup, 1986). Differential extinctions of populations, species, or clades result because some taxa are more affected than are others. Extinction processes (see Section 8.7) thus permit the continuation of some populations or lineages and cause the termination of others. Thus, taxa which survive major extinction events may provide as much information about those events as do their victims (Jablonski, 1986a; Lewin, 1986). The probability that a species will become extinct is said to be independent of the age of its clade (Van Valen, 1973), although Bambach (as cited in Lewin, 1985a) suggests that clades late in their history are generally more extinction prone than those early in their history.

Survival capability and extinction vulnerability of shallow-water, bottom-dwelling marine species are different during times of background and mass extinctions (Jablonski, 1986a). Traits favouring survival during background times include species-rich clades, constituent species with broad geographic ranges, and high dispersal ability of larvae; during times of mass extinctions those "clades with broad geographic ranges, regardless of the geographic range of the constituent species, were those more likely to survive" (Lewin, 1986). These trait differences do not represent dichotomous categories, but conceptual ends of a continuum.

Land plants seem to be less vulnerable than animals to extinctions by catastrophic events (Knoll, 1984). Extinction resistant traits in plants include their ability to survive defoliation and regenerate foliage from roots, rhizomes, seeds, and spores. These traits provide means for plants to propagate even though the above ground foliage may be destroyed. Thus, plants have the ability to survive episodes of severe biomass mortality (Knoll, 1984).

8.6.1 Extinction prone taxa

Some interrelated features which make species susceptible to extinction include: (1) large body size, (2) position in upper trophic levels within communities (i.e. carnivores), (3) diet or habitat specialists, (4) poor dispersal abilities, (5) restricted geographic ranges (i.e. endemics), and

(6) tropical distributions (Brown, 1971; Diamond, 1984a; Stanley, 1984; Jablonski, 1986a). Species with features 1, 2, and 5 generally have low birth rates (r), low population densities (N), and great disparity between r and N (Diamond, 1984a).

8.6.2 Extinction resistant taxa

In species, interrelated features which generally enhance survival and lineage longevity include: (1) small body size, (2) position in lower trophic levels within communities (i.e. herbivores), (3) dietary and habitat generalists, (4) good dispersal capabilities and ability to recolonize, (5) broad geographic ranges and cosmopolitan distributions (i.e. "risk of extinction decreases with area available to the population"), (6) longevity of individuals (i.e. "extinction rates decrease approximately linearly with generation time"), (7) large population size and density, (8) low metabolic rates, and (9) high birth/death ratios (Diamond, 1984a).

8.6.3 Living fossils

The term living fossil is applied to taxa that belong to extant clades, have long geologic ranges, often occur in low densities, and exhibit features regarded as primitive or generalized (Stanley, 1979). Examples (and known durations) include: horseshoe crabs (230 Myr), bowfin fishes (105 Myr), bairdiid ostracods (230 Myr), notostracan crustaceans (305 Myr), garfishes (80 Myr), sturgeons (80 Myr), snapping turtles (57 Myr), alligators (35 Myr), aardvarks (20 Myr), tapirs (20 Myr), and pangolins (35 Myr) (Stanley, 1979).

8.7 EXTINCTION CAUSES AND PROCESSES

A multitude of causes and processes have been proposed to explain why organisms become extinct, and the "abundance of explanations leads one to wonder that anything managed to survive" (Cassels, 1984).

It has been proposed that extinctions may result from chance or random causes (Raup, 1978, 1984), the implication being that extinctions are inevitable (Section 8.1). Yet all extinctions of populations and taxa surely have causes, even if these causes are not readily evident. The terms chance and random are thus misleading and should be deleted from the extinction dictionary. The term unpredictable is more explicit and may be substituted for chance and random as used by earlier workers.

There is no way to arrange all the proposed extinction trends, tendencies, correlations, and scenarios into a simple hierarchy; the categories I use below are by necessity representative but not mutually exclusive. This arrangement results from the fact that many extinctions are produced by an array of interrelated processes, many of which are simply links along an extinction continuum (Section 8.8). In other cases the precise cause of extinction can be securely identified. Many of the extinction processes discussed below are conjectural and should accordingly be considered with critical awareness.

8.7.1 Pseudoextinctions (anagenic, phyletic or taxonomic extinctions)

When a taxon evolves into another of equal or higher rank it 'appears' that the ancestral taxon becomes extinct and the descendant taxon originates. Yet there is no 'real' extinction or origination, only evolutionary change within a single lineage. For example, *Homo habilis* is apparently the direct ancestor of *H. erectus* (Johanson and Edey, 1981); and the large extinct kangaroo *Macropus titan* of the Australian Late Pleistocene is probably the direct ancestor of the smaller living *M. giganteus* (Marshall and Corruccini, 1978) – this exemplifies phyletic dwarfism (Marshall, 1984). Pseudoextinctions are present in most fossil data sets and they increase the number of apparent extinctions for any given time interval (Marshall, 1984). Multiple and synchronous pseudoextinctions in a data set will produce apparently high turnover rates and, assuming no real extinctions, the standing diversity will remain constant. Pseudoextinctions may account for many background extinction patterns and intensities in the fossil record, although this possibility has yet to be assessed.

8.7.2 Population extinctions

Fluctuations in populations can produce what have been called accidental, ecological, and/or population extinctions (MacArthur and Wilson, 1967; Marshall, 1984; Martin, 1984a). Local populations of a species may from time to time disappear (become extinct) and be re-established (originate) by immigration of individuals from surrounding areas. These sorts of extinctions are common on true (oceanic) islands (Diamond, 1984a) and on habitat islands on continents such as restricted ecologies on mountain tops (Brown, 1971; Patterson, 1984); they occur in ecologic time, and are difficult if not impossible to identify in the fossil record.

8.7.3 Coextinction, trophic cascades, domino effect

Extinction of a 'key' taxon from the base of a specialized food chain will predictably result in extinction of other 'dependent' taxa higher in the same food chain. In such cases, extinction of one taxon may be the direct result of extinction of another (Diamond and Case, 1986). As examples, predator–prey coextinction will predictably result when extinction of a predator (e.g. *Smilodon*) follows that of its principal or only prey (mastodon); scavengers and commensals such as carrion feeding birds may become extinct along with their ecologically-dependent partners, in this case, megafauna in North America; parasites may disappear along with their hosts; and herbivore extinctions may follow disappearance of particular plant species or communities (Marshall, 1984). Coextinctions or trophic cascades (Diamond, 1984c) are predicted by the impact theory (Section 8.7.10) in which extinction of plants produced a domino effect collapse of the food chain leading to extinction of herbivores and then carnivores in the Late Cretaceous (L. Alvarez, 1983).

8.7.4 Competition

Competition between species may occur when they share requirements for limited resources (Chapter 9). The replacement of four 'evolutionary floras' during the Phanerozoic is attributed largely to the competitive superiority of each successively younger flora (Niklas *et al.*, 1983; Knoll, 1984); biotic interchanges following continental suturing apparently resulted in some competition between interchange taxa (Knoll, 1984) described by Webb (1984a) as invasion-induced biotic turnover episodes (Section 8.7.7); competition and subsequent extinction may follow introduction of taxa into new areas (Diamond, 1984a; Sloan *et al.*, 1986) so that replacement occurs between vicars; and decrease in living or habitat space and overall productivity may promote competition for food resources and increase predation pressure (Stanley, 1984; Section 8.7.7).

8.7.5 Prey naïvety

The immigration or introduction of a predator into a fauna may result in extinction of its prey species. This may result because the prey species are initially naïve about predators in general or about a 'new kind' of predator in particular. If the predator does not quickly overexploit the prey species and cause its extinction, then the prey species may adapt to the predator and the two may coexist (Chapter 9). This prey naïvety has broad implications relevant to mixing of faunas. For example, continental suturing of the Americas and the invasion of placental carnivores into South America may have resulted in extinction of some naïve native South American herbivores (Marshall, 1977); cats and foxes introduced by man into Australia took a heavy toll of native herbivorous marsupials

(Diamond, 1984c); rats introduced onto oceanic islands which previously lacked rats and/or land crabs (the invertebrate ecological equivalent of rats) caused extinction of many species of birds (Diamond, 1984c, 1985); and the arrival of man-the-hunter into the New World and onto oceanic islands coincided with the disappearance of many native species (Martin, 1984a,b; Section 8.7.12).

8.7.6 Climate-related

Climatic and associated environmental changes have been credited with more extinctions than any other cause and were long regarded as the principal extinction process in the history of life. Speaking of such changes, Simpson (1953) notes, "the populations involved may not change sufficiently and adaptation may be lost; that, quite simply, is the usual cause of extinction." Examples include changes due to northward drift (i.e. latitudinal shifts) of India resulting in extinction of certain land plants on that subcontinent (Knoll, 1984); global cooling during marine regressions and concurrent changes in oceanic thermal patterns, salinity, or dissolved oxygen which are credited with most mass extinctions in marine realms (Fischer and Arthur, 1977; Stanley, 1984); global warming and increase in carbon dioxide produced a 'greenhouse' effect which caused Late Cretaceous extinctions (McLean, 1978, 1985); Middle Oligocene extinctions of North American land mammals are attributed to climatic and/or ecological causes (Prothero, 1985); faunal changes in Late Cenozoic land mammal faunas in South America may have been caused by environmental changes associated with uplift of the Andes (Webb, 1978; Marshall, 1981; Marshall *et al.*, 1982); and disappearance of many land mammals in the Late Pleistocene which may have resulted from abrupt climatic changes linked with glacial advance and retreat (Martin and Klein, 1984). Climatic change can directly affect taxa by causing intensification of seasonality such as drought, or by linking mechanisms which affect plant quality or gestation periods (Marshall, 1984). Habitat destruction by modern man represents a form of environmental perturbation (Section 8.7.12). The effects of climate on contemporary biotas produce range reductions or local extinctions (Chapter 6); however, "almost no modern cases exist of total extinction due clearly to climate" (Diamond, 1984c). Grayson (1984) has identified testable (falsifiable) attributes of the climatic paradigm for Late Pleistocene megafaunal extinctions.

8.7.7 Species–area effect

As discussed in detail in Chapter 4, diversity is determined in part by area. Extinctions will predictably result from a decrease of living space, both in area and number of habitats, due to intensification of competition and predation; decrease in size of populations – low species abundance produces fluctuations in sex ratios; and decrease in environmental heterogeneity and resources (Brown and Gibson, 1983; Stanley, 1984). Extinctions linked to species–area curves may result from habitat destruction and fragmentation on continents (Bakker, 1977; Lovejoy *et al.*, 1984); retraction of mountain top habitats due to climatic warming (Brown, 1971; Diamond, 1984a; Patterson, 1984); size reduction of islands due to rising sea levels (Diamond, 1984a); and restriction of living space for shallow water (shelf) marine communities due to regressions of epicontinental seas (Van Valen, 1984a). Following continental suturing, interchange of previously isolated biotas may result in turnover episodes as a mechanism for maintaining diversity equilibrium levels, so that the suturing of India with Asia may have resulted in land plant extinctions (Knoll, 1984), and the appearance of the Panamanian land bridge apparently resulted in high turnover rates in South American land mammals (Marshall *et al.*, 1982). The species–area effect is the most important unifying concept in understanding extinction dynamics because it is applicable to and apparent in both living and fossil biotas.

8.7.8 Volcanic eruptions

Volcanic eruptions can eliminate local biotas and produce regional extinctions of populations and endemic species. This process is evidenced by the recent eruption of Mt. St. Helens in the State of Washington (Chapter 15) and the 1883 eruption of a volcano on the island of Krakatau in the Sunda Strait. The latter lasted two days and ejected an estimated 10 cubic miles of debris into the atmosphere, causing hazy skies around the globe (L. Alvarez, 1983). Extensive volcanic emissions from restricted areas such as the Deccan Plateau in India (McLean, 1981, 1985; Zoller *et al.*, 1984) or oceanic spreading centres such as the mid-Atlantic ridge (Officer and Drake, 1985) may produce local concentrations of iridium (Section 8.7.10), but there is no secure evidence that they have a role in global extinctions (L. Alvarez, 1983; W. Alvarez, 1986).

8.7.9 Magnetic reversals

Periodicities of magnetic reversal intensities have been reported for cyclic intervals of 32 Myr (Negi and Tirwari, 1983), and 15 Myr (Mazaud *et al.*, 1983). Stimulated by these reports, Raup (1985a) analysed the earth's magnetic field record over the last 165 Myr and identified a periodicity of 30 Myr intervals. He concluded that these reversal intensities were accompanied by cosmic particle bombardment which he linked with mass extinctions, suggesting that reversals, meteorite impacts, and extinctions were possibly interrelated. Lutz (1985) demonstrated that Raup's 30 Myr signal is an artifact of the geologic time scale, hence an accident of the length of the record, leading Raup (1985b) to retract his view. Pal and Creer (1986) subsequently reported that reversal intensities do indeed increase at about 30 Myr intervals and these correlate with episodes of mass extinctions. The role, if any, of magnetic reversals in mass extinctions remains unresolved.

8.7.10 Impact theory

In the 1960s and 1970s suggestions (MacLaren, 1970) were made that one or more mass extinctions resulted from the impact of a large meteorite (asteroid or comet). It was not until L. Alvarez *et al.* (1980) presented credible evidence for such an impact at the Cretaceous–Tertiary boundary that this idea received serious consideration. The impact theory was originally formulated to explain extinctions at the end of the Cretaceous and has subsequently been suggested as an extinction mechanism for other periods of time. Evidence of an iridium anomaly (an unreactive element normally found in extremely low concentrations in the upper part of the earth's crust but in concentrations 10^4 times greater in meteorites) has now been found at more than 75 marine and continental Cretaceous–Tertiary boundary localities round the world (W. Alvarez *et al.*, 1984; W. Alvarez, 1986), in the Late Devonian of Australia (Playford *et al.*, 1984), and at the Eocene–Oligocene boundary in the Caribbean, Pacific, and Indian Oceans (Alvarez and Muller, 1984). The impact theory is regarded as testable (L. Alvarez, 1983): if it can be shown that the extinction was not synchronous with the asteroid impact, then the theory can be rejected.

The evidence for an impact at the Cretaceous–Tertiary boundary comes from knowledge of:

(1) high concentrations of iridium;
(2) osmium isotope ratios;
(3) shocked quartz which indicates environmental perturbation;
(4) particles of carbon soot; and
(5) sanidine spherules believed to be formed by impact melt

(Hsü, 1980; Smit and Hertogen, 1980; W. Alvarez *et al.*, 1984; Bohor *et al.*, 1984; W. Alvarez, 1986).

The Cretaceous–Tertiary meteorite was apparently

(1) a carbonaceous chondrite Apollo class asteroid (i.e. one which intersects the earth's orbit at its closest approach to the sun);

(2) had a diameter of about 10 km;
(3) impacted the earth at about 20 km s^{-1};
(4) possibly struck the ocean since, on land, it would have left a crater 100–150 km in diameter – though the resulting ocean crater has either escaped notice or perhaps was subducted along with 20% of the oceanic crust during the last 65 Myr;
(5) produced a tidal wave initially 5 km high that encircled the globe within hours and inundated all low lying coastal areas;
(6) released about 10^8 megatons of kinetic energy (about 10^4 times more energy than that which would be produced by the simultaneous explosion of all existing nuclear weapons); and
(7) produced a dust cloud which settled to form the iridium rich clay layer which is now the Cretaceous–Tertiary boundary (L. Alvarez, 1983; W. Alvarez, 1986).

The Cretaceous–Tertiary asteroid impact may have produced effects analogous to a 'nuclear winter' (Ehrlich *et al.*, 1984) and created killing mechanisms that included

(1) an initial increase in ocean temperatures of several degrees;
(2) dust and debris ejected into the atmosphere darkened the earth by "shutting out" the sun, causing a sharp and significant drop in temperature, followed by a domino effect (Section 8.7.3) collapse of the food chain due to termination of photosynthesis which resulted in death of marine algae and land plants leading to starvation and death of herbivores and hence carnivores;
(3) after dust particles settled a "greenhouse" effect was produced by vapour in the atmosphere and temperature rise to produce a "sweltering heat";
(4) "radiant energy in a rising fireball would go through the atmosphere and fix a lot of the nitrogen to make enormous amounts of nitrogen oxide" (nitric acid) producing acid rain; and
(5) radiation from the fireball and cloud of rock vapour, and falling hot ejecta ignited vegetation, producing wildfires that killed terrestrial organisms and formation of a large soot content at the boundary (Silver and Schultz, 1982; L. Alvarez, 1983; W. Alvarez, 1986).

It is important to note that not everyone accepts the impact scenario and alternative explanations for the Late Cretaceous extinctions have been offered (see above; Archibald and Clemens, 1982; Clemens, 1982; Sloan *et al.*, 1986).

8.7.11 Comet shower hypothesis

The comet shower hypothesis (Hut *et al.*, 1985) predicts a high influx of comets over a one to three Myr period that could produce "stepwise mass extinctions" (W. Alvarez, 1986). "If the comet shower hypothesis is confirmed, it will offer a satisfactory resolution of what has appeared to be an irreconcilable conflict between the gradualistic and catastrophic views of mass extinction. Each extinction event in a stepwise mass extinction would be the catastrophic result of a single impact, but the clustering of several of these events over 1–3 Myr would produce an apparently gradual mass extinction" (W. Alvarez, 1986). This hypothesis is testable.

8.7.12 Man-related

Man has been directly or indirectly responsible for the majority of vertebrate extinctions in recent times and an excellent overview of killing mechanisms produced by modern man is given by Diamond (1984c). The first six of the mechanisms listed below were also probably operative in prehistoric times:

(1) *Overkill*, whereby extinctions result from hunting at a rate beyond a population's reproductive capacity. An extreme form of overkill is blitzkrieg, the "sudden extinction following initial colonization of a land mass inhabited by animals especially vulnerable to the new human predator" (Martin, 1984b; Section 8.7.5). The three ingredients of blitzkrieg are: rapid deployment of human populations into an area not previously inhabited by man; man's possession of a big-game-hunting technology; and the synchronous extinction of megafauna resulting from hunting by humans (Marshall, 1984).

Two types of blitzkrieg are recognized: direct blitzkrieg in which man-the-hunter is the sole cause of extinction of unstressed megafauna, and associated blitzkrieg in which man-the-hunter and concurrent natural climatic change contribute hand-in-hand to megafaunal extinctions (Marshall, 1984). The blitzkrieg model was originally developed for the continental land mammal fauna of North America and subsequently applied to other land masses; the theory is reportedly testable (Martin, 1984a,b) although some workers (Grayson, 1984) have challenged this claim. Diamond (1984c) notes that there are no records of overkill of deep sea vertebrates with the possible exception of the Atlantic Grey whale, and that extinction of Steller's sea cow, Japanese sea lion, and Caribbean monk seal "took place on land or in shallow water at their breeding grounds."

Diamond (1984c) recognizes six categories of overkill: (a) hunting for meat (Steller's sea cow, Passenger Pigeon); (b) hunting motivated by economic value of inedible body parts (furs, sea otters; skins, crocodiles; feathers for bedding, Great Auk; oil, whales); (c) cultural value of body parts (rhinoceros horn, display plumes of birds); (d) protection of gardens (Carolina Parakeet) and domestic stock (lions, wolves); (e) capture for pets (parrots, primates, tropical fish); and (f) hunting for sport without economic motive.

Marshall (1984) recognizes two categories of overkill in addition to blitzkrieg: innovated overkill which requires the combination of man's long-established presence in an area (i.e. Africa, Asia, Europe), natural climatic change stressing megafauna, and man's innovation of a big-game-hunting technology; and attritional overkill which requires man's previous presence in an area, man's manipulation of his environment and alteration of ecologies through farming practices, occasional hunting, and the slow disappearance of megafauna. Two types of attrition are: indirect, which involves man-the-farmer and habitat manipulator (activities which stress populations of native animals), with occasional hunting; and competitive, which can result from man's competitive exclusion of other animals due to overlap in diet preference, feeding strategies, or habitat utilization.

(2) *Habitat destruction* that can result from: (a) land clearing for agriculture, grazing of stock, or deforestation for use of timber; (b) browsing and grazing by introduced animals such as rabbits, goats and sheep; (c) drainage of swamp lands for use in agriculture and housing; and (d) fire.
(3) *Introduction of predators* (see Section 8.7.5) resulting in: (a) predation on eggs and juveniles; (b) predation on adults; and (c) over exploitation of prey species.
(4) *Introduction of competitors*, for example, the dingo apparently outcompeted the thylacine and Tasmanian devil on the Australian mainland.
(5) *Introduction of diseases*, for example, the introduction of chestnut blight to North America from Europe resulted in decimation of the American Chestnut.
(6) *Trophic cascades* (Section 8.7.3).
(7) *Chemical pollution*.

A different perspective of man-related extinctions is provided by Diamond (1984c) who groups them into three variable categories, quoted below:

(1) *Man-linked independent variables*, specifying man's arsenal of mechanisms for extermination: introduced grazing and browsing animals (leading to habitat destruction); crop farming (leading to habitat destruction); introduced cursorial predators (especially cat, fox,

mongoose, mustelids); introduced pigs (acting as predators on naïve ground animals, and also leading to habitat destruction); dogs (possibly increasing hunting efficiency, especially for hunters without firearms); weaponry (firearms, or quality of stone weapons visible archaeologically).

(2) *Island- or continent-linked independent variables* that facilitate man-linked extinction: lack of prior exposure to man (hence fauna naïve to human hunters); lack of native flightless mammals or land crabs (hence small birds, reptiles, amphibia, and terrestrial invertebrates susceptible to commensal rats); lack of native cursorial mammalian predators (hence native birds and small mammals susceptible to introduced cursorial predators; may also make fauna more naïve to human hunters); long arid seasons (hence habitats especially susceptible to destruction by fire).

(3) *Dependent variables*: widespread destruction of habitat (leading in turn to possible extinction of any species); widespread extinctions of megafauna; widespread extinctions of small mammals (as in the European extinction wave in Australia, and the West Indies, but not in the late Pleistocene waves in Australia and the New World); widespread extinctions of small volant birds (as in the Polynesian and European waves in Hawaii, but not in the European wave in Australia and North America).

8.8 CONCLUSIONS

Extinction is only one part of the evolutionary process (origination is the other) and serves the primary role of making room for the origination of new species; it is thus an integral part of the process of renewal of life on Earth. The majority of extinctions represent crises in the history of life and offer two possibilities to biological organisms: disaster for the victims owing to extinction of some populations, species, or clades; and opportunity for the survivors, presenting the chance for high origination rates and adaptive radiations (W. Alvarez, 1986).

The extremes of the extinction continuum are represented by background and mass regimes. "Currently evolutionary history is formulated almost exclusively in terms of . . . background times." Yet "mass extinctions play a larger role than is generally appreciated in creating opportunities for faunal change; removing dominant taxa and thereby enabling other groups – previously unimportant but with traits enhancing survivorship during mass extinctions – to undergo adaptive radiations." It is "the alteration of background and mass extinction regimes that shape the large-scale evolutionary patterns in the history of life" (Jablonski, 1986a).

8.8.1 Role of extinction in biogeography

Extinction has played a decisive role in biogeography by influencing the distribution of organisms over time and in space. In turn, many aspects of the biogeography of organisms have determined or contributed to their susceptibility to various extinction processes. Thus, in order to study and understand patterns and processes of extinction it is necessary to consider aspects of the biogeography of the organisms.

Effects of extinction on biogeography

Over time, extinctions serve as filtering processes which permit the continuation of some populations or lineages, and cause the disappearance of others. Extinction episodes predictably result in decreases in diversity and simplification of biotic communities. Extinction processes thus influence the historical development of biotas and monitor diversity patterns through time. In space, taxa with certain kinds of biogeographic distributions (e.g. restricted geographic ranges, endemics, small population densities) are particularly prone to extinction. Biogeographic patterns may also be determined by extinctions which produce range reductions or disjunct distributions. Extinctions related to climate may produce range displacements in which taxa become extinct in one area but survive by migrating to another (see p. 149). The disappearance of local populations may be permanent, or repopulation may occur from other areas depending upon the proximity and access of potential repopulating sources.

Effects of biogeography on extinction

The size of a geographic area may influence extinction rates: large areas (continents) usually have lower rates than small (island) areas. Extinctions linked to species–area curves may result from a decrease in living space, for example from habitat destruction and fragmentation on continents; retraction of mountain top habitats due to climatic warming; reduction in size of islands due to rising sea levels; and restriction of living space for shallow water marine organisms due to regressions of epicontinental seas. Extinctions may also be associated with a decrease in environmental heterogeneity or resources or both. Taxa in areas susceptible to environmental change will predictably be more extinction prone than those which are not. Movements due to plate tectonics have contributed to extinction by causing latitudinal shifts in continents such as India and hence climatic change; continental suturing, for example, India with Asia, North with South America, has resulted in mixing of biotas, permitting invasions of taxa and potential extinction due to prey naïvety and competition. Taxa which belong to species poor families, are dietary and/or habitat specialists, of large body size, have a position in the upper trophic level within communities such as carnivores, have low population densities, or poor dispersal abilities, are more extinction prone than taxa which do not have these features. Coextinction (trophic cascades, domino effect) will predictably occur in areas with specialized food chains. Lastly, endemics which occur near volcanically active areas will have a higher risk of extinction than those taxa in other areas.

8.8.2 Commonality in divergent extinction paradigms

Extinctions may be viewed conceptually and operationally as constructive, non-constructive, or inevitable. Some extinctions may result from known processes and in these cases a cause–effect relationship may be securely established. Other extinctions, particularly episodes (see p. 234), are not so 'black-and-white' but may be additive, include multiple, possibly diachronous, components and may fall into a 'grey' category. Episodes may encompass both gradual and catastrophic components, and direct, indirect, immediate, ultimate, earthly, or extraterrestrial causes (W. Alvarez *et al.*, 1984; Stanley, 1984; W. Alvarez, 1986).

Some extinction episodes are best explained by multidimensional paradigms (Marshall, 1984) even though these are the least falsifiable (Martin, 1984a,b). The two best examples are those at the ends of the Cretaceous and Pleistocene, namely Late Cretaceous mass extinctions and Late Pleistocene megafaunal extinctions in North America. These two extinction episodes demonstrate a commonality by sharing fundamental conceptual and operational attributes. For example, both episodes reportedly include: (1) gradualistic (climatic) and catastrophic (asteroid, man) components that apparently result in smearing of patterns; (2) the gradualistic components involve climatic change; and (3) the catastrophic components apparently dealt the *coup de grâce* to stressed populations (Guilday, 1967; Stanley, 1984), produced coextinctions (L. Alvarez, 1983; Marshall, 1984), and are regarded by their avid proponents as falsifiable – if it can be shown that the extinctions were not synchronous with the appearance of the asteroid or man, then the impact and blitzkrieg theories can be rejected (L. Alvarez, 1983; Martin, 1984a,b). Grayson (1984) opposes this view.

The Late Cretaceous mass extinction may have coincided with changing conditions of climate and temperature, and increased seasonality may have been produced by regression of epicontinental seas, concurrent shifts in oceanic circulation patterns and water chemistry (W. Alvarez *et al.*, 1984; Stanley, 1984; Sloan *et al.*, 1986), and possibly impact of an asteroid (L. Alvarez *et al.*, 1980). The ultimate effect of an asteroid impact on a biota "may depend on conditions of diversity and robustness in which various groups find themselves at the moment of impact" (W. Alvarez *et al.*, 1984). The concurrent climatic impact may have stressed the biota and the asteroid impact may simply have dealt the final blow to taxa already in a stressed state (Stanley, 1984). Thus, some of the extinctions recorded in the Late Cretaceous may have occurred irrespective of the effects produced by an asteroid impact. As noted by Archibald and Clemens (1982), "the . . . store of data indicates that the biotic changes that occurred before, at, and following the Cretaceous–Tertiary transition were cumulative and gradual and not the result of a single catastrophic event." Yet this interpretation does not negate the possibility that an asteroid impact may have occurred and produced a catastrophic event that resulted in extinction of taxa that might otherwise have survived the reputed climatic changes.

In the Late Pleistocene of North America the "large mammal taxa did not simply vanish from a scene of ecological composure, but instead disappeared during a time of great ecological ferment when biotas were adjusting, dissolving, and reforming under new-climatic parameters to emerge into the Holocene in greatly altered aspect. All this ecological fermentation served as a backdrop behind the spread and increasing technological sophistication of man-the-hunter. Hunting pressure becomes just one of many interrelated 'causes' acting in concert to drive the species to extinction, none of which could be singled out as the ultimate, or even the primary cause" (Guilday, 1984).

8.8.3 Final note

Many aspects of our current understanding of extinction in the fossil record have been gleaned from knowledge of marine animals. The data set of these organisms encompasses all of Phanerozoic time and the patterns they display have come to be regarded as a 'standard' for all life. Recent studies of land plants and non-marine tetrapods display extinction patterns that contradict, in part, those of marine animals (Fig. 8.1a–c). The three data sets are thus not totally in agreement and it is becoming evident that all life on earth may not have 'danced to the same tune' as did the marine taxa. Had land plant or non-marine tetrapod data sets been analysed initially in the same detail as marine animals, then our perception of extinctions during Phanerozoic time may have been different from what it is today. Contemporary workers must be attuned to this feature and receptive to patterns that do not simply 'confirm' conventional wisdom about Phanerozoic extinctions based on the 'standard' of the marine record.

ACKNOWLEDGEMENTS

I thank W. Alvarez, J. Betancourt, J. Brown, A. Burgess, R. Craw, G. H. Curtis, J. Diamond, K. W. Flessa, P. S. Giller, D. Goujet, D. Grayson, C. Lindquist, J. Lynch, P. S. Martin, A. A. Myers, C. Otts, C. Patterson, B. Rosen, P. Taylor, and S. D. Webb for helpful discussions and/or comments on various aspects of this work.

CHAPTER 9

Ecological interactions

T. W. SCHOENER

9.1 INTRODUCTION

The central objective of ecology is often stated as understanding the distribution and abundance of organisms. As such, ecology can be seen to have an intimate association with biogeography. Yet ecology by definition deals with processes in ecological time, i.e. in the order of one to one hundred generations or years. This time-scale is shorter term than that of interest to most biogeographers, and in fact most of the other chapters in Part III deal with longer term phenomena. Nonetheless, it is often plausible to view such long-term phenomena as the sequential action of short-term ecological processes. For example, analyses of differences between sympatric species, classically the province of community ecology, are in fact equivalent to analysis of the degree of overlap of geographic distributions according to the ecological properties of their species. Thus an understanding of how communities are structured and function is important in biogeographical studies.

The ecological processes relevant to determining species distributions can be given the following classification. Firstly, physical processes, those resulting primarily from the action of weather, geologic, hydrodynamic and similar factors, can be distinguished from biological processes, those resulting primarily from the action of organisms. Biological processes can be labelled intraspecific (effected by members of the same species), and interspecific (effected by members of other species).

Ecologists investigate the relative importance of physical and biological processes, and examine intraspecific versus interspecific processes (Schoener, 1987). The importance of the latter has, in particular, engendered debate, a debate that at times has been so acrimonious that it has frightened some away from the field and evoked a wonderment in others as to what could possibly be so controversial. We are not yet extricated from this imbroglio, and many of the controversies concern the fundamental variable of biogeography, the species distribution. Hence this chapter must meet the controversies head-on, at least in places. I hope that curiosity about how things have become so extreme and about how this antagonistic era might be terminated will sustain the reader through what will at times seem tedious and convoluted material; I also hope that participants in the controversies will realize that only simplification can lead to intelligibility in a book of this sort.

9.2 BACKGROUND

9.2.1 Kinds of ecological interactions

The principal kinds of ecological interactions between species are competition, predation and mutualism. All can be defined as population phenomena, as follows: determine the effect of species 1 on the population growth rate and/or population size of species 2; is it positive (+) or negative (−)? Do the same for the effect of species 2 on 1. Then examine the pair of signs so produced. If it is (− −), competition occurs. If it is (+ −), predation occurs (with + being the predator population). If it is (+ +), mutualism occurs. In addition to these definitions, mechanistic definitions have also been proposed, as follows.

Competition has been traditionally divided into exploitative, involving the deleterious effects of resource depletion and thereby being indirect, and interference, involving a variety of direct effects, including fighting with, injuring and even consuming individuals of the other species (e.g. Park, 1962). Schoener (1983a) proposed an alternative taxonomy. In this scheme, consumptive competition designates consumption of resources that deprives other individuals of those resources, whereas pre-emptive competition occurs when a unit of space is passively occupied by an individual, thereby causing other individuals not to occupy that space before the original occupant disappears. Four kinds of purely interference competition are also distinguished: encounter, chemical, territorial and overgrowth.

Predation can also be subdivided (Toft, 1986). Predation proper occurs when an organism kills and eats an animal. Herbivory occurs when an organism eats a plant or plant part. Parasitism occurs when an organism eats a part of another organism but does not consume the entire organism. Obviously, by this latter definition herbivory and parasitism overlap (Toft, 1986), and in older works parasitism has been restricted to the case where the organism being affected is an animal.

Mutualism has not undergone similar mechanistic subdivision, but can be divided into facultative (one or more species unnecessary for persistence) and obligatory (one or more species necessary for persistence).

We might require that both the 'sign' and 'mechanistic' criteria be met in order for an interaction to be identified (see especially the recent paper by Abrams (1987)). Where there is mechanistic ambiguity, as for example were individuals of one species to sometimes consume individuals of another – e.g. flour beetles (Park, 1962), lizards (Schoener *et al.*, 1982) – the interaction is called competition if the weight of all effects of species 2 on species 1 is negative (so the population effect is −−) and called predation otherwise.

This simple scheme is subject to the following complications. Firstly, in practice reliable information on only one of the populations may exist, so that the other's sign must be inferred. For example, suppose in an experiment we show that species 1 affects species 2 negatively, but we have no corresponding reverse experiment. Then the relation could, by the above logic, be predation or competition. By convention, we usually call the relation competition if the species appear to be on the same trophic level and predation if they appear to be on adjacent trophic levels.

Secondly, the above definition considers only pairwise combinations. When more than two species exist in a system, and some species are linked indirectly (say species 1 affects species 2, and species 2 affects species 3), then direct effects of one species on another need to be distinguished from indirect effects via intermediary species: indirect effects may also be considered to include physical factors acting as 'species' (Schoener, 1987). An intriguing indirect effect is Holt's (1977) 'apparent competition', in which, because of a reduction in abundance of prey species A, a predator species decreases in abundance and thereby leads to an increase in the abundance of prey species B. Unless otherwise designated, we shall use the 'direct' definition here, with the exception that competition for resources that are themselves species, which is really an indirect effect, will be considered as simple competition along with the other direct effects.

Thirdly, terms like 'competition' and 'predation' are sometimes used to describe effects on individuals rather than populations. Usually the two would be equivalent when all effects on apparently reproductive individuals are determined. But often they are not; e.g. one might have information on how removal of a species affects the somatic growth rate of individuals of another species; this may or may not imply an effect on the population of the latter. Because such information is so common, it is probably best to extend our definition of interactions to individual effects. Thus, for example, competition for resources is broadened to include the situation in which an individual could have produced more offspring had other individuals not used the resources needed for this additional reproduction (Schoener, 1977).

Of the three principal kinds of interactions, competition and predation have received by far the most attention, and both have been repeatedly implicated in experimental studies of particular systems (see

Section 9.2.2). Mutualism is less well-studied; pollination appears to be the most common direct mutualism, but other types may be more common than the data presently indicate (Boucher, 1985). Other types of interactions, in which 0 (no effect) is substituted for one or the other species' sign, are sometimes distinguished. For example (+0) is called commensalism. Consideration of zero-type interactions requires us to specify the degree to which an effect is so small that it is classified as zero, a specification that in fact is generally not done in practice. We do not consider zero-type interactions further here. The possible roles of species interactions over the longer term as potential causes of extinction are discussed in Chapter 8.

9.2.2 Ascertaining the existence and nature of ecological interactions in the short term

Typically, biogeographers frame hypotheses that cannot readily be tested using experimental manipulation. Indeed, because of the necessarily observational nature of most biogeographic investigations, some feel that too many unfalsified possibilities exist for ever achieving unambiguous explanation of most biogeographic phenomena. This hard-line opinion would, of course, place biogeographical research in an inferior position with respect to other scientific endeavours, a rather dismal and in my opinion unwarranted view (see also Chapter 2). On the other hand, because of the inferential nature of many biogeographic studies, especially with regard to ecological interactions, we must establish that such interactions do occur in places where they are expected and can be investigated directly. Fortunately, many experiments have now been done in natural systems that *in toto* serve this function. Recent reviews of literature exist for field experiments on interspecific competition (Connell, 1983; Schoener, 1983a) and predation (Sih *et al.*, 1985).

The upshot of these reviews is that competition is found very often where it is looked for; differences between conclusions are mostly quantitative and result from whether the whole (Schoener) or a selected (Connell) literature was surveyed, whether (Connell) changes in population density were used as the primary criterion for designation of competition and whether (Connell) one-species-one-experiment studies were excluded (Schoener, 1985; Ferson *et al.*, 1986). Experiments detect predation even more frequently than competition, and in comparisons the effect of predation is the greater. It can plausibly be argued that systems expected to show an interaction are differentially studied (Connell, 1983; Schoener, 1985), so interactions are more likely to be found. On the other hand, because of their laborious nature, field experiments often have low power (Schoener, 1986d), so that figures for the systems examined might in fact be higher were power greater. Whatever the case, the experiments have shown conclusively that competition and predation often occur in systems where they are expected *a priori*, and that these systems span (but not equally) most habitat types and kinds of organisms. The fact that competition or predation is not found all the time is quite in accordance with ecological theory, which, for example, expects size (Connell, 1975; Schoener, 1974c), trophic position (Hairston *et al.*, 1960; Menge and Sutherland, 1976) and degree of niche separation (Roughgarden, 1979; Pacala and Roughgarden, 1985) to affect the intensity of competition. In short, we certainly need not be defensive about hypothesizing species interactions as causes of biogeographic patterns, although of course as precise an evaluation as possible needs to be made in each case. How this can be done is discussed in Sections 9.2.4 and 9.2.5.

9.2.3 The theory of ecological interactions

This section explores how, in theory, species are expected to coexist or not, in some broad area, in the face of ecological interactions.

Population dynamics of species in communities

Competition. The simplest approach to modelling the effects of competition is that which uses differential or difference equations such as the Lotka–Volterra model or its alternatives (Hassell and Comins, 1976; Schoener, 1976a). The two-species version of the Lotka–Volterra model gives three kinds of outcome: (1) the species coexist at a fixed 'equilibrium' value; (2) one species competitively exterminates the other, the identity of that species depending on initial population sizes of the two species (a 'priority effect' or 'saddle'); and (3) one species is dominant and always competitively exterminates the other. Alternative models can give more complicated outcomes, including several equilibria rather than a single one. Lotka–Volterra competition between three or more species is also more complicated (Strobeck, 1973) and includes the possibility of limit cycles (the species cycle in abundance in a pattern that repeats itself with time) or other cyclical behaviour (Gilpin, 1975).

Potentially, these models might seem a good basis for understanding the distribution of interacting species over space: species disappear in units of space where competitive extermination occurs and, where coexistence occurs, abundances diminish as a function of the degree to which they are reduced by competing species. Roughgarden (1979) has exploited this approach, but as he notes, it is exact only with the assumption of zero migration.

When migration is substantial, things are not so simple. The models just discussed assume spatial homogeneity and a closed system, i.e. the entities of interest interact as if in a well-stirred soup. While islands may often approach these assumptions, gradients of intersecting species ranges over continental scales would seem far removed from them. What is the effect of migration into and out of the arenas of competition? Theoretical work on this question is limited and is mostly quite recent. Two conceptualizations predominate, that picturing space as divided into patches and that picturing space as a continuous gradient.

One approach, using patches, keeps track of the abundances of each species in each patch. In simple two-species-two-patch competition models with priority effects, migration mitigates the effect of competition (Levin, 1974); this kind of result can be generalized to any number of patches and species (Levin, 1978), providing a basis for distributional heterogeneity of species as an equilibrium condition. When migration is modelled simply as an input of immigrants into a closed system (Schoener, 1976a), mitigation of competition again results. Yodzis (1986), using a related approach, shows that maximum diversity increases when priority effects, as opposed to dominance, are more probable.

A second approach using patches has as variables the fraction of patches containing particular species (Cohen, 1970), but technical flaws in some models (Levins and Culver, 1971), have marred their initial development (as pointed out by Slatkin, 1974 and Levin, 1974). Their main result is that coexistence can be facilitated by the environment being broken up into patches; in particular, a species inferior in competition but superior in migration can coexist with the reverse kind of species provided the rate at which patches become empty via extinction is sufficiently high.

Slatkin (1974) derived other, less intuitive conclusions. First, unlike the spatially homogeneous Lotka–Volterra system, no priority effect can exist, so that species cannot exterminate one another by virtue of an initial preponderance in abundance of one or the other species in the total set of patches. Second, species too similar in their migratory and competitive properties cannot exclude one another.

Both these results have been shown to be sensitive to certain assumptions (Hanski, 1981, 1983); most importantly, when the time-scale of local population processes (births and deaths) is fast relative to that of regional changes (extinctions and immigrations for patches), the original conclusions apply, whereas if not, they must be broadened. Hanski argues that the original class of models apply mostly to 'fugitive' species, and for more permanently occupied patches, which are those closer to the interests of biogeographers, the 'comparable-time-scale' theory is more appropriate. To demonstrate the plausibility of comparable time-scales, Hanski (1982) collected data on extinction rates of species averaged over all patches as related to the fraction of patches occupied. A negative correlation was generally found, presumably because the latter variable is positively correlated with mean population size – large population sizes reduce extinction probabilities (Toft and Schoener, 1983; Schoener, 1986c).

Using the theoretical machinery on patches just discussed, Hastings (1987) has shown that fugitive-type systems cannot produce very strong negative associations between species even with moderately strong competition.

The second predominant conceptualization deals with competition along continuous spatial gradients in which migration is a key feature. Shigesada *et al.* (1979) considered species for which undirected migration increases with the number of individuals of the same and different species ('repulsive forces'), and for which directed migration increases as favourable habitat elsewhere increases. When interspecific repulsion is high, non-overlapping distributions result. Shigesada *et al.* (1979) also analysed the continuous analogue of priority-effect competition which Levin (1974) analysed for discrete patches. Results were similar: with suitable parameter values, migration was shown to stabilize an unstable competitive situation. Shigesada and Roughgarden (1982) extended this approach and pointed out the similarity of spatial gradient analysis to analysis of resource utilizations. Theoretical machinery transfers from one to the other when dispersal rate is high relative to population growth rate, so that one can assume that movements of individuals will allow them to select the most favourable positions along a habitat-axis dimension, even when the latter does not have a continuous distribution in real space. Both types of migration incorporated into the model of Shigesada *et al.* (1979) favour coexistence in this situation.

Pacala and Roughgarden (1982) modelled continuous movement toward and through an environment consisting of two habitats; invasion of the environment succeeds or not depending on the size of the 'suitable' habitat, defined as a habitat that can be invaded in the absence of migration effects. In this theoretical analysis, migration can allow a species to invade the environment even if both habitat types cannot be invaded in the absence of migration effects, and migration can prevent invasion even if both habitat types can be invaded in the absence of migration effects. Dispersal of the resident causes the first effect, dispersal of the invader causes the second effect. In these models, dispersal is in part a cost, in that it weakens the ability of a species to persist where it already occurs.

The niche overlap approach is taken further by Brew (1984), who examines competition in 'coarse-grained' habitats, defined as habitats whose encounter rates are under the control of individuals through their active choice (see also Rosenzweig, 1979). The approach allows a variety of models to be used for the actual mechanics of competition, and like the model of Shigesada and Roughgarden (1982), assumes that individual responses are rapid enough, relative to population processes, that favourable positions along spatial gradients can be selected effectively instantaneously. A major result is that ecological separation along coarse-grained dimensions (macrohabitat) implies strong competition, whereas the opposite holds for fine-grained dimensions (e.g. prey size). These results agree quite nicely both with previous theoretical intuition and with the outcome of field experiments (Schoener, 1974a, 1983a). Except for the applications discussed by Brew, the theory of competition along continuous gradients has received little testing so far.

Predation. Population-dynamics models of predation have a similar distribution of assumptions to those of competition. The most sensible closed-system, two-species model is one in which prey grow logistically in the absence of predation and in which a predator's consumption ability saturates with increasing numbers of prey. Such models give a stable equilibrium point or a limit cycle, depending upon parameter values (Roughgarden, 1979). Thus in predation models, unlike competition models, cyclical coexistence is possible (and likely) with two species. Models having three levels, with two predator-prey pairs, also exist. Ludwig *et al.* (1978) present a model of birds, herbivorous insects and trees which gives two stable equilibria, one at a high herbivore level (an outbreak) and one at a low herbivore level.

As in the case of competition, models of predation in systems with spatial components exist for both patchy and continuous environments. A nearly universal trend in such models (but see Allen, 1975; Hilborn, 1975) is that explicit inclusion of space stabilizes predator–prey interactions that might be unstable in the simple models just discussed (McMurtrie, 1978; Hastings, 1978; Levin, 1978). For example, Hastings (1978) constructed a model whose variables are the fraction of patches occupied by predators, prey, or neither; he concluded that spatial fragmentation is a stabilizing factor given sufficiently high dispersal in both predator and prey. In addition to greater persistence of species, dispersal can reduce oscillations in predator–prey models (McMurtrie, 1978).

Mutualism. Development of population-dynamics models for mutualism is more inchoate than that for competition or predation (Boucher, 1985). Models are mostly non-mechanistic, and results turn *inter alia* on whether mutualism is facultative or obligatory. Spatial heterogeneity models for mutualism do not seem to exist.

Niche dynamics of species in communities

The preceding section is limited to population-dynamics models, whose output is the set of species persisting in the system and their abundances or frequencies of occurrence. Another class of models characterizes the system by the ecological niches of its component species. Coevolutionary versions of such models follow changes in traits of particular species that are produced by the interaction itself acting as a selective force. A typical trait is the size of a trophic structure, e.g. bill length in birds, which can be correlated with the prey-size distribution (a one-dimensional version of the niche).

Community patterns in such traits have long been of interest to ecologists, especially with regard to whether or not they reflect the action of competition (Schoener, 1986a). These concepts are much less developed for the action of predators (Ricklefs and O'Rourke, 1975) and mutualists (Templeton and Gilbert, 1985), so I concentrate here on competition.

The principal issue has been the minimal degree of similarity, or limiting similarity, that species can have and still coexist. Models of limiting similarity have been constructed for ecological time, i.e. when limiting similarity is exceeded species cannot invade a community or cannot persist there in the face of invasion, and for evolutionary time, i.e. species can change their traits by evolution, possibly allowing them to persist in the community. Limiting similarity is measured in units of d/w, where d and w are properties of the ecological niche, as follows. Define the niche as the utilization distribution of a particular species population over one or more resource dimensions (Schoener, 1986a). For one dimension, d and w, assumed constant, are the distance between peaks of adjacent utilizations and the standard deviation of the utilization, respectively.

Initial models (MacArthur and Levins, 1967; May and MacArthur, 1972) suggested a limiting similarity of $d/w \sim 1$. However, the value varied in subsequent ecological models as a function *inter alia* of niche shape and how competition and environmental variation were incorporated (Roughgarden, 1974a; Roughgarden and Feldman, 1975; Turelli, 1981). Certain evolutionary models (e.g. Rummel and Roughgarden, 1983; Taper and Case, 1985) but not others (Abrams, 1986) generally predict larger values for d/w than ecological models (see Section 9.2.4). The upshot of all this theory is that, while a limiting similarity should exist for any particular case, its value will vary between cases, and this value may be very small (Abrams, 1983). In more biological terms, the effect of competition should be to produce a minimal difference between niches (and by implication, correlated morphological traits) within a community, but its value may well be specific to types of species or communities.

Niche dynamics within particular species

The above discussion centres on collective properties of species in communities, for example, minimum size differences between adjacent species on a resource dimension. Another approach is to conceptualize changes in traits of a particular species in communities having different sets of competitors or predators. The traits can be morphological (e.g. bill size) or behavioural (e.g. habitat, prey types selected), and we can consider the population mean, variance or other properties of such traits, as well as (for behavioural traits) the individual mean and variance. Theoretical treatment of these topics can be arranged under three headings.

Character displacement. This is the situation in which two species differ more in some character where they overlap geographically than where they are allopatric (Brown and Wilson, 1956). Certain models of coevolution discussed in the last section are actually models of character displacement, such as Roughgarden's (1976) model, allowing only niche position to evolve, and Slatkin's (1980) model, allowing both niche position and between-phenotype variation to evolve. Case's (1982) model allowed for niche position and within-phenotype variation to evolve. This was extended by Taper and Case (1985) to allow the evolution of between-phenotype variation; moreover, resource dynamics are incorporated explicitly, rather than assuming static resources, as the earliest models do. In outputs from this later model, divergence frequently occurs, in contrast to Slatkin's (1980) model (Matessi and Jayakar, 1981). The model also predicts that d/w (while remaining large) should decrease as species number increases, that within-phenotype variation should be larger than between-phenotype variation, and that resource utilization separation should be smaller than trait separation. All predictions are consistent with available data. Milligan (1985), in a very general analysis, points out that divergence is more likely in models where species utilizations match the resource spectrum relatively imprecisely (Roughgarden, 1976); but not Slatkin (1980); the latter analysis also points to biological differences between species in resource exploitation favouring character divergence. Finally, a recent model by Abrams (1986) also explicitly includes resource dynamics and additionally includes other density-dependent processes. Convergent changes are achieved more easily than in most previous treatments, and the situation in which *both* species become more similar in sympatry than in allopatry can result. Abrams' paper also contains an extensive review of earlier theory.

Habitat shift. Habitat shift occurs when utilizations shift along a habitat-niche dimension, such as feeding height. Obviously, limiting-similarity and character-displacement models apply here, although details may vary. Models in behavioural time have also been used to describe habitat shift, namely the cluster of effects known as the compression hypothesis (MacArthur and Wilson, 1967; Schoener, 1974b). This predicts that habitat utilizations will shift away from one another in the initial, behavioural phase of two species becoming sympatric, which may represent a net contraction of the range of habitat types utilized. I also predict that in contrast prey types will not diverge, and the range of prey types taken will typically expand. Notice the difference between this model and coevolutionary models, which predict that both prey types and habitat types should diverge over evolutionary time. Schoener *et al.* (1979) modified this conclusion for species whose ranges overlap only along their boundaries; then the behavioural predictions should apply over evolutionary time, because migration should keep species traits in their original, non-coevolved state.

Niche breadth. Utilizations such as those describing prey size or foraging height can be characterized by their variances as well as means, and this can give some measure of niche breadth. A similar characterization applies to corresponding morphological traits. Much theoretical treatment of these variances is discussed under the previous two headings and in Roughgarden (1972).

9.2.4 Methods of analysing distributions of species at the community level

The simplest definition of an ecological community is a set of species in some place. Limits on the species and the place are, in this definition, both arbitrary. The sort of theory developed in the last section deals with sets of ecologically similar, sympatric species. Such species sets can be called 'guild' communities (Root, 1967; species consume the same general kinds of resources) or 'similia communities' (Schoener, 1986b; species have various ecological similarities, including consumption of the same general kinds of resources).

Considerable methodology has recently been developed to assess the degree to which such communities are structured, and in particular, the potential role of ecological interactions. Because such communities represent the collection of individual species whose distributions overlap at some location, they can be considered point samples of species distributions with respect to one another. Analysis of the properties of such communities is thus analysis of the degree of overlap of species distribution, the basic biogeographic unit. Such properties will have important implications in the establishment success of new species in the community (Chapter 1) and will be pertinent to discussion of speciation (Chapter 7) and extinction (Chapter 8).

In 1959 Hutchinson listed seven sets of sympatric congeneric species ranked according to size of some trophic structure, in which the ratio of the larger to smaller member of adjacent species tended toward a constant value (about 1.3). Hutchinson's data and certain subsequent data (Schoener, 1986a) suggested that trophically similar species in the same place may be evenly spaced on a logarithmic scale along certain niche dimensions (such as prey size), perhaps minimally separated as would be expected from limiting similarity ideas. In turn, the inference might be made that this pattern resulted from interspecific competition thereby affecting the number of species that can coexist within an area.

Are these patterns deterministic and how does one evaluate them statistically? Clearly one method would be to test whether the empirical mean ratio is significantly different from 1.3, using a one-tailed *t*-test or some equivalent. Critics might caution, however, that the value 1.3 is itself induced from the data, so that a one-tailed expectation is invalid, i.e. 1.3 is not really 'fixed'. The evaluators, however, might counter that they are really testing a limiting-similarity model, in which 1.3 is the theoretically predicted ratio. Critics might counter again, however: theory predicts values of limiting similarity that vary widely with biological details of the system (see p. 263), and these have not yet been established.

Partly to avoid such entanglements, the 'null' or 'neutral' (Caswell, 1976) model, which in its idealized form is independent of any particular theory and free of biological assumptions, was proposed. Roughly following Harvey *et al.* (1983), such a model is a logical or mathematical device that yields expectations about some characteristic for the situation in which one or more specified processes or phenomena are absent. A venerable example of a null model in population biology is the Hardy–Weinberg equilibrium; this model gives genotype frequencies in the absence of selection, mutation, gene flow, genetic drift and assortative mating. A null-model approach to Hutchinson's size data would ideally specify the ratios expected in the absence of interspecific competition, or more informally and more generally, were the community structured 'at random.'

An example of a null-model treatment of size data, applied to the Galapagos finches (which comprised part of Hutchinson's data set) is that of Strong *et al.* (1979). The general method is to combine all taxa, such as species, in all communities together to form a species pool, then select taxa 'at random' from this pool, using rules that preserve certain features of the biological situation (for example, the same species cannot occur twice at a site). Real communities are then compared to random ones by conventional statistical tests that ask, for example, whether real communities have significantly larger size ratios. This method is, in my opinion, the best generic one for dealing with size data, yet criticisms of it are numerous (Sections 9.3.2 and 9.3.3). But critics of null models mostly do not wish to do away with the technique altogether. Rather than viewing null models as straightforward, impartial statistical tests, however, they view them as containing major assumptions that must be justified biologically (Schoener, 1982; Colwell and Winkler, 1984).

As discussed below (Section 9.3), null models have been applied to numerous community characteristics, including species lists, species sizes, species habitats, and species-to-genus ratios. The above viewpoint will perhaps bring more perspective to what are often very heated arguments about the applicability of null models.

9.2.5 Methods to detect interactions by analysing given species in various portions of their ranges

Tests for changes in traits of individual species from one place to another are much more straightforward than those just described for community characteristics, because only one species at a time is considered. For example, rather than asking whether all species in a community, taken together, have bill-size ratios significantly larger than expected, we ask whether a particular species has a larger bill in the presence than in the absence of another species. To analyse comparisons such as character displacement or habitat shift, a simple t-test or non-parametric analogue for a central tendency variable is appropriate. To test hypotheses about niche breadth, an F-test on variances or some analogue is appropriate.

Notice that this methodology, by definition, does not allow us to say anything about a 'typical' or 'representative' species; rather, a particular species is focused upon *a priori*. In order to combine data from different species to conclude something about a representative species, we need either to use a null-model approach or have an expectation of exactly which species are expected to show an effect and which not. For example, in a study of habitat shift in Caribbean *Anolis* lizards, Schoener (1975) compared the habitat of each species population occurring with a putatively competing species to that of each population occurring alone. The proportion of combinations where shift was in the expected direction was compared statistically for groups designated as expected competitors versus groups designated as expected non-competitors. Care was taken to eliminate non-independent comparisons. This procedure allowed hypotheses such as "larger species are more likely to cause habitat shift" to be tested (Section 9.3.4).

The individual-species method is thus able to make substantial headway in testing alternative hypotheses for size and other changes. For example, species may differ in size in sympatry but not in allopatry for at least two reasons. One is that there is selection in sympatry for those individuals of each species different from individuals of another species, say because of competition. A second is that both species vary clinally (Grant, 1972), and that the variation is parallel, leading to greater

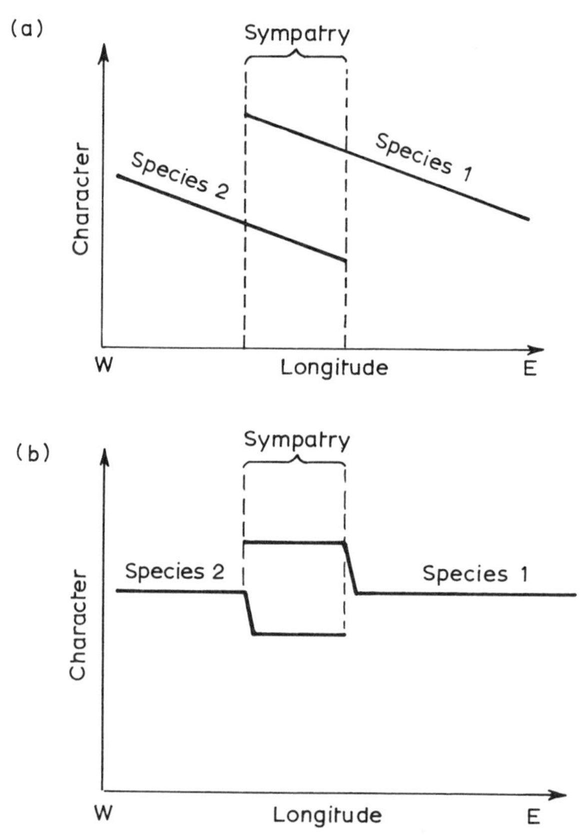

Figure 9.1 Alternative explanations of differences in sympatry. If two species have characters like size graded in space (a *cline*) then they may automatically have different characteristics in the zone of sympatry (a). If differences at sympatry are solely the result of character displacement, characters should be distributed in space as in (b). (Colinvaux, 1986, reprinted by permission of John Wiley and Sons, Inc.)

differences in sympatry than allopatry (Fig. 9.1). Techniques to discriminate such hypotheses range from fitting spatial regressions to mapped data, to more complex multivariate regression techniques (Dunham *et al.*, 1979). A second example concerns habitat shifts, for which differences in habitat availabilities may have to be factored out before the competition hypothesis can be evaluated (Schoener, 1975).

9.3 COMMUNITY CHARACTERISTICS

9.3.1 Number of species

General introduction

The number of species in a community has always been of great interest to biologists. Various trends in species number, such as those related to latitude, area and isolation, are covered in detail elsewhere (Chapters 3, 4 and 15). Here I focus on how species interactions are expected to affect those trends.

Latitudinal diversity gradients

It is well established that in many taxa, the number of species increases, often precipitously, toward the tropics, and this phenomenon doubtless has multifactorial explanations (Pianka, 1983; Chapter 3). Of most direct concern to us here are the predation and competition hypotheses.

The predation hypothesis assumes that predation is greater in tropical than temperate zones. Based on the work of Paine (1966) and others, predation is assumed to increase the number of prey species, primarily by reducing the density of competitive dominants amongst the prey species. As is illustrated by Lubchenco's (1978) data on marine algae and their molluscan herbivores, very severe predation reverses the effect. That an intermediate amount of predation increases species number has also been suggested by certain theoretical models (Hastings, 1978). Few data are available to test the assumptions of the hypotheses in relation to latitudinal diversity gradients; however, Haigh and Orians (MacArthur, 1969), in a study of a latitudinally wide-ranging species, the red-winged blackbird (*Agelaius phoeniceus*), showed that predation on nestlings was greater in Costa Rica than in Washington state.

The competition hypothesis assumes that competition is greater in the tropics and predicts that this should lead to greater species numbers there. Unfortunately, neither the assumption nor the conclusion is supported by much evidence. While it is plausible that competition might be driven upwards in the tropics because greater climatic stability allows populations to reach carrying capacity more often, a greater predation rate would counteract this effect. Almost no field experiments on competition (Schoener, 1983a) or predation (Sih *et al.*, 1985) have been done in tropical regions, so comparisons are premature at this time. In addition, no theory apparently exists that predicts a greater species diversity over ecological time with increasing competition except Yodzis's (1986) work, which predicts increasing diversity with increasing number of priority effects. In standard models, severe competition exterminates species (Section 9.2.3). The competition hypothesis should be laid to rest or greatly modified. In contrast, the idea that tropical species can be more specialized, so that w is reduced and species can be more tightly packed, is plausible to the extent that tropical stability has allowed such specialization.

Another hypothesis, greater structural heterogeneity in the tropics, is in part related to the importance of species interactions, as follows. A greater number of prey species can lead directly to a greater number of predator species by allowing more taxon specialists, or indirectly by allowing more structural variation in prey characteristics such as size, thereby increasing the width of the available resource spectrum. Arnold's (1972) study of snakes exemplifies the first possibility: in a multivariate analysis of snake species number as a function of latitude and of prey species number, he found the latter to be a better predictor than the former, and latitude dropped out of a multivariate analysis. Insect sizes and insectivorous bird species number illustrate the second possibility: tropical arthropods have a greater variance in size than temperate arthropods (Schoener and Janzen, 1968), and tropical insectivorous birds overall show a comparably greater variance in bill length, mostly due to the existence of large-billed species (Schoener, 1971).

Species–area and species–isolation relationships

Attempts to explain the often very strong relation of number of species to the area containing those species (the species–area relationship) vary widely (Williamson, 1981; Chapter 4); some believe many of its properties are statistical artifacts, while others have proposed a highly deterministic causal structure. Adopting the second attitude, Schoener (1976b; Wright, 1981) incorporated competition explicitly into a derivation of a species–area relationship, and found that it depressed the slope of the species–area curve, i.e. it diminished the steepness of the relation. Moreover, as would be expected from a model such as MacArthur and Wilson's (1967), with a finite number of species in the source, P, the slope is not constant but asymptotes: for the interactive version, the slope z ranges between 0.5 and 0 and depends only on $\hat{S}/P$, where $\hat{S}$ is the equilibrium number of species. This model has not been explicitly tested very often, but when it has, agreement has been extremely good, e.g. for birds (Rusterholz and Howe 1979; Martin 1980, 1981a,b).

Theoretical models of how predation might affect the form of the species–area curve do not exist. However, observations on lizards preyed upon by rats (Whitaker, 1973) and orb spiders preyed upon by lizards (Toft and Schoener, 1983; Schoener, 1986c) show that the species–area intercept is substantially lower in the presence of predators. Moreover, scatter in the species–area relationship is substantial in both cases; this is to be expected when predation reduces population sizes of most species of prey, as stochastic extinction of the latter is then more likely. The role of predators (including man) in extinction processes is further discussed in Chapter 8.

Competition and other interactions have not been explicitly incorporated into species–isolation models, and perhaps it is generally sensible not to do so. A survey by Schoener (1986c) contrasts kinds of species more likely to be controlled by dispersal factors with those more likely to be controlled by factors on recipient islands such as resource abundance. Arthropods, which fall more often into the first group, show relatively stronger isolation effects and relatively weaker area and habitat effects than do vertebrates, which fall more often into the second group.

9.3.2 Species lists

Diamond (1975) proposed a set of 'assembly rules', outlining a rather deterministic structure for the birds of the Bismarck Archipelago, a group of islands adjacent to New Guinea. These rules imply that only a subset of possible species combinations can occur; the biogeographic implications being that not all range expansions and introductions are possible. Among other things, Diamond found that certain combinations of species – e.g. cuckoo-doves, gleaning flycatchers – occurred on the islands much less than expected 'by chance'. These data were argued to implicate competition in determining these species' distributions. Diamond's method for analysing such 'forbidden' combinations was to treat each guild separately and compute the frequency of particular co-occurrence combinations if the species were randomly distributed over the islands and the number if each of a set of types of islands containing species was fixed and so on (the number of each of a set of types of islands containing species A was fixed, and so on). While this analysis is correct as far as it goes (Section 9.5.2), the problem is that many possible species sets could have been examined, and for any significance level, a certain number of these are expected to show non-random, negative associations 'by chance'. Thus the guild-by-guild analysis may not hold overall.

Connor and Simberloff (1979) criticized the method and used null models to test (for all species from some taxon and archipelago taken two or three at a time) whether or not a greater proportion of combinations existed than would be expected. They concluded that the forbidden combination rule describes a situation "which would for the most part be found even if species were randomly distributed on islands".

This analysis, in turn, engendered substantial counter-rebuttal from Diamond and Gilpin (1982; Gilpin and Diamond, 1982). They pointed out among other things that Connor and Simberloff's compilations for West Indian birds and bats (but not New Hebridean birds) supported more species being negatively associated than expected by chance. Despite superficially similar observed and expected values, such tendencies were statistically significant, with P values ranging down to $< 10^{-8}$! For further debate see Strong *et al*. (1984).

Gilpin and Diamond (1982) proposed their own method of analysing a collection of communities, which centres on degree of association between pairs of species. The observed and expected distributions of such associations for the Bismarck Archipelago turned out to be significantly different ($P < 10^{-8}$) due mostly to an excess of positively associated species but also to an excess of negatively associated species. New Hebridean birds showed only the former excess. The former would be expected biologically were there certain types of habitats available in greater-than-threshold abundances on some islands but not others; islands with ponds, for example, can have ducks, but those without ponds typically do not. This illustrates the importance of correcting for habitat differences whenever possible before doing analyses of species co-occurrences as well as habitat and other niche shifts (Sections 9.2.3 and 9.4.2). Gilpin and Diamond (1982) point out that, even for the negatively associated species, a variety of factors may determine their distributions (Section 9.5), e.g. preference for different sized islands, different invasion rates into the archipelago leading to incomplete allopatry, and interspecific competition. Similar multiple interpretations can be given to positive associations. A more detailed observational approach, or an experimental approach, is needed to discriminate between pairs of species. The observed and expected distributions of impossible for island bird species, the former would in most cases seem the more practical route.

Discussion so far has been limited to analyses of similarities and differences in species distributions. The inverse approach is to analyse similarities and differences in species lists of islands. A similar, although so far less extensive, set of arguments and counter arguments exists for island list analysis as for analysis of species occurrences (Connor and Simberloff, 1978; Wright and Biehl, 1982; Simberloff and Connor, 1984; Hamill and Wright, 1988). This methodology has some of the problems just reviewed for species occurrences and in addition has its own special problems. One of the latter is that islands may share the number of species expected from some null model, yet the species themselves may have quite non-random distribution. This suggests that island similarity may be too abstracted a quantity to be very useful for analysing species' relationships. In addition, the effect of interactions such as competition is sometimes counter-intuitive (Hamill and Wright, 1988).

In this section we saw how attempts to examine all the closely related species of interest in the area, rather than a single or few associations averted the *ad hoc* singling out of those cases having the most negative associations. We also saw, however, how inclusion of vast numbers of species, i.e. all birds, gave expected and observed distributions seemingly close to one another, even though the latter were sometimes statistically very distinct. The inclusion of species not expected to interact by any reasonable ecological premise has been labelled 'dilution' by Diamond and Gilpin (1982). Most participants in the controversy agree that dilution must be avoided, but exactly how is far from resolved. Careful formation of *a priori* hypotheses, including advance specification of ecologically similar species, is obviously desirable. Yet hypotheses are seldom framed in total ignorance of results, so that even this advance specification will be biased to some degree. Clearly, however, investigation of theoretically likely and unlikely cases for species interaction, specified beforehand, does have strong merits, and I illustrate this approach in Sections 9.4 and 9.5.

9.3.3 Species sizes

One characteristic that may determine which species can coexist in some regions is their difference in size. Two major approaches toward analysing size differences within communities dominate the recent literature.

One involves testing sets of species one set at a time, analysing, say, a single Galápagos finch community. This method (Simberloff and Boecklen, 1981) is substantially weaker than the second one and has so many statistical difficulties that it is not treated further here (Colwell and Winkler, 1984; Schoener, 1984; Tonkyn and Cole, 1986).

The second approach involves treating, in some aggregate way, sets of species whose size differences are of interest, say analysing all communities of Galápagos finches using the method of Strong *et al.*

(1979) discussed in Section 9.2.4. They concluded that "actual communities on individual Galápagos islands show no tendencies of character displacement, compared to null communities . . .". Both Hendrickson (1981) and Grant and Schluter (1984), however, were able to find more cases where sizes were significantly different from random expectation than did Strong *et al.*, although the number was far from overwhelming. After various other tests and evaluations, Simberloff (1984) concluded "On balance, there are aspects of Galápagos avifaunal taxonomy, morphology, and coexistence that are inconsistent with simple models of randomness and independence, and are consistent with competitive hypotheses, though for the latter two traits one must note that there are at least as many aspects that are consistent with random, independent models."

The Strong *et al.* (1979) method of generating null communities has been criticized on a variety of grounds, from conceptual through statistical to biological (Hendrickson, 1981; Case, 1983; Colwell and Winkler, 1984; Grant and Schluter, 1984; Schoener, 1984). One criticism is that some studies include species ecologically dissimilar in ways other than size, say in habitat, which is similar to the 'dilution' criticism discussed above. Another criticism (Colwell and Winkler, 1984), is that if past competition among the species of an archipelago were to have exterminated certain species too similar to the others, then random selection from the remaining, relatively different species would underestimate the role of competition (see also Case, 1983). Despite these criticisms, the Strong *et al.* procedure or some analogue is the most objective and powerful way known to analyse size differences in the species aggregate.

In this spirit, various studies like those of Strong *et al.* (1979) have been performed. Schoener (1984) tried to avoid the 'dilution' problem while retaining a sample size with reasonable statistical power, by forming all possible groups of 2–5 species from the world's accipiter-like hawks. Size ratios were calculated for each, and the resulting null distributions of ratios were compared to those of actual communities. Figure 9.2 shows how real ratios of pairs contrast with those expected under the null model, and in most cases real ratios were significantly larger. In a second study, using among other methods the Strong *et al.* procedure, Case *et al.* (1983) found greater than expected size differences between West Indian birds. A third, very recent study, by Eldridge and Johnson (1988), analysed North American shorebird data; again, real ratios were significantly larger than random. In addition, a striking tendency toward ratio constancy existed: as many as 8 unique species pairs have bill-size ratios between 1.2 and 1.3 (Section 9.2.4.)! Conclusions from all these studies are conservative, in that possible past competitive extermination cannot be taken into account,

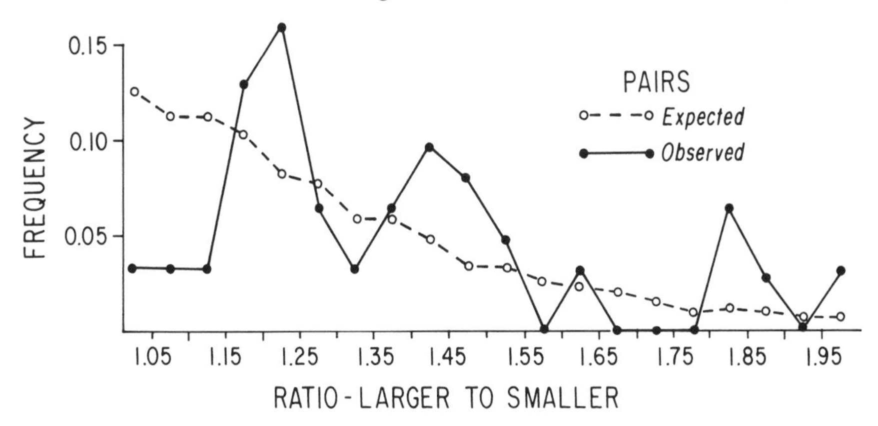

Figure 9.2 Expected and observed distributions of size ratios for pairs of *Accipiter* species. (Schoener, 1984.) (Reprinted by permission of Princeton University Press.)

yet size differences are still significantly larger than expected. In these studies, however, especially in the first and third, the likelihood of competitive exclusion of some species everywhere would seem relatively small, given the huge and often continuous areas involved.

The likely severity of the Colwell–Winkler effect was evaluated in a study on the *Anolis* lizards of the Lesser Antilles (Schoener, in press). On these islands, species pairs always have large ratios, so that only low power tests (Simberloff and Boecklen, 1981) could conclude that they are not significantly different from random, and indeed, the Strong *et al.* (1979) procedure concludes the opposite. The data are intriguing in another way, however. If, as postulated by Williams (1972), the ancestral source island of northern Lesser Antillean *Anolis* was Puerto Rico, we can then construct null communities from all Puerto Rican species taken two at a time and again compare these to actual Lesser Antillean communities. Interestingly, the mean ratio of the null distribution so produced is about the same as or actually greater than that from the Strong *et al.* procedure. The variance of the former distribution is always substantially greater: very large and very small Puerto Rican *Anolis*, which together generate very large size ratios, do not occur on the small islands at all, perhaps not so much because of their poorer dispersal abilities as because such specialists are not favoured there. In short, the Colwell–Winkler effect is counterbalanced by a tendency for extreme sized species not to occur on islands.

An additional possible problem with the ratio analysis of Strong *et al.* is that redundancy may exist when combining communities to make a statistical comparison (Schoener, 1984; Hopf *et al.*, in prep.). For example, a particular species may be very widespread and different in

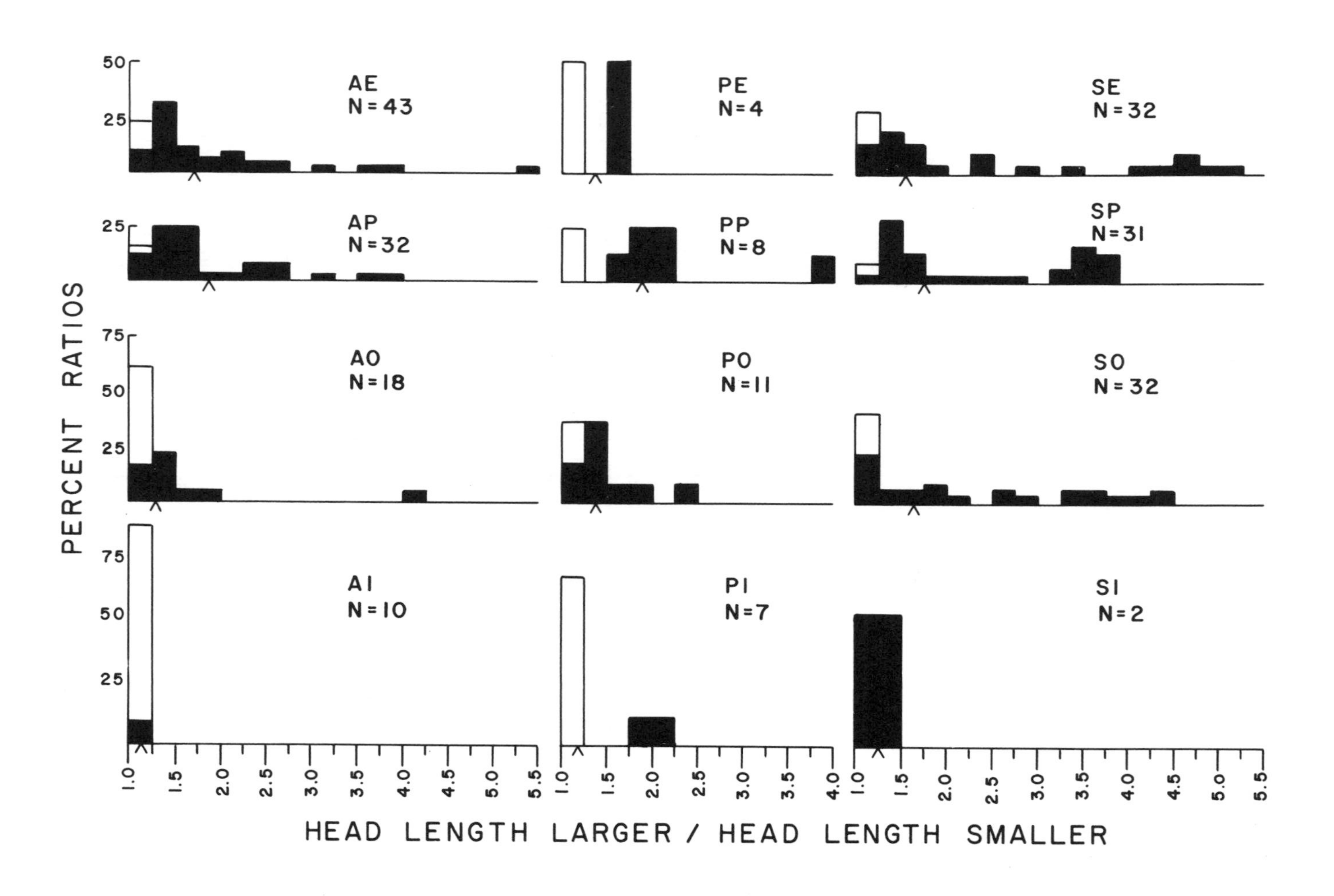

AE
N=43
PE
N=4
SE
N=32
AP
N=32
PP
N=8
SP
N=31
AO
N=18
PO
N=11
SO
N=32
AI
N=10
PI
N=7
SI
N=2
50
25
75
PERCENT RATIOS
1.0
1.5
2.0
2.5
3.0
3.5
4.0
4.5
5.0
5.5
HEAD LENGTH LARGER / HEAD LENGTH SMALLER

size from all the others. This species will repeatedly generate large size ratios, yet the reason for its widespread distribution may not be because it is so different in size from the other species that it can coexist with all of them (although, of course, it may be). Redundancies can be viewed as biologically important, in the sense that they might represent repeated 'trials' for persistence under competition. On the other hand, the 'trials' may not have been independent in reality, in which case the redundancies are biologically real and need to be dealt with statistically. Hopf *et al.* (in prep.) have proposed some novel methods to do this.

An alternative method to Strong *et al.* (1979) for examining the hypothesis that species whose distributions overlap are more different in size than are non-overlapping species de-emphasizes the species pool (Schoener, 1970). Bowers and Brown (1982) examined all possible pairs of rodents in three North American desert systems and classified each according to two properties: whether or not geographic ranges overlap and whether or not body-mass ratio exceeds 1.5. Geographic overlap was examined on both a local and regional scale, using a Monte Carlo simulation, in which a null distribution was generated *ad hoc* for each case. Results showed that granivorous species of similar size occurred together less than expected by chance, local differences were more pronounced than regional ones (presumably because the latter had pairs with substantial habitat segregation), and when all other rodents were combined with the granivorous ones, thus increasing 'dilution', the first tendency disappeared statistically.

In an analysis of *Anolis* lizards (Schoener, 1970), each pair of species was designated allopatric, parapatric or sympatric (see Chapter 7 for definitions), and each was also designated as having structural habitats (perch height and diameter, primarily) identical, largely overlapping, peripherally overlapping or discrete. This gave twelve categories, and frequency histograms of size ratios were computed for each (Fig. 9.3). It was first postulated that similarity in structural habitat implies similarity in size. To test this, species with little or no geographic overlap (and hence no competitive pressures) but varying degrees of structural habitat overlap were compared. The result was that the greater the structural habitat similarity, the greater the size similarity (read down the allopatric column in Fig. 9.3). Next, to show that species overlapping

Figure 9.3 Frequency histogram for ratios of head length (larger to smaller) for pairs of *Anolis* species on Cuba. Caret is median ratio. Two-letter abbreviations are as follows: First letter – A = allopatric, P = parapatric, S = sympatric; Second letter – E = structural habitat exclusive, P = structural habitat adjacent, O = structural habitat overlapping, I = structural habitat identical (Schoener, 1970). (Copyright 1970 The University of Chicago.)

in geographic range are more different in size from those not so overlapping, distributions for allopatric pairs similar in structural habitat were compared to distributions for sympatric pairs similar in structural habitat (read across the third row in Fig. 9.3). Again, the latter had larger ratios, as expected from competition. Such comparisons were mostly statistically significant, but significance was judged by chi-square tests, not as in Bowers and Brown (1982) by adjusting for partial dependence of pairs. The dependence problem here, however, is complex, as sizes were recomputed for each species for each case of range overlap, so that geographic changes in size were taken into account. Bowers and Brown, in contrast, always used the same size for a given species. It is unclear how to reduce dependence in the lizard data, but given Bowers and Brown's results, little reason exists to suppose that conclusions would be changed were some kind of Monte Carlo simulation attempted.

9.3.4 Species' habitats

Although, as we have seen, substantial attention has been given to null models of species' sizes within communities, virtually no parallel studies exist for species habitats. Again, the issue arises as to which species can have overlapping ranges, now considered as a function of habitat similarity. The only such study of which I am aware again deals with *Anolis* lizards (Schoener, in press). Satellite islands of the Greater Antilles contain up to four species, and each species can be identified with a species from the hypothesized source island, one of the Greater Antilles. Each species can be placed into one of seven major structural habitat categories. Using the classification of Rand and Williams (1969) and Williams (1983), the number of species in each structural habitat category can be determined for each source island (Table 9.1). Assuming

Table 9.1. Number of species of *Anolis* lizard in various structural habitat categories for the three largest islands of the Greater Antilles (Schoener, in prep.)

	Cuba		*Hispaniola*		*Puerto Rico*
Trunk-crown	2	7	4	7	2
Giant	5		3		1
Twig	3		4		1
Trunk-ground	8		7		3
Grass-bush	11		10		3
Trunk	5		6		0
Semi-aquatic	1		1		0
Miscellaneous*	2		2		0

* *Rock faces, rock outcroppings on the ground*

that species occur on satellite islands without regard to their structural habitats, the distribution of structural habitats on source islands can be considered null, from which random communities can be constructed to compare with real ones. Using only those real communities that are hypothesized to have been formed from independent dispersal events, real communities generally had species more different in structural habitat than expected by chance. This conclusion, however, was statistically sensitive to exactly how the species pool was constructed; using structural habitat categories never found on satellite islands blurred differences, for example. A similar procedure was used to evaluate size differences of species on satellite islands; no tendency for them to be larger than random existed. These results back up earlier observations that *Anolis* of satellite islands tend to partition resources by structural habitat rather than prey size, in contrast to the Lesser Antilles (Schoener, 1970; Williams, 1972; Section 9.3.3).

9.3.5 Community convergence

General considerations

Analyses of community convergence ascertain to what degree widely separated communities, e.g. those on different continents, are similar in various properties, including some discussed above: species number, species relative abundances, kinds of trophic specialists, niche breadths and overlaps, total biomass and so on. When convergence occurs, the existence of strongly deterministic factors, sometimes involving species interactions, is supported.

Meaningful studies of convergence must satisfy two requirements. First, the species must be relatively distinct phylogenetically; otherwise parallel, rather than convergent, forces will be involved. Second, the communities must exist under comparable conditions or convergence will be impossible. Satisfying these requirements, especially the second, on a finite planet, is not easy and they sometimes must be partly compromised in order that a study can be performed at all. The most extensive studies of community level convergence as opposed to convergence at the individual species level deal with three kinds of organisms – lizards, birds and terrestrial plants, discussed in turn below.

Lizards

Fuentes (1976) performed the first detailed study of lizard convergence, comparing saurofaunas of Chilean and Californian chaparral. 'Control' sites were coastal sage and montane forest, which border chaparral on either side of an altitudinal transect. Major convergences were found in microhabitat, activity time and food type, both with respect to total community range and individual species' utilizations (the latter identify 'ecological equivalents'). In all these respects, chaparral sites were more similar between the two continents than sites with different vegetation from the same place. Crowder (1980) later challenged Fuentes' interpretation, concluding from a null model that similarities within chaparral relative to other sites were entirely due to similar vegetational availabilities within chaparral sites. Fuentes (1980) replied among other things that species do not randomly select habitats as in Crowder's null model.

In contrast to Fuentes's conclusions, Pianka (1986) in a survey of the lizards of three warm deserts – Australia, North America, and Africa – was more impressed by lack of convergence than the reverse. As Table 9.2. shows, deserts differed strikingly in, *inter alia*, number of genera and species (not only of lizards, but also of other vertebrates), and average niche overlap for several separate dimensions. Deserts were similar in number of lizard families and average niche breadths. Pianka noted, that the three deserts do differ in certain major climatic features, perhaps affecting ability to converge.

A third lizard example, this one supporting major convergence, is *Anolis* of the Caribbean (Williams, 1972, 1983). Particular 'ecomorphs', defined as species similar in size and structural habitat, evolved repeatedly from different stocks; all species save one on Jamaica, for example, are monophyletic. The two largest islands, Hispaniola and Cuba, are especially striking in that they both have 34 species of *Anolis* and closely similar numbers of species within each structural habitat category (Table 9.1).

Birds

Cody's (1974) studies are the most extensive concerning convergence of bird communities. He found that grassland communities converged strikingly in species number and ecological type: 10-acre plots contained 3–4 passerines, a large vegetarian 'grouse-like' species, both a long and short billed wader, and 2–3 raptors. Unlike Pianka's (1986) lizard example, mean niche overlaps between typical grassland communities were quite similar: contrasting Kansas with Chile, overlap in habitat was 63 and 60%, in feeding height 78 and 89%, and in feeding behaviour 18 and 21%, respectively. Mediterranean systems in Chile and California also showed a great deal of convergence, paralleling the conclusions of

Fuentes. Despite major differences in family composition, virtually all 'full' members of one community had an ecological equivalent in the other; 20 pairs in all were so designated. In a later study, Mediterranean sites were expanded to include Sardinia and South Africa (Cody and Mooney, 1978). Convergences were then much weakened, especially for the latter locality where certain floristic features differed strongly from the others. Relatively weak convergence was also found in boreal forests (Cody, 1974). A later, also quite extensive, study of Mediterranean systems in three regions – California, Chile, and France – failed to support ecomorphological convergence according to the criterion that differences between a temperate control region and any of the three Mediterranean regions should be greater than between any two of the latter (Blondel *et al.*, 1984).

Schluter (1986) invented a more precise technique different from those previously used (Blondel *et al.*, 1984) to study convergence and applied it to finch communities. The method uses analysis of variance to determine whether the effect of 'habitat type' is statistically significant; if so, convergence is deemed as occurring. Substantial convergence was concluded for four of the five characters studied.

Working on a much broader spatial scale, Lein (1972) developed an index of similarity in trophic type and applied it to zoogeographic (Sclater) regions. While taxonomic affinity declined with increasing distance between regions, trophic similarity was highest among regions of similar climate, for example Neotropical and Ethiopian, regardless of distance: values were typically near 90%.

Terrestrial plants

While finding convergence at the individual functional level, Mooney and Cody (1977) were unable to find much in community level properties of plants of Chilean versus Californian Mediterranean sites. Climatic differences were hypothesized as accounting for this result. For example, thermal regimes are more moderate in Chile, and plants show less seasonality there; powerful chronic winds occur in California, causing a complex of fire-resistant plants to evolve. These results, which differ from those found with lizards and birds, suggest that plants are more sensitive to gross climatic differences than are vertebrates.

In contrast to the previous study, Orians and Solbrig (1977) found major convergences in the number of species, total plant biomass, life-form types, shrub height, spacing and density (in some cases as related to other variables such as soil texture) in warm desert plants of Arizona and Argentina.

Table 9.2. Major differences and similarities among continental desert-lizard systems (Pianka, 1986)

	North America	*Kalahari*	*Australia*
Thermal climate	Colder, shorter growing season (especially in north)	Warmer	Warmer
Precipitation	Great Basin: distributed fairly evenly over year Mojave: winter precipitation Sonoran: bimodal annual rainfall, with peaks in both late summer and midwinter	Summer rains	Summer rain, variable annual precipitation
Higher taxa (number of lizard families)	5	5	5
Number of lizard genera	12	13	About 25
Species diversity			
Lizards	Low (4–11 species)	Medium (11–18 species)	High (15–42 species)
Snakes	High (4.5 species)	Low (2.2 species)	Medium (3.6 species)
Birds (all)	Low (7.8 species)	Medium (22.8 species)	High (28.3 species)

Insectivorous ground-feeding birds only	Low (about 2–3 species)	High (7.3 species)	Medium (4.8 species)
Small mammals	High (5.3 species)	—	Low (1.5 species)
Insect food resource diversity	High (8.7)	Low (4.4), dominated by termites	Intermediate (6.6)
Microhabitat resource diversity	Low (2.8)	High (6.8), variable	High (7.0)
Lizard niche breadths			
Dietary	1.1–7.3 (4.4)	1.1–8.2 (3.9)	1.0–9.4 (3.8)
Microhabitat	1.0–3.9 (2.2)	1.0–6.0 (3.4)	1.0–6.4 (3.0)
Proportional utilization coefficients	Fit decaying exponential	Fit decaying exponential	Fit decaying exponential
Estimated electivities	Fit decaying exponential	Fit decaying exponential	Fit decaying exponential
Niche overlap			
Diet	Intermediate	Highest	Lowest
Microhabitat	All-or-none	Skewed toward low values	Skewed toward low values
Overall	Highest	Intermediate	Lowest
Guild structure	Little evident	Present	Conspicuous
Morphospace	Intermediate	Smallest	Largest

Other organisms

Few studies of convergence as detailed as those above exist for other organisms. Patrick (1961) found strong similarities in the number of species of freshwater diatoms per area of substrate when comparing rivers between north-eastern and south-eastern US and even Amazonian regions. Anurans of Argentinean and Arizonan deserts showed substantial convergence in ecological properties (Orians and Solbrig, 1977). In contrast, few convergences among insect groups were found for warm deserts by Orians and Solbrig (1977) or for chaparral by Mooney and Cody (1977). Mammalian convergence is also substantially more equivocal in the two kinds of vegetation (Mares *et al.*, 1977; Cody *et al.*, 1977).

Conclusion

In summary, about as many cases of convergence as lack thereof seem to exist, and the validity of some claimed cases has been questioned. Part of the problem is lack of climatological (or for animals, floristic) equivalence between regions and part is simply obtaining statistically powerful sample sizes from a single planet. Additionally, the usual discordance in methodology has now arisen in this area, possibly contributing to the differences in conclusions between studies.

9.4 SPECIES' CHARACTERISTICS

9.4.1 Character displacement

Character displacement has been discussed in Sections 9.2.3 and 9.2.5. Some of the more famous cases occur in Galapagos finches (see below), already discussed in relation to aggregate size differences within communities (Section 9.3.3). Character displacement is simply the evolutionary part of the process by which such differences are generated. The list of well-documented cases of character displacement is not extensive, but it is growing.

Two species of burrowing lizards (*Typhlosaurus*) have different body sizes and food sizes from each other in sympatry, whereas the species that also occurs allopatrically has intermediate body and food sizes (Huey and Pianka, 1974).

Two species of particle-ingesting gastropods (*Hydrobia*) differ in shell size in sympatry but not in allopatry (Fenchel, 1975). These body-size differences are associated with differences in particle-size utilization (Fenchel, 1975; Schoener *et al.*, 1986; but see Levinton, 1987).

Ficken *et al.* (1968) showed that a species of warbler (*Dendroica*) has a longer, more attenuated bill in the presence but not in the absence of a related species; this bill enables prey to be probed from loblolly pine cones. Grant (1972) suggests this may be a secondary result of parallel clinal variation rather than competition (as in Fig. 9.1 above), but his argument is less convincing than for the similar case of the Asian nuthatch (*Sitta*), for which he was unable to find a consistent ecological consequence of bill-size differences (Grant, 1975). In another set of cases the tables are turned; ecogeographical rules in certain mammals have been re-interpreted in terms of character displacement (McNab, 1971). In a multivariate study, Dunham *et al.* (1979) pitted geographical variables against characteristics of co-occurring species as predictors of character change in catostomid fishes. The latter were found much more important than the former.

Fjeldså (1983) recently published a very extensive study of character change in grebes (Podicipedidae): 2925 specimens were measured and 47 000 prey were identified for species occurring in northern Europe, Columbia, Peru, Patagonia and Australia. In all examined cases, where two closely related species overlapped broadly, one or both diverged in bill morphology. Where data were available, these morphological changes were usually accompanied by changes in food type. An example is given in Fig. 9.4; note that *Podicips grisegena* displaces considerably in bill length where it overlaps with *P. cristatus*; the genus has a practically worldwide distribution.

A number of cases of displacement are known from Galapagos finches (*Geospiza*), and their ecological significance understood, although several cases of sympatry in these species have apparently not resulted in displacement (Grant and Schluter, 1984; Schluter *et al.*, 1985).

Numerous cases of character change, some involving displacement, are known in *Anolis* lizards, a genus for which, virtually without exception, differences in morphology are accompanied by differences in structural habitat and/or prey size (Schoener, 1977). Striking changes in morphology and pattern accompany the major changes in habitat, as discussed in Section 9.4.2. Cases of divergence and convergence in size (Schoener, 1970) are consistent with a model (Schoener, 1969; see also Abrams, 1986) of optimal size, based on feeding considerations. Convergences are also possible with most of the models of displacement discussed in Section 9.2.3.

In summary, character displacement is not as uncommon as has recently been claimed. Nonetheless, only a few studies have systematically searched for such cases among all possible opportunities for a given group (Schoener, 1970; Dunham *et al.*, 1979; Fjeldså, 1983; Grant and Schluter, 1984), or have carefully evaluated alternative explanations

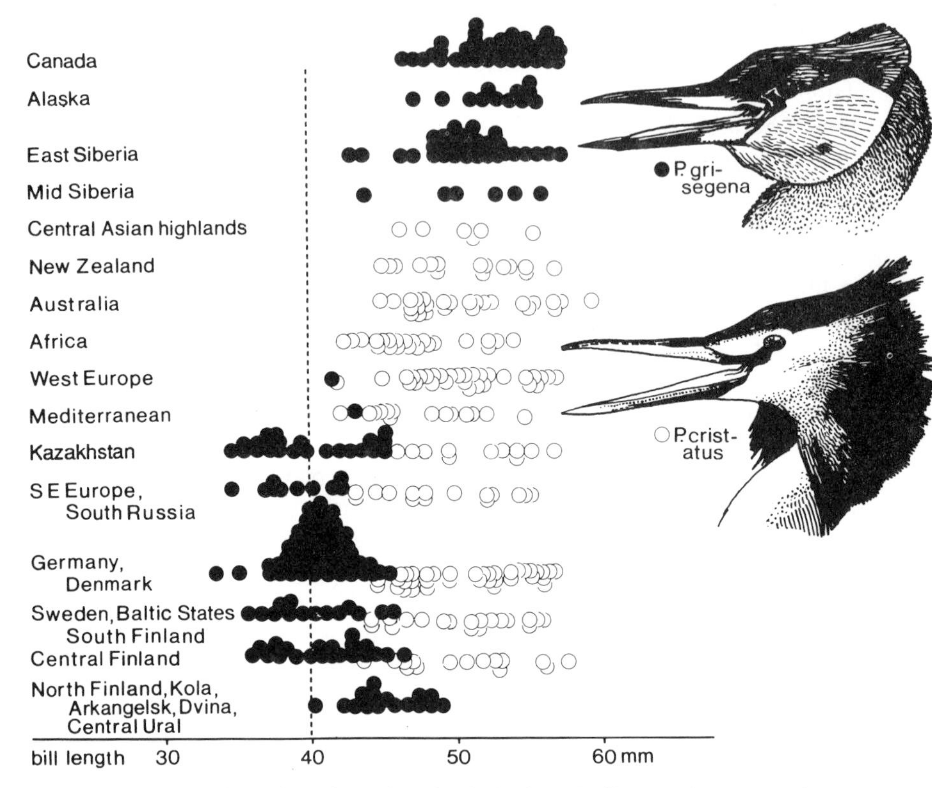

Figure 9.4 Bill lengths of Red-necked Grebes *Podiceps grisegena* and Great Crested Grebes P. *cristatus* from different geographic areas. (Fjeldså, 1983).

such as geographical variables (Grant, 1975; Dunham *et al.*, 1979) or variability in resource availability (Schoner, 1975; Schluter *et al.*, 1985).

9.4.2 Habitat shift

Habitat shifts (Section 9.2.3) have been observed in a variety of organisms. In Tasmania, where taxa of birds specialized to feed on trunks are absent, species from other avian taxa shift into this zone (Keast, 1970). In New Guinean (Diamond, 1970) and New Hebridean (Diamond and Marshall, 1977) birds, the commonest type of shift is spatial: in altitude, in vegetation type and in foraging height. Shifts in diet are much rarer, as predicted by the compression hypothesis (Section 9.2.3). In certain of these species, as well as ant species (Wilson, 1961), shifts from low to high altitude or from secondary to primary forest are especially common, as would be expected for colonizers of remote islands. Other examples are in Williamson (1981).

Whilst response to interspecific competition is one hypothesis for habitat shifts, changes in resource availability is a second (Section 9.2.4). Only a few studies have carefully taken into account such variability. Schoener (1974a, 1975) measured vegetation changes from one locality to another in the study of structural habitat shift mentioned earlier (Section 9.2.4). Habitat shifts in apparent response to competition were both enhanced and obscured by inter-locality vegetation differences. This study also showed that: (1) the more similar that species are in other niche dimensions such as food size or climatic habitat, the more likely they are to shift structural habitat in response to one another, and (2) smaller forms tend to shift more in response to larger forms than vice versa. The second result is explainable on the basis both of consumptive competition (larger forms eat more) and interference competition (larger forms win in aggressive encounters) (Schoener, 1983a). In a study on the two species of North American tanagers *Piranga*, Shy (1984) showed by multivariate analysis that the species change habitats in response to regional vegetational changes. Given these changes, the species' habitats still differ more in sympatry than in allopatry. Shy postulates that the species coexist by habitat shift, maintained in part by interspecific aggression; observations of aggressive response by one species to vocalizations of the other species support this view.

Habitat shifts have also frequently been produced experimentally by manipulation of putative competitors (Schoener, 1983a), although mostly over smaller spatial units than are typically of interest to biogeographers.

9.4.3 Niche expansion

In contrast to habitat or niche shift, unambiguous examples of niche expansion as plausibly affected by interspecific competition are rare (Beever, 1979). Crowell (1962) studied three species of birds that were much denser on species-poor Bermuda than on the species-rich mainland. Bermudan populations showed no new foraging behaviour, and their niche widths, if anything, were slightly smaller than on the mainland. The barred antshrike *Thamnophilus doliatus* is the most abundant insectivorous foliage-gleaning bird in second growth on both Trinidad and Tobago, but only on the latter does it extend into closed rain forest. Keeler-Wolf (1986) proposed that diffuse competition keeps it out of rain forest on Trinidad because (1) it is typically absent from rain forest when many similar species co-occur, (2) the biomass of similar birds is higher in rain forest on Trinidad than Tobago, and (3) foliage arthropods are significantly less abundant in Trinidad rain forest. The gastropod *Conus miliaris*, virtually the only representative of the genus on Easter Island, has an expanded diet and possibly depth range but not microhabitat range (Kohn, 1978). Certain *Anolis* lizards show a relatively small climatic breadth in species-poor communities (Schoener, 1977), but major exceptions exist (Hertz, 1983). The evidence for prey-size expansion in *Anolis* is equally mixed (Roughgarden, 1974b; Schoener, 1977). The most striking example supporting niche expansion is perch height in *Anolis sagrei* (Lister, 1976); temperature and prey-length breadth also increased with decreasing number of species. Finally, Schoener *et al.* (1979) tested the compression hypothesis (Section 9.2.3) on a biogeographic scale with data from two species of Kalahari lizards (*Mabuya*) having peripatric distributions (see Chapter 7 for details). The expectation that food types decrease in breadth from sympatry to allopatry while habitat types increase was realized in 7 out of 8 cases.

9.5 COMPLEMENTARITIES IN SPECIES DISTRIBUTIONS AND ABUNDANCES: BRIDGING THE COMMUNITY AND INDIVIDUAL-SPECIES APPROACHES

9.5.1 Characterization of the species distribution

Rather than focus on species distributions as they combine at the community level, we now focus on the distributions *per se*. When a species range is fragmented, either because it is already insular or is amenable to reasonable subdivision, its distribution can be characterized by properties of the fragments on which it is present or absent. A straightforward way to do this uses occurrence sequences (Schoener and Schoener, 1983b), as follows: order islands along some quantitative scale, say from small to large area (Fig. 9.5). Label each island as to whether a given species is present (P) or absent (A). Examine the sequence of P and A for regularity with some simple statistical technique, such as the Mann-Whitney U-test. The greater the separation of P and A, the better is the associated island variable at predicting the given species' occurrence. For example, suppose all islands below a certain area do not have a given species, whereas all above that area do. A precise threshold then exists with respect to the variable 'island area', so that in this sense island area is an excellent predictor. Fig. 9.5 gives occurrence sequences with respect to area for three species from the same archipelago: a lizard species is predicted best while a migratory bird species is predicted least well. Occurrences can also be investigated using logistic approaches, which allow several island variables to be considered at once (Schoener and Schoener, 1983b; Adler and Wilson, 1985).

A second method of characterizing occurrences uses Diamond's (1975)

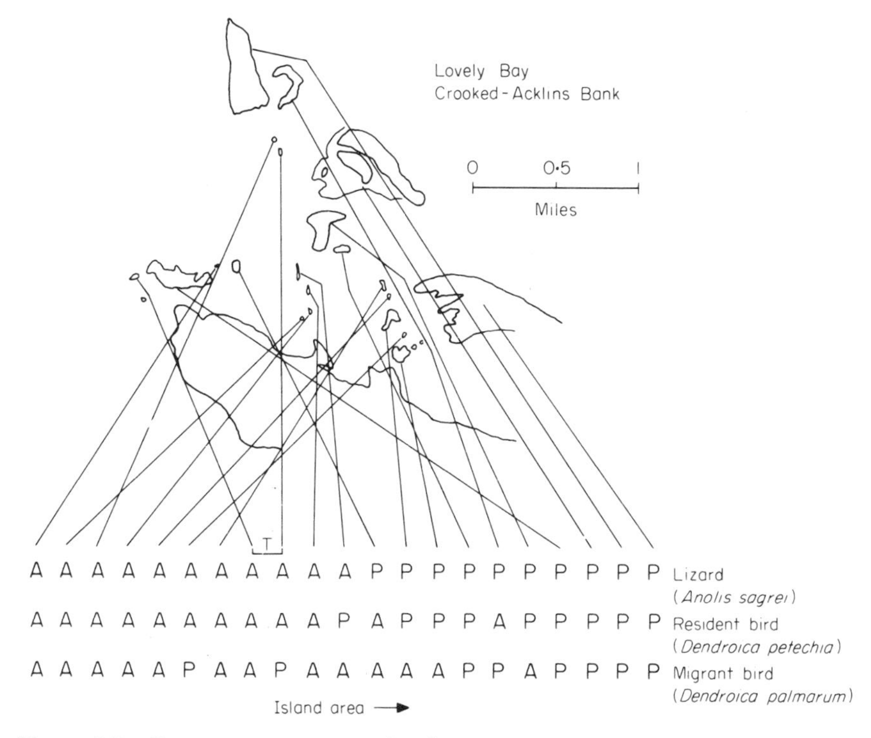

Figure 9.5 Occurrence sequences for three species in the Lovely Bay archipelago. Islands are ranked in order of increasing area from left to right. A = absent. P = present. T = having the same area. Notice that the lizard *Anolis sagrei* has a perfectly ordered occurrence sequence with respect to area. The resident bird *Dendroica petechia* has a somewhat less highly ordered occurrence sequence. The migrant bird *Dendroica palmarum* has the occurrence sequence closest to haphazard. Examples are representative of the degree of regularity shown by the three kinds of species. (Schoener and Schoener, 1983b).

incidence functions. Here, the fraction of islands having a given species is plotted for each of a set of categories indexing some island variable. In contrast to occurrence sequences, incidence functions do not directly lead to significance tests.

Using occurrence sequences, Schoener and Schoener (1983b) found that most resident species of Bahamian birds and lizards had very precise thresholds with respect to certain island variables, particularly area and abundance of foliage in various height layers. Cole (1983a) found a similar tendency for ant species of Florida mangrove islands. When all the species in some group have precise, possibly different, occurrence thresholds with respect to an island variable, occurrence distributions are nested: all islands with n species have the same

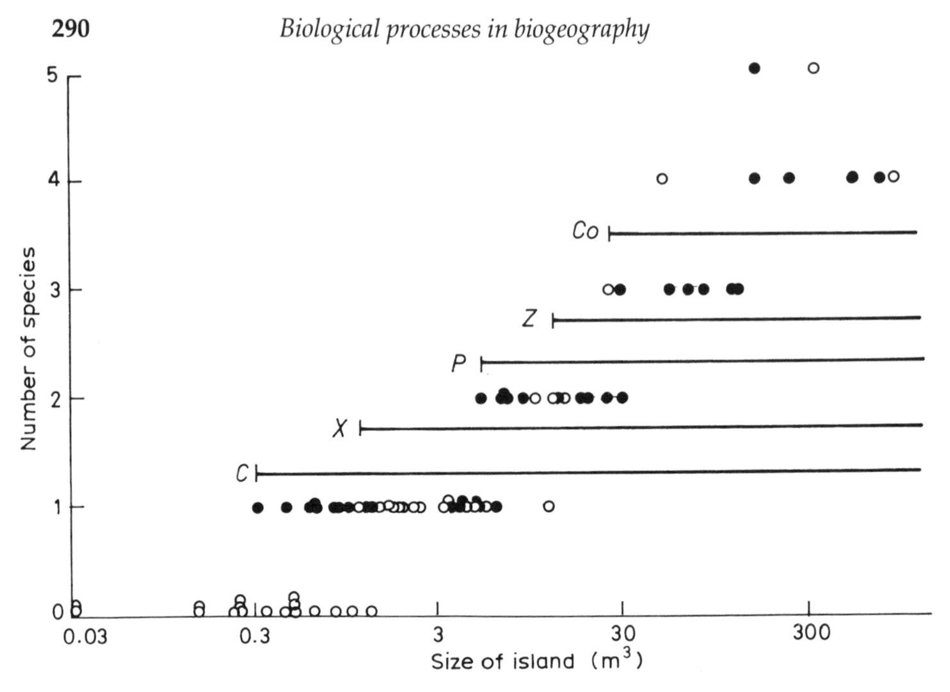

Figure 9.6 Species number-island size relation for mangrove ants. The 81 points represent individual mangrove islands. • indicates the presence of C. *ashmeadi*. ○, excluding islands without species, indicates the presence of *X. floridanus*. The bars indicate the range of island sizes on which the five major species of mangrove ants are found. (Cole, 1983a).

species, and any species on an island with n species also occurs on islands with greater than n species. Fig. 9.6 shows Cole's data for ants, which have highly nested distributions. Nested distributions in themselves do not implicate ecological interactions in their aetiology; on the contrary, each species may be responding to a different threshold habitat factor independently of the other species. (A nested pattern, however, is strongly consistent with a high degree of resource partitioning, which itself may have resulted from past competition, although not necessarily on the archipelago of interest.) On the other hand, the boundaries, say in units of island area, between n- and (n + 1)-species islands may be affected by present-day interactions, but this can only be determined by experiment. Intriguingly, Cole (1983b) did just that for the Florida ants; he found that certain species could live on smaller islands than they normally inhabited, when other species were experimentally removed from those islands.

9.5.2 Distributional complementarities

Complementarities in distribution within an archipelago contrast sharply with the nested distributions just discussed. Such complementarities by themselves comprise the most convincing kind of observational evidence for negative species interactions. When the species involved are trophically similar, competition is strongly implicated, especially if islands are identical in habitat and area, so that differences in the kind of island preferred are ruled out. 'Checkerboards' (patterns in which islands having one or the other species are spatially highly interdigitated; Diamond, 1975) are even more suggestive of competition, because they rule out the possibility of the two species having disjunct distributions simply because they invaded the archipelago from different directions. When jump dispersal rates are estimated to be high or islands are large, complementarities in occurrence are unlikely, because at least some individuals of each species are likely to be everywhere (Toft *et al.*, 1982). Under such conditions sufficiently strong competition should still produce complementarities in abundances, although complementarities may be difficult to detect if habitat types vary between the islands.

A number of striking cases of complementary distributions are known. Most examples involve birds, and these are illustrated in Diamond (1975) and Schluter and Grant (1982). Fig. 9.7 shows complementary distributions for eight species of grass finches (*Lonchura*) on New Guinea; here, actual islands are not involved but rather the species occur in zones of midmontane grassland. Most sites support only one species (Diamond, 1975). Complementary distributions have also been reported for lizards on actual islands (Radovanović in Nevo *et al.*, 1972; Kiester, 1983; Schoener, 1986c), for salamanders in stream sites (Means, 1975), and for snails in ponds (Lassen, 1975). These studies suggest competition because the species involved are ecologically similar (although habitat differences must always be ruled out (Section 9.3.2). However, Lomolino (1984) documents a case of complementarity between two species of mammals, the herbivorous role *Microtus pennsylvanicus* and the carnivorous shrew *Blarina brevicauda*, occurring in the Thousand-Island region of North America. The shrew preys mainly on immature stages of the vole, often wiping it out quickly, but the vole is the better disperser.

The simplest way to evaluate distributional complementarities statistically is to classify all islands into a contingency table (Edmondson, 1944). Two species give a 2×2 table with 4 boxes: both species present, both species absent, species 1 present but species 2 absent, and species 2 present but species 1 absent. A simple chi-square test of association can

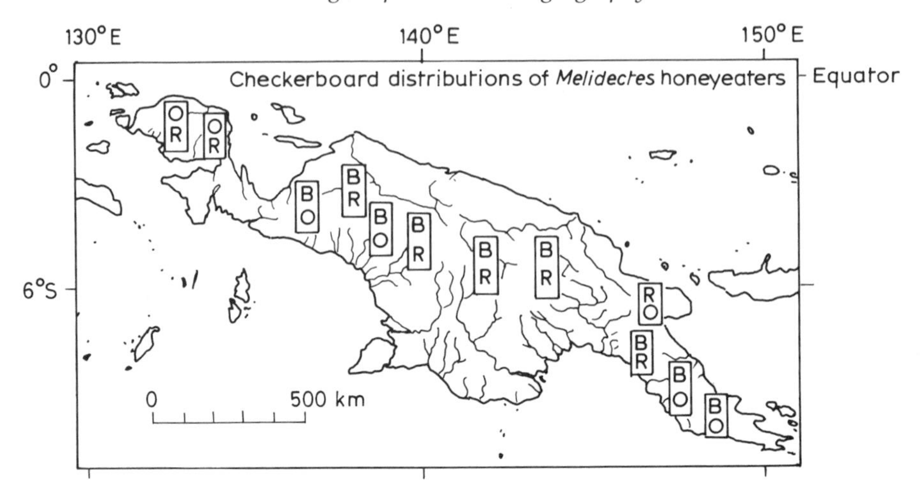

Figure 9.7 Compound checkerboard exclusion: distributions of three *Melidectes* honey-eaters. O, *Melidectes ochromelas*; B, *M. belfordi*; R, *M. rufocrissalis* super-species. Most mountainous areas of New Guinea support two species with mutually exclusive altitudinal ranges. At each locality depicted on the map of New Guinea, the letters above and below indicate the species present at higher and lower altitudes, respectively. The identity of the locally missing third species is subject to irregular geographical variation. (Diamond, 1975).

then be performed. Cohen (1970) detailed the methodology and also gave a theoretical justification that showed how competition can produce a negative association. The method has been used on a small scale by Strong (1982), who found no significant associations among pine beetle species on *Heliconia*. It was used on a larger spatial scale by Toft *et al.* (1982) for subarctic ducks on ponds; some significant complementarities occurred, preferentially among species most overlapping in habitat and phenology. As they discuss, the method is very sensitive to how islands are determined as absent-absent, rather than being simply unsuitable.

When three or more species are included, multiway-contingency analysis is appropriate, and the number of kinds of interactions increases. In Whittam and Siegel-Causey's (1981) study of Alaskan seabird colonies, the only interactions suggesting competition were three-way; positive two-way interactions were typical. Obviously, as more species are added to the set under consideration, contingency models become impractical; one can then use methods such as Diamond (1975) originally used or modifications (Sections 9.3.2).

A completely different method, the variance-ratio test (Schluter, 1984; McCulloch, 1985) is also available for large groups of species. It

Table 9.3. Ecological processes which may result in a positive or negative association among species, with emphasis on interspecific interactions (Schluter, 1984)

Association	*Interaction*	*Example of process*
Negative	Competition	Interference or resource mediated, producing occasional exclusion or negatively covarying population sizes.
Negative	None	Different resource requirements or climatic tolerances with negatively covarying resource/climate states.
Negative	Mutualism	Resources compete and are used exclusively by species. Local overharvest of some resources decreases abundance of consumer species, but density of remaining consumer species increases due to release of resources from competition (modification of Vandermeer 1980).
Negative	Predation	High predator densities produce a local depression of prey and result in negatively covarying population sizes.
Positive	Competition	Population sizes of competing species fluctuate in unison in response to fluctuations in their identical, limiting resources.
Positive	None	Common response to changes in climate or in supply of unlimiting resources.
Positive	Mutualism	Species enhance each other's survival probabilities.
Positive	Predation	Predator abundance fluctuates in positive response to variation in prey abundance.

computes the ratio of the variance in total species number in samples (e.g. islands or other spatial units) to the sum of the variances of the individual species' occurrences. Systematic examination of Bahamian orb spiders by this method showed them to be significantly positively associated in many cases (Schoener, 1986c). As Schluter (1984) pointed out, a variety of processes can lead to negative or positive associations. Table 9.3 lists the possibilities. Moreover, differences in sampling techniques for species or sites may also be responsible for non-zero associations (McCulloch, 1985), and sample sizes needed to detect competition where migration between sites is great, may have to be large (Hastings, 1987; p. 260).

9.5.3 Complementary abundances

As discussed above, competition between species, particularly in situations of high dispersal, is more likely to lead to complementary abundances than complementary occurrences. Negatively associated abundances can be examined with the variance-ratio test (Section 9.5.2), or one can examine plots of one species' abundances against that of another. Due to sites having neither species and because of other factors giving nonlinearities, such plots are very difficult to evaluate statistically (Schoener, 1986c). Scattered examples of statistically significant negative associations exist: 1 of 13 beetle species in Strong's (1982) study; several duck species in Toft *et al.*'s (1982) study, but no orb spider species in Toft and Schoener's (1983) study. Using the variance-ratio test on both occurrences and abundances, terrestrial vertebrates are significantly more negatively associated than arthropods in Schluter's (1984) data (Schoener, 1986c). The most convincing evidence for complementary abundances are so-called 'density-compensation' studies.

9.5.4 Density compensation

Competition theory predicts that, the smaller the number of competing species, the larger a typical species' density should be (Section 9.2.3). Comparisons of particular species' densities between communities with varying numbers of species (typically performed between islands and mainlands) are called density-compensation studies (Wright, 1980; Williamson, 1981; Faeth, 1984). The term 'density compensation' actually refers to the degree to which the summed species densities in a species-poor community equals the same quantity in a species-rich community (MacArthur *et al.*, 1972).

A variety of processes other than competition could cause island species' densities to be relatively large. One is a relatively smaller predation intensity on islands (MacArthur *et al.*, 1972; Case, 1975). A second is a more benign insular climate (Case, 1975). A third is the 'fence' effect, in which individuals are trapped within insular bounds and so unable to emigrate to marginal habitats (Krebs *et al.*, 1969; MacArthur *et al.*, 1972).

Studies differ in the degree to which density compensation is found. Summed densities may not be equal between depauperate and rich environments, even though individual species in the former have, on average, larger densities. Sometimes 'excess compensation' occurs, in which summed densities are higher in depauperate regions. Simple observational studies of densities by themselves are typically not so good for implicating particular mechanisms because of the large number of consistent explanations (Wright, 1980; Faeth, 1984). For example, degree of density compensation increases with number of species in the depauperate communities (because more species are then able to exploit a greater fraction of the available resource spectrum), as well as with the similarity of habitats from depauperate and rich sites (because species will then be better adapted to habitats in the former). Wright (1980) was able to implicate both variables statistically and account for a respectable 65% of the variance in density-compensation data.

In an ingenious effort to get at density compensation experimentally, albeit on a small scale, Faeth (1984) transplanted individual oak trees into isolated sites, leaving bunched trees as controls. Isolated trees eventually lost some species of leaf-mining insects relative to controls, but no evidence of density compensation or habitat shift was found. Possibly the diminution in species number was not sufficient to give a detectable effect, but the result also accords with proposals that interspecific competition is low in herbivorous insects (Hairston *et al.*, 1960; Lawton and Strong, 1981).

9.6 CONCLUSION

It would be marvellous if, at the end of a chapter documenting so many unresolved issues, one could reassuringly claim that most controversies can be settled by some vastly improved methodology. While a number of my colleagues view field experimentation as such a panacea, I fail to see how it is practical for testing typical evolutionary hypotheses. For example, the possibility of past competition as a causal agent is sometimes disfavoured to the point of mild ridicule, perhaps because it is largely impossible (by definition) to test experimentally and so subverts a purely experimental programme. Yet past competition has happened, and it might have happened often. We will stumble blindly through nature if we ignore sometimes probable explanations because they fail to fit our narrow methodological protocol. Indeed, Chapter 8 documents extinction hypotheses based on competition; Chapter 2 also discusses the interplay between historical and ecological factors within a biogeographical systems approach.

This is not to diminish the absolute usefulness of experiments for biogeographic purposes broadly conceived (Chapter 15). At the finest level, theoretical results on how spatial fragmentation affects the nature of interactions such as competition and predation need to be worked into our field experiment designs. Natural immigration and emigration are often more appropriately measured than ignored, and as dispersal nearly always stabilizes ecological interactions in theory, we must revise empirical expectations when measured dispersal is high. At a broader level, many ecological experiments are of peripheral interest to biogeographers because they deal with the detection of an interaction and not with its ultimate effect on community composition, namely which species persist or become extinct. Although difficult, manipulations using entire isolated communities, such as those of small islands, test hypotheses about the outcomes of interactions and so should be pursued as much as possible. Natural experiments (Diamond, 1986a) are also useful in this regard (Chapters 3 and 15).

If manipulations are agreed in principle to be incapable of resolving all major issues, we must then recourse to observational methodology as a second line of attack. In this chapter, such methodology is divided into community and individual-species approaches.

The advent of null models has helped make community approaches more rigorous. As reviewed, such models have drawbacks. They can be in part circular in the sense that they use information from communities, after some interaction may have had an effect, to deduce features of pre-interaction communities. Null models sometimes have low power for detecting the effect of an interaction; large sample sizes may be required. Even if a community's structure were distinguishable from 'randomness', alternative hypotheses for the direction of difference are sometimes so numerous as to offer little progress – the community view may be so distant that mechanisms are invisible.

The individual-species approach would seem to solve some of these problems. Because numbers of individuals, rather than numbers of species or communities, are the variables of interest, sample size is not so much of a problem. A close look at individual species can also reveal mechanisms, in part simply by direct observation. In studying individual-species phenomena such as character displacement, one should not be discouraged by the claim that the phenomenon is rarely found. Implicit here is perhaps the idea that it has been looked for many times, but the fact is that systematic studies of these phenomena are rare, and that character displacement and habitat shift (but not niche expansion) are often found when looked for. A more individualistic approach to community convergence, at least at the biome-type level, will probably often be impossible because biomes are too rare; it may, however, be quite profitable at a finer level, e.g. for replicated evolutionary radiations of some non-vagile group on different islands.

As discussed above, however, the individual-species approach has a major drawback: it often concentrates where a phenomenon is expected. This, of course, is what community approaches explicitly try to avoid. At least two ways to coalesce individual-species studies into a broader picture might alleviate the problem. First, an aggregate significance level can be obtained if probabilities are combined according to Fisher's method (Sokal and Rohlf, 1981). Care must be taken that all cases are independent; here the redundancy (Section 9.3.3) is not biological but statistical and therefore never desirable. Second, an *a priori* characterization can be made of cases expected and not expected to show some effect; then contingency analysis can be performed to obtain an aggregate significance. Again, independence is required. Examples of the second approach are given in Sections 9.4.1 and 9.4.2. In both approaches, an overall assessment is eventually arrived at, but unlike community methods, it is constructed from below, accumulating individual case studies. Perhaps these fine-scale, biologically oriented approaches are our best hope for resolving many of the controversies that now fracture biogeography.

PART IV

Biogeographic reconstruction

Introduction

The starting point for any reconstruction of biotic histories is an analysis of the distributional ranges of the taxa under study. Analyses are generally based on extant distributions but may utilize distributions at any geological horizon (fossils). Synoptic approaches, used in some methods of reconstruction, involve analyses of distributions at several horizons, one of which may be the present. The available data base for these analyses is a compilation of collection localities of the taxa concerned. Analyses based on 'absence' data are constrained by the fact that apparent absences may in fact be pseudo-absences (the taxon occurs in a locality in which it has not yet been collected) or temporal absences (the taxon did occur in a locality in the past but no longer does). Past distributions are even more difficult to reconstruct than modern ones due to incompleteness of the fossil record, decreased taxonomic precision associated with incomplete remains and the fact that the site of collection of a fossil is not necessarily its site of fossilization (due to earth movements) nor its site of existence in life (as a result of movement of its remains between death and fossilization). Nevertheless, despite these deficiencies, distribution patterns based on collection data are all that can be used to construct hypotheses of biotic history which in turn can be tested as more distributional data become available.

Small-scale patterns of distribution are largely governed by ecophysiological processes. Large scale disjunct distribution patterns, on the other hand, result from dispersal or vicariant events and many historical biogeographers attempt rigorously to determine the comparative roles of these two processes. Others, however, consider that since the processes by which distribution patterns are formed are not known, only hypothesized, the incorporation of process hypotheses into analyses introduces a subjective element. They therefore restrict their analyses to biotic distribution patterns. This is a major point of contention between 'evolutionary' cladists and 'pattern' cladists and we will return to this controversy later. If a choice must be made between dispersal and vicariance, biogeographers generally resort to the criterion of 'parsimony' which states that the simplest sufficient hypothesis is to be preferred even though others are possible. If a large number of unrelated taxa show closely similar distribution patterns involving two or more disjunct areas, it is parsimonious to hypothesize that they represent a vicariated assemblage of organisms, since it is less likely that such a variety of unrelated taxa, presumably with differing dispersal capabilities, could independently come to occupy such similar (congruent) disjunct ranges. In the analysis of congruent distribution patterns taxa are generally unweighted, that is to say, all congruences are considered to be of equal importance. It may be, however, that certain congruences should be considered as having more information content than others. Edmunds (1981), for example, has drawn attention to the importance of co-adapted species pairs. It is unlikely that the dispersal capabilities of each member of a co-adapted pair, unless congeneric, will be closely similar, so their occurrence together, in disjunct populations, should provide a good test of vicariance. Single unsupported divergent patterns, on the other hand, could be explained equally plausibly as a result of their prior dispersal, or of vicariance. In a historical context, dispersal is at best shown only to be equally as likely as vicariance in any given case, whereas vicariance can be supported more rigorously. This in no way refutes dispersal as an important process in pattern forming, but doubtless explains why most attempts at historical biogeographic reconstruction in this decade have been based on vicariance rather than dispersal models, the latter often being relegated to the status of 'Just so' stories. This is not the case where distributions over more recent (ecological) time are considered. Dynamics of colonization patterns on newly created or recently disturbed habitats and species richness patterns on 'true' or 'habitat' islands have been modelled using a dispersal-based hypothesis, the MacArthur and Wilson (1967) theory of island biogeography. The model is based on an equilibrium between immigration via dispersal and extinction rates, but affected by island size and degree of isolation (in both space and time) from source pool areas. The model of island biogeography and more recent refinements of it have been tested in a wide range of environments, using a variety of taxa and experimental systems which are outlined in Chapter 15. Some aspects of the theory are open to other interpretations (Chapter 4) but it remains one of the few models in biogeography amenable to rigorous experimental testing.

A prime objective of vicariance biogeography is the identification of distributional fragments of a previously continuous biota and their subsequent reassortment or integration in a modern biota. Fragments which have resulted from vicariance must be discerned from the background noise of distributions which result from jump dispersal of the biota since the vicariant event. Since jump dispersal has at least the potential to obliterate vicariant patterns, vicariance biogeographers work on the assumption that it is too rare to conceal major disjunctions. Vicariance biogeography, of course, accepts range expansion as a normal process or 'means of survival' (Croizat, 1958), but their methodologies do not require consideration of this in the development of their hypotheses.

In response to changing physical and biotic conditions, populations of organisms undergo periods of expansion and contraction in their geographic ranges and fluctuations in population size over time. Populations do not, however, exist in isolation but form part of interacting complexes of species populations within communities. It may be difficult rigorously to delimit communities, since species distributions, especially in the case of plants, often integrate according to underlying ecophysiological optima. Nevertheless, certain 'key species' mould the habitat for other members of the community e.g. through space occupation by intertidal barnacles, or domination of habitat structure by corals or forest trees. Small-scale disturbances caused by biotic (trophic interactions) or abiotic (e.g. fire or flood) factors can lead to local distributional changes within a community through their action on the key species. Larger scale tectonic, eustatic, climatic and oceanographic (TECO) events, again through their effects upon these key species, can result in expansion or contraction, or in range shift, of whole biotopes. In the case of range shift, individual species may move out of phase with each other and with the key species, resulting in time lags before reassembly of the biotopes (Connor, 1986). Patterns of species richness in contemporary communities have been hypothesized to result from these changes in the geographical extent of biotopes (Chapters 3 and 10). In periods of contraction, biotopes may become vicariated into isolated units. Rosen (1984) has demonstrated, with his 'three-island' model, the importance of the pattern of distribution of oceanic islands in producing high diversity in oceanic areas. According to the model, varying barrier intensity as a result of TECO events can result in periods of allopatry when speciation occurs, followed by periods of sympatry, leading to increased diversity as a result of the prior allopatry. Where episodes of allopatry are of insufficient duration, or where there are insufficient environmental differences in the allopatric areas (founder effect notwithstanding) for full speciation to have occurred, initial divergence may later be exaggerated by ecological interactions during sympatry (Diamond, 1986b). Another model has been developed to explain high diversity in broadly homogeneous (at least in regard to key species and climate) ecosystems. As biotopes contract during periods of climatic deterioration (for those particular biotopes), organisms stenotopic to the contracting biotopes become isolated in allopatric populations within the boundaries of the contracted biotope. Such biotopes have been termed 'refugia' and, as described above, are hypothesized to result in high rates of speciation and hence high diversity, in the integrated ecosystem. It is argued that if the apparently uniform biotope was once vicariated into regions of refuge, areas of endemicity in the geological past should still be recognizable today, provided that sufficient time has not passed for complete homogeneity of the biota to have occurred through the expansion of ranges of the refugial components. Thus, patterns of endemicity need to be observable, although the resolution of refugia and suture zones (Chapter 10) will depend upon the scale employed. Studies of widely different taxa may result in the recognition of more and more refugia or subrefugia. This complicates, although does not necessarily refute, the refugial model. Taxa used in framing refugial hypotheses must be stenotopic, since eurytopic taxa may not have been restricted in the past to biotopes similar to those in which they now exist, and may not, therefore, have been faithful to them. Most refugial models have been developed in relation to Pleistocene glaciations, where it is clearly implicit that refugia are contemporaneous with each other and with the glaciation event. These assumptions have been questioned (Connor, 1986; Chapter 10). Strategies for testing the refugial model, with particular reference to neotropical rainforests, are discussed in Chapter 10.

The broad field of vicariance biogeography has itself become vicariated by the development of different methodologies which result from two divergent philosophical approaches (Fig. IV.1). One school of approach is founded on the phylogenetic taxonomy of Hennig (1966), the other on the panbiogeography of Croizat (1958). Hennig's major contribution was the development of a method of analysing taxa with respect to shared characters which have been derived from an ancestor common only to themselves (shared derived character states or synapomorphies). The result of such analyses is the development of branching diagrams (cladograms) which represent the hierarchical distribution of synapomorphies and hence the phylogenetic relationships of the taxa studied. Cladistics is based upon the assumption that speciation is overwhelmingly dichotomous and not polychotomous (Fig. III.4, p. 162) or reticulate. In addition, it should be noted that our interpretation of the speciation process in a historical sense is restricted to phenotypic data since the genotypes of extinct ancestral taxa are largely unknown (but see recent work on gene amplification and DNA sequencing, carried out on 17–20 Myr old fossil leaf samples (Golenberg *et al.*, 1990)). The phenotype of a taxon results from the interaction of the genotype with the environment. Lack of a character state in a given taxon (usually morphological, but can be behavioural or physiological) may imply absence of the relevant genetic instructions, but could also result from lack of the appropriate environmental stimulus for its expression, e.g. application of ecogeographical rules (Lane and Marshall, 1981), or colouration in Lepidoptera (Myers,

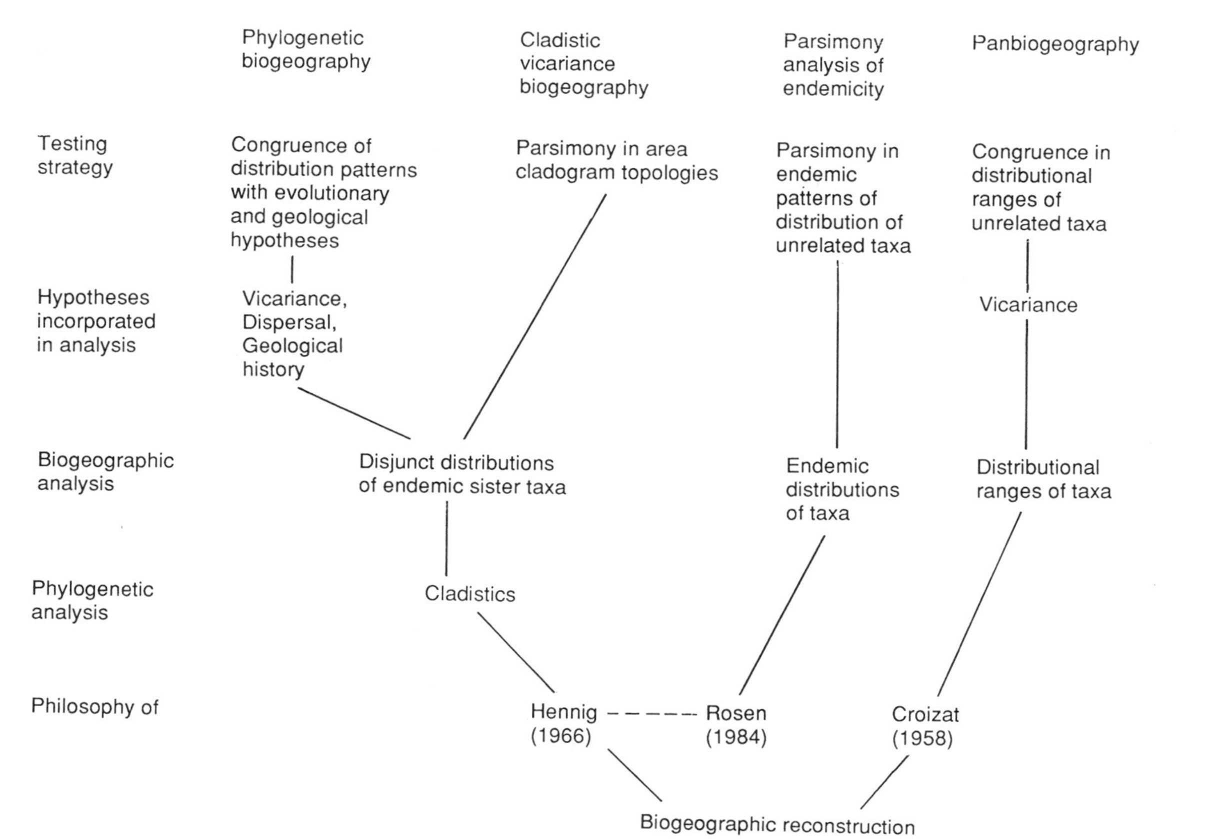

Figure IV. 1 Pathways for biogeographic reconstruction using different analytical techniques.

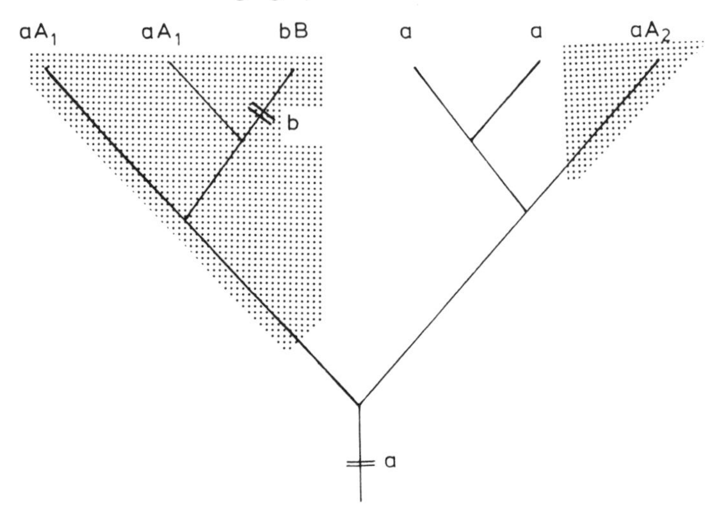

Figure IV. 2 Hypothetical effect of disjunct, convergent environmental conditions on phenotypic expression of genotypes. a = latent genotype, A_1-A_2 = phenotypic expression of genotype 'a' in spatial and/or temporal disjunction, b = modified genotype 'a', B = phenotypic expression of genotype 'b'. Stippled areas = similar, disjunct environmental conditions.

1977). In cladistics, the occurrence of a character state in an incongruent position in a cladogram is usually taken to indicate either convergent evolution or a misinterpretation of the generality of the character state. It could also, however, represent re-emergence of a previously latent character state (parallel evolution) as a result of convergence in environmental conditions (Fig. IV.1; Brundin, 1986). Notwithstanding these potential falsifiers, cladistics still offers the best available tool for determining phylogenetic relationships.

Wallace (1860) pointed out that the natural sequence of species by affinity was also geographical. In the same way that higher taxa contain nested sets of lower taxa, so the geographical range of a higher taxon has nested sets of lower taxa within its boundaries. It is not surprising then that Hennig (1960) himself realized that cladistic methods could be applied to biogeography. Once a phylogenetic cladogram has been developed, the substitution of areas in which the taxa occur, for the taxa themselves, leads to an 'area cladogram', expressing the hierarchical arrangement of the areas, with the dichotomies representing the sequential breakup (literally in the case of tectonics, effectively in the case of sea-level eustasy and other barrier formation) of the areas concerned. A hypothesized reconstruction of earth history is then possible. Although Hennig clearly recognized the potential of his method (cladistics) for biogeographic analysis, it was left to Brundin (1966), who studied the relationships of chironomid flies in the Southern Hemisphere, to apply the method in detail. On the basis of this work, he produced a hypothesis for the sequential breakup of a former southern land-mass, Gondwana. Brundin terms his method 'phylogenetic biogeography' emphasizing its close adherence to the evolutionary approach of Hennig. Analysis of both vicariance and range expansion are included in the method, the latter being determined by Hennig's often criticized 'progression' and 'deviation' rules (Chapter 11). The relative age of a given group (determined by its position in the hierarchy) is related to the ages of barriers and hence maximum age of vicariated subgroups. The method is, therefore, not independent of geological hypotheses and will presumably have greater resolution the more recent the formation, and the more accurate the dating, of the barrier. The methodology is detailed in Chapter 11.

More recently, a form of cladism sometimes termed 'pattern cladism' or 'transformed cladism' has developed, largely through the work of Nelson (1979), Nelson and Platnick (1981), Patterson (1981a), Humphries (1983), and Humphries and Parenti (1986). Pattern cladistics differs from Hennig's original concept by removing its dependence upon evolutionary theory, and in this way claiming greater objectivity (a claim which is not uncontroversial). Pattern cladistics is free from *a priori* commitment to any geological or geophysical theory (as also is panbiogeography). D.E. Rosen (1978) has maintained that "a geological area cladogram neither tests nor in any way affects the generality of a biological area cladogram. A geological area cladogram differing from a biological pattern does not refute the pattern, it is simply irrelevant to it, because it contains no explanatory information regarding the biological pattern". The empirical basis of pattern cladistics is the formation of groups of sister taxa by the most parsimonious distribution of characters. This is extended to biogeography as the formation of consensus area cladograms by the substitution of sister taxa with the areas in which they occur. When real data are analysed, different hierarchical patterns of area relationship can result, and the criterion of parsimony allows a choice to be made between competing area cladograms. The main emphasis among pattern cladists in a biogeographical context has been the development of an array of analytical techniques (component analysis) to resolve a large range of conflicting patterns parsimoniously. Chapter 12 reviews the main methods, discusses their efficacy and outlines some recent refinements to the technique.

In all cladistic methods, randomness in distribution patterns is held as a null hypothesis which is refuted by finding area cladograms which are replicated with greater frequency than could be accounted for by chance alone.

Recently (Rosen, 1984; Rosen and Smith, 1987) a new analytical method based on parsimony has been developed, the data base of which is the distribution of endemic taxa in the area studied. Unlike other cladistic methods, the phylogenetic relationships of the endemic taxa are not considered, but shared presences of taxa within an area (endemicities) are treated in the same way that synapomorphies are treated in cladistics. The method, known as parsimony analysis of endemicity (PAE) is described in Chapter 14.

In obviating the need for any prior phylogenetic analysis of the taxa involved, and in being therefore a primary biogeographic method, PAE resembles panbiogeography. It differs fundamentally, however, in the way it uses the base data. PAE looks for the most parsimonious clustering of endemicities, while panbiogeography looks for the repeated congruence of 'tracks'. Tracks are lines on a map joining the modern and fossil centres of distribution of isolated (disjunct) populations of a species or of higher taxa. Tracks are oriented in terms of the sea or ocean basins that they cross or circumscribe. A number of criteria for track construction are applied and are described in Chapter 13. When track patterns are repeated many times by unrelated organisms, the combined tracks (generalized track) are assumed to represent a once continuous biota which has become vicariated. A diagnostic base-line for the generalized tracks analysed is then hypothesized as a summary of vicariance episodes. Convergences of generalized tracks are assumed to represent tectonic convergence where fragments of two or more ancestral biotic and geologic worlds meet and interdigitate in space and time. Panbiogeographers argue that since taxa are spatiotemporally bounded entities, the spatial element is an important component of the evolutionary process and that phylogenetic relationships cannot be determined by purely typological methods (in the sense that taxa have ontological status as classes that can be defined in terms of characters, e.g. synapomorphies that are both necessary and sufficient for membership). With regard to a spatial component at least, there is some agreement with the philosophy, if not the method, of phylogenetic biogeography. Both stress the importance of the spatiotemporal element of the speciation process, whereas cladistic vicariance biogeographers do not incorporate hypotheses of evolutionary process into their methodology. Panbiogeographers stress the important reciprocal relationship between distributional and phylogenetic data. Chapter 13 provides examples of how biogeographic analyses can generate phylogenetic hypotheses, test established ones, and help to choose between conflicting hypotheses. Panbiogeography has been criticized for being broadly descriptive and not amenable to rigorous statistical testing. Craw (Chapter 13), in response to this criticism, elaborates a method developed by Page (1987) involving graph theory, where drawing a track is equivalent to finding the minimal spanning tree that connects the occurrences of a taxon. Rigorous testing of panbiogeographic hypotheses by using connectivity matrices (Page, 1987) is then said to be possible.

It is clear from the foregoing that there are fundamental differences between the cladistic and panbiogeographic methods, between the evolutionary (Hennigian) and transformed cladistic methods and between all of these methods and those of parsimony analysis of endemicity (Fig. IV.1). Whilst attempts have been made to merge Croizat's panbiogeography with Hennigian cladistic techniques, the two approaches to biogeographic analysis are philosophically quite incompatible, and the underlying philosophy of the former has often been misinterpreted. All cladists are agreed that no meaningful biogeographic analysis can be carried out without a thorough prior phylogenetic analysis. Panbiogeographers on the other hand believe that a prior phylogenetic analysis is not essential and claim that a primary panbiogeographic analysis can in fact help to elucidate phylogeny. Parsimony analysis of endemicity is also performed without a prior phylogenetic analysis, but since the method is based on endemicity alone no inference on phylogenetic relationship can be made. Phylogenetic biogeographers (*sensu* Brundin) believe that process and pattern must be considered together and stress that the process of speciation results in orderly and decipherable space occupancy by taxa through time. With this, at least, panbiogeographers would doubtless agree. Pattern cladists, on the other hand, argue that all that is known is pattern and that anything else is purely hypothesis. They reason, therefore, that pattern analysis is the only objective method in biogeography. It should be borne in mind, however, that no existing method of reconstructing earth history from biological data is truly objective since none can rigorously test either the nature or timing of geological events. When a biological area cladogram has been produced, it still has to be compared with an independent palaeogeography on a narrative basis (Myers, 1988; Chapter 12).

THE FUTURE

It is perhaps useful here to return to the problem of scale first introduced in Chapter 1. Do the various methods described in this section work equally well at all scales? The different methods have not yet been rigorously applied to comparable spatial, temporal and taxonomic scales, nor has each method been systematically applied to different scale situations. Chapter 13 cites a specific application of panbiogeography to a small-scale area (New Zealand) on a small temporal scale (post-Pliocene) and small taxonomic scale (species). In this example, intraspecific (not sister species) distributions are being examined and the question of whether or not a prior phylogenetic analysis should be carried out clearly does not apply. It is doubtful whether a cladistic analysis would be appropriate anyway in this case, since only two areas are involved (southern North Island and northern South Island). A minimum of three areas is needed for a cladistic analysis. The same chapter considers generic distribution in Australia, the Moluccas, New Zealand, New Caledonia and New Guinea as well as more general distributions in the Southern Hemisphere on a broader time-scale. How effective the panbiogeographic method is, at higher taxonomic levels, in the absence of prior phylogenetic analysis is a matter of controversy. Phylogeny can never be known, only hypothesized. Phylogenetic reconstructions derived from cladistic methods or panbiogeographic methods are not open to independent testing except by congruence with the results of other methods. Ignoring the spatial element of the evolutionary processes on the other hand would seem to lead to the loss of potentially valuable information. As emphasized in Chapter 13 a primary spatial analysis can reveal phylogenetic relationships hitherto obscured. PAE methods are too recent for assessment of their applicability, but they appear to be most useful at large spatial scales (e.g. Pacific ocean, Tethys), and large time-scales.

The various philosophical approaches to historical biogeography have been surrounded by outright and often bitter controversy. These controversies have led to such a watershed that impartiality becomes virtually impossible. We have, of necessity, expressed some bias by our selection of cladistic vicariance biogeography (both Hennigian and transformed cladistic-based methods) and panbiogeography, and our rejection of dispersal-centred historical biogeography. Both cladistic vicariance biogeography and panbiogeography have been criticized, but much of the criticism has been aimed at the various *ad hoc* accretions with which the competing methods have been surrounded as they have been modified or expanded. In our view, the rigorous cladistic core of cladistic vicariance biogeography and the important primary spatial base of panbiogeography form an essential groundplan for an interpretation of the space-time evolutionary framework of historical biogeography. A multiplicity of methods is in itself useful as it provides an enhanced possibility for testing hypotheses through congruence of independently derived reconstructions of biotic and geologic histories. We would not exclude parsimony analysis of endemicity as a broadly applicable biogeographic method of reconstruction which has a particular application to fossils although it is not yet fully tested.

CHAPTER 10

Refugia

J. D. LYNCH

10.1 INTRODUCTION

Biogeographic refugia, if we can identify them, provide indirect evidence of events in the past which have moulded the shape of the present. Generally, refugia are geographic areas whose limits are distinct. They are envisaged as biogeographic arks protecting their passengers from certain extinction during some series of environmental disturbance events. Thus, they tend to be small areas, often tucked into a corner of some larger ecogeographic unit identified today. They also serve another implicit function, as causal elements in the speciation process (Chapter 7). What makes refugia of great interest is the fact that they involve a large subset of a regional biota. These arks are not conceived to explain the survivorship of a single taxon. Rather, they provide a means of explaining large scale survival – and, of necessity, demand pattern as a consequence. In this chapter, the Pleistocene forest refugia hypothesis is examined as a model of these arks.

The tropical rain forests of South America exhibit a biotic diversity unparalleled by any other ecogeographic unit in the world. The rich and varied fauna and flora had scarcely been sampled before Charles Darwin skirted the fringes of this immense forest in 1832. It began to be discovered in the middle of the nineteenth century by Henry Bates, Richard Spruce, and Alfred Wallace. Those initial forays only whetted the appetites of other biologists who raced headlong into this biological wellspring over the next century. Although there are endemic species of animals in virtually every phytographic formation (e.g. see Müller, 1973, for neotropical vertebrates), ecologists have long noted that with the structural complexity of rain forests there is at least a proportional animal diversity. Dry tropical forests may be green much of the year but they lack the structural complexity and the biotic diversity associated with tropical rain forests.

In the late 1960s, biologists had documented much of the diversity of the neotropical forests, so much so that these forests and their biota assumed a nearly mythical quality. This diversity stood in sharp contrast to what passed for diversity elsewhere and could be subsumed into many general ecological descriptions of the planet. The diversity revealed by Bates, Spruce, and Wallace became one extreme in a latitudinal diversity gradient. The facts about the biota of these great forests became mingled with anecdotes and all fitted rather nicely into an image of these forests as a sort of living biotic museum whose diversity was a product of the climatic stability enjoyed over great expanses of time. Many an ecological theory appeared, based at least in part on this tranquil, ancient, and diverse biota where competition had moulded a highly structured community (Chapter 3: Chapter 9).

In 1969, Haffer shattered the stillness with the publication of his paper on the Pleistocene refugia of lowland forests. He argued that the diversity was a product of spatial heterogeneity as well as a series of contractions and expansions of the forests during the Pleistocene. Although the forests appear at present to be an essentially unbroken expanse of multi-storied rain forest, in the recent past it had existed precariously as a series of isolated forest islands in a sea of grasslands. With each glacial advance in the North Temperate regions, the rain forest shrank into fragments; with the retreat of the glaciers, the fragments enlarged, fused, and extended over outlying areas. The organisms who called the forests home were herded along into new territories or into crowded refuges with the vagaries of climatic cycles (Fig. 10.1). During contractions, while isolated geographically, these forest-dwellers underwent speciation, and when their ranges expanded with the expanding forests, they came into contact with their now differentiated kin forming various contact zones but retaining their separate identities. Over the next few years, the Pleistocene neotropical lowland forest refugia hypothesis gained extensive support.

The rain forest refugia hypothesis would be important in nearly any context in biogeography. However, the introduction above is intended to highlight one of the serious problems in biogeography. Haffer made no test of the prevailing 'explanation' for tropical diversity. He simply provided a very different 'explanation' for the same body of facts augmented by some observations that he had made on birds. The prevailing explanation was not discarded so we now have two mutually exclusive explanations both somehow preserved, if not as equals, as having some truth. Such opposing explanations should remind us of the dangers of entertaining unfalsifiable hypotheses or of being content with a theory having much 'explanatory' power. The widespread popularity of the hypothesis advocated by Ab'Saber, Brown, Haffer, and Simpson should not be taken for evidence that it is true. If it is true, it should be capable of withstanding critical testing. This chapter is about the hypothesis and the efforts to falsify it.

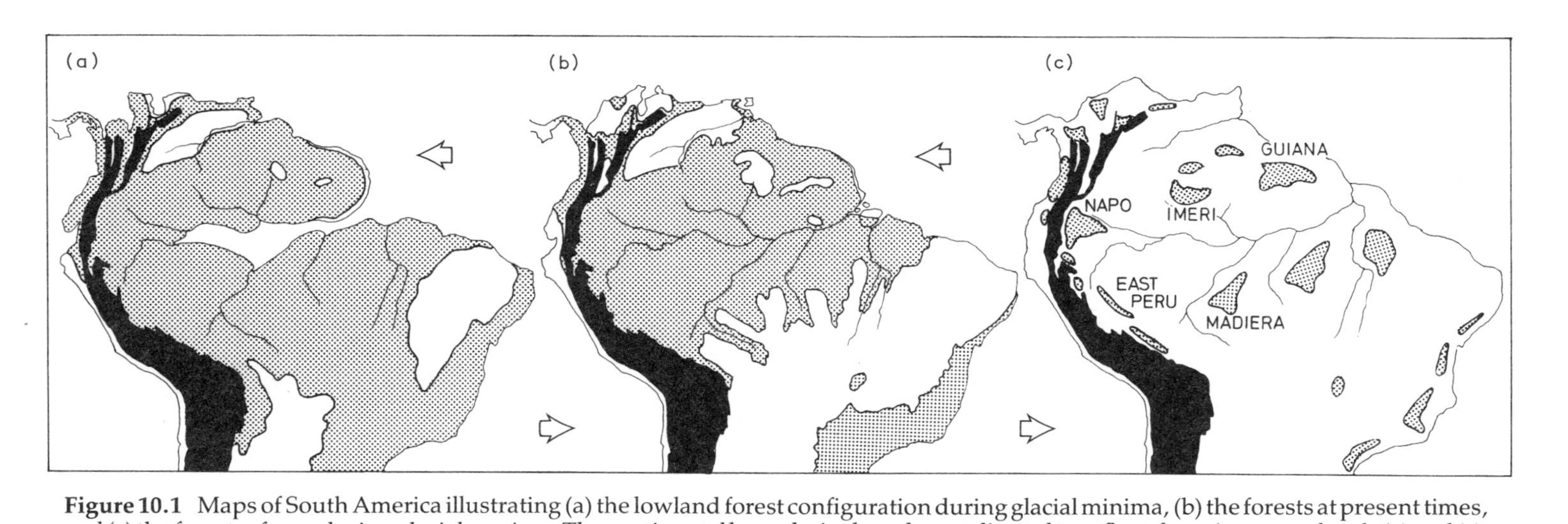

Figure 10.1 Maps of South America illustrating (a) the lowland forest configuration during glacial minima, (b) the forests at present times, and (c) the forest refuges during glacial maxima. The continental boundaries have been adjusted to reflect changing ocean levels (a) and (c). The Andes (above 1000 m) are indicated as solid black. In all maps, the general outline of the Amazon drainage is preserved so as to allow comparisons. Adapted from various sources, chiefly Dixon (1979), Haffer (1969, 1974, 1979), Hueck (1966), Lynch (1979), Schmithusen (1968), and Simpson and Haffer (1978).

10.2 THE PLEISTOCENE RAIN FOREST REFUGIA HYPOTHESIS

10.2.1 Contrasting proposals

Haffer's (1969) proposal of the refugia hypothesis came about the same time that Müller (1973, 1974) proposed his centres of dispersal. The two proposals may be profitably contrasted in that each utilized some of the same data. Müller's centres required the presence of one or more endemic species per centre but did not require any necessary connection between centres. Species with recognized subspecies were of some value to Müller if there was an endemic subspecies found in the centre. However, species spread over two or more centres without showing differentiation into subspecies were irrelevant data, serving neither as corroborators nor as falsifiers. Müller's major concern (1973) was the determination of centres of origin once the centres of dispersal had been identified. The attention to centres of origin perhaps explains Müller's emphasis on the presence of monotypic genera and of endemic genera (having two or more endemic species) within a dispersal centre. The only objective is to demonstrate that the area (centre) is actually distinct. This use of level of differentiation characterizes a very different sort of biogeography from that which Haffer addressed (descriptive rather than explanatory and predictive).

Haffer's proposal did not utilize unique taxa. He did require endemicity to be present (otherwise refugia and contact zones would never be recognized from distributional data) but coupled it with the requirement that sets of species be employed. His scheme differed from that of Müller in: (1) not using species of monotypic groups, (2) in requiring that the groups providing the endemic species be made up of no fewer than two species, and (3) in containing two levels of generality rather than one, so addressing two dimensions (time and space) rather than one (space). One other apparent advantage was that Haffer could use the data provided by a set of two species, neither of which was endemic to what would subsequently be identified as separate refugia. His advantage derived from his use not only of occurrence but of the range limits. These range limits were the subset involved in the transition from one taxon to another (but not the transition from one taxon to the absence of any member of the set).

The two approaches can be most effectively distinguished by contrasting how they utilize the same data sets. Within the forests of eastern Ecuador and adjacent countries, one finds many species of the leptodactylid frog genus *Eleutherodactylus*. Most of the species found in this wet forest are restricted in their distributions to the upper Amazon Basin. Consider first the distribution of a small leptodactylid frog (Fig. 10.2a).

Müller would use this as a species endemic to the Amazon centre. However, this species belongs to a clade otherwise distributed in montane environments in the Andes of Colombia (Lynch, 1981). Another species, *E. nigrovittatus*, having a similar distribution (Lynch, 1979) to that in Fig. 10.2a (except that it is confined to Müller's Napo subcentre) does serve to outline much of what Haffer (1969) associated with the Napo refuge, but the species cannot apply in Haffer's scheme because it has no other lowland relatives, repeating the 'pattern' seen in the other species (Fig. 10.2a). Its nearest relatives live in the Andes, mostly at elevations above 1000 metres (Lynch and Duellman, 1980).

Each approach would utilize the data in Fig. 10.2b,c. The data are not of actual species but represent a polytypic species (Fig. 10.2b) having subspecies associated with four different refuges (dashed lines). For Müller, the level of differentiation (failure to speciate fully) would be evidence for grouping the four areas. For Haffer, the important data are that the forms have differentiated into geographically distinct forms and that the suture zones (zones of intergradation) are defined. The other data set (Fig. 10.2c) is based on a subset of the *Eleutherodactylus conspicillatus* group (Lynch, 1986). The five species are allopatric and each is associated with one or more postulated refuges. Within so large a genus as *Eleutherodactylus*, having over 400 recognized species (Frost, 1985), there are many subunits. However, if the genus is monophyletic, there can be only one centre of origin. It is not clear that Müller actually assigned that centre of origin to the West Indies (Müller, 1973), but he treated species of the genus as independent elements, exactly as one would do if the centre of origin were distinct from the centre of dispersal. Had Müller known that these five species formed a monophyletic group within the genus he would have treated them in much the same fashion as would Haffer; however, Müller utilized polytypic species preferentially in order to develop his affinity groups (sets of dispersal centres). For Haffer, these five species would be used in recovering a pattern of refuges. These data would be factored in with other acceptable data such as sets of species having two or more representatives in lowland tropical rain forests – each species being found in one or more postulated refuges. A monotypic genus which is relatively widespread, *Lithodytes* (Fig. 10.2d) ought not contribute to either hypothesis yet was identified as a faunal element of the 'Guyanan centre' by Müller (1973) as was the large toad, *Ceratophrys cornuta* (Fig. 10.3). Haffer would not use *Lithodytes lineatus* (Fig. 10.2d) because it is monotypic and would presumably not utilize the subgenus *Stombus* (Fig. 10.3) because only one species occurs in rainforests. However, the genus *Ceratophrys* does have a species confined to the rain forests of

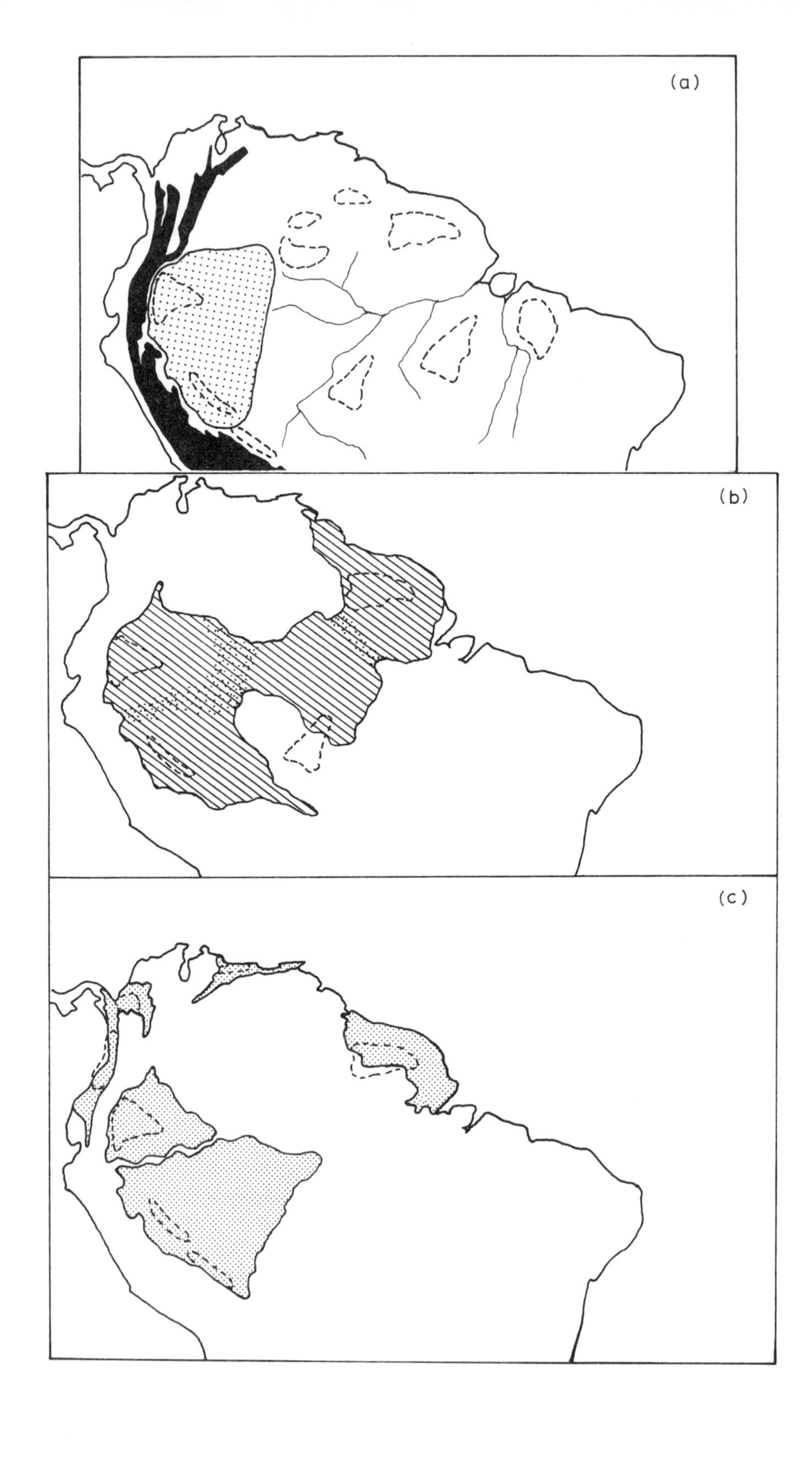
(a)
(b)
(c)

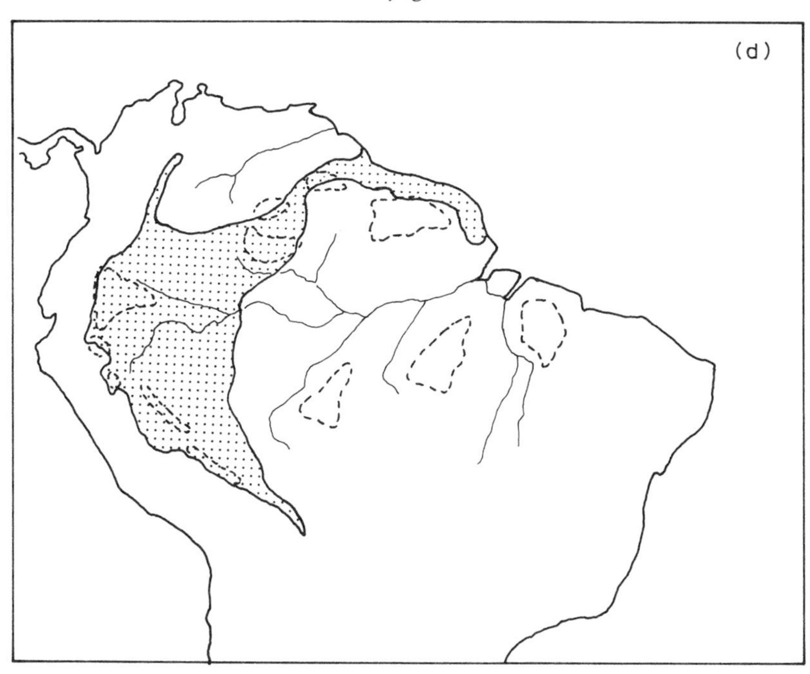

Figure 10.2 (a) Distribution of *Eleutherodactylus sulcatus* in upper Amazonia. The species is endemic to the Napo-Ucayali refuge area but has no lowland relatives. Its nearest relatives are species confined to Andean forests (Lynch, 1981). Postulated refuges are indicated as areas formed by dashed lines. (b) A hypothetical distribution of a species having four subspecies. The subspecies are more or less associated with four different refuges (dashed lines). (c) A hypothetical distribution of a superspecies having five included species, each restricted to forested habitats. Two species have parapatric distributions (upper Amazon), otherwise the superspecies is distributed with sharp disjunctions. Each species is associated with one or more refuges (dashed lines). (d) Distribution of *Lithodytes lineatus*, a frog, family Leptodactylidae. Postulated refuges are indicated with dashed lines. After Lynch (1979).

south-eastern Brazil (Lynch, 1982) and thus Haffer might well have utilized the genus even though only two named species have distributions in rain forests (as does one unnamed taxon). For purposes of hypothesis formulation Haffer would have ignored the four species of *Ceratophrys* distributed in more xeric habitats. Both Haffer and Müller would ignore a species having a distribution away from forests (even when its distribution corresponds to the edges of forests) in terms of formulating a hypothesis about forests.

Haffer (1969) used Chachalaca and Toucanet birds to illustrate the

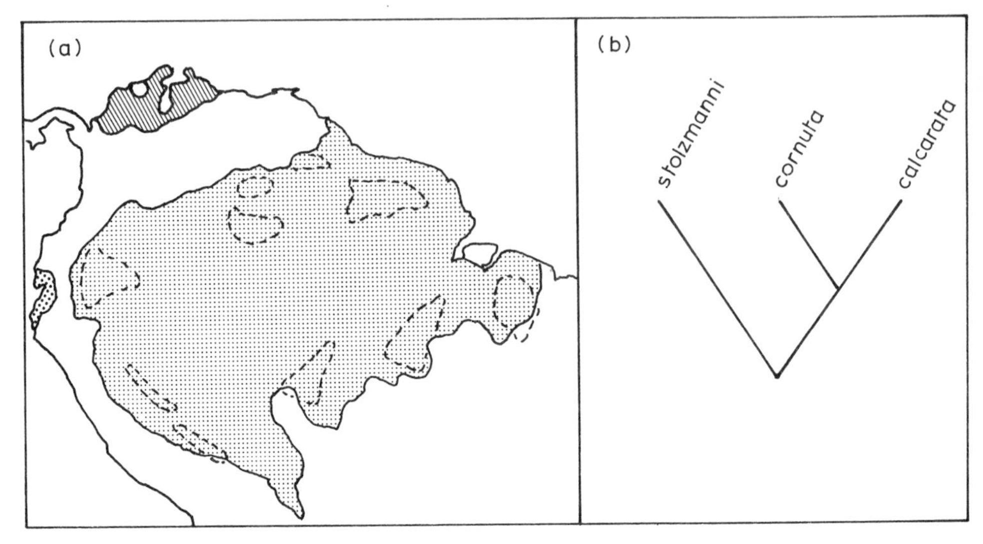

Figure 10.3 (a) Distribution of species of the subgenus *Stombus* (genus *Ceratophrys*). The largest distribution (*C. cornuta*) is for a species distributed over most of the Amazonian rainforest. Postulated refuges are indicated with dashed lines. (b) Cladogram of relationships within *Stombus* (*C. calcarata* is the northern species and *C. stolzmanni* is the western one, both confined to xeric environments). After Lynch (1982).

coincidences of distributional patterns between data sets (Fig. 10.4). Under Haffer's hypothesis the number of refuges required equals the number of allopatric species. However, that number varies with each data set. For example, if we examine his *Ortalis* data (Fig. 10.4a), we find that *O. guttata* embraces three sets of refuges (Napo, eastern Peruvian, and Madeira-Tapajós) whereas *O. motmot* embraces two (Imerí and Guiana). His *Selenidera* data (Fig. 10.4b) call for a different combination such that *S. culik* is associated with the Guiana refuge and *S. nattereri* with the Imerí refuge. The Napo refuge is occupied by a single species (*S. reinwardtii*) as are the eastern Peruvian refuges (*S. langsdorfii*). These two groups of birds have three (*Ortalis*) or five (*Selenidera*) Amazonian species but, taking the two groups simultaneously, we require six refuges to accommodate all the data. It appears that as more relevant data are added to the system, Haffer's scheme reflects more subdivisions (refuges) whereas Müller's scheme reflects fewer divisions (centres of dispersal).

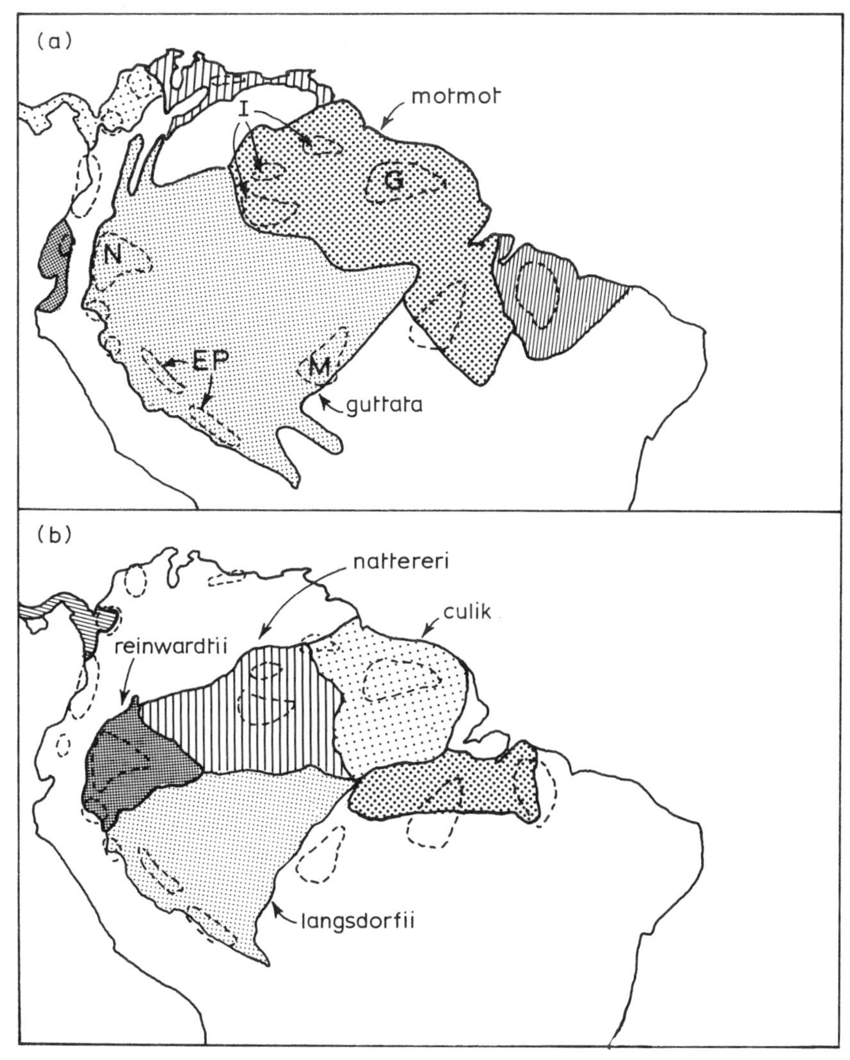

Figure 10.4 (a) Distribution of a superspecies of birds, genus *Ortalis*. (b) distribution of another superspecies of birds, genus *Selenidera*. After Haffer (1969). Postulated refuges are shown with dashed lines (N = Napo, E.P. = East Peru, M. = Madeira, G. = Guiana, I = Imerí).

10.2.2 Components of the rain forest refugia hypothesis

The chief difference between the Haffer and the Müller models in terms of primary distributional data is that the former made constant use of the implicit phylogenetic hypotheses (i.e. species relationships) embedded in the classificatory data employed. It was this use of information from phylogenetic groups of taxa that enabled Haffer to address the time dimension. Anyone using only an endemic taxon can accomplish nothing more than the identification of the area occupied by that taxon; no time component is implied, no matter how many endemic taxa are cited. By restricting himself to the members of a superspecies, Haffer called upon a phylogenetic diagram having two levels – one for which the superspecies is a component, and the second for which members of the superspecies are components. Nowhere in Haffer's published analyses is there evidence that he employed more complex phylogenetic diagrams which would specify more steps in a temporal sequence (Fig. 10.5b). Haffer appears to have always used the minimum information contained in the hierarchy of a classification – all the members of one of his sets of species are envisioned as evolving simultaneously (Fig. 10.5a). It is thus not surprising that his model does not reach farther back into time than the most recent global event – the model is predisposed to be a Pleistocene model. However, if we did not know about the Pleistocene but were aware of some other set of events having a global impact, Haffer's model would appear to correspond to those events.

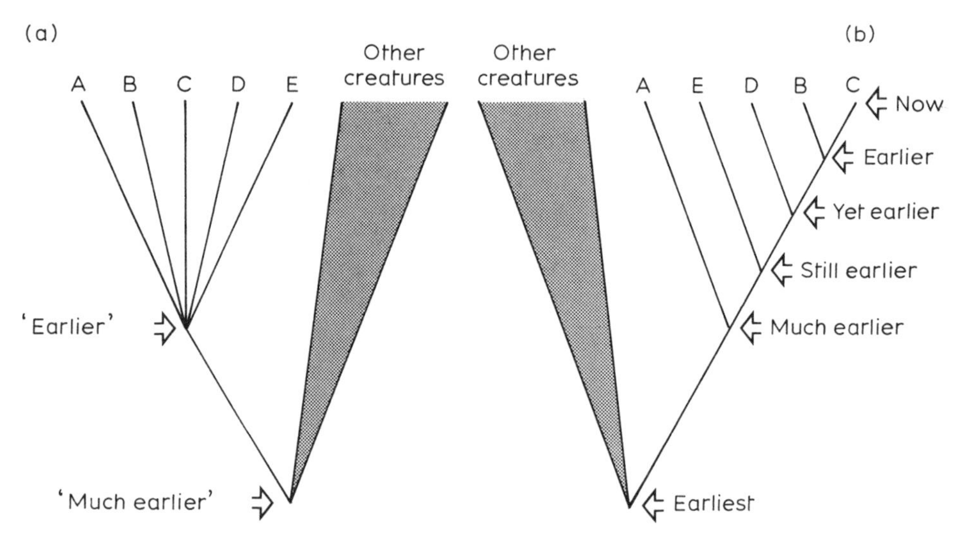

Figure 10.5 (a) The form of the implicit phylogenetic hypotheses employed by Haffer. The superspecies, containing species A, B, C, D, and E, forms a subset distinct from all other taxa and two time periods are defined. (b) A fully resolved phylogenetic diagram wherein five time periods are defined – more information is contained within (b) than in (a).

Haffer did not commit the common 'error' of assuming that the differentiation of a set of subspecies was younger than a set of species. He used the coincidences of the patterns of their distributions in order to define the pattern of endemism; the first component of the hypothesis meant that Haffer would later need to 'explain' the different rates of evolution (or speciation) seen among the representatives of the pattern of endemism.

The general distributional pattern discovered by Haffer was more than simply a sum of the 'centres' of endemism because Haffer also made use of the transition zones between member pairs. These transition zones were zones of intergradation (secondary contact in Haffer's terms) when the taxa were conspecifics or were the narrow zones of sympatry (suture zones) in the cases of differentiated (parapatric) species (see Chapter 7 for details). The repetition in the pattern of endemism seen in distinct superspecies and genera of birds was thus reflected in the presence of core distributional areas being more or less congruent and in the presence in the interspaces of transition zones which were more or less coincident.

At this stage of the analysis one might well ask whether or not one should be impressed by these coincidences. Haffer never explored this question in print.

The second component of the hypothesis followed after Haffer's original (1969) model was proposed. It was not in fact different, it only seemed different. This component was based on the presence of organisms having different capacities and requirements from those of birds but exhibiting distributional patterns congruent with them. Many of these data are summarized and added to in Simpson and Haffer (1978), Duellman (1979), and Prance (1982a). Much of the success of the model is that the pattern first detected in birds was 'repeated' in several other groups of organisms. Haffer guessed that the probability that this was due to chance was too low to worry about and argued that if there is a single pattern, then we require a single, sufficient, cause. That cause could not be idiographic simply because such divergent sorts of organisms exhibited the same pattern.

The third component of the hypothesis arose from his search for a sufficient cause for the distributional pattern when Haffer in effect argued that the distributional data are informative of history. What he did was to make the assumption that the presence of the organism antedated the existing topography, in other words, that the organisms demonstrated rain forest fidelity. This assumption (A) means that past changes in the geography will be imprinted in the matrix of the organism (in the form of changes recognized by the taxonomist). The opposite assumption (B) seems equally risky but is not. It would hold that the organism was somewhere else when the changes in the landform occurred and the organism has subsequently dispersed onto the landform. Assumption (B) – dispersal – is less desirable because it reduces rather than increases the potential number of falsifiers of the hypothesis.

Haffer (1969, 1974) and Simpson and Haffer (1978) noted that the Amazonian forests were not as uniform as they appeared to the casual observer, but differed in terms of rainfall – the basis of the fourth component of his model. Haffer noted that the areas of highest precipitation were generally concordant with the areas of endemism in his pattern and that the transition zones in the pattern of endemism were generally concordant with the areas having less precipitation.

Various rain forests in lowland South America are exceptionally wet (up to 10 000 mm rainfall p.a.) whereas others are comparatively dry (as little as 1500 mm rainfall p.a.). In addition, the seasonality of rainfall also determines whether or not an area will support a rainforest (Richards, 1952). For his sufficient cause, Haffer hypothesized that a universal reduction in the amount of precipitation over the landform under consideration (now supporting rain forest) would result in a retraction of rain forests to a system of isolated islands in a sea of non-forest habitats. Such a system of rain forest isolates would provide the geographic isolation necessary to permit the development of differentiated populations and reproductive isolation. The refugia model was thus articulated into a global water cycle influenced by the waxing and waning of glaciers during the Pleistocene, providing the fifth component of his hypothesis. Haffer offered a model that was sufficient to produce a series of allopatric/parapatric species distributed through the now contiguous rain forests.

Each of the above five components were what Haffer termed indirect evidence. In the late 1970s, Ab'Saber (1977, 1982) provided what Haffer has termed direct evidence (component six) in the form of xeric fossil soils within existing forests and in some of the areas for which Haffer had predicted loss of rain forests during dry cycles. Some of these were dated by ^{14}C as being no older than 8000 years old.

10.2.3 Summary

The explicit claims of the model are: (1) that the speciation events are Pleistocene in age, and (2) that the speciation events of different members of a postulated refugial community are coincident in time and space (i.e. are due to a simultaneous cause).

This model, like any testable model involving a series of areas of endemism, does not predict that speciation must occur when refugia are separated nor does it predict that extinctions must occur in the 'compressed' communities 'forced' into the postulated refuges (Chapter 8).

If the model made those predictions, it would be much easier to test. Those are possible outcomes of the model, but not necessary outcomes, for a particular monophyletic group. In explaining the model, Haffer has suggested that species of larger organisms would show less effect, or would require more time to reach detectable effects, than would species of smaller organisms (for which Haffer assumes that evolution proceeds at a more rapid rate).

The components of the Pleistocene forest refugia model are therefore as follows:

(1) At least two sets of species-level taxa showing coincidences of distributional areas and of the suture zones between included members.
(2) Other groups of organisms having distinct dispersal powers or natural histories (biologies) in contrast to those seen in the organisms of component (1) but exhibiting the same pattern of distribution as detected in component (1).
(3) Forest fidelity, that is, the organisms identified as components (1) and (2) are organisms unique to tropical lowland rain forest. Forest fidelity also reflects the assumption (A) that dispersal can be ignored (see above).
(4) Precipitation component, which is related to the above points because rainfall is required to sustain a tropical rain forest. Beyond that, this component reflects the variation in rainfall amounts across the rain forest habitat and is a geographic pattern of variation.
(5) Global hydrological cycle. As water is sequestered in glaciers, there is a reduction in the amount of rainfall at each locality; when glaciers are reduced, the freed water floods the hydrologic pool and rainfall amounts are increased.
(6) The direct evidence of fossil soils.

Haffer's model works if the following conditions are met:

(a) The populations 'trapped' in the forest refugia are in fact trapped.
(b) Alterations of the total amount of water available for precipitation do not alter other climatic features such as wind patterns.
(c) Changes in the precipitation base are simultaneous (if not, then refuges are no longer the same ages).
(d) Dispersal is invoked only to account for movements within habitats and never between habitats (dispersals across barriers are denied).
(e) The time available is sufficient to permit speciation to such a level that upon meeting, daughter species serve as self-barriers and form narrow zones of overlap.

The model is so attractive that it seems to be necessary. However, our greatest fear as biogeographers is that we may confuse sufficiency with necessity. That the model seems overwhelmingly corroborated is no assurance that it is correct or even that it comes close to representing reality. Unfortunately, the model is almost intuitive in that it appears self-evident. This has led many observers to interpret their data in terms of Haffer's model (Duellman, 1972). These 'additional' data sets do not provide evidence for the model nor do they serve to test it. If the model is a scientific one (i.e. a testable model), we should first be able to state what sorts of data would be sufficient to falsify the hypothesis. Until such specification is made, we cannot be certain that the hypothesis is testable.

Not every tropical biologist has approached Haffer's model with either approbation or enthusiasm. Several have challenged it in the sense of proposing that their data sets showed that the hypothesis was either partially or entirely false. Others have been less direct and have argued that some other explanation (not specified) was probably true. Direct and indirect assaults have been made on the model. These are dicsussed below.

10.3 TESTING STRATEGIES

The refugia hypothesis consists of a pattern and a process. Attacks made upon the model may be divided conveniently into pattern attacks and process attacks. Pattern attacks dispute the very existence of a pattern, obviating any attention to process, whereas process attacks question the necessity of the process or its sufficiency to produce the specific pattern. For example, if the same pattern emerges from two different processes, then we are left with only the descriptive value of the pattern. There is no necessary cause for the observed pattern, making it a trivial coincidence possibly due to chance alone.

10.3.1 Is there a pattern?

Components (1) and (2) of Haffer's model are the most attractive elements in the model. Under the best of circumstances, such data would resemble those in Fig. 10.6a. Each of the five exclusively rain forest species is distributed such that a single hypothesized refugium is included within the distribution area of the species. If we have several such groups, each containing at least three species, each will conform to the model and be corroborative. At some point, we will be impressed by the model's 'explanatory' power, in that it seems capable of providing a simultaneous explanation for so many diverse bits of data. So long as we

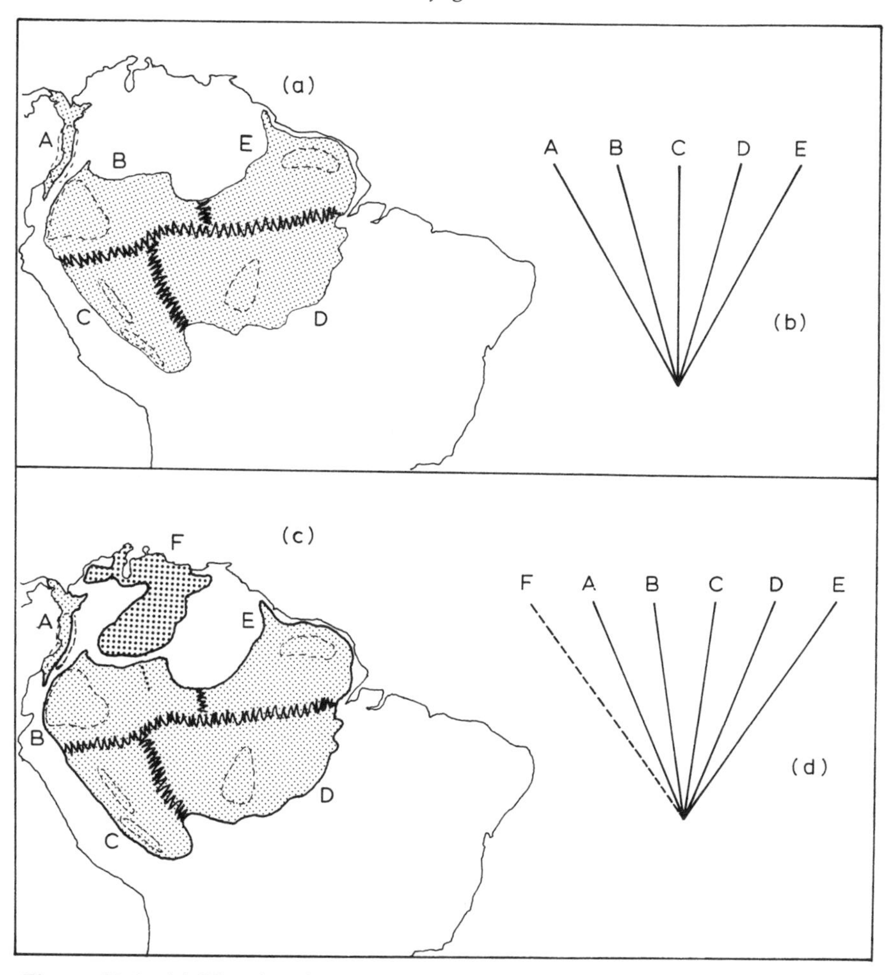

Figure 10.6 (a) The distributions of five species of a monophyletic group in agreement with the Pleistocene forest refugia hypothesis (contact zones indicated with zigzag lines). The non-restrictive phylogenetic hypothesis (b) is in the form used by Haffer. (c) The same group of organisms but with a sixth species, F (more dense stipples), added whose distribution is in open habitats. If the hypothesis of relationships (d) is non-restrictive, this taxon does not cause the group (ABCDE) to disagree with the refuge hypothesis.

have at least three species, and each is restricted to the forested environments, the data support the model crafted from the set of species providing the greatest fragmentation.

When we examine the phylogenetic hypotheses implicit for each group of organisms, we again find that they are entirely compatible (Fig.

10.6b). Every hypothesis simply states that all of the included taxa of the monophyletic group share a common ancestor. Whether the hypothesis is for two or six species makes no difference. There is no way in which phylogenetic hypotheses may disagree with one another (simply because each phylogenetic hypothesis is unrestrictive, i.e. unfalsifiable). Such phylogenetic hypotheses 'require' fragmentation(s) to be simultaneous and thus require some global (or simultaneous) force.

But what if one or more species have distributions entirely outside the rain forest habitat? In Fig. 10.6c, I have used the same distributional diagram as in Fig. 10.6a except that I have included a taxon 'F' whose distributional area consists of the llanos (non-forested environments in north-eastern Colombia and adjacent Venezuela). If we compare the phylogenetic hypothesis for this group (Fig. 10.6c) with the combined phylogenetic hypotheses in Fig. 10.6b, we discover that the presence of a non-endemic species outside the forests has no impact – that is, we are able to secure our refugial hypothesis without hint that there is a grave problem. What has been overlooked is the fact that we can no longer argue that the group exhibits forest fidelity because we have a single species (from group 10.6c) that is a non-forest species. The same result is obtained no matter how many non-forest species are included in the monophyletic group (Fig. 10.7). However, of greater consequence is the realization that a non-exclusive forest group cannot challenge the apparent coincidence of the phylogenetic hypotheses (the coincidental set can be termed the area hypothesis). The area hypothesis is merely a non-restrictive hypothesis linking a group of areas of endemism (Nelson and Platnick, 1981). The non-forest species seem irrelevant to the refuge hypothesis.

However, returning to Fig. 10.6a, our assertion that the group is forest-adapted follows only from the observation that in this monophyletic group no non-forest member occurs. If we have a group with only one non-forest member (Fig. 10.6c), we may no longer assert that the group is forest-adapted (in the case of Fig. 10.7, the group has as many forest species as non-forest species). Of course, we might assert that the forest dwelling species are more closely related to one another than any is to a non-forest member. However, our hypothesis is then endangered because we are making an assumption that is in fact necessary to salvage our hypothesis. In the existing refugial literature one encounters many examples of taxa like those seen in Figs. 10.6b and 10.7, such as the tree frogs of the *Ololygon rostrata* group (Duellman, 1972); the Jacamar birds, of the *Brachygalba albogularis* superspecies, and two groups of Toucans, the *Pteroglossus aracari* superspecies and the *Ramphastos discolorus* superspecies (Haffer, 1974); and the poison-dart frogs, *Dendrobates auratus* species group (Lynch, 1979). The authors

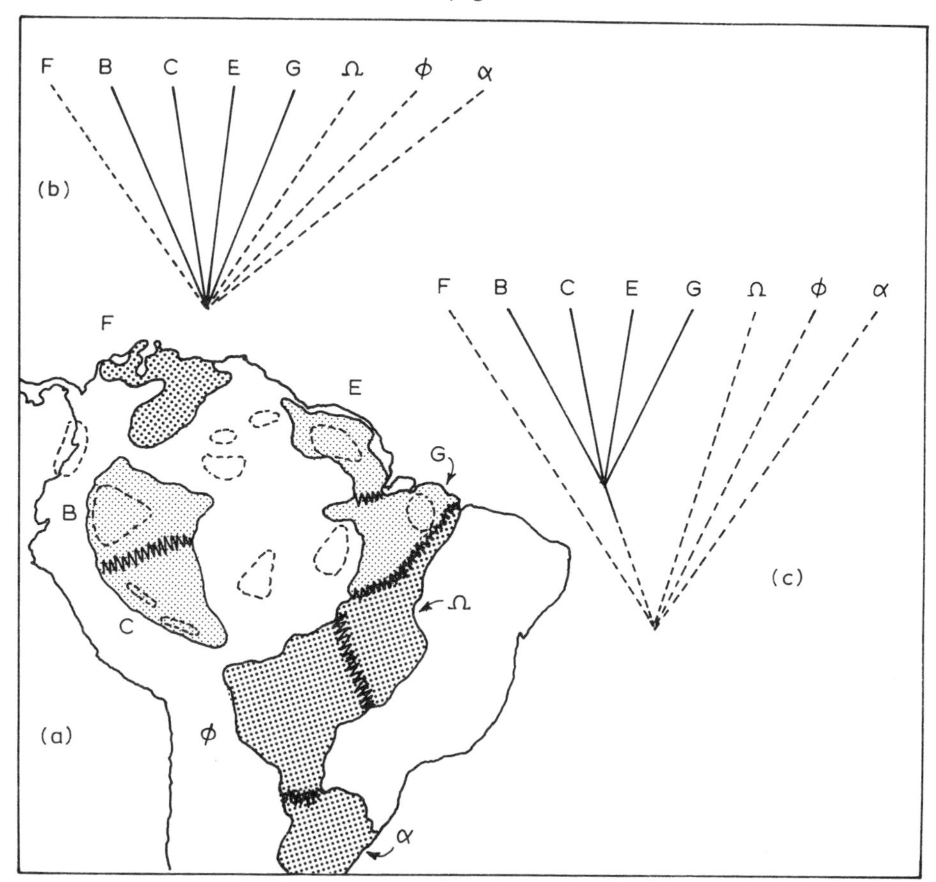

Figure 10.7 (a) Another hypothetical group having four forest species (BCEG) and four species distributed in non-forest habitats (F, Ω, ϕ, and α). The non-restrictive phylogenetic hypothesis (b) is compatible with the refuge hypothesis. If the forested species are each other's nearest relatives, the phylogenetic hypothesis can still be non-restrictive (c) but assumes a very different form than (b).

reporting these make no reference to the existence of any serious problem in their use (with the notable exception of Heyer and Maxson (1982) and see below). Likewise, they make no assertions that the forested members of a set are more closely related to one another than any one is to a non-forest member of the set. Their failure to comment on this problem suggests that they were unaware that it was a problem for the forest fidelity component.

However, if we assume that the forest-dwelling species of a group (which is not exclusively forest-dwelling) are more closely related to one another than any is to a non-forest member (the solution apparently made by those authors reporting such data sets), then the phylogenetic hypothesis required is different (Fig. 10.7c) from that for Fig. 10.7b. The difference is significant, since it essentially tells us that the four non-forest members of the set are not members of the set (the set is actually the subset BCEG). We have ignored any non-forest dwelling taxa of the group and, consequently, find that our data fit rather nicely with the general area hypothesis.

In order to avoid the problem identified in the preceding paragraph, we need to take the phylogenetic hypothesis data seriously. We cannot employ it some of the time and ignore it at other times. If the group in question, for example, that in Fig. 10.7 is monophyletic, then this group of species cannot be used to support the general area hypothesis or the Pleistocene refugia hypothesis, as component three (forest fidelity) must be discarded from the model. Thus the model fails. On the other hand, if we ignore those groups having some forest and some non-forest members, we are guilty of selectively sampling the world in order to accumulate data sets consistent with our hypothesis.

If we have restrictive phylogenetic hypotheses, we should be able to examine more carefully the question of forest fidelity. We might discover that it is an illusion. In Fig. 10.10 for example, we might find that each forest species of the set is most closely related to (a sister species of) a non-forest species. In that case, we would not be inclined to entertain a hypothesis that explained things about forested environments.

Furthermore, if we have restrictive hypotheses, we might discover that the pattern is not a pattern at all but is an illusion, a true coincidence.

Asking the question 'is there a pattern?' challenges a wide range of corroborative data. Those data were published as though they constituted confirmations of the general prediction of the Pleistocene forest refugia hypothesis. Given the hypothesis, the distributions of the species of an unanalysed group of organisms should be predictable, if Haffer's hypothesis is true. There is an error in this line of logic which is most easily seen by examining the testing power of sets of data, given the hypothesis. The hypothesis of Haffer (1969) recognized nine refugia (three trans-andean and six amazonian). Even with taxa which exhibit sharp forest fidelity, not every group will have cis and trans-andean representations, nor have an endemic species correlated with each postulated refugium. Groups having at least two species found exclusively in forest habitats will however confirm the hypothesis at some level because the suture zone between those members will be near one of the more general suture zones.

The confirmatory nature of these data holds up until there are more than three trans-andean taxa or more than six cis-andean taxa. Once our group consists of more taxa (and more suture zones) than there are refuges, the data begin to challenge the model. The challenge is asymmetrical however, because the challenge requires not rejection of the hypothesis, but an increase in its complexity, i.e. an increase in the number of postulated refuges beyond the nine proposed originally.

Haffer's original proposal of nine refuges may be compared with his paper published in 1982 in which he accommodates the data from other studies of many different organisms. One can see the emerging conflict by contrasting Haffer's (1969) system of nine refuges with Brown's (1977) system of 60 or so sub- and semi-refuges based on studies of nymphalid butterflies. The refugium hypothesis, when based upon the distributions of endemic taxa, increases in complexity (number of and sizes of the postulated refuges) as we consider sets of allopatric species having small individual distributions. The greater complexity of the hypothesis as advocated by Brown and his co-workers may reflect nothing more than that insect species (or subspecies) generally have smaller distribution areas than do most vertebrates.

How does one decide which data are useful in describing the general pattern and which are too detailed to contribute usefully to the general pattern? Success in pattern recognition appears to demand that we do not use all the relevant data.

Cladists have also challenged the hypothesis from the perspective that the unrestrictive phylogenetic hypotheses implicitly employed by Haffer (1969, 1974, 1979, 1982) are not useful. Consider a simple system in which there are but three postulated refuges (Fig. 10.8). For this hypothesis there exist three groups of organisms, each having an endemic species in each of the refuges. As long as we employ unrestrictive phylogenetic hypotheses, the three data sets agree with one another and with the hypothesis of simultaneous refuge formations. The two challenges by cladists are given below (both derive from the insistence by cladists that phylogenetic hypotheses be fully resolved).

Non-synchrony

If fully resolved, a three species phylogenetic hypothesis (and a three area biogeographic hypothesis) requires one speciation (or fragmentation) event to be younger than the other. This challenge is not necessarily serious except that if we have a means of aging speciation events, we will be able to falsify the claim of Pleistocene age. In order to preserve the hypothesis of Pleistocene events being mainly responsible, we must concede that some distributions (not specified) might be older than others (Haffer, 1981 including published discussion with Tattersall).

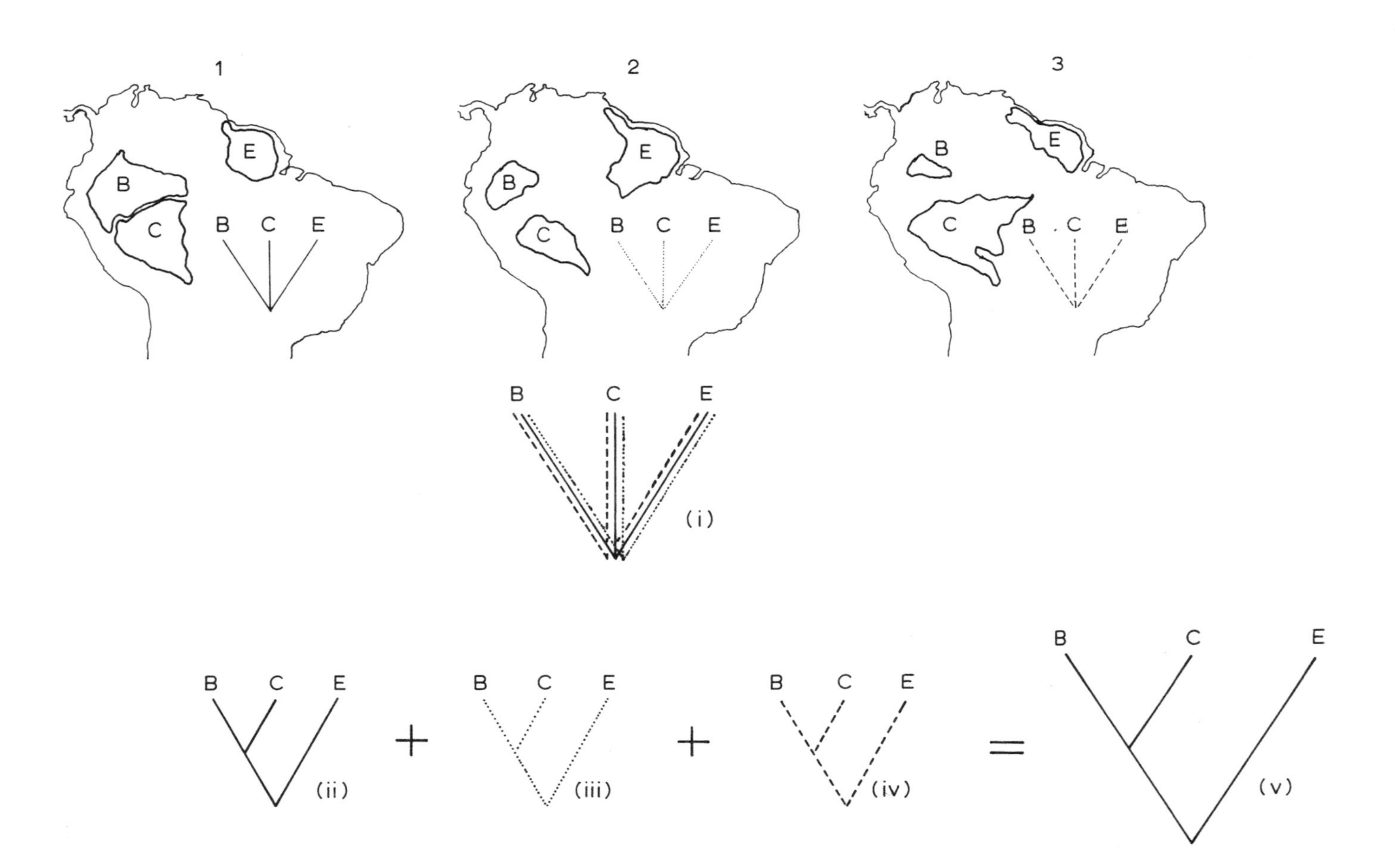
1
2
3
(i)
(ii)
(iii)
(iv)
(v)

Absence of a pattern

The contributing data sets (Fig. 10.8) have the potential of being incompatible in formation (Fig. 10.9). Three simple groups, each having a different set of area relationships, allows the mistaken impression of a pattern when only noise exists. Haffer apparently did not realize that if his hypothesis was true, then a powerful (and independent) means of testing it existed in the developing literature of cladistics. Without concern for what the sequence of refuge formation might be, systematists have the capacity to test (in the sense of to falsify) the hypothesis because the systematist is not limited by how many or how few species are contained in the group. The systematic prediction is that if there is a pattern, then all member elements exhibiting that distributional pattern must exhibit the same area cladogram. This prediction is valid without reliance on which area cladogram might be selected. In the example cited (Figs. 10.8 and 9) there can be three area cladograms and only three – if there is a pattern, then only one area cladogram will be found in the three groups of species having representatives in the three areas. If there is not a single cause to the distribution, i.e. if the 'pattern' (Fig. 10.8) is an illusion, we expect lack of agreement among phylogenetic diagrams (Fig. 10.9).

10.3.2 Unnecessary predictions – a false test

Perhaps the most common error that might be made in testing the Haffer hypothesis is to view a widespread species as evidence that the hypothesis must be false. Haffer never claimed that the postulated fragmentation must cause changes in the subpopulations inferred to have been separated. Thus, Lynch's (1982) observation that *Ceratophrys cornuta* (Fig. 10.3) exhibits no geographic variation is not sufficient evidence to falsify the hypothesis. Haffer (1969) suggested that larger forms would evolve more slowly than smaller forms and thus might be expected to show less evidence of speciation. He did not deal explicitly with this generalization but included it as a descriptive qualifier.

Figure 10.8 Three groups of organisms (1, 2, and 3) each having an endemic species in each of the areas B, C and E. For each group of organisms, the phylogenetic hypothesis is non-restrictive and the three sets of data have coincidental patterns of relationship and distributions (i). If these organisms had actually experienced simultaneous speciation events such that the fragmentation of area E from B and C occurred before the fragmentation of area B from C, then the individual data sets (ii, iii, iv) would combine to form a single area – cladogram (v) where areas B and C form a subset of the area BCE.

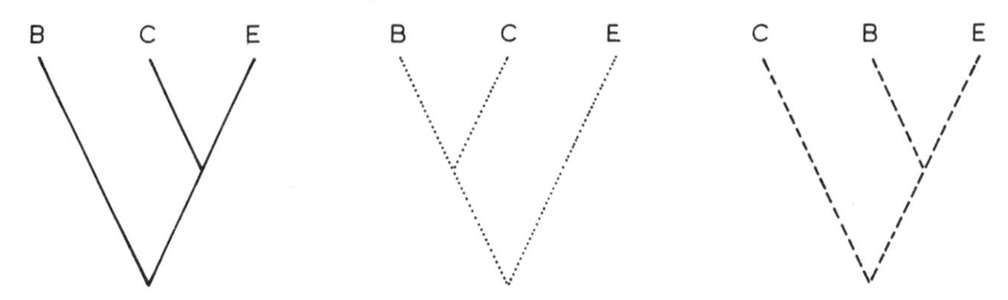

Figure 10.9 Given three areas there are three possible ways to group them. Each of the three is incompatible with the other two. These are the restrictive hypotheses contained within the non-restrictive hypothesis in Fig. 10.8 (i).

According to Haffer's model, a fragmentation event (or series of events) where a single forest (and distribution) is broken into two forests (and distributions) can result in no accumulated differences in subpopulations, in recognizable subspecies, full species or even genera. These different expectations are permissible because Haffer's model does not require a constant rate of evolution for all members of each forest fragment. That he imposed no such requirement makes the hypothesis more attractive but it also makes the hypothesis more difficult to test.

A similar effort to render the hypothesis deterministic is seen in Endler's (1982a) suggestion that when conditions ameliorate and the populations confined to refuges begin to expand toward one another, they will disperse at equal rates. He also suggested that the overlaps between vicars should be greater in more rapidly evolving species (such as butterflies) than in less rapidly evolving species (such as birds). Had Haffer made explicit claims that the model was deterministic, these might be reasonable suggestions. However, Haffer made no such claims. Endler (1982a,b) argued that data from African birds and butterflies were not consistent with the 'predictions' and claimed that therefore the hypothesis deserved less attention. Mayr and O'Hara (1986) defended Haffer's hypothesis by arguing that the African data are either consistent with the Haffer proposal or are too imprecise to allow challenge to the hypothesis.

10.3.3 Testing by consilience

Haffer was not the first author to evoke refugia as biological arks. There is a long tradition in biology for what might be called refugia hypothesizing. Many authors identified refugia from glaciation in the northern hemisphere. Haffer (1974) provided a brief review of earlier refuge models. The 'success' of Haffer's model is reflected in the neotropical model being essentially synonymous with refugial models.

Taking Haffer's model, based as it was on the biota of the neotropical rain forest biome, one may test it by asking what demands it makes on the world outside this biome. The predictions for other biomes, including non-tropical ones, are testable. This method of testing is neither deterministic nor corroborative, rather it works by consilience. If Haffer's hypothesis is true, then we should be able to organize a broad array of data sets from other biomes.

This approach is suggested by Haffer's (1982) world map of lowland forest refuges during glacial maxima. Testing by consilience led Endler (1982a,b) to examine the bird distributions in Africa. Other studies approaching the problem from this perspective include Graham (1982) and Toledo (1982) for Mexican areas. The predictions for these areas are not different from the predictions for the neotropical rain forest biome. Many of the drawbacks to serious testing for the neotropics will also be drawbacks elsewhere. It is not until the data are needed that we begin to appreciate just how scarce are species-level cladistic analyses. This is no less a problem for North America and Europe than for South America.

10.3.4 Falsification with biological clocks

Many cladists were doubtful of Haffer's hypothesis because it was so exclusively Pleistocene. Even with a fully resolved phylogenetic hypothesis, all one can do is argue that some events are older than others (Fig. 10.5). Having access to relative time is much better than having no notions whatsoever about time. There is no question that Pleistocene events have shaped current distributions. The charge by most cladist-critics has been that not all distribution patterns are of equal age (i.e. Pleistocene). So long as one side asserts "all are of Pleistocene age" and the other asserts that "not all are of Pleistocene age", we have a stand-off with little or no opportunity for resolution. The techniques of vicariance biogeography explored by D. Rosen (1975, 1978), Platnick and Nelson (1978), and Nelson and Platnick (1981) (Chapter 12) allow correlation of 'suture zones' and geological events (which may be dated independently of the organisms postulated as affected), but such correlations require selecting from an unknown number of geological events which may correlate with the suture zone. Even knowing that the event must be in a particular sequence does not allow much improvement in precision.

Heyer and Maxson (1982) challenged the Haffer model through use of a biochemical biological clock (Wilson *et al.*, 1977; Radinsky, 1978; Carlson *et al.*, 1978). Their immunochemical dendrograms suggest that the speciation events separating the species distributed within the forests are much older (Pliocene to Oligocene) than the Pleistocene age postulated by Haffer.

Heyer and Maxson specifically sought to compare their data for frogs of the *Leptodactylus pentadactylus* group (animals mainly of the forest 'domain') with those for frogs of the *Leptodactylus fuscus* group (animals of open domains). These authors use the term 'domain' in the sense of environment to which the organism is adapted (such terminology parallels Haffer's use of 'fidelity'). For purposes of dating speciation events, they used (as a measure of the molecular clock) a conversion of 1.8 immunological units equals one million years (Carlson *et al.*, 1978).

Two populations of *Leptodactylus pentadactylus*, one from Panama and the other from Amazonian Peru, differed by 8–9 immunological units, but are morphologically indistinguishable (Heyer, 1979). These populations are associated with very different Pleistocene refuges. As Heyer and Maxson point out, for most frog species examined by Maxson and her co-workers, within-species variation in immunological units is of the order of 0–2 units. *Leptodactylus knudseni* differed from *L. pentadactylus* by about 15 units; *L. knudseni* is a species of the Napan forests of eastern Ecuador. Except for the variation within *L. pentadactylus*, most of the speciation events are dated as Miocene or Oligocene. The earliest dichotomy is between *L. laticeps* and the other five species – this event is dated as Early Eocene or Late Paleocene.

Two comments are required here. The first concerns the biochemical data. One of the problems with the technique is that the results can be reported to imply something (Fig. 10.10) which isn't true. The within-species data report average or 'consensus' results. In each case (*fuscus* and *pentadactylus*), Heyer and Maxson had inconsistent results. The two samples of *L. pentadactylus* gave distances of 0 and 8–9 units (these average as 4; Fig. 10.10). A result of 0 might have been compatible with the hypothesis the authors sought to test. A result of 4 or of 8–9 (some 7–16 Myr) is not consistent with a Pleistocene scenario. Their data for *L. fuscus* yielded results of 0 and 14 units (average 7; Fig. 10.10). Results of 13–25 Myr are not compatible with a Pleistocene scenario but less than one million years (0 units) is compatible. These were the only two species for which the authors had more than one sample so we are unable to decide how firm the other dates might be. However, even with such qualifications, the data appear quite inconsistent with a Pleistocene age. If Heyer and Maxson are correct (or even have a correct order of magnitude), then the forces operating during the Pleistocene (component five of the model) must be viewed as having no connection between the distributions, and some other explanation must be forwarded to explain the presence of distributional patterns.

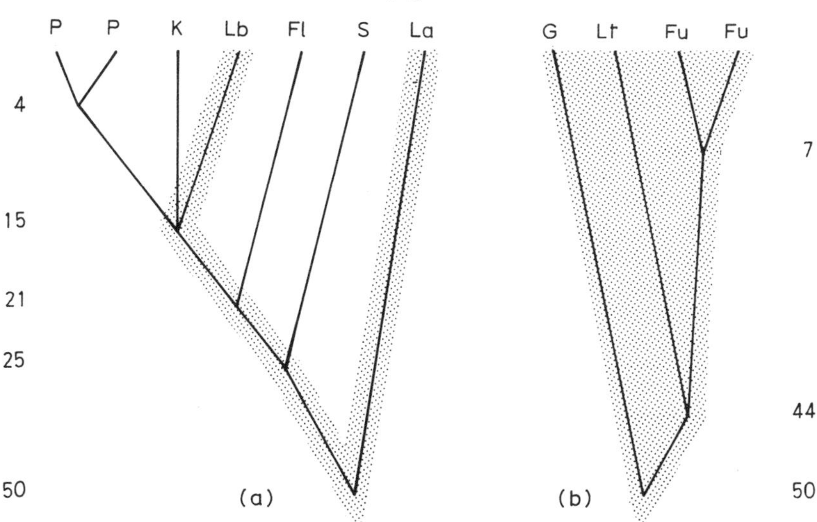

Figure 10.10 (a) A phylogenetic hypothesis taken from the dendrogram of the *pentadactylus* group of *Leptodactylus*. The lineages embraced by stippling are for frogs found in open (non-forested) habitats (this is thought to be the ancestral habitat for these frogs). (b) A hypothesis taken from the dendrogram of the *fuscus* group of *Leptodactylus*. The entire clade is considered to be from open habitats. Numbers at branch points refer to the distances (expressed in immunological units) from the branch point to the present. After Heyer and Maxson (1982).

The second comment about this falsification set concerns the forest fidelity of the organisms (Heyer and Maxson, 1982). They pointed out that the *pentadactylus* group is not confined to forested domains. Both *L. labyrinthicus* and *L. laticeps* are found in open (non-forested) domains and *L. pentadactylus* occurs in forest, forest-edge, and open formations but its distribution is essentially coincident with the distribution of rain forests. They characterized the *fuscus* group as one of open domains but pointed out that *L. fuscus* is distributed over both forested and open domains although it is associated most strongly with open formations.

Lynch (1979) argued that the reproductive mode of these frogs was not consistent with adaptation to forested environments. Frogs exhibiting the aquatic foam nest show reduced levels of endemicity in neotropical rain forests and there is a higher than expected frequency of non-endemics among frogs having this reproductive mode. If these frogs are not forest-adapted, what expectations might we have for the ages of their speciation events? Should we be surprised that those events are not Pleistocene events?

The approach taken by Heyer and Maxson (1982) appears to be a fruitful one. If biological clocks can be identified, they provide a decisive means of testing the ages of postulated events. However, the vehicle chosen by Heyer and Maxson does not appear to have been appropriate.

10.3.5 Falsification with systematic hypotheses

If we can successfully identify species which are actually forest-adapted, then the use of biological clocks such as that advocated by Carlson *et al.* (1978) and Wilson *et al.* (1977) would be of immense value in testing the Haffer hypothesis and separating those events of Pleistocene age from those that are much older.

The identification of forest-adapted organisms is also critical to another testing strategy, which broadly challenges the hypothesis by trying to discover if there is one and only one pattern in the relationships among the various sets of organisms which exhibit the same geographic pattern. As mentioned above in the description of the hypothesis, one of the serious drawbacks to Haffer's model is that he employed non-restrictive hypotheses (Tattersall, 1981).

Once restrictive hypotheses of relationships are brought to bear, the refugial hypothesis is sustainable only if all sets of taxa exhibiting the geographic pattern exhibit the identical pattern of branching sequences (Fig. 10.8).

Unfortunately, the general scarcity of restrictive phylogenetic hypotheses (especially those including species-level resolution) has prevented the articulation of a general pattern (assuming that a pattern does in fact exist).

There is no reason why the postulated refuges must be of the same ages. Given the geographic patterns inherent in precipitation in South America today and in the past (Haffer, 1969, 1979; Brown, 1982), one would not expect simultaneous separations among the postulated refuges achieved at glacial maximum. Minimally, one expects those proto-refuges furthest from montane regions to be the first isolated and those nearest to structural features which hold rainfall to be the most resistant to fragmentation (Ab'Saber, 1982). If the Pleistocene changes in the hydrologic cycles somehow caused simultaneous fragmentations, we would not expect systematic analyses to result in fully dichotomous diagrams. However, non-dichotomous diagrams are achieved in various manners. One way is by employing data above the species level, securing only superspecies and taxa of like rank. This is a very different method of discovery from 'concluding' that because the data are incompatible we must 'accept' a consensus diagram (one that is non-restrictive). If evolution actually produces non-dichotomous branching, then our expectations must be that each species shows unique exclusive traits (autapomorphies) and that we do not find shared traits except at 'higher' levels, such as superspecies, species groups, genera, or families (Eldredge and Cracraft, 1980; Wiley, 1981).

Few systematic challenges have been made against the Haffer hypothesis. Lynch (1982) employed the toads of the genus *Ceratophrys* as a test and discovered that the forest-dwelling species were not nearest relatives. However, *Ceratophrys* is like *Leptodactylus* (see p. 335) in being of questionable status as a forest-adapted group: most species are denizens of open habitats and the species appear to have reproductive biologies consistent with adaptations to xeric environments. Weitzman and Weitzman (1982) attempted to test the hypothesis using gasteropelecid and lebiasinid fishes distributed within Amazonian forests. They developed a species-level hypothesis for gasteropelecids of the genus *Carnegiella* but were unable to relate the distributional patterns to those identified by proponents of the Pleistocene forest refugia hypothesis. A similar lack of geographic coincidence was seen in the lebiasinids of the genus *Nannostomus*, for which a resolved phylogenetic hypothesis could not be obtained (Weitzman, 1978). Each fish group 'suffered' from extensive sympatry. The lack of obvious allopatry also posed a difficulty for Duellman (1982) and Duellman and Crump (1974) in their analyses of the forest-dwelling treefrogs of the *Hyla leucophyllata* group (five species are sympatric in the upper Amazon Basin).

As Duellman (1982) and Heyer and Maxson (1982) have emphasized, there are many frog groups having distributional patterns macroscopically consistent with the Haffer hypothesis but whose reproductive biologies reveal adaptations to non-forest environments. So long as we have evidence that the organisms are not forest-dwellers of necessity, we should be cautious about the meanings attached to the phylogenetic hypotheses generated for those organisms. The most easily accessible evidence that the group is not appropriate might well be the presence of one or more species within the group having a non-forest distribution. Obviously, more precise information would derive from autecological studies of the species under consideration. A great deal of data are available and those data raise questions about the efficacy of some often-used groups (e.g. a non-forest species of one of Haffer's (1974) demonstration groups, the toucans). Many of the butterfly distributions lend themselves to some sort of support for a refugial hypothesis but others suggest that it is premature to suggest that all of these examples are forest-adapted species (Brown, 1982; Turner, 1982).

Haffer (1969) originally mentioned nine refuges. If there were only seven, there would be 10 395 different ways that they could be linked together into testable hypotheses (Figs. 10.8 and 9 deal with the much

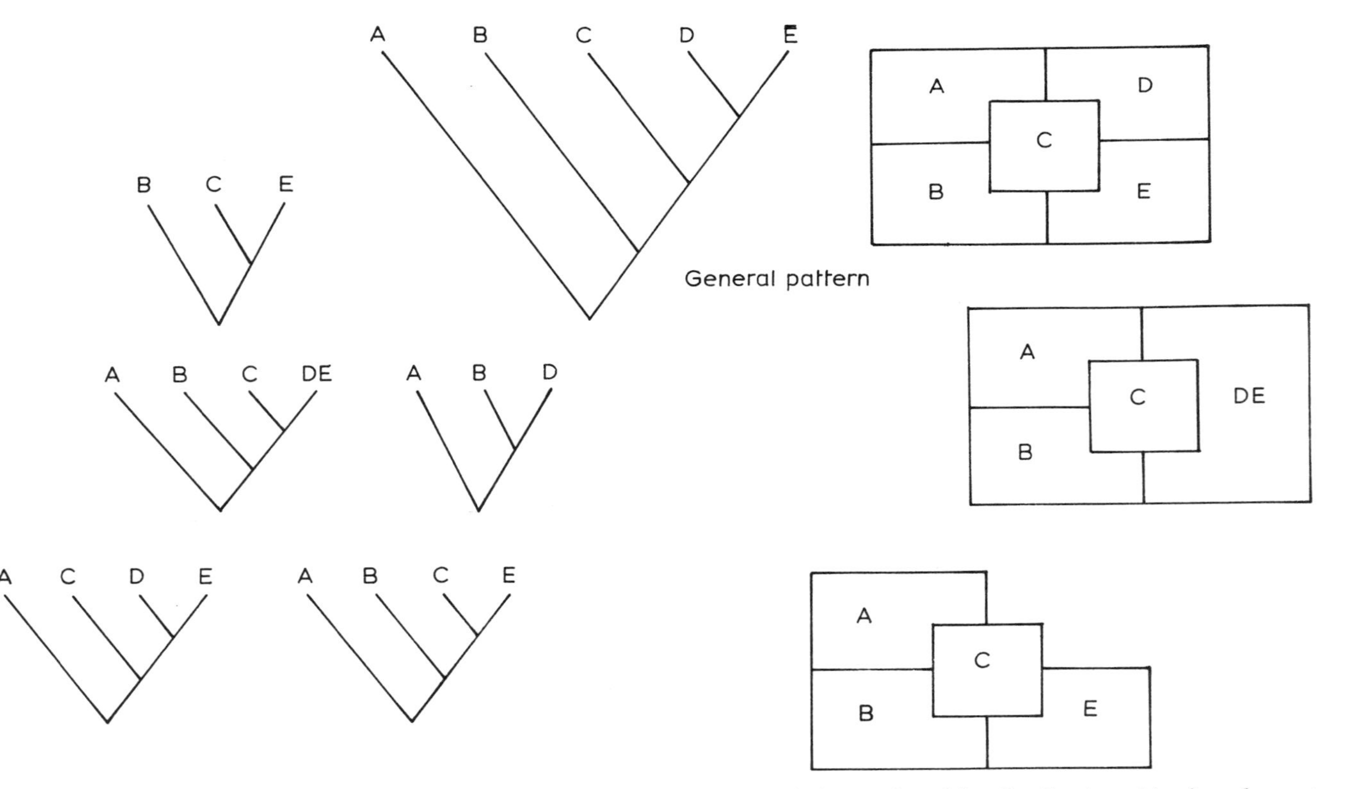

Figure 10.11 The general pattern (hypothesis) for areas (the five element cladogram) and for distributions (the five element rectangle). The smaller cladograms and the other distributions are among the many compatible data sets which corroborate the general hypothesis.

simpler system of only three refuges where there are three different ways that they can be related). This exceeds by some orders of magnitude the number of groups of organisms that have so far been studied by those investigating the refugia problem. If we consider the more complex system of refuges such as that proposed by Brown (1977, 1982), the number of possible testable area cladograms becomes astronomical (Felsenstein, 1978).

The lack of available phylogenetic hypotheses and the intimidating power of the number of possible diagrams when considering many refuges means that we cannot identify the pattern of the hypothesis nor can we test it at this time. However, we can examine the process whereby the Pleistocene hypothesis can be falsified using phylogenetic information and explore the consequences of falsification.

Consider a general hypothesis for five areas (Fig. 10.11). There are 105 different, fully resolved, possible patterns (Felsenstein, 1978). Individual data sets, compatible with the general hypothesis, may assume many forms depending on whether or not speciation was complete or whether extinction events occurred during forest contractions (Fig. 10.11). Some of the geographic combinations (not illustrated in Fig. 10.11, but imagine the pattern if the populations in areas A and E had failed to speciate, or had been merged subsequently by introgression) initially appear to be falsifiers but are in fact compatible (Nelson and Platnick, 1981).

However, a contrary data set (Fig. 10.12) is sufficient to falsify the general hypothesis. Firstly, let us consider the case in which we have species present in each area (Fig. 10.12(i)). This cladogram specifies that the species in areas A, B, C and D are each other's nearest relatives, a result inconsistent with the general hypothesis (Fig. 10.11). The clado-

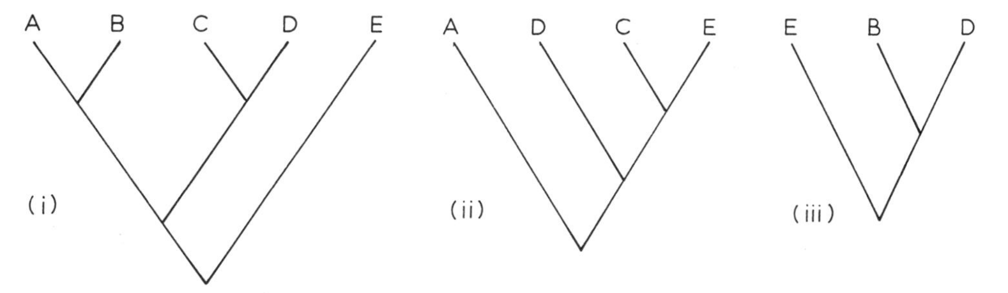

Figure 10.12 Three falsifiers of the general hypothesis (Fig. 10.11). The five taxon diagram (i) is entirely incompatible, the four taxon diagram (ii) is partially incompatible, and the three taxon diagram (iii) is entirely incompatible with the general hypothesis, but each is a falsifier of it.

gram also specifies a sequence of three fragmentation events (E from ABCD, followed by AB from CD, followed by A from B and C from D) completely different from those specified by the general hypothesis. Secondly, consider a different sort of falsifier (Fig. 10.12(ii)), which may be of more general occurrence. This cladogram of four taxa (in areas A, C, D and E) is partially consistent with the general hypothesis (Fig. 10.11) in that the relationship between A, D and E is correct as is that between A, C and E. However, the relationship of the species in areas C and D is reversed. Lastly, consider the case of three taxa in areas B, D and E (Fig. 10.12(iii)). This cladogram is inconsistent with the general hypothesis (Fig. 10.11) and serves as a falsifier of it. Many other falsifiers having at least three species in three areas are possible but these examples serve to provide cases of differing diversities.

Unfortunately, there are several 'solutions' to the detection of falsifiers. The least attractive is the claim that they don't matter in the face of all the supportive data that went into the production of the general hypothesis. It is not uncommon to encounter someone who treats exceptions (falsifiers) as though testing of scientific hypotheses was an exercise in statistics where a certain frequency of incompatible results are shrugged off. At another level, such contrary cases are likely to be dismissed as demonstrations that 'the exception proves the rule.' Exceptions prove the rule only in the proper use of the phrase, that is, they prove the rule to be false.

A second 'solution' is that the contrary systematic hypothesis is false. This is always possible and whether or not it is true depends upon systematic analyses rather than biogeographic or ecological analyses. Proper attention is more likely to be paid to a systematic hypothesis which 'challenges' a popular or generally held more inclusive hypothesis (such as the Pleistocene forest refugia hypothesis). The proponents of the refugial hypothesis would presumably be interested to discover that the contrary systematic hypothesis is false, and should lead the way in calling for its falsification.

A third 'solution' is to provide some plausible explanation for the contrary data. The most frequent evasion of falsification is to claim that the contrary pattern is produced by one or more dispersal events. For example, consider the smallest falsifier (Fig. 10.12(iii)). This data set can be forced into consistency by claiming that the species present in area E actually originated in area A and dispersed to area E before it became extinct in area A. Even a complex contrary data set (such as that in Fig. 10.12(i)) can be explained away by creative use of dispersals and extinctions. It can also be explained away by using a combination of extinctions and missing species but all such 'explanations' are remarkably uneconomical. Examples and terminology are given by Nelson and Platnick (1981).

The fourth 'solution' is to argue that the falsifiers reflect a second process. This, however, denies that the pattern has a single cause and asserts simple coincidence. This last 'solution' was explored recently by D. Rosen (1985) using the potential contradictions between geological data sets and phylogenetic (or systematic) data sets. Rosen's analysis is called for if and when 'corroborated' falsifiers exist for each general area cladogram.

10.4 CONCLUSIONS

The past twenty years have been active and exciting years in biogeography. Haffer's (1969) proposal of Pleistocene refuges revitalized interest in historical biogeography. Although non-historical explanations remained popular (MacArthur, 1972), the developing field of vicariance biogeography promised a possibility of separating ecological (non-historical) from historical explanations. Ecological explanations (Endler, 1982a; MacArthur, 1972) 'seek' to minimize the contributions of systematic biology whereas historical explanations (Haffer, 1969; Ball, 1975; Platnick and Nelson, 1978; Rosen, 1978; Nelson and Platnick, 1981; Wiley, 1981) seek to emphasize the centrality of systematic data in biogeographic formulations and use ecological explanations secondarily.

The Haffer refugial hypothesis was proposed first for Amazonian forest birds but rapidly gained widespread support. The hypothesis has been extended even to non-forest systems (de Granville, 1982; Huber, 1982). Part of the attractiveness of non-forest environments may derive from their comparatively impoverished biotas, e.g. savanna vertebrates (Webb, 1977, 1978). The reduced diversity may preclude pattern decomposition. One of the serious problems arising from the success of the forest refuge hypothesis is that as more and more groups of organisms are added to the data set, the number of required refuges increases to such a point that enough refuges are present to absorb any distribution pattern within the forests; expressed in another way, the 'pattern' becomes lost in its necessary detail.

The problem of pattern decomposition arises from using discontinuities in organism distributions in order to recognize the approximate locations of refuges. Very different approaches also have been made. Vanzolini and Williams (1970) utilized reduced within-species variability in order to identify core areas within the distribution of a forest-dwelling lizard, *Anolis chrysolepis*. The approach taken by Rosen (1978) was to identify areas of endemism by mapping areas of geological activity (events which might produce range disjunctions) and then selecting the undisturbed areas as potential areas of endemism. He kept his biological data independent of his geological data until he was prepared to ask if centres of endemism (based on fish distributions) coincided with geologically stable areas. A parallel approach may be possible in the Amazonian forests by using the clues in fossil soils (Ab'Saber, 1977, 1982). If we could identify a system of stable areas through detection of soil studies over the areas now forested, we could then ask about the coincidences of distribution patterns of organisms without being caught up in the necessity of positing at least one refuge for each species and binding ourselves to an allopatric speciation model – although not necessarily that advocated by Mayr (1954).

Our capacity to test the refuge hypothesis is seriously hampered if we do not employ scepticism in accepting data sets consistent with the model. As Duellman (1982), Heyer and Maxson (1982), and Lynch (1979) point out, many organisms have distributional patterns only loosely associated with the pattern we are studying. Unless some autecological data are entered into our biogeographic equation, we run a risk of confusing results of some other process with the results we are presumably studying.

Systematic data can be used to test the area hypothesis (the system of postulated refuges). Although it would be desirable for some other source to identify the sequence of fragmentation, a sequence can be formulated using systematic hypotheses (Platnick and Nelson, 1978; Rosen, 1978) and then other systematic data can be brought to bear to test the generality of that formulation. Systematic data which are contrary to the sequence called for under the general hypothesis would stand as falsifiers of that general hypothesis.

Temporal estimates can be gathered by merging dated geological events with an area cladogram (Rosen, 1978) but such estimates are only correlations. Direct dating methods, such as the albumin clock (Wilson *et al.*, 1977; Carlson *et al.*, 1978), provide a distinct method of testing the refugial hypothesis by challenging its temporal component.

If refugia are shown to have multiple causes or different time-scales, then either there is no pattern to be explained (that is, the distributional data constitute noise) or some other analytical method is called for. A potential analytical method might be the one recently discussed by Rosen (1985) in his paper examining the tensions that exist between biological and geological data when one set is considered to have primacy.

The scepticism of workers such as Endler, Heyer, Lynch, Maxson, and Weitzman is necessary if we are ultimately to 'discover' that the refugial model is true. Their scepticism is broad-based, ranging from doubts as to whether ecological and historical factors can be separated to questioning whether there is a single pattern as opposed to a mixture of superimposed patterns. No amount of corroboration serves to test the model nor does it improve its scientific credibility.

CHAPTER 11

Phylogenetic biogeography

L. Z. BRUNDIN

11.1 INTRODUCTION

To do him full justice, Joseph D. Hooker has to be designated as the real founder of causal historical biogeography. In 1843 the young Hooker returned to England with a large plant collection and varied experiences from the 'Antarctic' voyage of the two ships *Erebus* and *Terror* under the command of James Clark Ross. During that lengthy voyage, Hooker had been able to study the floras of Tierra del Fuego, southern Australia and Tasmania, New Zealand and several islands in the southern temperate and subantarctic zones. The results were published in the *Flora Antarctica* (1844–47), *Flora Novae Zelandiae* (1853–55) and *Flora Tasmaniae* (1855–60). Although mainly taxonomic, these works contain introductory essays which are of great biogeographic interest.

From the results of his comprehensive taxonomic analyses of the southern flora, Hooker (1853) concluded that: "Enough is here given to show that many of the peculiarities of each of the three great areas of land in the southern latitudes are representative ones, effecting a botanical relationship as strong as that which prevails throughout the lands within the Arctic and Northern Temperate zones, and which is not to be accounted for by any theory of transport or variation, but which is agreeable to the hypothesis of all being members of a once more extensive flora, which has been broken up by geological and climatic causes." That South Africa was also involved is indicated in the Flora Tasmaniae, which discusses the relationship between the floras of South Africa and south-west Australia. Before formulating his general explanation of the history behind the circum-Antarctic vicariance patterns displayed by endemic groups now separated from one another by wide stretches of ocean, Hooker made a careful investigation of the dispersal mechanisms of the taxa and felt convinced that dispersal was not significant. He was also deeply impressed by the fossilized gymnosperm wood (*Araucaria, Podocarpus*) on the barren Kerguelen, a place that had demonstrated drastic climatic changes at high southern latitudes since the Tertiary. It seemed clear to Hooker that he was dealing with an old southern flora that had once formed a complete counterpart to the northern continents; but was now broken up into fragments and strongly depauperate because of the marked deterioration of the climate which resulted in glaciation of the centrally positioned Antarctic continent and destruction of its presumably temperate flora.

Hooker's conclusions and unprejudiced approach stand out as a remarkable achievement. By studying "the botanical affinity" between the different southern continents on the basis of the apparent relationships of genera in two or more of these continents, Hooker applied the principles of what is now called vicariance biogeography. On biological grounds he was keen enough to draw the conclusion – independently of the geologists – that the southern continents must formerly have been directly connected to one another. In other words, he foresaw the former existence of what is now called Gondwanaland.

The biogeographic theories of Hooker signified the start of a period of lively interest in the history of life in time and space that still continues. Darwin (1859) and Wallace (1876) appeared as the leading opponents of Hooker, both of them presupposing that the main geographic features of the globe had been stable. On the basis of the great dimensions of the northern continents and their rich fossilized vertebrate faunas, Darwin and Wallace proposed a dominant centre of evolution in the north with the southern continents as independent receivers of a primarily northern biota, "while any intercommunion between themselves has been comparatively recent and superficial, and has in no way modified the great features of animal life in each" (Wallace, 1876). Apparent trans-Antarctic disjunct relationships were, according to Darwin and Wallace, consequences of trans-oceanic long-distance dispersal; the real scope of the southern problems was underrated by both men.

It seems justifiable to say that until about 1970, historical biogeographers were divided into two camps, both very active, one representing the Hooker tradition, the other that of Darwin and Wallace. The touchstone has always been an explanation of the many great disjunctions between closely related groups. The views of the two camps were on the whole incompatible, and it is clear that the position that is taken becomes decisive for the judgment of the history of life in time and space. However, the dominance of the Darwin–Wallace tradition could not conceal the general want of convincing arguments that marked biogeographic discussions.

While it seems clear today that many of the views and conclusions of Darwin and Wallace and their followers have to be laid to rest, and that of Hooker (1853) was right in principle, there is irony in the fact that

Hooker (1882) adopted the opposite theory concerning the history of the southern biota (Turrill, 1953). An obvious need was a method of phylogenetic reconstruction to determine the hierarchy of vicariant species and species groups occurring in the actual areas of endemism.

Historical biogeography and systematics in general entered a new era when Hennig (1950) published 'Grundzuge einer Theorie der phylogenetischen Systematik' with its carefully thought out principles and methods for the reconstruction of monophyletic groups. This provided historical biogeography with a method, the application of which could

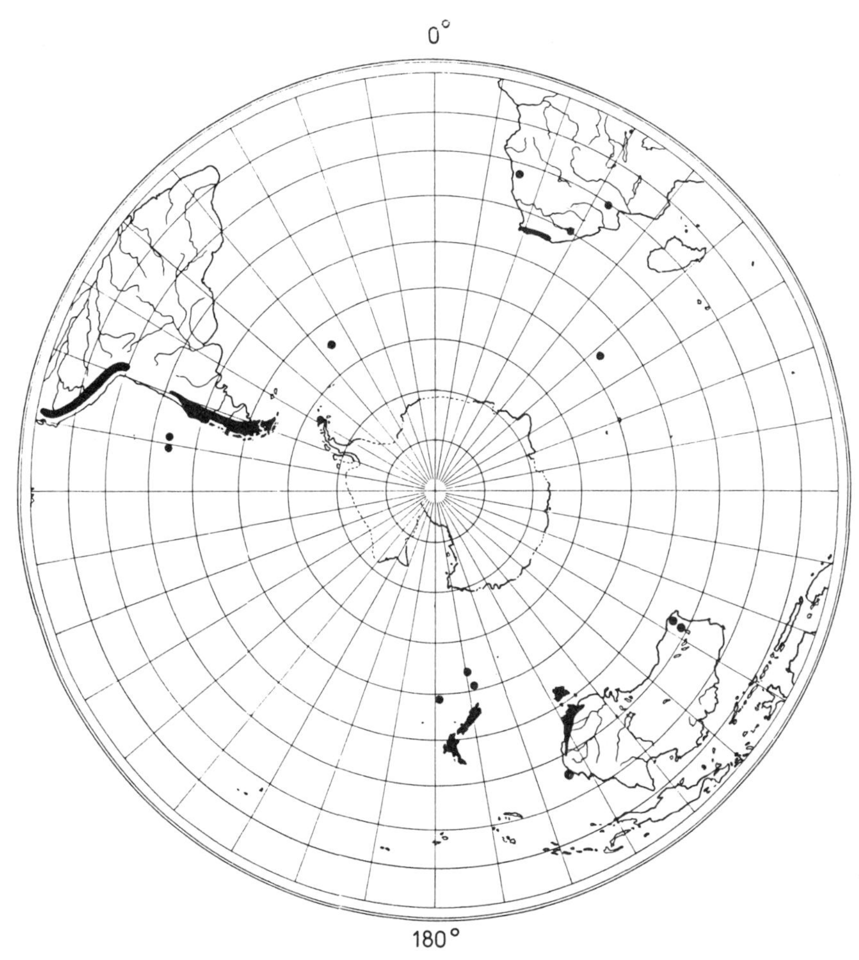

Figure 11.1 The combined Southern hemisphere distribution of the subfamilies Podonominae, Aphroteniinae and Diamesinae (Diptera: Chironomidae). (From Brundin, 1966.)

release biogeographic research from its marked dependence on speculation. From this basis and by reference to the distribution of chironomid midges (Fig. 11.1), Brundin (1966) attempted to provide an adequate explanation of the history behind the drastic intercontinental vicariance patterns around Antarctica, which has been such a controversial matter since the time of Darwin, Hooker and Wallace. It seemed clear from this analysis that the chironomids were an animal group that was better suited for biogeographic analysis than most others. The family is old, with a history going far back into the Jurassic (see below, p. 367); it is well diversified, comprising about 5000 species, and it is well represented in all freshwater biotopes from the Arctic to the Antarctic. Chironomids are abundant and easily collected. Furthermore, the imagines as well as the pupae and larvae are rich in morphological structures which facilitate phylogenetic analysis. Collections were made in andean South America, from Peru to Tierra del Fuego, in South Africa, Australia and Tasmania, New Zealand and New Caledonia during 1953–63. The phylogenetic biogeographic analyses were concentrated on three representative subfamilies of Chironomidae: Podonominae, Aphroteniinae and Diamesinae. The members of these groups are principally inhabitants of running waters and are more or less cold-stenothermal. Furthermore, the three subfamilies apparently originated in the south temperate zone, were lacking in the tropical lowlands and had an amphitropical (bipolar) distribution. This restriction of the study kept the extent of the analyses within reasonable limits, in spite of the more general need for consideration of global distribution patterns.

When the results were published in 1966, the conclusions about the nature of trans-Antarctic relationships were based on 23 cases of southern intercontinental sister-group connections. Ten patterns connected New Zealand with South America, evidently via West Antarctica, and ten more connected East Australia and Tasmania with South America, evidently via East Antarctica. South Africa was also involved, but in a more general way. The pattern is shown in Fig. 11.2 which, in the terminology of Nelson and Platnick (1981), is a combined taxon-area cladogram. The fact that the hypothetical reconstruction of the hierarchies and vicariance patterns of the three chironomid subfamilies could be interpreted as a lucid exposition of a particular historical process in time and space was very satisfying. The combined result can be put into a single diagram (Fig. 11.2), where the stepwise subordination of monophyletic groups in time demonstrates simultaneously their history in space and hence also the sequence of paleogeographical events which separated the areas of endemism inhabited by these groups. In other words, the biological data gave concordant evidence of the former existence of a large southern continent that had been broken up

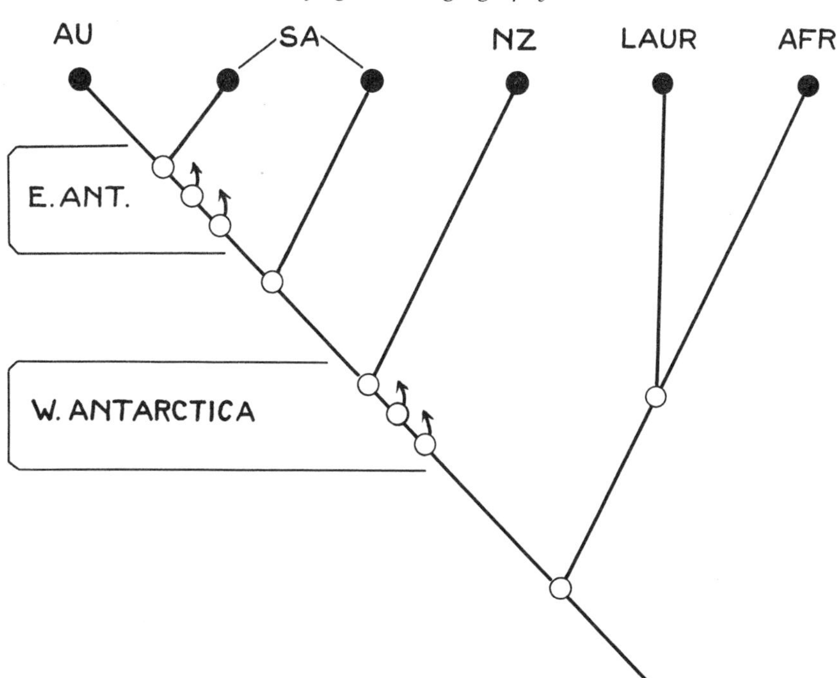

Figure 11.2 Taxon-area cladogram showing the connection between phylogenetic relationship, relative age, geographical vicariance and area relationships in chironomid groups of southern Gondwanic origin. Circles with attached arrows indicate the multiple occurrence of congruent trans-Antarctic sister-group relationships. The different evolutionary and biogeographical role played by East and West Antarctica after the separation of South Africa (AFR) from the other southern lands is indicated (modified from Brundin, 1970).

according to a sequence beginning with a separation of South Africa from East Antarctica, followed by a break in the connections between New Zealand and West Antarctica and later by a separation of Australia and East Antarctica and South America and West Antarctica. It seemed evident that a definitive opening of the Drake Passage had been preceded by a separation of East Antarctica from the Patagonian block along the margin of the Falkland Plateau.

Geological corroboration of these results appeared during the 1960s from new data gained by deep sea drillings, an activity which led to a rapidly increasing willingness among geologists to accept the theory of plate tectonics and continual displacement. The new geophysical results demonstrated the former existence of a great southern continent, Gondwanaland, that had been broken up according to a sequence that supported the biologically based conclusions. This reciprocal confirmation from life and earth history supported the far-reaching capacity of the applied phylogenetic approach.

11.2 PHYLOGENETIC BIOGEOGRAPHY

Division of phylogenetic biogeography into vicariance and dispersal biogeography is appropriate because they symbolize two different processes (Chapter 1), but dispersal analysis is not possible without preceding application of the methods of vicariance biogeography. The term phylogenetic biogeography is here preferred to cladistic biogeography because 'cladistic' symbolizes both the actual reconstruction of nature's hierarchy, and a somewhat reductionist attitude that in its 'transformed' variant tries to be as independent as possible of an evolutionary theory (Nelson and Platnick, 1981; Patterson, 1982; Chapter 12). In contrast to this view, I feel free to look upon 'phylogenetics' and phylogenetic biogeography from an evolutionary background. I thus perceive phylogenetic biogeography as a study of the history of life in time and space through simultaneous development and integration of an evolutionary theory beyond neo-Darwinism that answers the requirements of the phylogeneticist and phylogenetic biogeographer.

Croizat (1964) summed up the relationship between what he considered to be the two principal processes in the development of biotic distribution patterns as follows: (1) immobilism (vicariant form-making) and mobilism (dispersal) constantly alternate and interplay; (2) vicariant form-making excludes mobilism; and (3) vicariism arises by local form-making during immobilism within ranges that may have been acquired in an ancestral stage by mobilism. The causal explanation of the above can be expressed as follows: (a) development of barriers causes vicariance and leads to allopatric speciation; (b) disappearance of barriers allows range expansion (Brundin, 1981); and (c) a third process, though of minor importance, is jump dispersal over pre-existing barriers which may also result in speciation through allopatry.

11.3 VICARIANCE BIOGEOGRAPHY

11.3.1 The hierarchy of life

Since new species arise by successive cleavage of ancestral species, the result of the evolutionary process has been an immense hierarchy of groups within groups, where each group includes an ancestral species. The viability and creative force of the genetic material of individual populations and species has been kept up by gene recombinations and discrete mutations, while macromutations have probably marked the origin of new species (Schindewolf, 1950; Lovtrup, 1982).

Without any preconceived notion, one has to admit that the immense comprehensiveness of the hierarchy of life proves that the first organisms which occurred on the globe must have been provided with great developmental possibilities. An adequate term for these possibilities is 'evolutionary potentials'. It is now important to perceive that the perpetual hierarchical development of new groups within the frame of older groups has resulted in a correspondingly increased channelling of the primary evolutionary potentials. The evolutionary potentials have never been greater than in the first species occurring on the globe and never less than in species existing today. Processes giving rise to the ancient divisions into 'kingdoms' and 'phyla' will never be repeated because of lost potentials caused by the perpetual stepwise subordination of the evolving groups and their increasing specialization combined with increasing channelling of the mutations. The members of the monophyletic groups show characters that are variations on group-specific themes, demonstrating the directed constraints on mutations by the specific basic design of the genotype of each stem species. Among hundreds of adequate cases close to hand, reference may here be made to the genus- and species-specific variations of the fore-wing markings of the noctuid moths (Lepidoptera) comprising 15 000 species, where all variations lead back to a basic pattern that is distinctive for the family Noctuidae. Confronted with cases like this, the phylogeneticist is hardly inclined to look upon mutations as governed by pure chance (Brundin, 1968, 1972a, 1986). When the mutations are conceived as being more or less channelled or directed, the evolutionary process as a whole seems to fit in well with our present conception of the history of life in a panbiogeographic perspective. There has always been a close integration between the history of life in time and space and the history of the earth. Evidently the underlying dynamic process has been orderly and kept together by chains of given factors and their causal connections, which the scientist has to identify and explain. Attempts to explain certain events and phenomena as governed by chance are meaningless.

11.3.2 Methodological approach

Since the principal task of the vicariance biogeographer is the analysis of the vicariance patterns displayed by monophyletic groups, it is evident that all attempts to explain the history of these patterns must start with a reconstruction of the hierarchies formed by the actual groups. The appropriate tool is Hennig's method which requires determination of sister-groups on the principle of synapomorphy (Chapter 12). Cladistic analysis is best undertaken by starting with a small species group that is obviously monophyletic due to striking characters that may be synapomorphies. The stepwise establishment of sister-groups simultaneously indicates the degree of relationship of areas of endemism occupied by the taxa under analysis. The cladistic analysis should be continued until a hierarchy shows a vicariance pattern of global extent. Three-taxon, three area statements are generally the most basic viable units of biogeographic analysis. The results can be demonstrated in a simple taxon-area cladogram (Fig. 11.2). The generality of a three taxon-area statement has to be tested by extension of the analysis to other groups occurring in the same area. There are, however, certain limits set for the individual biogeographer with respect to the comprehensiveness of the project that is planned and possible to accomplish. The suitability of the group that is chosen for study and analysis is important. The accessibility of different groups to a reconstruction of their hierarchy is different. Lower invertebrates with very simple organization are generally difficult objects while, for example, many insect groups are comparatively well suited for the purpose. Especially valuable as biogeographical indicators are groups of reasonable size showing a well developed intercontinental diversification that can readily be subjected to analysis by one specialist. One example of this is given by the chironomids described earlier, where a fraction of the family shows 20 trans-Antarctic sister-group connections between South America and Australia/New Zealand; a number of connections comparable to those shown by birds and mammals together. Even if certain groups are exceptionally profitable in this context, there is still a very long way to real synthesis, and it is somewhat unrealistic to formulate the biogeographical method following Croizat *et al.* (1974): "Operationally, we consider that biogeographical investigation begins with the determination of general patterns of vicariance, and the determination of the geological changes that caused them". One cannot of course determine general patterns without previous time-consuming phylogenetic analyses of sequences of particular groups; and this type of work is still in its infancy. Croizat's panbiogeographic work (1952, 1958, 1964) is unique and his thoughts and conclusions have deeply influenced biogeographical thinking, but he worked on other prerequisites. He relied on distributional data published by taxonomists and on plant and animal groups whose monophyletic status is often questionable. Croizat had a fine intuition, but his conclusions about the history of many groups must certainly be verified by critical application of Hennigian methods.

For a more detailed treatment of the methods of hierarchical reconstruction and vicariance biogeography than that which can be given here, the reader is referred to the works of Nelson and Platnick (1981),

Wiley (1981) and Ax (1984). The role of true parallelisms in their quality of being 'incomplete synapomorphies' has been discussed by Brundin (1976) and Saether (1983).

11.3.3 Power of explanation

Generalized tracks

On the basis of his wide sweep through the taxonomic literature and its immense supply of distributional data, Croizat (1958, 1964) summarized his results in a method of graphic representation termed generalized tracks. He marked these tracks as lines on the map, where each line symbolizes the congruent course of many subordinate lines, each of them connecting the ranges of the vicariant components of a particular, supposedly monophyletic group. According to Croizat, a generalized track symbolizes the distribution pattern of an ancestral biota before it subdivided (vicariated) into descendant biotas and whose components have had a common history in time and space (Chapter 13). The fact that Croizat was able to show that biotic distribution patterns are on the whole non-random and fall into a limited number of tracks, whose courses on the map indicate drastic geographical changes since the early Mesozoic, meant a great step forward.

In a criticism of vicariance biogeography, Ball (1975) raises the question of what generalized tracks really mean and how they can be explained. He concludes that: "As statistical measures of overall similarity of disjunct biotas they mean about the same as measures of overall similarity in systematics". D. Rosen (1975) insists that the groups whose distributions make up a generalized track must be monophyletic, but Patterson (1981a) remarks that this requirement "seems equivalent to pheneticists' insistence that the attributes they select must be homologous". Patterson (1981) considers Ball's comment a fair assessment of the meaning of generalized tracks and says that "although Rosen added cladistic systematics to the method, its central concept – the generalized track – remains a phenetic concept. As such, I do not believe it can lift biogeography out of the narrative phase or meet the need for correspondence between systematics and biogeography". To me this discussion appears rather strained. To the extent that the discussion bears upon Croizat's method, Ball and Patterson seem both to be right, but when Patterson refers to the cladistic method applied by D. Rosen (1975) his conclusion is hardly defendable. Rosen's method requires: (1) that establishment of a generalized vicariance pattern presupposes reconstruction, based on synapomorphy, of the hierarchy involved in the vicariance pattern of every monophyletic component; (2) that the

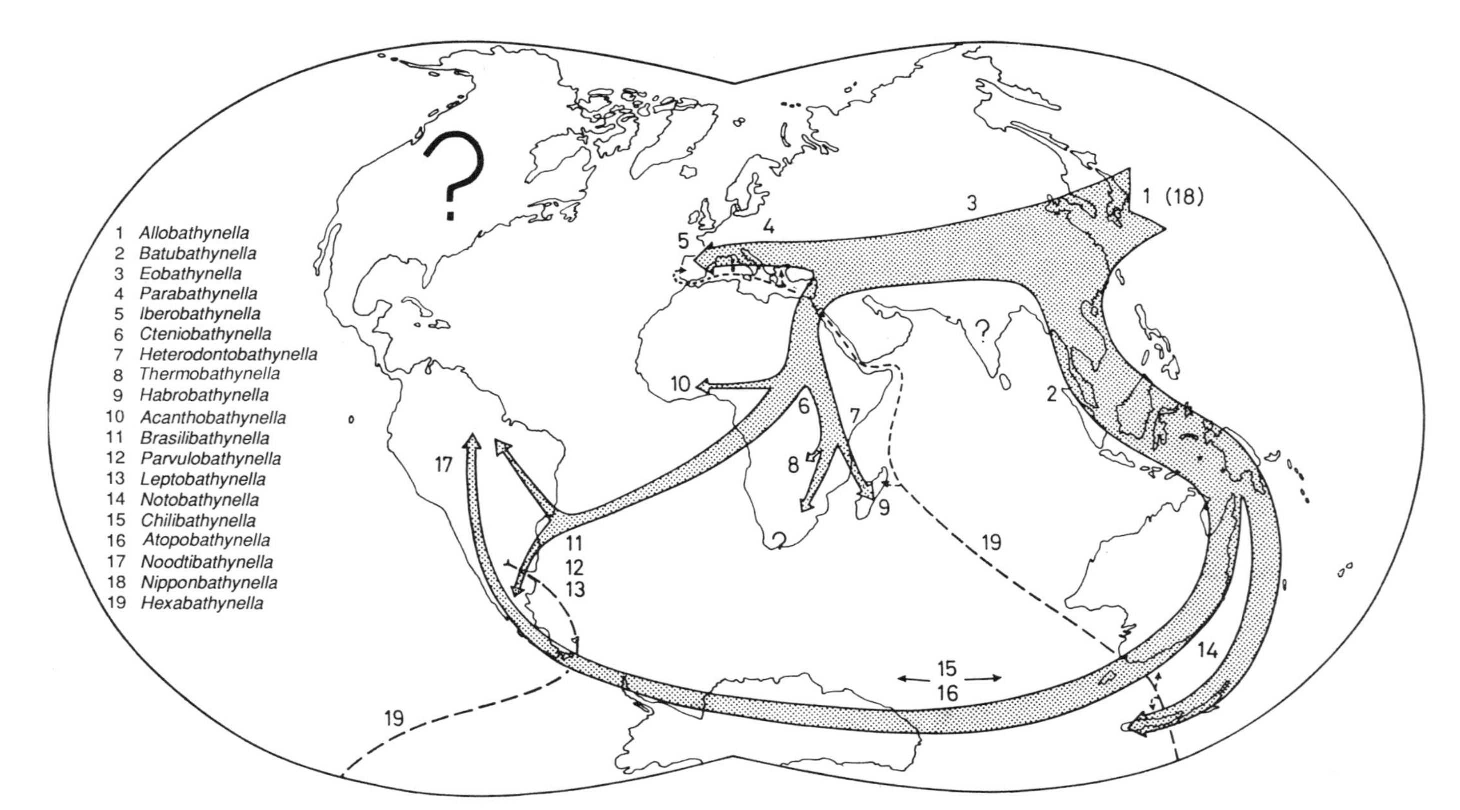

Figure 11.3 Dispersal routes of the crustacean family Parabathynellidae according to Schminke (1973). (By permission of Franz Steiner Verlag, Wiesbaden.)

hierarchies formed by other groups (components) are congruent with the first, and (3) that the hierarchies involved express meaningful patterns of area relationship. The whole can be presented in one taxon-area cladogram, where the different vicariance events are consequences of the development in time of particular barriers. A generalized track will then stand out, not as a phenetic concept, but as the expression of a unique biotic genealogy confirmed by multiple reciprocal illumination from life and earth history.

There are of course pitfalls. If our analysis of vicariance patterns is restricted to a special region, there is a risk that we do not detect the composite nature of an apparently homogeneous generalized track. One example is offered by the biotic connections between temperate South America, East Antarctica and Australia. Among chironomid groups of Gondwanic origin as already mentioned, there are ten sister-group connections, and in these cases the Australian groups stand out as morphologically apomorphic in relation to their South American relatives. However, although the subterranean crustaceans of the family Parabathynellidae show similar sister-group connections between Australia and South America (Chile), the situation is just the opposite (Fig. 11.3), indicating that the Chilean groups are apomorphic in relation to their closest relatives in Australia, and still more so in relation to their far-off relatives in Malaysia and East Africa (Schminke 1973, 1974). According to Schminke these small ground-water crustaceans had their origin in East Asia, and one main branch was able to disperse southwards. The New Guinea sector probably served as a forking point for dispersal, partly through eastern Australia to Tasmania, partly along the Inner Melanesian Arc to New Zealand, a process that resulted in sister-group relationships across the Tasman Sea which are lacking among the analysed chironomid groups of southern origin. From Australia, members of the genera *Chilibathynella* and *Atopobathynella* could use continuous land for further dispersal via East Antarctica to Chile, where they are represented by the most apomorphic species. Further research will probably show that several other groups of northern origin display a concordant pattern.

The pattern just discussed demonstrates that groups of different origin, and hence with a different prehistory, can appear quite erroneously as members of the same track over wide distances. The general need for a global perspective is again underlined.

Vicariance patterns as paleogeographic indicators

It is now generally accepted that the theory of plate tectonics is of basic importance to biogeographers in their attempts to explain the history of major vicariance patterns. The reconstructions of Gondwana and Laurasia and their relative positions as suggested by geophysicists are, however, rather divergent in certain respects, demonstrating that there still are several unresolved problems. There are also some reconstructions in the plate-tectonic literature which cannot be accepted by the biogeographer. One is the tremendous disjunction that appears between Gondwana and Laurasia and forms the eastern gap of an oceanic Tethys Sea that would have existed before the fragmentation of Gondwana. Faced with Schminke's reconstruction of the history of the Parabethynellidae (a group of Upper Paleozoic age (Noodt, 1965) that required continuous land connection to explain range expansion), Carey's (1958) theory of an expanding earth is a possible solution of the Tethys problem (Brundin, 1975). This theory is now becoming more acceptable. According to Shields (1979) there is "geological evidence for a narrow, epicontinental Tethys Sea", which seems very promising from a biogeographic point of view. The greatest riddle of historical biogeography today is the history of the Pacific island biotas and the wealth of probable east-west amphi-Pacific sister-group relationships. The recent development of the theory of an expanding earth seems to offer a radical solution, by implying that the Indian and Pacific Oceans once were more or less closed and became open in the Upper Jurassic some 155 Myr ago. Furthermore, most Pacific islands and island groups are now believed to rest on basements that are continental fragments (Shields, 1979). A further development of the theory of an expanding earth may provide a satisfactory geological-geographical background to the biogeographic problems. There are still, however, few representative plant and animal groups with Pacific or amphi-Pacific distributions that have been subjected to cladistic analysis, a deficiency that needs rectifying.

A test of the reliability of different attempts by geophysicists to reconstruct Gondwanaland is possible on purely biological grounds as exemplified by the problem concerning the mutual positions of Africa, Antarctica and South America. In most reconstructions, the western portion of East Antarctica is positioned close to Mozambique, while Madagascar takes a more northern position with its mid portion opposite Zanzibar (Fig. 11.4). Consequently the Antarctic Peninsula points to Cape Town. Considering the generalized biotic track that connects Australia with Patagonia via East Antarctica, this particular reconstruction is unrealistic. The reconstruction by Tarling (1972) fits the biogeographic data best (see p. 347). In Tarling's reconstruction (Fig. 11.5) we find direct contact between Patagonia and East Antarctica at the margin of the Falkland Plateau. A consequence of this arrangement is a more southern contact between East Antarctica and Africa and a more Southern position for Madagascar, off Natal.

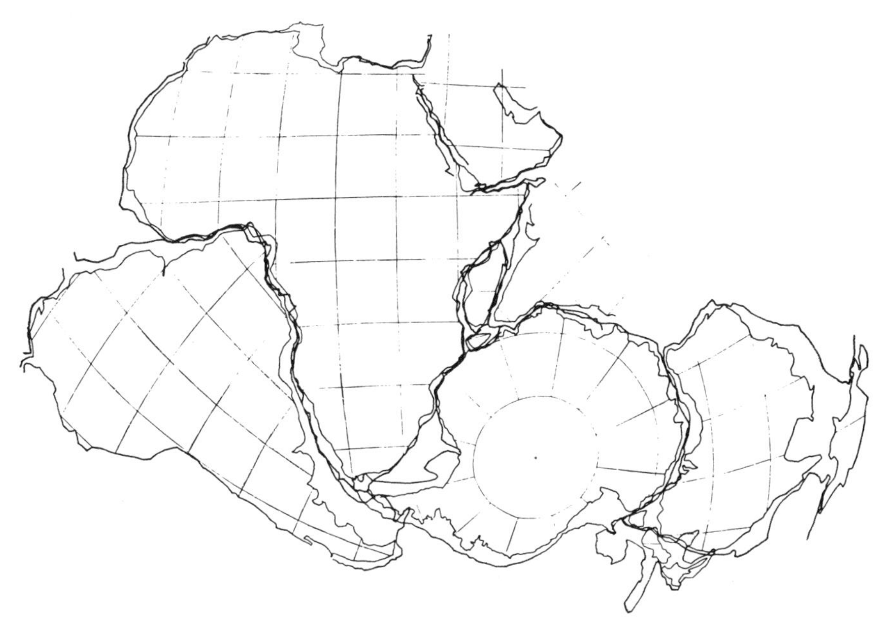

Figure 11.4 Reconstruction of Gondwanaland by Smith and Hallam, 1970. (By permission of the publishers of *Nature*.)

Figure 11.5 Reconstruction of Gondwanaland by Tarling 1972. (By permission of the publishers of *Nature*.)

11.4 DISPERSAL BIOGEOGRAPHY

11.4.1 The dispersal phenomenon

The most difficult and controversial matter within historical biogeography is the role of dispersal. The principal reason for this difficulty has always been the production of convincing arguments from the available data. Since Darwin, Hooker and Wallace, discussions have concentrated on the provocative disjunctions among the isolated subgroups of the many groups now separated by wide expansions of ocean. Hypotheses of jump dispersal (including wave dispersal by sea currents) based on investigations of dispersal mechanisms, or of supposed land bridges by those not willing to accept random dispersal, were the main arguments. The emergence of modern vicariance biogeography and the theory of plate tectonics has apparently put an end to former exaggerated ideas about the role played by jump dispersal over pre-existing barriers.

Two of the main hypotheses of historical biogeography are as follows: appearance of barriers means vicariance, disappearance of barriers means dispersal. In this case dispersal means biotic progression in space, step by step, without jumps over pre-existing barriers, for which the term range expansion is used. The best known examples of range expansion are given by the biotic invasions of northern Europe and northern North America in the wake of the retreating ice caps at the end of the Pleistocene. Pollen analysis, ^{14}C dating, and a wealth of subfossil plants and animals allow biotic progression in space during the last 10 000 years to be followed in great detail. In Scandinavia the last apparent major range expansion was marked by immigration of the spruce (*Picea abies*) and its accompanying biota, which advanced from the northeast, north of the Gulf of Bothnia, and stopped in Blekinge and the northern part of Scania, the southernmost provinces of Sweden, only about 200 years ago (Berglund, 1966; Digerfeldt, 1974).

Nelson and Platnick (1981) have pointed out that "dispersal is vicariance in disguise . . . the reason is that the postulated dispersal takes place prior to the appearance of the barrier and prior to the fragmentation of the range of the ancestral species. The effect of the postulated dispersal is only the creation of primitive cosmopolitanism (a requirement of the vicariance model)". There cannot be any disagreement with this postulate, disregarding the unnecessary inset of the word 'only'. Indeed, the achievement of cosmopolitanism by dispersal of ancestral species must have been a fundamental prerequisite for the widespread occurrence of harmoniously diversified biotas, except in areas that offer extreme conditions of life or that have long been isolated. Vicariance and range expansion have constantly alternated, bringing about intermittent interchange between biotas of different parts of the world.

11.4.2 Methods and their explanatory power

On the assumption that sympatry between closely related groups indicates range expansion of previously allopatric populations (rather than coevolution), and by stressing that this type of dispersal only played an important role prior to the appearance of barriers (vicariant events), some biogeographers prefer to restrict investigations of the history of life in time and space solely to analyses of vicariance patterns.

At this point, phylogenetic biogeography, as I understand it, departs from the position of pure vicariance biogeography. Even though traditional dispersal biogeography has shown its weakness by restricting itself to a narrative stage, there is no reason to go to the opposite extreme by admitting only vicariance analysis and attaching the label *noli me tangere* to the problem of range expansion. There are methods available for determining the former occurrence of certain range expansion events and their direction, area of origin and minimal age of origin. These include: reconstruction of hierarchies for taxa of global proportions; assessment of the position of particular groups in the hierarchy, indicating relative ages; the relative apomorphies of the vicariating subgroups; consideration of the history of barriers, their appearance indicating the minimum age of the range expansion; and the maximum age of the vicariated subgroups. Two examples may demonstrate the explanatory power of the methods at our disposal (Brundin, 1966, 1981). In Bolivia, Peru and Ecuador, the Andean mountain streams at an altitude of 1700–4850 m are inhabited by several endemic chironomid species which have their closest relatives far to the south, in southern Chile and the Andean parts of Patagonia. Within the genus *Podonomus* alone there are ten such examples. My reconstruction of the chironomid hierarchies involved has shown that these chironomids are comparatively young offshoots of an old Jurassic biota with multiple trans-Antarctic sister-group connections with south-east Australia and New Zealand which now shows a marked diversity in southern South America ranging from the Valdivian to the Magellanic regions. The chironomids of the tropical High Andes are more or less apomorphic in comparison with their close relatives in the south and are now isolated from southern species by a broad barrier: the desert zone of northern Chile and southern Bolivia (Fig. 11.1). This is a consequence of the cool Humboldt Current which presumably started to operate in connection with the opening of the Drake Passage and the subsequent development of a circum-Antarctic current system in the Oligocene which in turn led

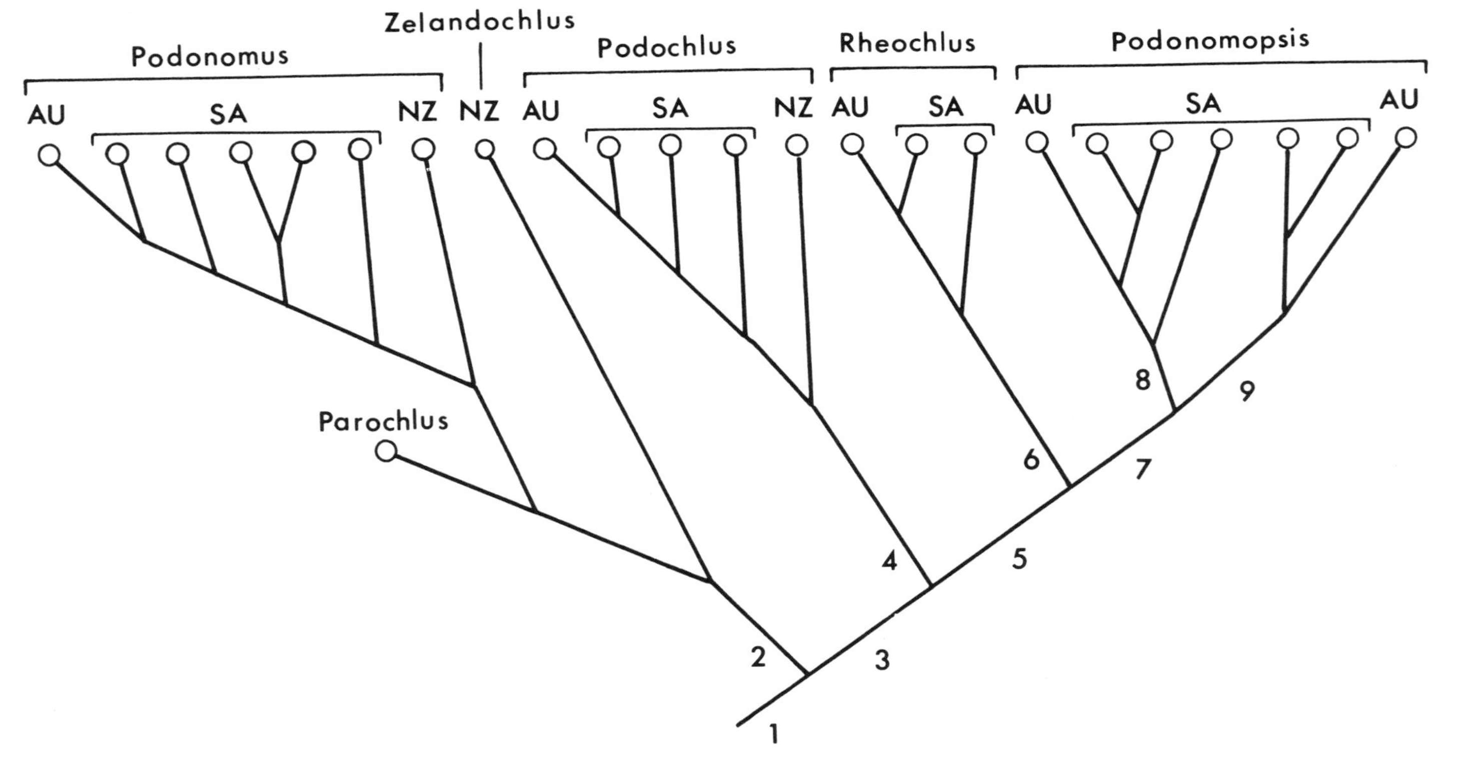

Figure 11.6 Cladogram of the Chironomid tribe Podonomini, with exclusion of the genus *Parochlus*. From Brundin, 1981. (By permission of Columbia University Press.)

to glaciation of Antarctica in the Miocene (Kennett *et al.*, 1974; Hammond, 1976). Andean orogeny began in the Upper Cretaceous and continued through several phases during the Tertiary, culminating in the Pliocene–Recent, an uplift which still continues today. Considering that these cool temperate groups are absent from tropical lowlands, there can be little doubt about the process that caused the above vicariance pattern. The rise of the Andes implies the disappearance of former barriers and possible range expansion of several elements of a cool-temperate biota northwards, along the Andean chain. Later isolation has evidently resulted in local speciation. It may be added that the actual cases exemplify multiple vicariance of closely related sympatric groups, but the pattern could still be explained without sympatry.

My second example of range expansion refers to a vicariance pattern already touched upon. There is good evidence for the presence of a generalized biotic track connecting South America with south-east Australia and Tasmania via East Antarctica. Among cool temperate chironomids there are ten examples of such sister-group connections. Edmunds (1975, 1981) presented four similar patterns among Ephemeroptera and one among Mecoptera, the sister group of Diptera, and there are many more examples from other groups, although few have been verified by cladistic analysis. Chironomids attract special interest because nine of the ten cases of trans-Antarctic connections between South America and Australia refer to groups which are sister groups among themselves (Fig. 11.6), a clear indication of multiple range expansion as exemplified by sympatry. This is emphasized further by the fact that the Australian members of the different groups are often found together in the same mountain stream. At this point the biogeographer has of course to ask what kind of range expansion has been involved? It can be observed that the Australian components comprise the total Australian representatives of the three chironomid subfamilies analysed (Brundin, 1966). Given that the ten Australian groups all stand out as comparatively young apomorphic offshoots of older hierarchies in southern South America and that they appear apomorphic by comparison with their closest relatives in the latter continent, it is logical to conclude that their occurrence in Australia is due to Gondwanic range expansion by ancestral species via (or from) East Antarctica before Australia became separated from East Antarctica in the Eocene.

The role of Antarctica in the history of trans-Antarctic groups can be assessed by extrapolation. There is no reason to doubt that before its glaciation Antarctica was an important centre of evolution, and also that it has served as an area of intermittent interchange of biotic elements of different origin. It seems conceivable that the present areas of endemism in temperate South America, South Africa, New Zealand and also in the subantarctic islands (see below, p. 363) mirror the general composition of the preglacial biota of Antarctica, not least with respect to the linking position of that continent.

The analysis of the two vicariance patterns given above shows that the use of meaningful data in combination with common sense may be sufficient for making judgments about the occurrence and nature of range expansion events. This offers hope, because it seems a sound approach and we need not fall into a trap and seek dispersal when none occurred, i.e. the 'Type-1 error' in the sense of Nelson (1974).

11.4.3 Controversies

The progression rule

As mentioned above, extreme vicariance biogeographers dismiss attempts to investigate range expansion events. This concentration on the static, 'immobilistic', portion of the history of life in time and space is in my view sterile. Significantly, Hennig's (1950) 'progression rule' is dismissed by Nelson (1974) and Croizat *et al.* (1974). The rule simply says that there is a parallelism between morphological and chorological progression, meaning that when a species extends its range by range expansion and a population becomes isolated in a new environment by the appearance of a barrier, and this event is followed by a speciation process, then the new species with novel characters will stand out as comparatively apomorphic in relation to the populations of the ancestral species that remain within the old range. The same will result from jump dispersal over a pre-existing barrier by a small founder population which leads to successful speciation after establishment. The progression rule can be conceived as the expression of a very normal evolutionary process in time and space. The two types of dispersal events discussed above confirm the action of the rule, as does the history of the family Parabethynellidae (see p. 353). Hennig's deviation rule which states that when a species splits, one of the two daughter species tends to deviate more strongly than the other from the common stemspecies, i.e. from the common original condition (Hennig, 1966) has also been heavily criticized. The two rules have been wrongly interpreted as representing simply a methodological approach. In reality these rules are expressions of life's own experiments and have as such been of fundamental importance in understanding the history of evolution (Brundin, 1972b). Some discussions of the progression rule have suffered from another misinterpretation. According to Croizat *et al.* (1974) and Ball (1975) the meaning of the rule would be that it is the ancestral populations that remain at, or near, the point of origin, while the derived forms migrate. This overlooks the fact that the progression rule refers to the results of range expansion not to the question of the relation between plesiomorphy/apomorphy and the aptness to disperse.

In this connection it seems appropriate to refer to Rosen and Buth (1980) who discuss progress in empirical evolutionary research. They point out that Buth's (1979) method, of using electrophoretic data for comparing sister taxa in adjoining areas, seems to provide a model of how genetic differences might arise allopatrically. "His data indicate that populations in habitats that are in some sense new compared with those of sister populations, possess a higher proportion of derived allelic characteristics. Because his comparisons involve sister-taxa that, having a common ancestor, must logically be of the same age, a difference in the proportion of their accumulated autapomorphies suggests that they have been affected by different factors which permit the accumulation or, as suggested below, even enhance the production, of new mutations and their phenotypes' (Rosen and Buth, 1980). This interpretation seems to explain the causal connections involved in the concordance between morphological and chorological progression.

Areas of origin

It has been mentioned several times above that the chironomid subfamilies analysed by me are evidently of southern, i.e. Gondwanic, origin. The reason for this conclusion is that South Africa, New Zealand and southern South America are involved in the basic branchings of the three hierarchies and that the vicariant sister groups of Laurasia stand out as morphologically derivative in relation to their southern relatives. But a closer estimate of the areas of origin of the three subfamilies cannot be made at present, a situation that is far from unexpected. We have to be content with 'southern Gondwanaland' as their area of origin. But even this implies a rather important position in connection with our efforts to reconstruct the history of life on a global scale.

As to the concept 'centre of origin' so disliked by many vicariance biogeographers, it should be realized that semantics are involved. 'Area of endemism' is a concept in common use among vicariance biogeographers. Such areas are evidently the places of birth of their endemic species, provided that these are not relict. When an area is distinguished by a marked concentration of endemic species, it can, as a rule, also be called a centre, or area, of origin. Not only continents, however, but also smaller areas of endemism like the old Cape fold belt of southern Africa, south-west Australia and New Caledonia, are rich in monophyletic groups of rather different rank whose hierarchies evidently have evolved inside these areas. This can be shown by phylogenetic analysis.

Pronounced areas of endemism also include south temperate South America and New Zealand, each with many species groups of their own. But the question of the area of origin of several of these groups raises a problem. The reason is the intermediate position of the Antarctic continent, almost a biotic blank because of its glaciation. The chironomid groups of southern Gondwanic origin discussed above may be taken as an example. They show the presence of several sister-group relationships between South America and New Zealand, evidently via West Antarctica. Since it is generally agreed that Antarctica had a comparatively congenial climate during the Mesozoic and Lower Tertiary, there is every reason to suppose that it was inhabited by a representative south temperate biota with good opportunities for speciation, not least in the mountainous, partly volcanic landscapes of West Antarctica. But the probable involvement of West Antarctica in the areas of origin of the ancestral species of groups with sister-group relationships between South America and New Zealand, is a matter that – lacking direct evidence – cannot be further resolved.

In retrospect, Croizat's violent critique of the conventional, aprioristic approach to disjunct distribution patterns as being the result of jump dispersal from particular centres of origin appears well justified. However, even Croizat could not do without the biological phenomenon behind the term centre of origin, but he found it so tainted that he consistently used other terms like 'baseline', 'hub', 'node', 'gate' and 'centre of endemism'. For example, according to Croizat (1952) the angiosperms started to disperse during Late Jurassic and Early Cretaceous from the Antarcto-Gondwanic 'baseline' in three main streams through three 'gates', the West Polynesian, the Magellanian and the African 'gates'.

11.4.4 Isolated island biotas: jump dispersal or vicariance?

The discussions of dispersal in this chapter have focused so far on range expansion in the absence of barriers. Jump dispersal or chance dispersal over pre-existing barriers is a different phenomenon whose influence on the development of disjunct distribution patterns has long been a matter of controversy. Yet the majority view now is that the influence of adherents of the Darwin–Wallace tradition has led to a markedly overrated role for jump dispersal. Recent research based on phylogenetic analyses seems to have confirmed the thesis of Croizat that the main patterns of geographic distribution are orderly and result from a close integration of life and earth history. But former mistakes are easily explained since many biogeographers were compelled to presuppose a mainly stable geography in the absence of a well documented alternative. However, in the face of appreciation of comprehensive tectonic activity and its biological consequences, biogeographers have been left behind, lacking hierarchical reconstructions of amphi-oceanic vicariance patterns and adjoining patterns displayed by now widely separated island biotas. This deficiency is especially felt with regard to the tropical zone. Croizat's investigations have indicated the scope of the problems waiting for solution in that zone. One example is the challenging problem raised by a vicariance pattern among plants joining the floras of Juan Fernandez, Marquesas and Hawaii (Skottsberg, 1956). It may be predicted that only a combination of the theories of plate tectonics and an expanding earth can give the necessary geographic-geological background for an understanding of such patterns. But what is needed first and foremost are hierarchic reconstructions.

A volcanic cone arising above the surface of the sea and increasing in height and extent will, however isolated, sooner or later receive a biota by jump dispersal (Chapter 15). The extension of the habitable area, the degree of development of different types of habitats and not least the establishment success of the arriving immigrants all become decisive in the results of the colonization. But in relation to larger continental areas, the biota of isolated islands will remain disharmonic and comparatively poor in species. However, if the biota of an island is of great age, it will often contain several endemic taxa (Chapter 5). Against this background, the history of the Pacific island biotas has long been explained by reference to jump dispersal by air or sea currents, predominantly from west to east, with the Indonesia–New Guinea–Solomon sector as the leading source area, but also with a certain influence from America and Antarctica. This explanation may seem convincing when reference is made to maps which show the most evident feature as "the successive diminution in representation (of taxa) with progression eastwards from Asia, producing a successively more disharmonic fauna from west to east" (Gressitt, 1961). However, there is a clear disharmony between this general picture and its causal explanation by more or less orderly jump dispersal via wind systems and sea currents, because these prevail in the opposite direction.

Even if our concept of the geologic-geographical history of the Pacific area is now in flux (p. 354), this circumstance need not be a direct hindrance to initial interpretation of the history of the isolated subantarctic island biotas. True subantarctic islands according to Gressitt (1970) include: South Georgia, Marion–Prince Edward, the Crozet Islands, Kerguelen, Heard and Macquarie. They are widely scattered around the Antarctic continent between 46° and 55°S. The fauna is poor in species but attracts great interest because of a generally marked endemicity that in the Crozet Islands, for example, is 50% (Gressitt, 1970).

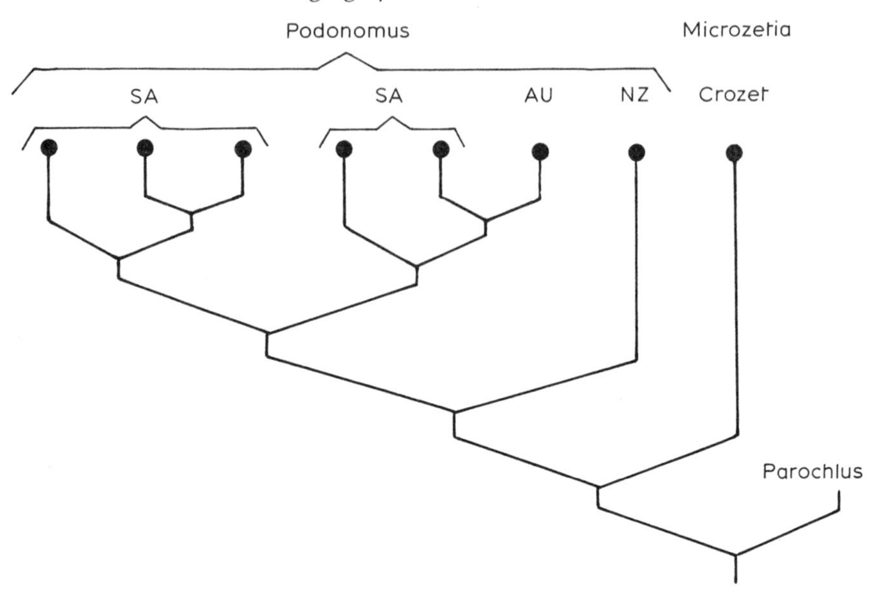

Figure 11.7 Cladogram of the chironomid genera *Podonomus* and *Microzetia*.

The mountain streams of Possession Island in the Crozet group are inhabited by two endemic chironomids, *Microzetia mirabilis* and *Parochlus crozetensis*, the latter recently described by Serra-Tosio (1986), and both belonging to the tribe Podonomini of the subfamily Podonominae. Analysis of the monotypic Crozet genus *Microzetia* indicates that it alone forms the sister-group to the strongly diversified genus *Podonomus* which is represented by 44 species in Andean South America, Juan Fernandez, New Zealand, southern Australia and nowhere else (Fig. 11.7). *Parochlus crozetensis*, also an old endemic taxon, is, according to the careful analysis performed by Serra-Tosio, the sister taxon of a major monophyletic aggregate formed by the *steineni*, *nigrinus* and *tonnoiri* groups of the genus *Parochlus* which are represented by 18 species in southern Andean South America down to Tierra del Fuego, South Georgia, South Shetland Islands, South-east Australia and Tasmania (Fig. 11.8).

The Crozet Islands extend over a distance of 120 km and are situated about half way between Madagascar and East Antarctica. The islands are volcanic and arise from a submarine pleateau. The other islands of the southern Indian Ocean, Marion–Prince Edward, Kerguelen and Heard, are similar. *Microzetia* thus stands out as the sister group of the genus *Podonomus*, the species of which are involved in trans-Antarctic

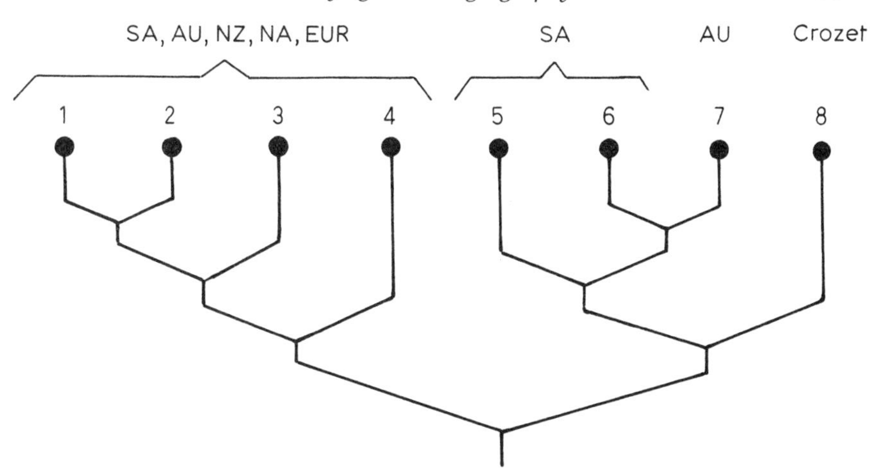

Figure 11.8 Cladogram of the chironomid genus *Parochlus* showing the position of *P. crozetensis*. (1) *chiloénsis* group (SA); (2) *squamipalpis* group (SA); (3) *conjungens* group (SA, NZ); (4) *araucanus* group (SA, AU, NZ, N. Amer., Europe); (5) *steineni* group (SA, S. Georgia, S. Shetland I.); (6) *nigrinus* group (SA); (7) *tonnoiri* group (AU); (8) *crozetensis* (Crozet I.).

sister-group relationships between South America and Australia and between South America and New Zealand. It has also been shown by geophysicists that New Zealand was separated from West Antarctica in the Upper Cretaceous. Therefore, there is good reason to conclude that the genus *Microzetia*, like *Podonomus*, is at least of Middle Cretaceous age. *Parochlus crozetensis* is evidently of about the same age as *Microzetia* – in spite of the fact that its sister group does not occur in New Zealand – because the other main group within *Parochlus*, of which *P. crozetensis* plus the three subgroups mentioned above are the sister group, is also represented in New Zealand (Fig. 11.8).

From this analysis, it seems justifiable to conclude that the Crozet Islands are the visible remains of a former 'Crozet Land' that was left behind as a continental fragment when East Antarctica, after its separation from South Africa at the transition between the Jurassic and Cretaceous, moved south, towards the South Pole. The later subsidence of the Crozet Land and the development of a hostile environment in connection with the deterioration of the climate and the glaciation of Antarctica, certainly implies much biotic extinction. There is reason to suppose that the other islands of the southern Indian Ocean had a similar history. *Microzetia* and *Parochlus crozetensis* stand out as true Antarctic elements. Therefore, the Crozet portion of the East Antarctic ranges of their ancestral species became cut off when the Crozet Land was left behind as a continental fragment. In cases like these we are evidently dealing with vicariance events, not preglacial jump dispersals. Among the subantarctic island biotas there are probably several more examples of biogeographic elements of the type demonstrated by the two chironomids of Crozet, though they cannot be clarified at present due to lack of cladistic information.

It seems probable that a certain portion of the subantarctic island biotas consists of post-Pleistocene immigrants that arrived by air or sea currents. Such elements should above all be widespread species which are known also to occur in a larger source area like the southern tip of South America. Also theoretically possible is the presence of widespread genera that are represented on the subantarctic islands by endemic species as a consequence of preglacial jump dispersal. A requirement for successful analysis will, however, be establishment of source areas based on cladistic arguments.

11.5 SIGNIFICANCE OF FOSSILS TO BIOGEOGRAPHIC HYPOTHESIS

Little by little some palaeontologists have perceived that Hennig's principles of phylogenetic systematics meant a revolution to their science. Working with extinct taxa and associated stratigraphic data, palaeontologists have long considered themselves to be in a favoured position when facing the task of reconstructing phylogenetic relationships. The general confinement of their studies to ancestor-descendant relationships has often led to construction of so called 'grade groups' which merely represented different evolutionary levels and not necessarily monophyletic groups. However, because we are dealing with a hierarchy, every fossil group must have a sister-group in the Recent biota. Hence a reconstruction of the phylogenetic relationships formed by the Recent taxa is a prerequisite for reliable placement of fossil taxa in the hierarchy. It is only through cladistic analysis of the Recent taxa that we can determine those apomorphic characters that decide the position of fossil taxa. Even if this insight means a methodological improvement, there are several factors restricting the exploitation of fossils for biogeographic purposes. Many groups are incapable of fossilization, while others have left only superficial impressions. Regrettable also is the poverty of the southern continents in fossiliferous terrestrial strata from the Jurassic and Cretaceous epochs, periods filled with events that have deeply influenced the history in time and space of plants and animals. Because of patchy documentation, the fossil record is at best able to supply only the minimum age, not the actual age of a plant or animal group (Hennig, 1966). But even such limited insight is possible only if the fossil is so well preserved that its identity and position in the hierarchy can be established on at least one autapomorphy of the basic design of a Recent taxon. Hence, fossils showing only relatively plesiomorphic characters are of little importance, while fossils with relatively apomorphic characters can be very important, because they not only prove the existence at a certain time of the group to which they belong but also the simultaneous existence of the sister group of that group and of older groups of the actual stem line. Further problems of using fossils with respect to the identification of extinction episodes are discussed in Chapter 8.

The great importance of a single, well preserved fossil is exemplified by chironomid midges. Darlington (1970) objected to my conclusions of 1966 by saying that the chironomids are too young to have been influenced by the breakup of Gondwanaland. However, the fortunate find (Schlee and Dietrich, 1970) of a fossil member of the subfamily Podonominae in amber from the lowermost Lower Cretaceous (Neocomian) of Lebanon, which is situated at the northern margin of Gondwana, nullified his objection. It could be shown by synapomorphies that the midge, *Libanochlites neocomicus* Brundin (1976), belongs to the tribe Boreochlini, where it is a member of the *Boreochlus* group that comprises the extant holarctic genera *Boreochlus* and *Paraboreochlus*. The sister-group of the comparatively apomorphic *Boreochlus* group is the strongly plesiomorphic *Archaeochlus* group of South Africa and south-west Australia (Brundin, 1976). The occurrence of *Libanochlites* in Lebanon 130 Myr ago not only supports the supposition (Brundin, 1966) that the history of the *Boreochlus* group includes old range expansion northwards, along the East African highlands, but proves that the history of the chironomid midges goes right back into the Jurassic.

A factor that probably reduces the indicator value of fossil findings is the different chances for fossilization in different types of environment. Sediments are predominantly deposited in the lowlands, and consequently fossilized plants and animals predominantly belong to the most common species of the lowlands. For obvious reasons, the chances of a stable deposition of sediments and preservation of fossils are much less in mountainous areas than in the lowlands. On the other hand, the chances for biotic isolation and speciation are comparatively high in mountainous areas. The eastern slopes of the tropical Andes provide a good example of this. Consequently a new group may carry on a marginal life for long periods without leaving any traces in the fossil record. Eventually, factors like the passing of a developmental threshold, extinction of dominant groups or the disappearance of abiotic barriers may open the gate to range expansion into the lowlands. This seems to be an adequate explanation of the sudden appearance in the fossil record of several new groups in different parts of the world. In other words, as stressed also by Croizat (1958), there may be a considerable difference between the age of fossilization of a group and its age of origin.

To biogeographers, the breakthrough in the theory of plate tectonics through increasing refinement of the timetable of continental displacements has increased insight into the ages of many plant and animal groups. If the range of an ancestral species is divided by a barrier, the age of the barrier will indicate the maximum age of two vicariant sister groups that evolved through allopatric speciation as a consequence of that barrier. The degree and type of involvement in the different phases of the process of continental fragmentations and collisions have indicated that many groups may be older than is supposed on the basis of fossil findings. As witnesses of the history in time and space of monophyletic groups, fossils and paleogeographic events stand out as complementary to each other (Chapter 14).

Finally, it is worth mentioning that not only vicariance biogeography (Nelson and Platnick, 1981; Patterson, 1981a) but also dispersal biogeography becomes comparatively insensitive to fossils if the conclusions are based on documentation of the sort I had at my disposal in 1966. I could then rely on several vicariance patterns of worldwide extent from different epochs demonstrating amphitropical and trans-Antarctic sympatry among closely related groups of southern Gondwanic origin including the deep involvement of southern Africa. With reference to these cosmopolitan patterns of different ages, I believe there is hardly any reason to suspect that the finding of fossils will be able to change the main observed patterns of vicariance and range expansion. Fossils will on the whole either confirm or complete the reconstructed patterns, which has been the case so far with the chironomids. But the history of the extinct stem groups of Chironomidae and other groups of the same or greater age will largely remain lost because of insufficient documentation. The history of many typical relict groups, is also out of reach.

11.6 CONCLUSIONS

A survey of the present status of causal historical biogeography shows that what is needed most is a greater knowledge of the hierarchy of life through application of Hennig's methodological principles on a far larger scale than has hitherto occurred. Considering that phylogenetic systematics has been at our disposal for more than 30 years, progress in our understanding of the history in time and space of different biotas and their components – the monophyletic groups – has indeed been rather limited, mainly because the necessary analytic work is very time-consuming. There has undeniably, been an unequal division between the efforts in reconstructing the history of actual groups and the great amount of philosophical, terminological and methodological discussions based on abstract groups. Ball (1975) summed this up well when he wrote the following: "Moreover, I wish to avoid the tendency, all too apparent in many recent contributions to debate on systematics, of arguing in a vacuum, using theoretical examples rather than real problems. The latter are never solved as neatly as the former, and it is illuminating in epistemological discussions to reveal everything, even the warts so beloved of Cromwell".

CHAPTER 12

Cladistic biogeography

C. J. HUMPHRIES, P. Y. LADIGES, M. ROOS and M. ZANDEE

12.1 INTRODUCTION

As witnessed from the other contributions in this volume, the theory and methodology of biogeography has received attention from a wide range of perspectives. Broadly, these approaches fall into two groups – those that provide explanations in terms of ecology and those that provide explanations in terms of biotic and earth history. Ecologists generally consider distribution patterns at various levels of organization, from species in habitats to species across ranges in ecosystems, and hence they consider biogeography as a subsidiary of ecology. Ball (1975) pointed out that some ecologists (MacArthur and Wilson, 1967) do not even distinguish between ecology and biogeography. We do not think that anyone would challenge the idea that general similarities occur between on the one hand the floras and faunas of the Arctic and the Antarctic and on the other between the rain forests of Africa and South America. However, when examined from a taxonomic point of view the species in the different areas are quite distinct from one another. In other words, species in similar 'niches' are different taxa which have had different histories. Historical biogeography focuses on evolution – change in time and space in the different groups of the taxa present in the areas of interest. Historical biogeographers attempt to explain the history of biotas by looking for general similarities in the individual histories of the component taxa.

The data for historical biogeography come from the distributional information in systematic monographs. As mentioned in Chapter 13, the major synthesis is found in the work of Croizat, in his book *Space, Time and Form* (1964). Croizat's contribution was to show that geological and climatological history of the earth and form-making in species are one and the same thing. The rekindled interest in Croizat's work (Croizat, Nelson and Rosen, 1974; Craw and Gibbs, 1984) came at the time of acceptance of plate tectonics, mobilist concepts of earth history and the ever widening use of cladistics for phylogeny reconstruction, especially with the use of the method outlined by Hennig (1966). Brundin (1966, 1972a,b; Chapter 11) perhaps most clearly demonstrated the practicalities of using cladistic classifications for biogeographic explanations. More recent studies have attempted to combine the principles developed by Hennig with those developed by Croizat. Examples can be found in the work of Cracraft (1975, 1982), Ball (1975), Rosen (1975, 1978, 1979), Nelson (1978), Patterson (1981a), Nelson and Platnick (1981), Wiley (1981), Nelson (1984, 1985), and Humphries and Parenti (1986). The association of Croizat with cladistics is anathema to both himself and his followers (Craw, 1983a, 1984a) because Nelson, Platnick and indeed Hennig are seen as authoritarian systematists rather than good investigative scientists (Heads, 1985). But it was through the work of Nelson, Rosen and Platnick that the works of Croizat were shown to be essential reading and through their work initially that Croizat's 'panbiogeographic method' was shown to lack a clear notion of relationship (Nelson and Platnick, 1981). Because historical biogeography is about the classification of areas, we hope to demonstrate here that a rigorous concept of relationship for areas can only come about by applying cladistic methods to biogeography and that the resolution of patterns is far greater with cladistic methods. Our aim in this chapter is to produce a short, but comprehensive, account of cladistic biogeography by describing briefly the recent refinements of methods beyond those described by Platnick and Nelson (1978), Nelson and Platnick (1981) and Humphries and Parenti (1986), and then applying these methods to a group of eucalypts from Western Australia.

12.2 CLADISTICS AND BIOGEOGRAPHY

A phylogenetic relationship can best be demonstrated with a rooted, branching diagram called a cladogram. In Fig. 12.1 the Magnoliidae are considered to be more closely related to the Gnetidae than to the Cycadidae, the Pinidae and Ginkgoidae since they have at least one common ancestor which is not ancestral to the latter taxa. Similarly, the Pinidae, the Gnetidae and the Magnoliidae form a strictly monophyletic group because they have a common ancestor not shared by any other taxon (e.g. the Cycadidae). Such groups are strictly monophyletic because they can be defined uniquely by particular character distributions (Fig. 12.1, characters 9–11 and 12–18). The task of all cladistic methods is to find the groups.

Hennig (1950, 1966) considered cladograms as common ancestry trees in which the monophyletic groups could be determined from the distribution of shared derived characters (synapomorphies) inherited from the most recent common ancestor (e.g. characters 12–18 in Fig. 12.1 for the Gnetidae + Magnoliidae). Relatively primitive shared characters (symplesiomorphies, e.g. Fig. 12.1, characters 1–8 for all taxa

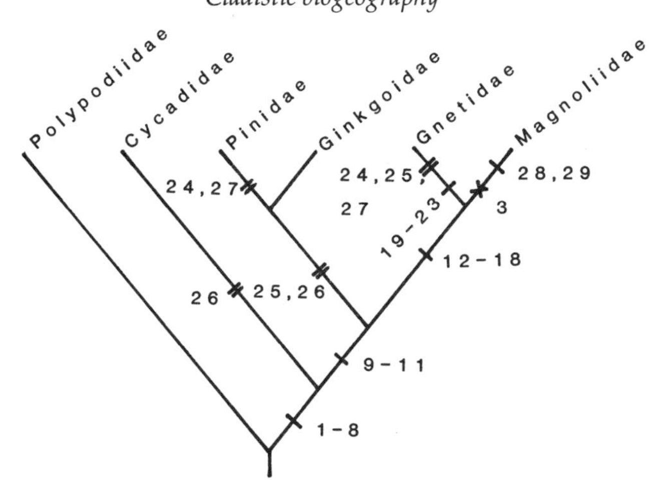

Figure 12.1 Cladogram of seed plants showing Hennig's definition of relationship. Characters are as follows: 1 – vascular cambium, 2 – eustele, 3 – embryogenesis with free nuclear phase, 4 – single functional megaspore mother cell, 5 – integument, 6 – micropyle, 7 – linear tetrad of megaspores, 8 – seeds, 9 – axillary branching, 10 – saccate pollen, 11 – platyspermic ovules, 12 – pollen with distal germinal aperture, 13 – loss of pollen sacs, 14 – siphonogamy, 15 – ovules borne in a cupule, 16 – megaspore membrane thin, 17 – uniovulate cupule, 18 – 'flowers', 19 – granular pollen wall, 20 – erect uniovulate cupule, 21 – unicupullate megasporophylls, 22 – vessels with porose perforation plates, 23 – 'flowers' with opposite-paired bracteoles, 24 – ovulate shoots aggregated into an inflorescence, 25 – sporophylls on short axillary shoots, 26 – pollen with distal germinal aperture, 27 – narrowly triangular awl-shaped leaves, 28 – leaves with essential oils, 29 – microsporogenesis simultaneous.

above the Cycadidae) do not help in recognizing the group Gnetidae + Magnoliidae. Similarly, the characters unique to a terminal taxon (Fig. 12.1, characters 19–23 and 27, 28–29), which Hennig called autapomorphies, are uninformative with respect to the formation of groups. Thus, we can summarize Hennig's view of similarity by saying that he recognized three types of resemblance – autopomorphy, synapomorphy and symplesiomorphy – all of which describe the status of any character in terms of a particular taxonomic problem. For example, in land plants (Crane, 1985) the seeds (Fig. 12.1, character 8) are autapomorphies when trying to determine the relationship of spermatophytes (e.g. subclasses in Fig. 12.1) with Algae and Bryophyta; they are symplesiomorphies when trying to determine the relationships of Magnoliidae and Gnetidae to Pinidae; and synapomorphies when trying to determine the relationships of Polypodiidae to the other five subclasses in Fig. 12.1.

From these concepts of character distribution, Hennig derived definitions for three types of groups:

(1) Monophyletic groups based on shared resemblance derived from a common ancestor (Fig. 12.1, characters 9–11 for the group including all seed plants except Cycadidae).
(2) Paraphyletic groups which are those that do not contain all of the descendants of a common ancestor. A paraphyletic group is one remaining after one or more parts of a monophyletic group are removed into a group of their own because they possess many unique characters. In Fig. 12.1 a paraphyletic group 'Gymnospermae' would result if the Magnoliidae were removed to a group of its own as a result of undue weight being applied to characters 28 and 29. The paraphyletic group 'Gymnospermae' is thus defined by not having the characters of the Magnoliidae. Other classic examples are the Algae, Invertebrata and Reptilia, none of which possesses defining characters.
(3) Polyphyletic groups based on resemblance due to convergent or independently derived characters not inferred to have occurred in the common ancestor (Fig. 12.1 a theoretical group combining Cycadidae with Ginkgoidae and Pinidae, determined by character 26).

In the real world character distributions do not always appear to fall into neat, congruent hierarchical patterns, but instead show general hierarchical patterns with numerous conflicting distribution patterns. For example in Fig. 12.1 these could be more characters like 26 which show conflicts with the distribution of characters 9–11. In fact, in most analyses it is possible to construct more than one cladogram from the same set of information. To distinguish between competing solutions, the principle of parsimony is applied (Farris, 1983). The empirical basis of cladistics is the formation of groups by the most parsimonious distribution of characters. The criterion of parsimony is a purely logical device used in the formation of hypotheses in science, which in systematics allows us to choose amongst alternative hypotheses of homology (characters) and alternative hypotheses of relationship (cladograms) by minimizing the number of *ad hoc* hypotheses of homoplasy (parallelisms, reversals and convergences). Farris (1986) states that, because the parsimony criterion is a logical criterion, it should not be equated with the idea that evolution is parsimonious. Parsimony both measures the ability of genealogical hypotheses to explain observed similarities and corresponds to Hennig's principle of synapomorphy. A genealogy that is consistent with a single origin of a character is able to account for all of those occurrences in terminal taxa as inheritance. Each additional requirement for a separate origin of a feature reduces the explanatory power of a genealogical hypothesis. Identification of the most explicit genealogical hypothesis is equivalent to finding the hypothesis that minimizes the number of multiple origins, namely the most parsimonious cladogram. It is no accident that parsimony has been independently applied by cladists, molecular biologists and language analysts. Biogeographical hypotheses based on the most parsimonious cladograms are also likely to be the most robust hypotheses.

Cladistics as described here really refers to the methodology of Hennig (1950, 1966). Although there is no space here to describe other approaches, Wagner (1961, 1980) also developed a method of phylogenetic inference, groundplan-divergence, which formed the inspiration for the Wagner parsimony method (Kluge and Farris, 1969; Farris, 1970). Other types of parsimony methods used in computer applications today include the Branch and Bound method (Hendy and Penny, 1982) and the three-taxon computations of Zandee (1985). Quite independent are the character compatibility methods of Estabrook *et al.* (1976).

Consider, for example, the five taxa in Fig. 12.2, ABCDE, and six characters (1–6). Character 1 identifies the group CDE. Other characters can relate to it in five different ways: (i) they may group CDE into the larger group ABCDE (e.g. character 2); (ii) they may characterize another group entirely, AB (e.g. character 3); (iii) they may characterize a subset DE of the group CDE (e.g. character 4); (iv) they may define the same group CDE (e.g. character 5); or, (v) they may characterize a conflicting

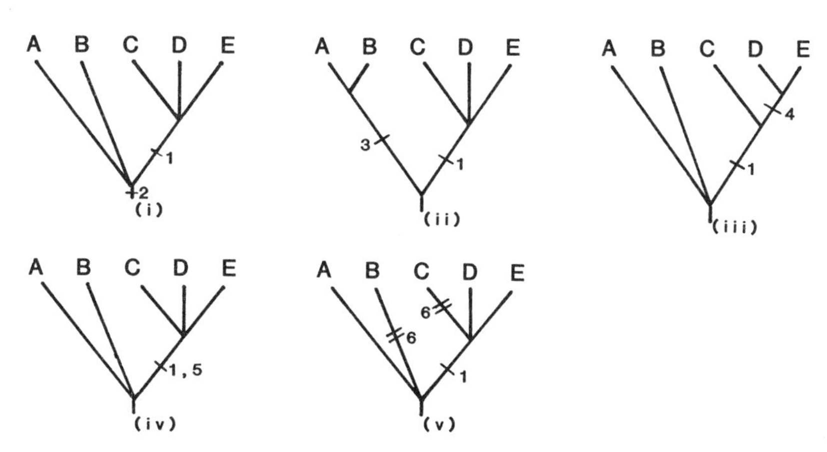

Figure 12.2 Cladogram showing five ways (i–v) that characters specify groups. Character 1 is a synapomorphy defining the monophyletic group CDE. Characters 2–6 can relate to character 1 in different ways. Character 2 supports a larger monophyletic group ABCDE, 3 supports the monophyletic group AB, 4 supports the monophyletic group DE, 5 defines the same group CDE and 6 suggests a conflicting group BC. (After Humphries and Funk, 1984).

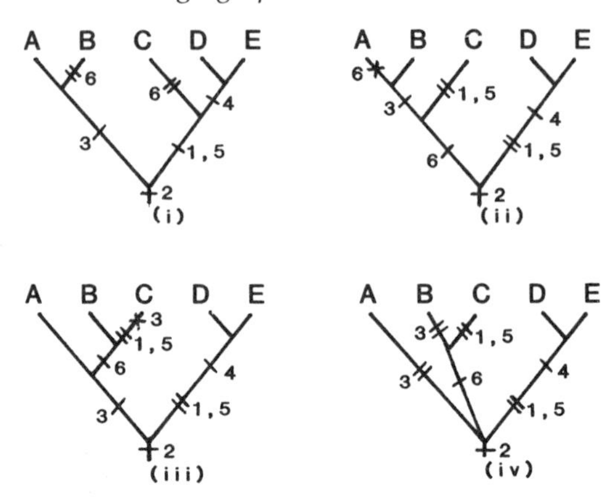

Figure 12.3 Four possible cladograms (i–iv) that can be constructed from the six characters in Fig. 12.2. (After Humphries and Funk, 1984).

group BC (e.g. character 6). If we accept the axiom that evolution has only one history, not all of these characters can be synapomorphies because there is some conflict. Fig. 12.3 shows the alternative classifications when all of the characters are considered together. It must be fairly obvious that some characters, such as characters 5 and 6, are very strong in the sense that they either fully corroborate or fully conflict with the original character 1 hypothesis. Characters 2, 3 and 4 lend only secondary support to character 1 by comparison. In the first cladogram, Fig. 12.3, (i) one potential synapomorphy is found to be false (character 6) which necessitates an *ad hoc* explanation of parallel evolution; in (ii) two potential synapomorphies are found to be false (characters 1 and 5) and one must be considered as a reversal (character 6), necessitating three *ad hoc* explanations, two of parallel evolution and one of reversal; in (iii) two potential synapomorphies are found to be false, characters 1 and 5, and character 3 is a reversal, necessitating three *ad hoc* explanations in total; in (iv) three potential synapomorphies are falsified (characters 1,3,5), necessitating three *ad hoc* explanations of parallel evolution. When counting up the steps, Fig. 12.3(i) requires the fewest steps (seven) and fewest *ad hoc* explanations (one), i.e. a total of seven steps, and is the one which gives most explanatory power to the data, since those cladograms in Figs. 12.3(ii), (iii) and (iv) each require nine steps (Farris (1983) gives an explanation of parsimony) or are not as informative with respect to the degree of resolution of the relationships. Consequently, 12.3(i) is the best defended hypothesis.

12.3 APPLICATIONS OF CLADISTICS TO BIOGEOGRAPHY

The earliest application of cladistics to biogeography was that of Hennig (1966) when he used a cladogram to determine the 'centre of origin' of a monophyletic group. Hennig used a concept known as the 'progression rule', which stated that the most apomorphic taxa were those furthest away from the centre of origin which contained the most plesiomorphic taxa. This approach is of little value as it has little predictive power to evaluate those instances of disjunct groups that owe their present distributions to vicariance. Hennig (1966) showed that there was a close relationship between species and the space each one occupies, and the best applications of Hennig's methods are found in Brundin (1966, 1972a,b, 1981; Chapter 11) and Ross (1974).

In our view, the breakthrough in the application of cladistics to biogeography came with the efforts of Nelson and Rosen in their interpretations of Croizat. Instead of a 'vacuum' theory of biogeography, whereby certain areas were originally devoid of taxa and were to be colonized later from other source areas with no relation to historical events, disjunct distributions could exist because of vicariance events. Ancestors originally occurred in areas where the modern taxa are present today, or the taxa we see today evolved *in situ* and the patterns we see were due to earth and biotic history being one and the same. To elaborate, dispersal models explain disjunction by dispersal across pre-existing barriers, whereas vicariance models explain them by the appearance of barriers fragmenting the former ranges of ancestral taxa (Chapter 1). However, the most important view that prevails now is the idea that dispersal patterns and vicariant patterns cannot necessarily be resolved decisively from one another. Instead, when faced with a particular distribution pattern, the main concern is whether it conforms to a general pattern of area relationships shown by other groups of taxa occurring in the same areas. In cladistic taxonomy, three-taxon statements are the most basic units used to express taxonomic relationships – cladograms indicating relative recency of ancestry in terms of the parsimonious distribution of synapomorphies – whereas in cladistic biogeography three-area statements are the most basic units for expressing area relationships, with area cladograms indicating the relative recency of common ancestral biotas.

The rationale for applying systematic theory to biogeography is based on the idea that phylogenetic reconstructions are essential precursors for historical interpretation. It is just as important to have a pattern to explain a process in biogeography as in systematics. By replacing taxa with areas, the principles of congruence and synapomorphy can be applied with the same rigour.

To obtain a precise cladogram of areas, based upon the biotic components within them, area cladograms are produced by substituting for taxa the areas in which they occur (D. Rosen, 1975; Platnick and Nelson, 1978). The generality of area cladograms can be examined by comparison with other unrelated taxonomic groups endemic to, or present in, the same areas. Corroboration of a particular pattern in a consensus cladogram or a general area cladogram (see Section 12.3.1), is the best general hypothesis that can be obtained to express the relative recency of biotas. Initial applications of this method (D. Rosen, 1975) encountered problems of incongruance and unresolved patterns in the consensus cladograms when two or more groups were considered together. Theoretically, it should be possible to connect every area of endemism in the world into one general statement of relationships no matter how complicated by dispersal and extinction. However, our perception of the world is made more opaque as a result of a variety of natural processes such as extinction, dispersal, and the generally restricted distributions of particular groups.

The most difficult problem for assessing biogeographical information is the choice of appropriate analysis. As was so clearly demonstrated by Croizat, biogeographic patterns repeat and they contain considerable redundancy. For this reason it is desirable to add the biogeographic information of one group to that of another, but the task is never made easy because each cladogram is never quite the same as another as a result of 'missing', 'redundant', 'unique' and partially overlapping areas. There have been two general approaches to solving the problem: component analysis and consensus trees (Nelson and Platnick, 1981), and the parsimony methods of Mickevich (1981), Brooks (1981, 1986) and Zandee and Roos (1987).

12.3.1 Component analysis

The purpose of component analysis (Nelson and Platnick 1981; Humphries and Parenti, 1986) is to determine a classification of areas even when the available biogeographic information is filled with seemingly unresolvable and conflicting patterns. Hypotheses of area interrelationships are most powerful when they are based on many different groups of taxa, each showing similar patterns of relationship for the areas in which they occur, but, as commonly happens, the different groups of taxa to be combined are wanting in one form or another. Component analysis attempts to provide the means of 'filling in' missing data and resolving conflicts so that a single, specifiable pattern can be derived from many different cladograms.

To describe the principles of component analysis, five different distribution patterns are given in Fig. 12.4. The basic steps include a code (usually a letter or a number) applied to the areas occupied by the taxa and substituted for the taxa onto the cladogram. Secondly, the cladograms are made comparable to one another by adjusting for missing information, redundant information, widespread distributions and unique areas. Thirdly, the different area cladograms are combined into a consensus tree (Fig. 12.5; Nelson, 1979; Nelson and Platnick, 1981).

Figure 12.4(i) shows a cladogram for four taxa a–d endemic to four areas A–D, and Fig. 12.4(ii) shows the corresponding area cladogram for the four areas, A,B,C,D and three components defined as follows. The area interrelationships are represented by one group-defining component (1) and two informative components (2,3). An additional definition is that a general component is one that is common to two or more cladograms. Fig. 12.4 (iii) shows a cladogram for three different taxa e, f and g, and the corresponding area cladogram (Fig. 12.4(iv)) for three of the same areas, A, C and D. There is one new group defining component (4) and one of the same informative components (3). When this pattern is compared with the four-area cladogram in Fig. 12.4(ii), area B is 'missing'. To combine these two area cladograms to give a consensus tree would require the addition of the missing area B to the area cladogram in Fig. 12.4(iv). Since we do not know the exact position of the 'missing' area B, there are five possibilities in which it can be placed: Fig. 12.4,(v–ix). The five new 'possible' cladograms offer three 'new' potential components (6, 7 and 8) that must be considered in a consensus of the two groups.

Figure 12.4(x) is a cladogram for six taxa (h–m) occurring in four areas: Fig. 12.4(xi). There is redundancy of information in the distributions of the taxa h, i, k and l that occur in areas A and C. The duplicated representation of areas A and C can be reduced to cladograms expressing each area and each component just once (Fig. 12.4(xii)), giving the same components as for the first group in Fig. 12.4(ii).

Figure 12.4(xiii) is a cladogram of three species, n, o and p occurring in four areas: Fig. 12.4(xiv). Widespread taxa such as n create a problem different from the one described for missing areas. The widespread occurrence of taxon n means in effect that the relationships of areas A and C to B and D are unresolved. Consequently, A and C can be alternatively represented at the basal node of the area cladogram – Fig. 12.4(xv) – and only two components, 1 and 6, can be expressed. We could make no assumptions and postulate that areas A and C are either related to one another or are wrongly considered as separate areas for some *ad hoc* reason. Species n might have failed to speciate in response to some vicariance event for example, but this rules out the possibility

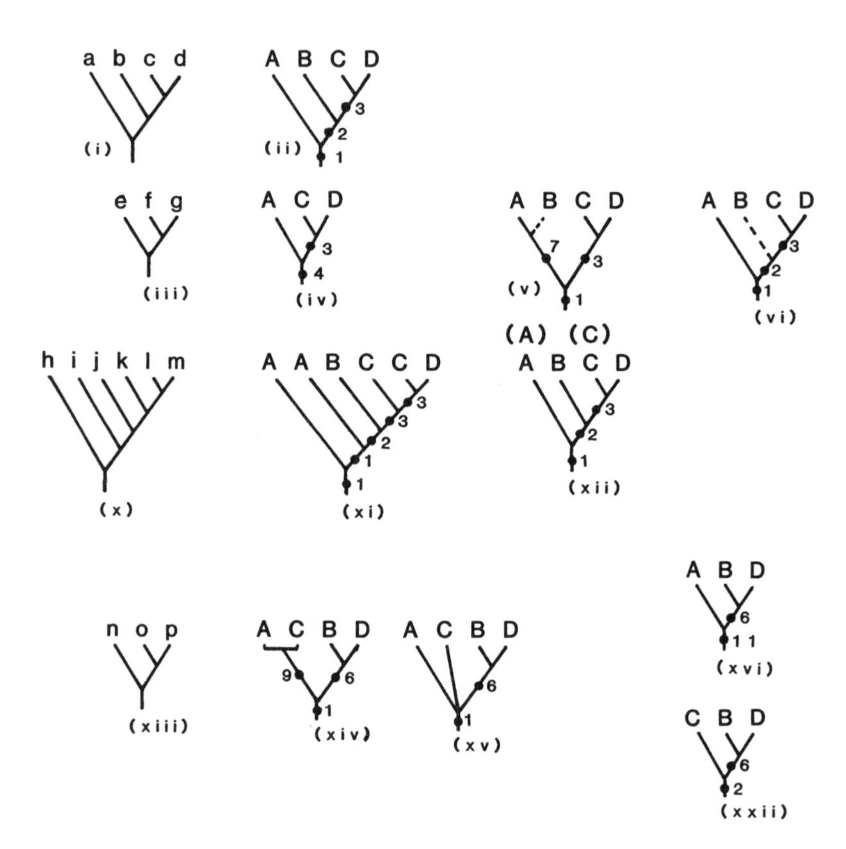
Taxa
Areas
a b c d
(i)
A B C D
3
2
(ii)
1
e f g
(iii)
A C D
3
4
(iv)
A B C D
7
3
(v)
1
A B C D
3
2
1
(vi)
(A) (C)
h i j k l m
(x)
A A B C C D
3
3
2
1
1
(xi)
A B C D
3
2
1
(xii)
n o p
(xiii)
A C B D
9
6
1
(xiv)
A C B D
6
1
(xv)
A B D
6
11
(xvi)
C B D
6
2
(xxii)

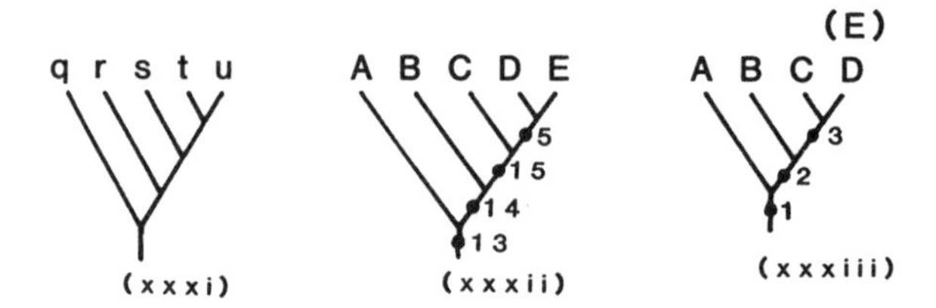
q r s t u
(xxxi)
A B C D E
5
15
14
13
(xxxii)
(E)
A B C D
3
2
1
(xxxiii)

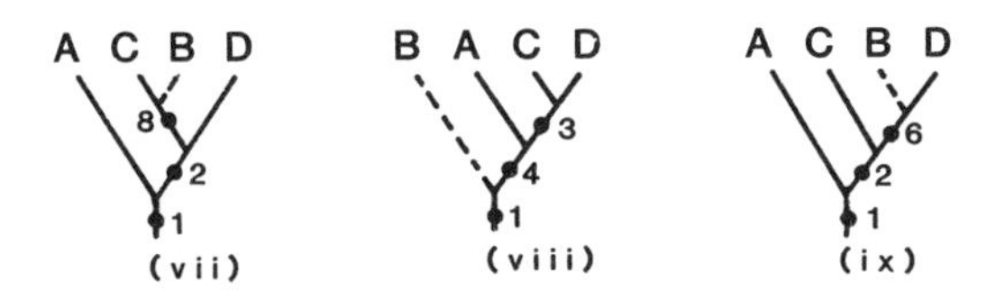

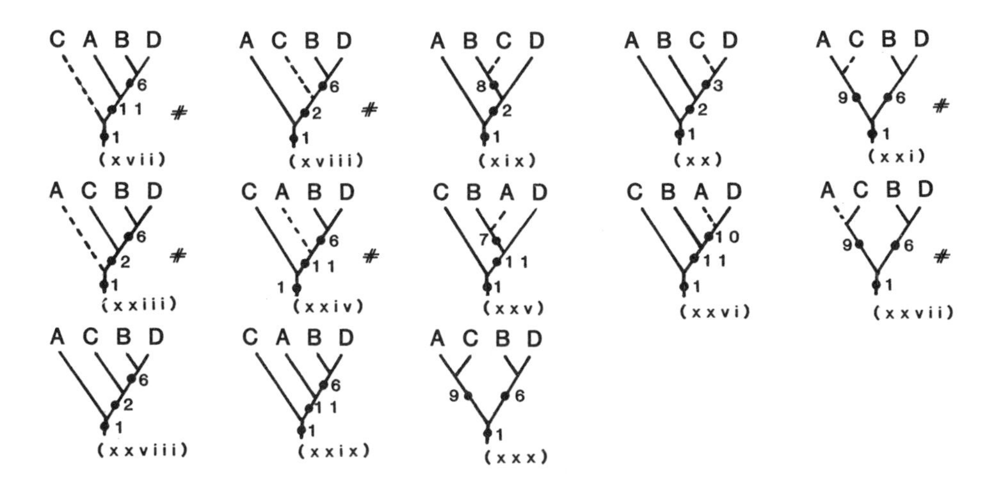

Figure 12.4 Patterns of distribution of species a–u for five hypothetical areas (A–E); (i–ii) taxa and derived cladogram for four endemic species; (iii–iv) taxa and derived area cladogram for three endemic species in three areas; (v–ix) possible components when cladograms for 'missing' area B are compiled; (x–xi) taxa and derived area cladograms for six taxa in four areas; (xii) cladogram corrected for redundant information; (xiii–xiv) taxa and derived area cladograms for two endemic (*o*,*p*) and one widespread species (*n*) in four areas; (xv) widespread pattern expressed as an unresolved area cladogram; (xvi–xxvii) possible components when areas A and C in (xv) are analysed under assumption 2 (see text); (xxvii–xxx) possible components when areas A and C in (xv) are analysed under assumption 1; (xxxi–xxxiii) taxa and area cladograms in five areas and 'corrected' for area E. (After Ladiges and Humphries, 1986). # represents equivalent cladograms from separate component analyses.

that there might be other reasons for the observed pattern. We could say instead that whatever is true of the relationship of the species n in area A (or in C but not in both) to o and p in areas B and D might not be true for species n in area C (or in A but not in both). Nelson and Platnick (1981) call this assumption 2. We could add that, because we do not know the reason for the widespread AC pattern (it might be due to dispersal, a failure to speciate or maybe there is no actual difference and A and C constitute one area instead of two), we should analyse the observed relationships separately. Consequently, with an unresolved cladogram of the type in Fig. 12.4(xv), we can consider two possibilities for component analysis; we can assume that the cladogram gives the correct relationship of A to B + D, with C unknown – shown in all possible positions in Fig. 12.4(xvi–xxi) – or we can assume that the cladogram gives the correct relationships of C to B + D, with A unknown, as shown in all possible positions in Fig. 12.4(xxiii–xxvii). This gives seven cladograms overall (since there are three in each row which are common to both analyses and these are marked with a #) and nine possible components (1–3, 6–11). Nelson and Platnick (1981) introduced another concept which they termed assumption 1. The widespread taxon n in areas A and C might be resolved into two or more taxa in the future simply in such a way that relationships with the taxa in areas B and D remain the same. In this case there are only three possibilities for component analysis: Fig. 12.4(xxviii–xxx).

The cladogram for five taxa q – u Fig. 12.4(xxxi), has introduced four new components (5, 13–15) because of the relationship of the unique area E in the area cladogram Fig. 12.4(xxxii). If we consider area E as a subunit of D because it is unique to this group, it is possible to adjust the cladogram so that it compares with the other groups: Fig. 12.4(xxxiii). Alternatively, we could consider E as missing in the other groups just as B is missing in Fig. 12.4(iv) compared with Fig. 12.4(ii).

To determine the area relationships in terms of all five groups of taxa, we need to sum the cladograms in Figs. 12.4(ii), (iv), (xi), (xiv) and (xxxii). There are two ways of doing this. We could assume that all of the components in the basic area cladograms are real and combine them into a single solution. This is shown in Fig. 12.5(i–vi). Four components are required for the resolution of the five taxa, but there are in fact ten (1–6, 9, 13–15) and there are overlaps and conflicts at every node. Components 3 (CD), 5 (DE) and 6 (BD) overlap as do also 2 (BCD), 4 (ACD) and 15 (CDE). Taking this naïve approach gives no resolved patterns (Fig. 12.5(vi)). On the other hand, it could be assumed that some of the components are false. Thus, if there is one history only, some of the implied components of each group which include or exclude components of other groups can be correct. Taking the examples of the taxa

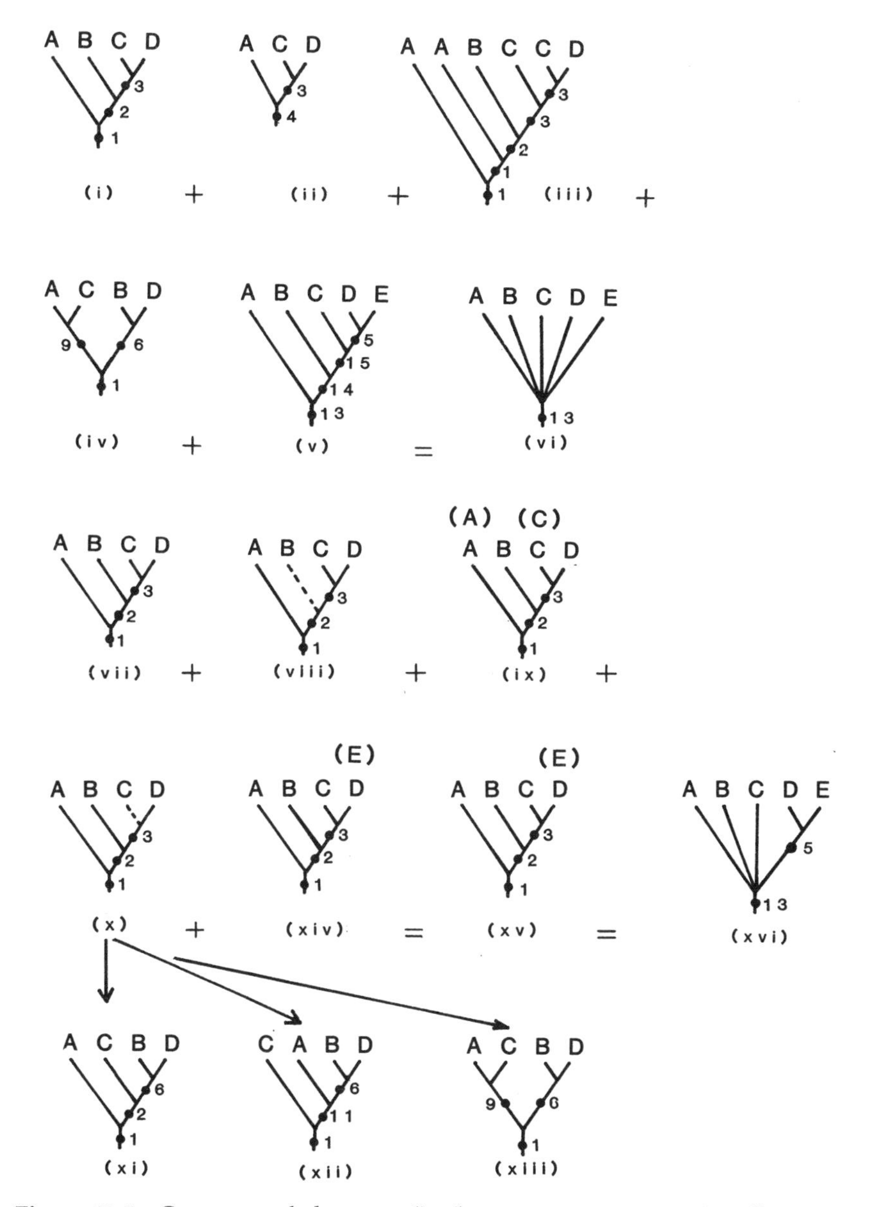

Figure 12.5 Consensus cladograms; (i–vi) consensus tree assuming all components as 'real'; (vii–xvi) consensus tree assuming only some of the components as real and analysed under assumptions 1 and 2 (After Ladiges and Humphries, 1986).

shown in Fig. 12.4(i), (iii), (x), (xiii) and (xxxi) there are 15 implied components. However, components 1, 2 and 3 occur in every taxonomic group. Component 5 is unique to the last group, Fig. 12.4(xxxii). Addition of the components (implied by Nelson and Platnick's second assumption) that are in common or exhibit congruency allows us to find a resolved consensus solution, shown in Fig. 12.5(xv). Alternatively, adding them together under assumption 1 (5(vii) + 5(viii) + 5(ix) + 5(xi) + 5(xii) + 5(xiii) + 5(xiv)) leaves A, B and C unresolved: Fig. 12.5(xvi). In conclusion, assumption 2 thus allows the maximum possible resolution of 'hidden' patterns that have been obscured by a variety of stochastic historical causes such as dispersal and extinction.

12.3.2 Nelson's problem

How effective is component analysis? Nelson (1984) provided an example with three cladograms that all shared some areas, but which also exhibited problems of widespread taxa, redundancy, 'missing' areas and conflicting information (Fig. 12.6). He said that this example was as opaque to intuitive appraisal as he could devise at the time. Nelson's problem can be expressed in terms of three-taxon statements. For the first cladogram, Fig. 12.6(i), of the six possible three-taxon statements derivable (Fig. 12.7(i–vii)) under Nelson and Platnick's assumption 2, only one, which suggests that B is much more closely related to C in relation to A, is a three-taxon statement: Fig. 12.7(iv). For the second cladogram, (Fig. 12.6(ii)), of the six possible three-taxon statements Fig. 12.7(ix–xiv)) there are only two that are informative, suggesting that B is more closely related to C in relation to D (Fig. 12.7(x)) or that C is more closely related to D in relation to B (Fig. 12.7(xiii)). Similarly, for the third cladogram (Fig. 12.6(iii)), there are two informative three-taxon statements (Fig. 12.7(xvii) and (xviii)). For the five informative three-taxon possibilities there are five possible positions into which can

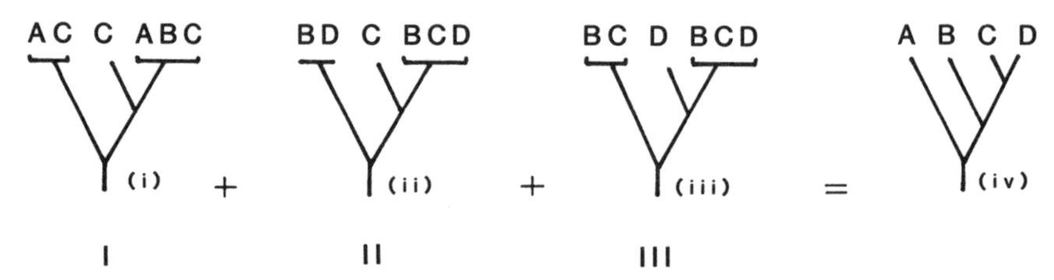

Figure 12.6 'Nelson's Problem'. A consensus combination of three area cladograms (I, II, III) involving lack of occurrence, widespread taxa, and redundancy, with a fully informative result. (After Nelson, 1984: Fig. 14.12, p. 288).

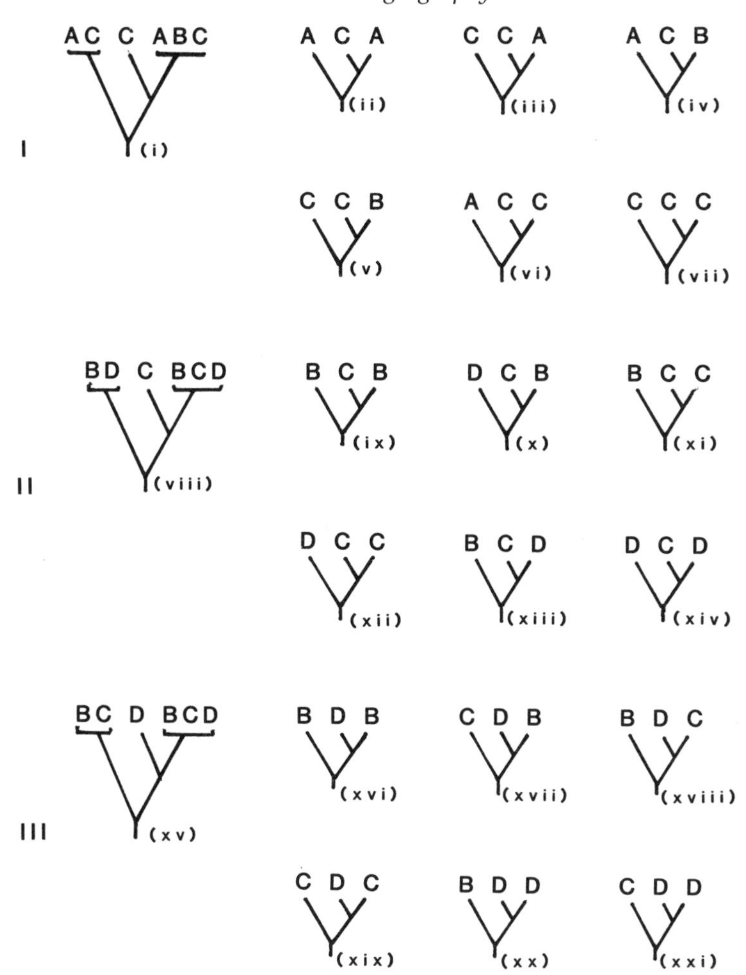

Figure 12.7 (i), (viii), (xv) – three area cladograms (I, II, III) involving widespread taxa in three areas; (ii–vii), (ix–xiv), (xvi–xxi) – possible three area statements from the given topologies under assumption 1.

be added the 'missing' areas. This then gives us $5^3 \times 4$ possible combinations for the consensus tree combinations (Fig. 12.8). In other words, adding in the 'missing' area into each three-taxon statement gives five extra possible combinations. Also, because there are two possible combinations for each of the original cladograms that excluded area A, there are four rows of possible additions. Three of four possible sets of 125 possible combinations give less than fully informative results. However, there is one solution that gives the result of Nelson, that C

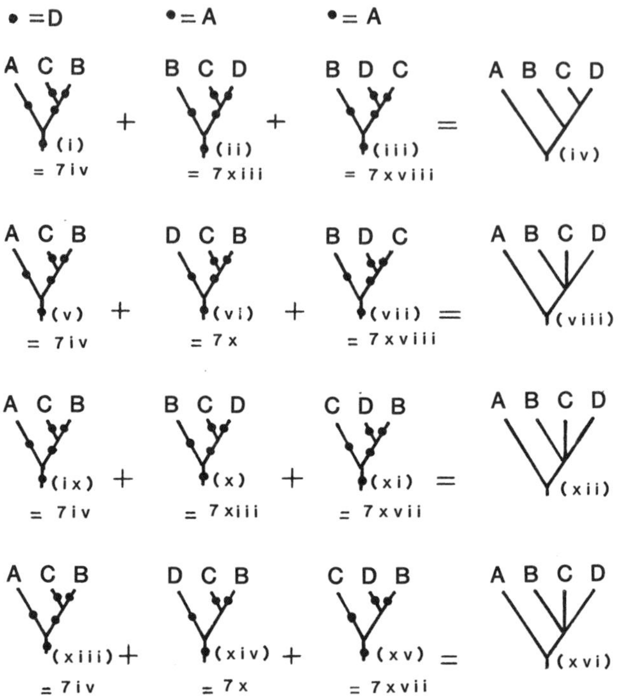

Figure 12.8 125 possible combinations for the five possible three-area cladograms derived from the original cladograms I, II and III (see Figs. 12.6 and 12.7).

and D are sister areas in relation to A or B and that BCD is a group relative to A (Fig. 12.8(i–iv)). Of the five possible solutions for inserting missing area D into Fig. 12.8(i), from which one can be selected as correct, the position of D as sister to C gives the nested set {A[B(C,D)]}. Similarly, of the five positions for area A in Figs. 12.8(ii) and 12.8(iii), that which groups A as sister to all three areas BCD gives the same nested set {A[B(C,D)]} and, hence, a resolved consensus tree.

12.3.3 Parsimony

Consensus trees have their own problems because they cannot be relied upon to be accurate summaries of logically different data sets. Although they can be regarded as the basis for generating general classifications from logically different character sets or taxonomic groups (Mickevich, 1978; Nelson, 1979; Nelson and Platnick, 1981), they are only summaries of congruent information in the cladogram topologies. Although consensus trees have a form of stability by recognizing the commonality or congruency of cladograms and providing an expression of maximum overall resolution, they do not take into account the relative strength of character data in the logically different character sets of taxonomic groups (Miyamoto, 1985). In other words, when consensus trees contain polychotomies they are not necessarily parsimonious representations of the data.

There are three other biogeographic methods which attempt to circumvent this problem and produce general area cladograms from some form of character data so that the principles of parsimony can be applied (Brooks, 1981, 1986; Mickevich, 1981; Zandee and Roos, 1987). The data matrices in all three methods are not based on the original characters, but on component data extracted from the cladograms to be combined. The quantitative procedure in all three methods is based on the notion that taxa or groups of taxa (components) can be considered as transformation series linking areas together into a biogeographic pattern. Any cladogram can be converted into a character matrix which can then be analysed. Data matrices from several groups can be combined into one matrix so that a general area cladogram is generated. As a test we have considered Nelson's problem using the methods of Zandee and Roos, and Brooks. Mickevich's (1981) method has yet to be fully evaluated and so the reader is referred directly to her paper.

Component compatibility

The method of Zandee and Roos has the advantage that the actual widespread distributions within the original cladograms can be simply coded into the data matrix. Zandee and Roos (1987) call this assumption O, since the original cladogram is considered to be the best estimate of the phylogeny.

As with the methods of Rosen, and Nelson and Platnick, the terminal taxa are replaced by the areas in which they occur. Each clade is then

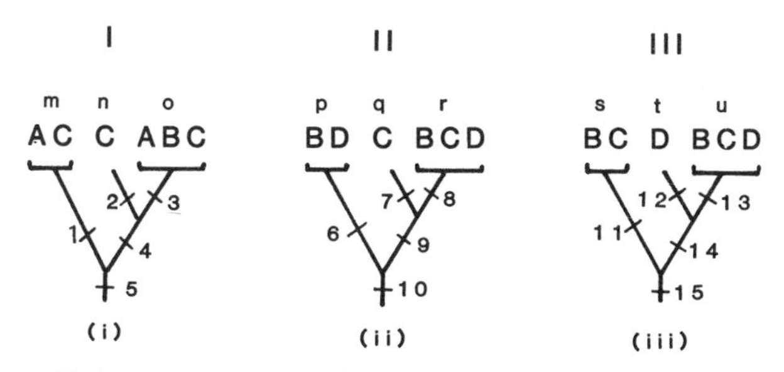

Figure 12.9 Cladograms I, II and III (see Figs. 12.6 and 12.7) coded for components (see Table 12.1).

Table 12.1. Binary matrix scored from component characters given in the three cladograms, I, II and III of Fig. 12.9

		I					*II*					*III*				
		Characters														
		1	2	3	4	5	6	7	8	9	10	11	12	13	14	15
Areas	A	1	0	1	1	1	0	0	0	0	0	0	0	0	0	0
	B	0	0	1	1	1	1	0	1	1	1	1	0	1	1	1
	C	1	1	1	1	1	0	1	1	1	1	1	0	1	1	1
	D	0	0	0	0	0	1	0	1	1	1	0	1	1	1	1

assigned a character (Fig. 12.9). The cladograms are then converted into data matrices (Table 12.1). The characters (components) are assigned in a form of additive binary coding (Farris, 1970).

The matrices can be analysed separately to produce individual area cladograms (Fig. 12.10). Within the data matrix there are seven partially monothetic sets (unique distributions of clades over areas; see Table 12.2) which can be arranged into five different cladograms (Fig. 12.11). Subsequently, all of the characters are put onto the cladogram, with those fitting the general area cladogram considered as support and those homoplasious characters requiring multiple, parallel or reversed steps considered as contradictory. The balance between support and contradiction serves as a measure of best fit with respect to the general cladogram and the data. Evaluation in terms of parsimony, the number of homoplasies minus support (Table 12.3), gives one possible area cladogram (Fig. 12.11(v)). Nelson's solution, (A(B(C,D))), is not recovered because there is no support for the clade C + D (Tables 12.1 and 12.2). Even if the algorithm had recovered it, putting characters 1–15 on Nelson's solution produces a cladogram that contains two extra steps compared with Fig. 12.11(v).

Analysing Nelson's problem under assumption 1 gives a total of 21 possible components for each of the three opaque cladograms (Fig. 12.9). The possible components for the first cladogram are given in Table 12.4. Analysing all of the components at once as single column

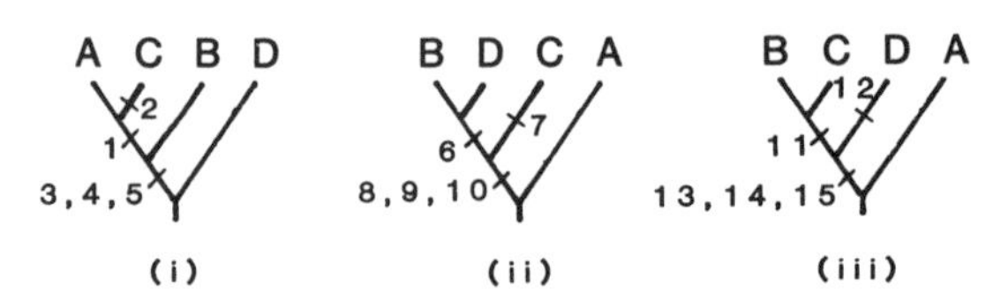

Figure 12.10 Area cladograms, derived by the method of Zandee and Roos from the data in Table 12.1, for each of the cladograms I, II and III.

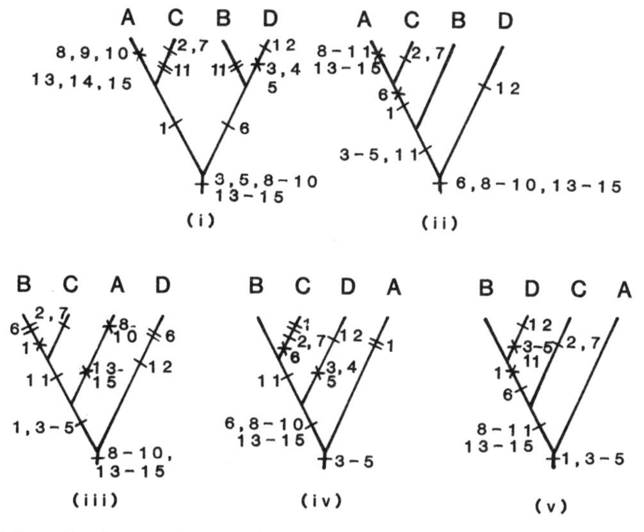

Figure 12.11 Maximum cliques (area cladograms) ordered from the monothetic data set in Table 12.2.

Table 12.2. Partially monothetic sets of areas (components) and corresponding columns ('characters', see text) for Table 12.1

		Characters					
Areas	C	2	7				
	D	12					
	AC	1					
	BC	11					
	BD	6					
	ABC	3	4	5			
	BCD	8	9	10	13	14	15

Table 12.3. Statistics for the area cladograms in Fig. 12.11: (i) homoplasies (multiple origins and reversals); (ii) support (single origins); (iii) homoplasies minus support (i–ii); (iv) number of steps

	I	*II*	*III*	*IV*	*V*
i	11	8	9	6	5
ii	14	15	14	14	15
iii	−3	−7	−5	−8	−10
iv	25	23	23	20	20

Table 12.4. Data matrix for cladogram I (see Fig. 12.7) scored for all possible components under assumption 1

		1	2	3	4	5	6	7	8	9	10	11	12	13	14	15	16	17	18	19	20	21
Areas	A	1	0	1	1	1	0	1	1	1	0	0	0	1	1	1	0	0	0	1	1	1
	B	0	0	1	1	1	0	0	1	1	0	1	1	0	0	1	0	1	1	0	0	1
	C	1	1	1	1	1	1	0	1	1	1	0	1	0	1	0	1	1	1	1	1	1
	D	0	0	0	0	0	0	0	0	0	0	0	0	0	0	0	0	0	0	0	0	0

entries in the data matrix leads to several resolvable solutions, the best of which are shown in Fig. 12.12. Similar analyses for the second and third cladograms are given in Fig. 12.12(ii–iv), giving two equally parsimonious cladograms for each. When the cladograms are analysed together the combined data matrix has 63 columns. The matrix yields eleven partially monothetic sets (components, Table 12.5) which gives eight general area cladograms including the original Nelson solution (Fig. 12.12(iii)). However, the best solution is that given in Fig. 12.12(iv) which is the same as the result given by assumption O. For this combined data matrix the balance between support and contradiction for the Nelson solution is two steps longer.

Under assumption 2 all components in the original cladograms, I, II and III are considered in dispute. A data matrix for analysis under assumption 2 should include all of the possible subsets of the areas of the range of the widespread species separately and in combination with all other clades, together with all of the combinations of the latter with missing areas (Zandee and Roos, 1987). As a consequence, the data matrices increase dramatically in size. For the first cladogram six resolved cladograms can be derived with one topology more parsimonious than the rest (Fig. 12.12(i)). Similar analyses for the other two cladograms give the results shown in Fig. 12.12(ii) and (iv) respectively, and all are similar to the results obtained using assumption O, but are again different from the Nelson solution. Analysing the data of all of the cladograms together gives a 4 × 174 character matrix. From this data set

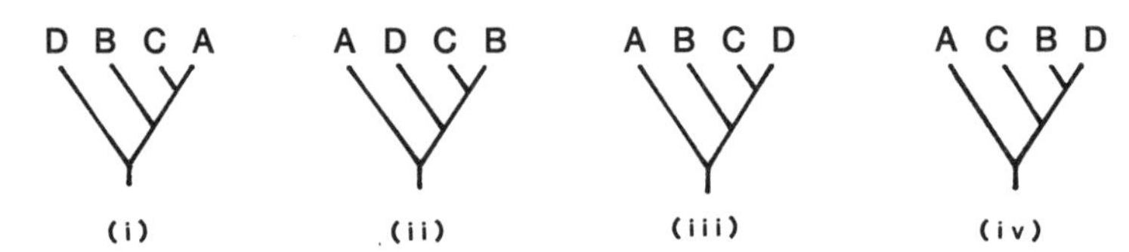

Figure 12.12 Solutions to analyses of cladograms I, II and III (Fig. 12.10) by the method of Zandee and Roos.

Table 12.5. Partially monothetic sets (components) for cladograms I, II and III analysed under assumption 1

A	7 13 (I)
B	11 (I) 7 13 (II) 7 13 (III)
C	2 6 10 16 (I) 2 11 17 (II) 6 11 (III)
D	6 10 (II) 2 6 10 (III)
AB	15 (I)
AC	1 14 19 20 (I)
BC	12 17 18 (I) 15 19 21 (II) 1 15 (III)
BD	1 14 (II) 14 19 20 (III)
CD	12 16 18 (II) 12 17 18 (III)
ABC	3 4 5 8 9 21 (I)
BCD	3 4 5 8 9 20 (II) 3 4 5 8 9 21 (III)

eleven monothetic sets can be derived (Table 12.5) which can be ordered into fifteen area cladograms, all of those possible for four areas. However, there is one most parsimonious solution in the final analysis (Fig. 12.12(iv)), again different from the Nelson solution as determined by the consensus method.

A quantification of component analysis

The method of Zandee and Roos differs from that of Nelson, not only because it discerns one extra assumption about the derivation of components, but also because it is based on different assumptions about the use of components in analysis. Interaction among components is less restricted, i.e. not confined to informative components and, as a consequence, more area cladograms can be found under the three assumptions (0,1,2) than is apparently the case using Nelson and Platnick's method. An alternative approach is to use a method similar to that outlined by Brooks (1981): apply Nelson and Platnick's assumptions and use a parsimony algorithm to analyse the data (in this case with the computer package 'phylogenetic analysis using parsimony' – PAUP; Swofford, 1984). The Branch and Bound algorithm in PAUP produces the most parsimonious cladograms from discrete character data.

Consider each of the three cladograms in Nelson's original problem (Figs. 12.7, 12.13). Cladogram I yields six possible three-area statements if we preserve the topologies but remove the conflicting areas under assumption 2 (Fig. 12.7(ii–vii)). The technique simply takes one area at a time from the terminal statements which then gives us six possibilities for further analysis. However, only one of these is informative (Fig. 12.7(iv)), as the remainder are simply one or two-area statements. Adding the missing taxa into one or two area statements recovers all of

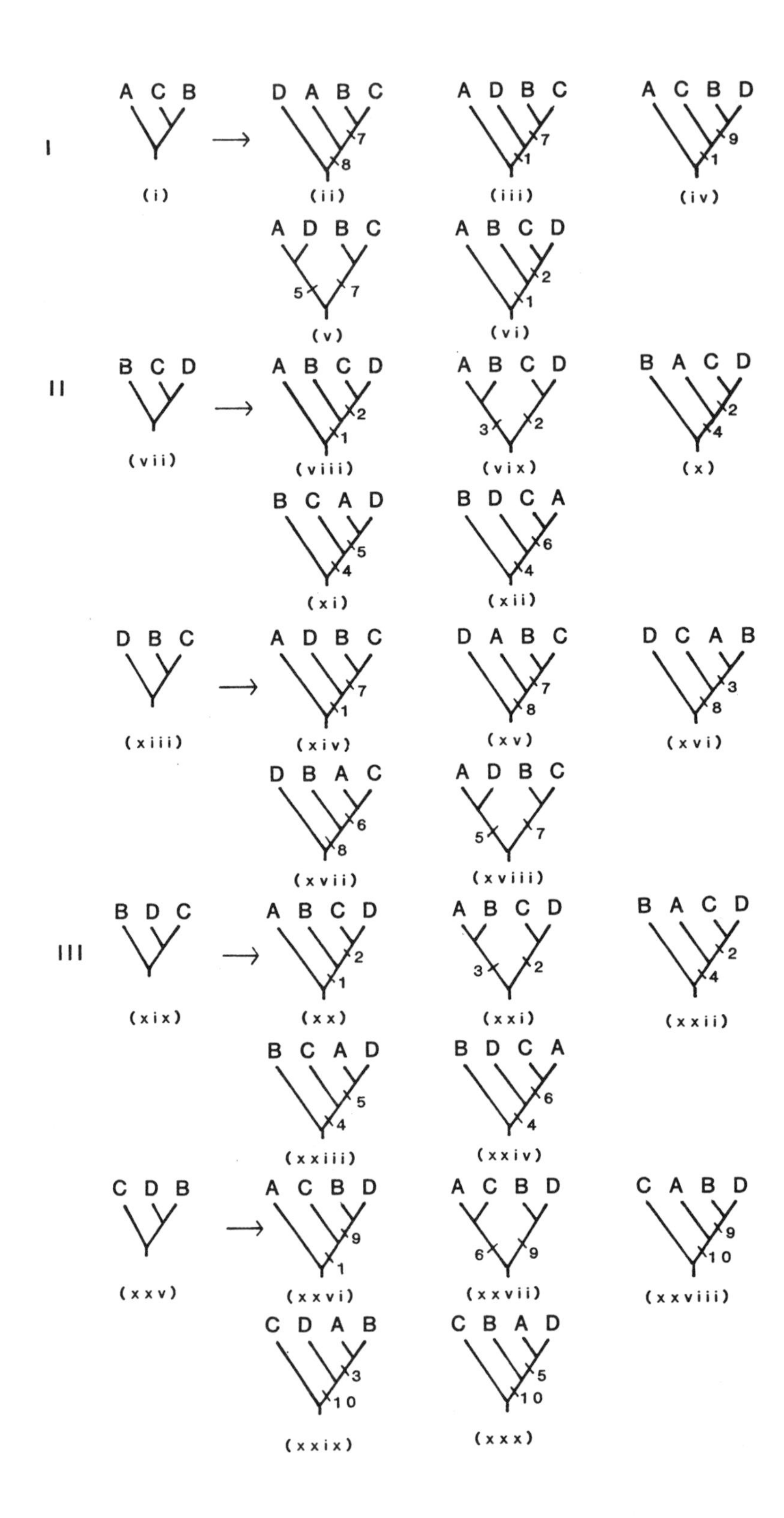

I
A C B
(i)
D A B C
7
8
(ii)
A D B C
7
1
(iii)
A C B D
9
1
(iv)
A D B C
5
7
(v)
A B C D
2
1
(vi)
II
B C D
(vii)
A B C D
2
1
(viii)
A B C D
3
2
(vix)
B A C D
2
4
(x)
B C A D
5
4
(xi)
B D C A
6
4
(xii)
D B C
(xiii)
A D B C
7
1
(xiv)
D A B C
7
8
(xv)
D C A B
3
8
(xvi)
D B A C
6
8
(xvii)
A D B C
5
7
(xviii)
III
B D C
(xix)
A B C D
2
1
(xx)
A B C D
3
2
(xxi)
B A C D
2
4
(xxii)
B C A D
5
4
(xxiii)
B D C A
6
4
(xxiv)
C D B
(xxv)
A C B D
9
1
(xxvi)
A C B D
6
9
(xxvii)
C A B D
9
10
(xxviii)
C D A B
3
10
(xxix)
C B A D
5
10
(xxx)

Table 12.6. Data derived from Fig. 12.13 as used in the PAUP analysis of the three cladogram problem outlined in Fig. 12.7 (see Fig. 12.14 and text for explanation)

		Characters									
		1	2	3	4	5	6	7	8	9	10
Areas	A	0	0	1	1	1	1	0	1	0	1
	B	1	0	1	0	0	0	1	1	1	1
	C	1	1	0	1	0	1	1	1	0	0
	D	1	1	0	1	1	0	0	0	1	1
Weight		7	7	4	6	5	4	6	4	4	3

the possible solutions and hence one or two area statements are totally uninformative. The next step is to add in the missing area D onto the informative cladogram (Fig. 12.13(ii–vi)) and, because we do not have a clue as to its correct position, there are five possible positions in which it can be placed (Fig. 12.13(ii–vi)). This exercise yields six different informative components. For cladograms II and III there are two informative three-taxon statements (Fig. 12.7(x), (xiii), (xvii) (xviii); Fig. 12.13(vii), (xiii), (xix), (xxv)) that can be derived from them. Consequently, there are ten possible cladograms for the cladograms II and III when the 'missing' area A is placed in all of the possible positions (Fig. 12.13(vii–xxx)). For all three original cladograms there are ten informative components represented in frequency from three (ABD) to seven times (CD); the full ranges are given as weights in Table 12.6. The weight value refers to the frequency of representation of components in Fig. 12.13. It should be noted that this data matrix contains all of the components also present in the data matrix derived under assumption 2 by the method of Zandee and Roos. In fact, these are all possible combinations of four items taken in twos and threes at a time, but with weights applied to match the frequencies in the 'possible' cladograms.

The data matrix was analysed with PAUP using the Branch and Bound option, and weights applied to the columns equivalent to the representation in Table 12.6. Two equally parsimonious unrooted trees were found with a length of 69 steps and a consistency index of 0.725. Note that parsimony in this context is taken to mean the total number of steps, while Zandee and Roos consider contradiction minus support to

Figure 12.13 Informative components for the informative three-area statements derived from cladograms I, II and III (Fig. 12.10) when combined with the missing areas (see Fig. 12.7): (i–vi) cladogram 1; (vii–xviii) cladogram II; (xix–xxx) cladogram III.

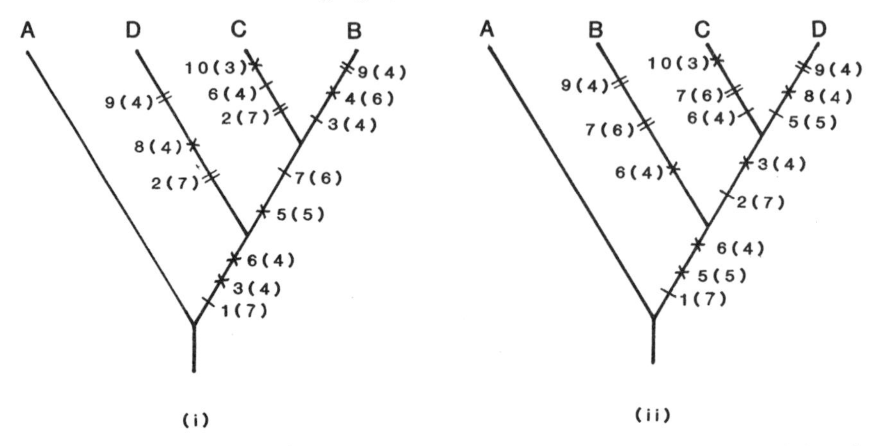

Figure 12.14 Solutions to analyses of cladograms I, II and III (Figs 12.6 and 12.9) by the method of quantified component analysis. Two equally parsimonious trees were obtained by the Branch and Bound algorithm in PAUP. Identical results were obtained with Lundberg rooting and rooting with area A.

represent parsimony. When rooted with a column of zeros (Lundberg rooting) or with A, the two cladograms were of the following configuration: Nelson's {A[B(C,D)]} solution and the {A[D(B,C)]} solution (Fig. 12.14). In conclusion, the method has the advantage of finding all solutions without the pitfalls of consensus trees and it utilizes the principle of parsimony.

12.4 CLADISTIC BIOGEOGRAPHY

Croizat's work (1964) contained two main conclusions:

(1) changes in earth history such as tectonic changes, rather than individual biotic histories, are the causal factors in evolution and distribution.
(2) areas that might be called biogeographic areas correspond not to the present-day continental configurations, but to fragments of continents and the intervening parts between them.

Croizat summarized all of the major patterns on biotic distribution that he could accumulate. His generalized track method (Chapter 13) indicated a high degree of congruence in widely different groups and he showed that there was an asymmetry between congruent and incongruent patterns. Congruence implies common history whilst incongruence invites a variety of different explanations. A recent analysis by Page (1987; Chapter 13) has shown that Croizat's tracks can be interpreted as minimum spanning graphs. Consequently, Page says that they can be considered as parsimonious graphical representations of area interrelationships. However, we disagree with Page because tracks are not based on analyses of character information but on a mixture of *ad hoc* criteria, and the problem with generalized tracks is that they are just that – generalized (Humphries and Parenti, 1986; Seberg, 1986). They lack resolution in terms of the taxa with which they deal and are not necessarily based on the empirical taxonomic information they are supposed to represent. We consider that cladograms are much better because they reflect the pattern of homologies as interpreted by the defensible criterion of parsimony and are the only means of determining resolution of strictly monophyletic relationships from empirical data. We demonstrate our own approach to biogeography by reviewing some recent work on the genus *Eucalyptus* (Myrtaceae).

12.4.1 Eucalyptus

The species of *Eucalyptus* informal subgenus *Monocalyptus* (Pryor and Johnson, 1971) have been the subject of intensive taxonomic study (Ladiges and Humphries, 1983, 1986; Ladiges *et al.*, 1987). The groups studied so far include the 'peppermints', 'stringybarks' and the Western Australian species. The latter group provides a good example of the fact that biogeography is only as good as the taxonomy upon which it is based. Just as polyphyletic and paraphyletic groups are not based on empirical evidence, area cladograms and biogeographic explanations derived from them have little or no connection to reality.

In the study of Ladiges *et al.* (1987), based on an analysis of 51 characters, the twenty-five species of monocalypts in Western Australia do not form a monophyletic group without the inclusion of at least three eastern Australian clades (Fig. 12.15) – indicated by *Eucalyptus planchoniana* representing the eastern ashes, *E. acmenoides* representing the stringybarks and *E. pilularis* representing the lineage leading to the peppermints. The stem species, *E. rubiginosa* is probably the sister taxon to the whole of the *Monocalyptus* group.

The areas in which the Western Australian monocalypts occur are given in Figs. 12.15 and 12.16. The cladogram in Fig. 12.15(i) gives just one of several solutions and a consensus tree for all of the solutions added together is given in Fig. 12.15(ii). This gives a precise definition of each area of endemism as determined by the taxa themselves. In other words, the areas are defined by the distributional boundaries of the taxa and, in the case of disjunctions within a species, these are given more than one letter code.

There are repeated patterns in the subclades, for example the eastern

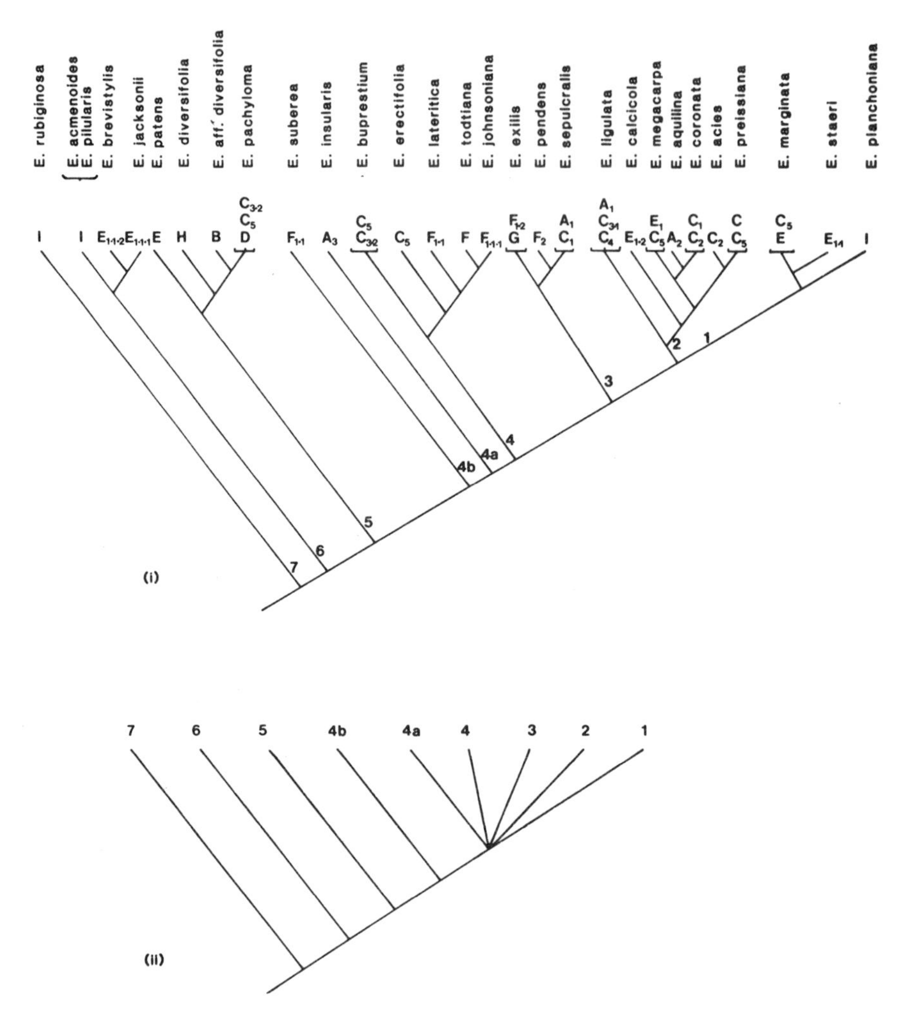

Figure 12.15 (i)Area cladogram for Western Australian *Monocalyptus* species; (ii) consensus tree for all 24 trees obtained in the analysis. Numbers refer to clades. The areas of endemism indicated by the letters are shown on Fig. 12.16 and described in Table 12.7 (after Ladiges *et al.*, 1987).

Figure 12.16 Distribution maps for the Western Australian taxa; (i) areas for subclade 1; (ii) areas for subclade 2; (iii) areas for subclade 3; (iv) areas for subclades 4 and 6; (v) areas for subclade 5; (vi) general map (showing combined areas of Western Australian taxa. WA – Western Australia, SA – South Australia, Vic – Victoria (after Ladiges *et al.*, 1987). Letters refer to areas of endemism for taxa in Fig. 12.15(i). See Table 12.7 for a description of areas and environments.

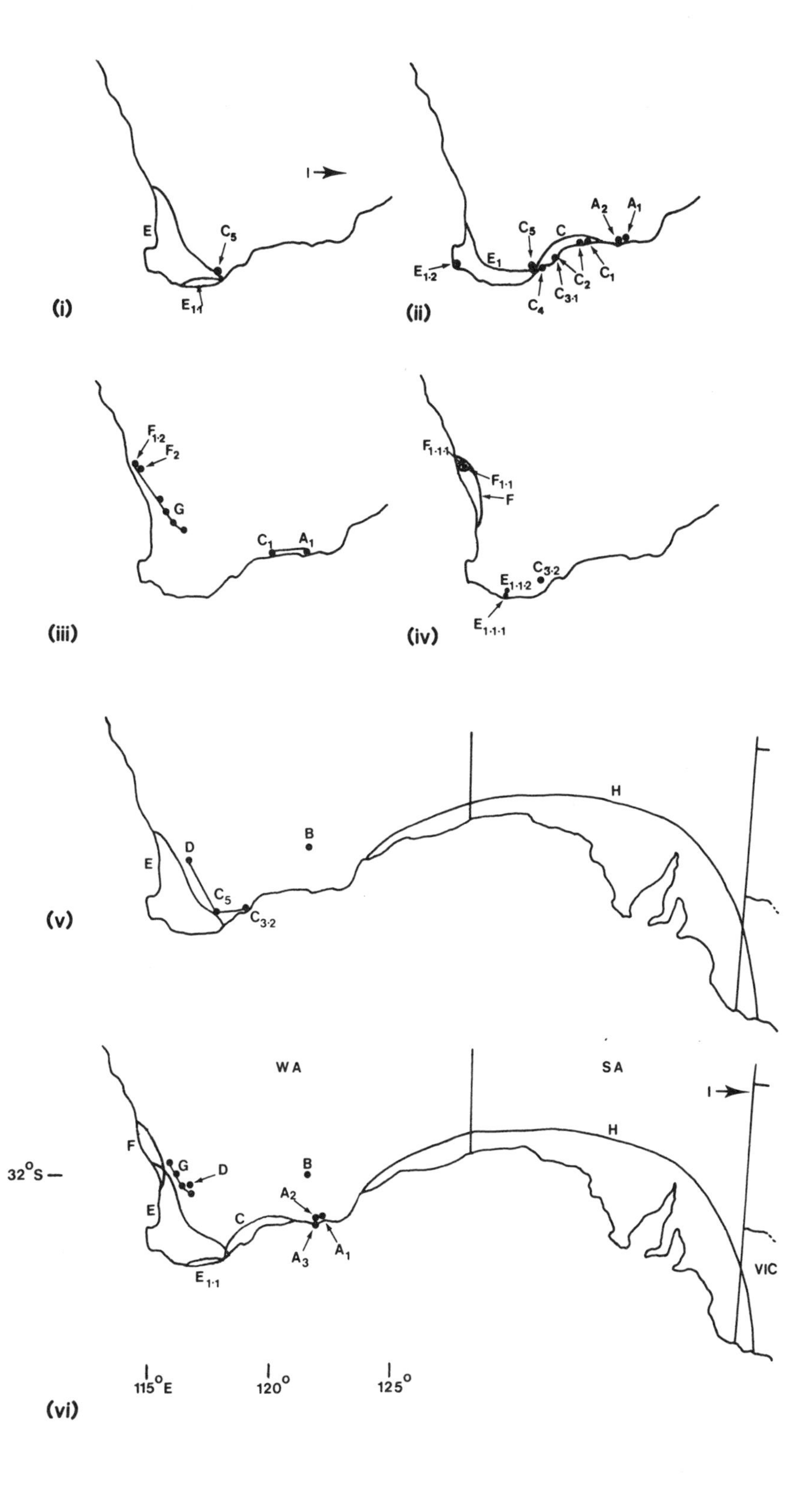

(i)
E
C5
E1·1
I
(ii)
E1·2
E1
C5
C
A2
A1
C4
C3·1
C2
C1
(iii)
F1·2
F2
G
C1
A1
(iv)
F1·1·1
F1·1
F
C3·2
E1·1·2
E1·1·1
(v)
E
D
B
H
C5
C3·2
(vi)
WA
SA
I
F
G
D
32°S
B
H
A2
E
C
A3
A1
E1·1
VIC
115°E
120°
125°

Table 12.7. Areas and environment of Western Australian taxa. General area I = eastern Australia

Species in major clades		Area and environment
(i)	*E. marginata*	Areas C_5 and E. Tall form with *E. patens* on well drained gravels, bounded by 625 mm isohyet; small tree and mallee form in Stirling Range and plains, on white sands
	E. staeri	Area $E_{1\cdot1}$. Gentle slopes near swamps and streams, coastal lowlands, sandy podsols to deep white sands
(ii)	*E. preissiana*	Areas C and C_5. Coastal, subcoastal lateritic or stony clay soil, 375–625 mm rainfall
	E. acies	Area C_2. Thumb Peak, Middle Mt Barren, Cheyne Beach
	E. megacarpa	Areas C_5 and E_1. Tree form on damp sands and podsols edging swamps and streams; mallee top of peaks in Stirling Range and in coastal heaths
	E. aquilina	Area A_2. Coastal granite
	E. coronata	Area E_1 and C_2. Coastal quartzite hills
	E. calcicola	Area $E_{1\cdot2}$. Coastal calcareous sands over limestone, limestone cliff outcrops
	E. ligulata	Areas A_1, $C_{3\cdot1}$ and C_4. Coastal sands and rocky slopes (middle to upper zone) in the Stirling Range
(iii)	*E. sepulcralis*	Areas A_1 and C_1. Emergent of inland heaths on stony to sandy soils on quartzite, foothills
	E. pendens	Area F_2. Emergent of inland heaths on lateritic sandhills
	E. exilis	Areas $F_{1\cdot2}$ and G. Emergent of heath on lateritic podsols, e.g. Boyagin
(iv)	*E. johnsoniana*	Area $F_{1\cdot1\cdot1}$. White sands over laterite
	E. todtiana	Area F. Coastal plains impoverished sands; sometimes impeded drainage
	E. lateritica	Area $F_{1\cdot1}$. Lateritic uplands on edges of outcrops and breakaway
	E. erectifolia	Area C_5. Lower slopes of Stirling Range
	E. buprestium	Areas $C_{3\cdot2}$ and C_5. Sand, sometimes over laterite, with *E. marginata* and *E. pachyloma* on middle to lower slopes of Stirling Range
	E. insularis	Area A_3. Largely on off-coast islands, base of granite rocks and cliffs
	E. suberea	$F_{1\cdot1}$. Lateritic uplands on edges of outcrops and breakaway; same habitat as *E. lateritica*
(v)	*E. patens*	Area E. Most soils with some clay content; usually not on laterite
	E. diversifolia	Area H. Coastal, shallow calcareous soils over limestone
	E aff. *diversifolia*	Area B. Granite outcrop near Norseman

Table 12.7. (*contd.*)

	Species in major clades	*Area and environment*
	E. pachyloma	Areas $C_{3\cdot2}$, C_5 and D. Sandy soils; lower slopes and foothills of Stirling Range with *E. marginata*, *E. staeri*, *E. buprestium* and *E. preissiana*
(vi)	*E. brevistylis*	Area $E_{1\cdot1\cdot2}$. Banks and adjacent areas of creeks in wet sclerophyll forest; with *E. marginata* and *E. megacarpa*
	E. jacksonii	Area $E_{1\cdot1\cdot1}$. Deep loams, 1250 mm annual rainfall in wet sclerophyll forest

area I occurs in more than one position in the area cladogram. Thus, if we are to conclude that the overall pattern is due to one history, then the repetitive area relationships in the cladogram mean that there were six partially sympatric subgroups affected by the same set of historical events that could have caused the patterns that we have today. There are nine major areas (areas A-I, Table 12.7, Figs. 12.15, 16) which can be divided up into 22 subunits to account for the individual species distribution patterns (A1, A2 etc.). Individual analyses of the six subclades (1-6 Fig. 12.15) gave only 3, 4, 5, 6 and 9 area solutions. Ladiges *et al.* (1987) combined the subclades using component analysis as if the sympatric and partially sympatric areas were describing the same thing (e.g. C1, C3.2) and the unique patterns were describing something different (e.g. H, B). The subclades were combined into a consensus tree using assumption 2. The 'best' (most parsimonious) solution gave seven resolved nodes (Fig. 12.17). Subunits of area C could all be treated as a single subunit, the southern sandplains, and it appears that area A1 and A2 are probably contiguous with C. Subunits within area E appeared to indicate three similar, but different, patterns (E1.1.1 + E1.1.2, E + E1.1 + E1.2, E1) as did the subareas of F, including G (F1.2 + G, F + F1.1, F2). F and G represent the drier, more northerly component of E and all three could be combined into a single area. The complete area–consensus tree, reduced to represent the broader areas (A,B,C etc.) as summaries of the lesser components, is given in Fig. 12.17(ii). To give some idea what this means in terms of a biogeographic pattern we have taken the eastern area I as the root of the consensus tree because the sister group to the informal subgenus *Monocalyptus* occurs in eastern Australia. Reading the consensus tree (Fig. 12.17(ii)) as an historical sequence the following geographical scenario, as presented in Fig. 12.18, can be suggested:

(a) The biota of southern and western Australia became isolated from that of eastern Australia: I + (A–H),

(b) A major division separated the southern and western taxa into three main areas: area E, now the *E. marginata* forests from north of Perth to Albany in the south; area FG, the northern sand plain and the western wheatbelt; and the remaining area of the south coast between the Stirling Range eastwards to Cape Nelson, Victoria: I + (E,E1) + (F,G) + (A1, A2, A3, B, C, D, H),
(b_1) Areas F and G became isolated into discrete areas of endemism: I + (E, E1) + F + G + (A1, A2, A3, B, C, D, H),
(c) Area E1, a coastal belt ranging from Perth southwards to Leeuwin and then eastward to the Albany district became a discrete area: I + E + E1 + F + G + (A1, A2, A3, C, D, H),
(d) The southern areas of Western Australia, including that between the Stirling Range, Cape Le Grand and north to Norseman, separated from area H determined by *E. diversifolia* which occurs between Madura, Western Australia and Cape Nelson, Victoria: I + E + E1 + F + G + H + (A1, A2, A3, B, C, D),
(e) The granitic outcrops where *E. aff. diversifolia* occurs, area B, separated from the remainder: I + E + E1 + F + G + H + B + (A1, A2, A3, C, D),
(f) The areas (A1, A2), A3, C and D became recognizable as distinct areas of endemism: I + E + E1 + F + G + H + B + (A1 + A2) + A3 + C + D.

12.4.2 A common history for areas and taxa?

Distributional congruence of the type found here suggests that the areas should share a similar sequence of climatic and geological events comparable with the taxonomic hierarchy.

The south-western and south-eastern biotas of Australia are now separated by a major climatic and edaphic barrier, the Nullabor plain, the limestone of which was formed by marine incursion. From the late Eocene through to the Mid-Miocene (50–15 Myr BP) there was a more or less continuous period of marine incursion into central Australia (Kemp, 1981; Nelson, 1981). In Western Australia, Late Eocene marine deposits occur as far inland as Norseman, and the limestones of the Eucla Basin (eastern Nullabor) reflect the marine transgressions of the Early Miocene (Kemp, 1981). During maximum marine intrusions in the Eocene, a mosaic of numerous islands would have flanked the south-western coastline of Western Australia (Hopper, 1979).

Eocene palaeobotanical data suggest that warm and moist conditions prevailed and were relatively uniform across southern Australia during that epoch. Nelson (1981) reiterated that marine incursion was the main

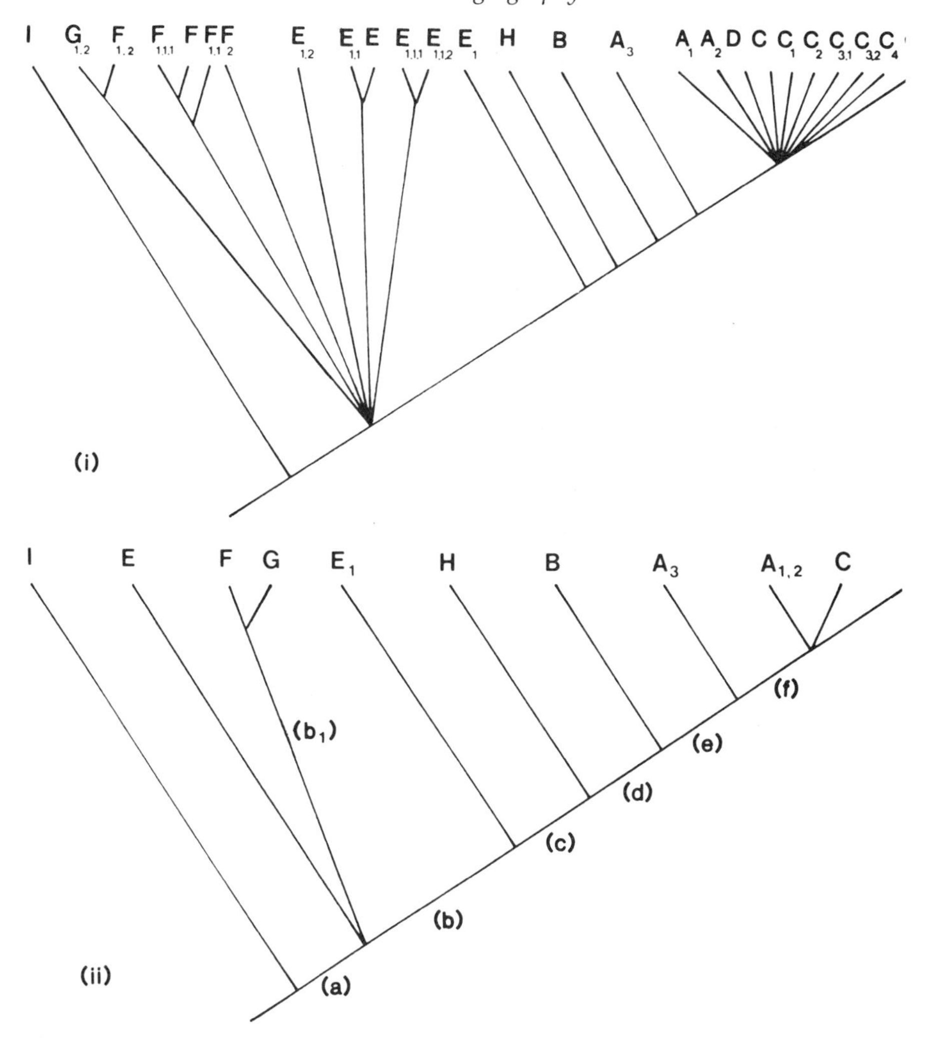

Figure 12.17 (i) Area consensus tree analysed under assumption 2 for Western Australian eucalypts; (ii) simplified consensus tree for the same taxa/areas (after Ladiges *et al.*, 1987). I = eastern Australia.

event that disrupted the flora. It could be argued that a corridor north of the Nullabor region could have existed, allowing connection between eastern and western floras. However, at the same time there was a trend towards a drier climate. During the Early Miocene, precipitation was abundant and laterization of the land occurred. Temperatures and precipitation then decreased in the Miocene, being associated with

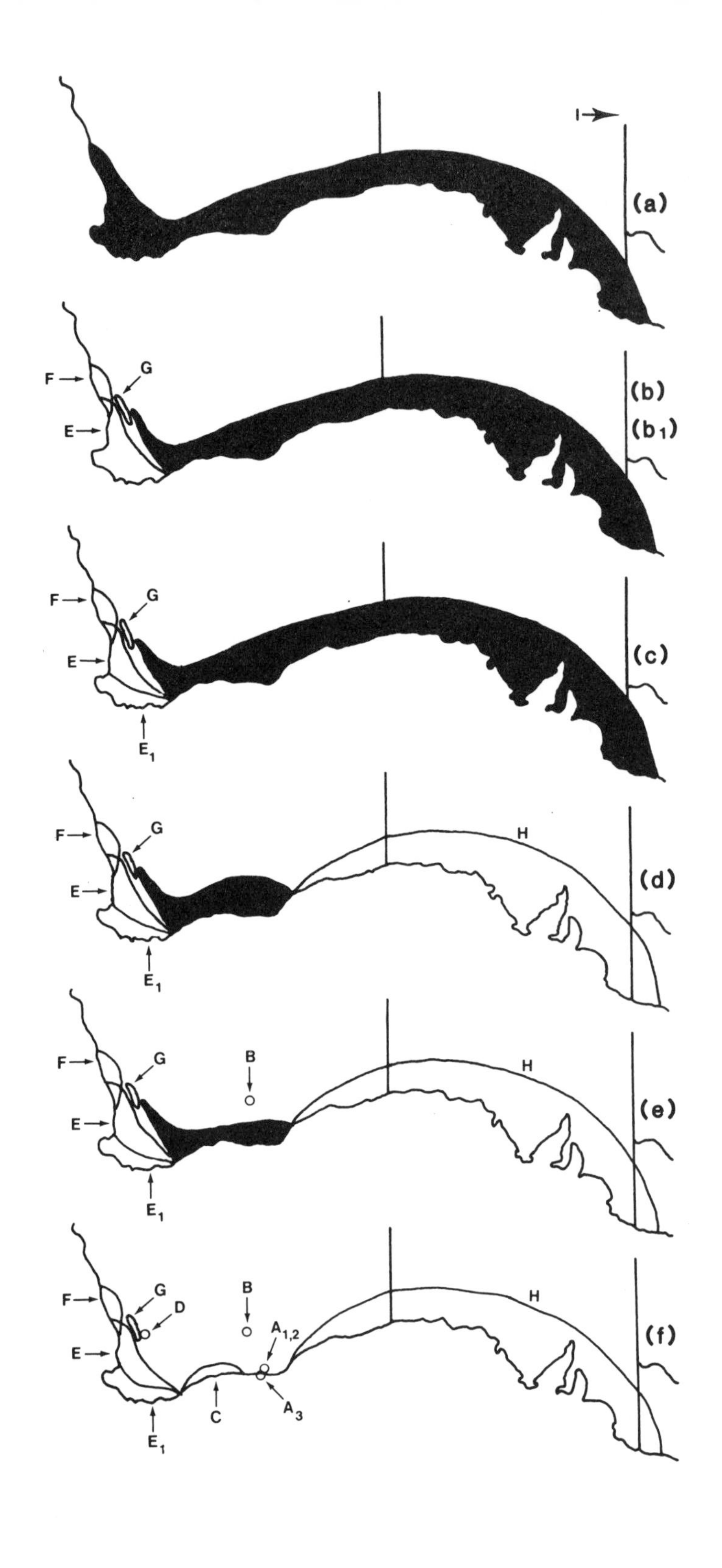

I
(a)
F
G
E
(b)
(b1)
F
G
E
(c)
E1
F
G
E
H
(d)
E1
F
G
B
E
H
(e)
E1
F
G
D
B
H
E
A1,2
(f)
A3
C
E1

Antarctic ice expansion and increasing aridity (Kemp, 1981). Nelson (1981) argues that, even if there had been climatic fluctuations, their magnitude and duration were insufficient to provide an adequate environment for a vegetation corridor north of the Nullabor plain.

Much of south-west Western Australia is a deeply weathered plateau rarely exceeding 500 m. Areas of relatively high relief are confined to a series of small ranges (formed of granite and quartzite) and remnants of the plateau along the south and western margins. Within the south-west, the plateau appears to have been uplifted and coastal rivers have dissected the area in the Tertiary and the Pleistocene; plateau margins today have relatively rugged and diverse topographical features (Hopper, 1979). Species such as *E. suberea* and *E. lateritica* occur on the lateritic breakaway and outcrops whilst *E. pendens* (area F2 in Fig. 12.16(iii)) and in *E. exilis* (areas F1, 2, G in Fig.12.16(iii)) extend onto deeper sandy podsols overlying the laterite.

According to Hopper, the south-west today has three main rainfall zones, two of which include monocalypt species. These are the high rainfall zone (800–1400 mm p.a.) and the forested region of the far south-west (area E), and secondly the transitional zone (300–800 mm p.a.) of the woodlands, mallees and heaths (areas F, G, D etc.) inland of the high rainfall zone and abutting the arid zone. Within each zone, soil characteristics are very important in determining vegetation patterns, and landform dissection and edaphic diversity are greatest in the transitional zone. The '*sepulcralis*' and '*todtiana*' groups occur on the deficient soils, whilst *E. patens*, *E. brevistylis* and *E. jacksonii* are restricted to the fertile soils with higher clay content. *E. calcicola* (area E1.2 in Fig. 12.16(iv)), *E. ligulata* (areas A1, C3.1, C4 in Fig. 12.16(ii)) and *E. diversifolia* (area H in Fig. 12.16(v)) are confined to calcareous soils over limestone. *E. diversifolia* is the only *Monocalyptus* species that occurs on the Nullabor limestone – emergent and dry since the Mid-Tertiary (Nelson, 1981).

Although not undertaken in a strictly rigorous manner, this geological and ecological account is consistent with the events in the consensus area cladogram. It points to the Tertiary as the time of divergence for the major lineages. The oldest divergences (Fig. 12.18(a)) between the west and the east (I) appear to be correlated with Eocene–Miocene marine transgressions and drying of the climate. Further speciation in the main clades can be interpreted in terms of climatic, topographical and edaphic patterns. Although our account has been expressed wholly as a vicariance scenario, there are alternatives which could be considered equally

---

Figure 12.18 Pictorial representation of proposed sequence of historical events. The areas are present-day distributions of taxa (after Ladiges *et al.*, 1987).

plausible. The problem with all scenarios is that they make speculations beyond the empirical data at hand. For some of the Western Australian eucalypts it is possible that some of the species boundaries that we see today were once quite different. For example, the limestone soils of the Nullabor plain have been formed as the sea retreated. Afterwards, the vegetation from surrounding areas dispersed into new terrestrial habitats. Thus, consider *E. diversifolia* which determines area H in Figs. 12.17(ii) and 12.18. Range expansion by *E. diversifolia* into the Nullabor is a very plausible possibility and still consistent with other parts of the scenario that we have described already. However, support for speciation during the Tertiary is found in Western Australian anuran frogs (Roberts and Maxson, 1985). The fact that there were at least three east-west patterns in the Western Australian monocalypts (Fig. 12.15; clades 1, 5 and 6) means that there were at least three monocalypt taxa before the Miocene.

12.5 CONCLUSIONS

The pursuit of a cladistic method in biogeography and the application of cladograms to solve biogeographic questions has given greater resolution than hitherto achieved. Because cladograms have a high empirical content with the greatest explanatory power for a given matrix of systematic characters, they are easier to falsify than other types of hypotheses. The developments of the techniques of area cladograms, consensus trees and component analysis, has opened the way for other methods, such as described here, which can apply the principle of parsimony to generalized area cladograms as well as to the individual cladograms upon which they are based. The pioneering work of Croizat applied track analysis to the problems of form, time and space, but little attention, other than the principle of congruence, was given to the principles of taxonomy so crucial to an understanding of pattern. We consider that the combined elements of resolving power and more robust hypotheses of relationship will eventually lead to a greater understanding of the general patterns that Croizat so painstakingly put together with his 'panbiogeographical method'. By removing spurious patterns caused by para- and polyphyletic groups, and by undertaking more rigorous studies of those regions, such as Australia in particular and the Pacific in general, eventually a resolved solution for the whole world will be achieved.

ACKNOWLEDGEMENT

We thank Ed Wiley for critically reading the manuscript.

CHAPTER 13

Panbiogeography: method and synthesis in biogeography

R. CRAW

13.1 SPACE-TIME AND BIOGEOGRAPHY: PHILOSOPHICAL CONSIDERATIONS

If philosophy is the devil's whore, as Martin Luther once quipped, then biogeography and biological systematics are fast becoming Old Nick's bordello. Advocates of the new biogeography, ecology and geography (Haines-Young and Petch, 1980; Nelson and Platnick, 1981; Simberloff, 1983) attempt to lay the foundations for these disciplines within the twin philosophical strictures of Popper's (1959, 1972) epistemological prescription for critical rationalism and falsification. This attempt to enlist philosophy as the paid companion of these sciences traps them within traditional binary oppositions; e.g. science versus non-science; falsifiable and testable hypotheses versus narrative, 'Just so' scenarios; ecological versus historical biogeography; pattern versus process; and vicariance versus dispersal, which serve to polarize debate (Chapter 2).

An important trend in modern philosophy has been a movement away from conceiving philosophy as providing the methodological and epistemological foundations for science. Instead philosophy plays an interpretative role; it is an aid in the search for meaning and understanding of the sciences rather than an absolute arbiter of knowledge and truth (Feyerabend, 1978; Rorty, 1979; Bernstein, 1983).

How might philosophy then illuminate rather than limit current debates in biogeography? One possibility is through a consideration of the differing conceptions of space within the three competing research areas currently recognized within historical biogeography: centres of origin/dispersal, vicariance cladistics and panbiogeography.

Geographers have identified two conceptions of space relevant to scientific spatial analysis: absolute and relative space (Gatrell, 1983). The concept of absolute space is the view that space is a fixed, immovable container with properties of its own, independent of and prior to the objects which fill it. In biogeography the dispersalist approach including island equilibrium theory (Chapters 1 and 15) can be directly equated with this view since taxa originate in fixed centres of origin (containers) from which they migrate outwards to other areas; the distance between, and size of, source and receiver areas is considered of prime importance (MacArthur and Wilson, 1967; Briggs, 1984).

The alternative relative concept of space treats it as a system of relations between objects. These objects are identifiable on the basis of internal qualities alone, prior to, and independent of, spatial and temporal properties. Space is nothing more nor less than the relationships between these objects. Vicariance cladistics (Chapter 12) can be associated with this view of space because spatial area relationships are established directly from form cladistics (taken here in its broadest sense to refer to the structural, functional and behavioural aspects of organisms) with no consideration of either spatial or temporal aspects of taxa (Nelson and Platnick, 1981, 1984; Humphries and Parenti, 1986). Vicariance biogeographic analysis proceeds from taxonomic analysis and "(biogeographic) area relationship is but the sum of relationships of endemic taxa" (Nelson, 1984).

A refinement of the relative space concept has been developed by Strawson (1959) who maintains that identification of objects rests ultimately on their location in an unitary spatiotemporal framework. Knowledge of spatiotemporal location and relations is just as important to the individualization and identification of objects as is knowledge of those objects themselves. This view of space can be directly associated with panbiogeography where biogeographic analysis establishes spatiotemporal homologies or relations between taxa (Craw, 1983a, 1984a) and these spatiotemporal homologies inform and in turn direct form cladistic studies of taxa (Croizat, 1961, 1968). The relations between biogeographic methods and metaphors, and concepts of space have been explored by Craw and Page (1988).

Recently, philosophers of biology (Hull, 1976, 1978) and systematists (Ghiselin, 1974, 1981; Hennig, 1966, 1984) have promoted the view that species and higher taxa are individuals and historical spatiotemporally bounded entities with a definite beginning and end in space-time rather than timeless definable abstractions, types or classes. This view was advanced earlier this century by Townsend (1913), Rosa (1918) and Woodger (1952), and regarded as a truism by Croizat (1978). Elsewhere, Croizat (1961) proposed a natural concept of a species "as an aggregate of forms having shared common history within a standard biogeographic range". In contrast vicariance cladists regard species as sets (Nelson, 1984), or types (Nelson and Platnick, 1981), with an identity independent of space-time. But if species and higher taxa are not classes but individuals contingent upon their spatiotemporal location then the relationship between sysematics and biogeography is somewhat different than has hitherto been suggested or suspected.

13.2 PANBIOGEOGRAPHY AND PHYLOGENY

Panbiogeography is a biogeographic method that focuses on spatiotemporal analysis of the distribution patterns of organisms. It is not a phenetic method in that Croizat was interested in the specialized similarity between distribution patterns of a variety of organisms, clustering these according to the diagnostic spatiotemporal components that his method identifies (Craw, 1983a). This is quite a distinct approach to phenetic biogeography which investigates the similarities between biotas of different geographic areas in terms of the numbers of taxa in common: Page (1987) gives an extensive comparison of panbiogeography, phenetic and cladistic biogeography. Panbiogeography has been misinterpreted as a phenetic biogeographic method requiring the addition of cladistic methodology and the conversion of the generalized track into the area cladogram before its full potential could be realized and communicated as the new methodology, namely vicariance cladistics (Nelson and Platnick, 1981; Patterson, 1981a; Humphries and Parenti, 1986). Ball (1981) has claimed that "Croizat . . . had little interest in the theory of systematics" while McDowall (1978) has asserted that "Croizat . . . failed to recognize the significance of phylogenetic relationship to biogeographic analysis". These views run counter to Croizat's perception of the origin of panbiogeography (Heads, 1984; Heads and Craw, 1984)

Discontented with his phylogenetic systematic work on angiosperms Croizat began to develop methods of analysis (to be described in Section 13.3) that dealt directly with the spatiotemporal aspects of taxa rather than approaching them strictly from the viewpoint of taxonomic relationship. This immense project (Croizat, 1952, 1958) involved the graphical analysis of the geographic distribution patterns of literally hundreds of animal and plant groups culminating in the opening chapters of his major reassessment of angiosperm systematics and phylogeny, the *Principia Botanica* (Croizat, 1961; detailed review by Heads, 1984). In this volume Croizat turns on its head the traditional argument that accurate biogeographic analysis depends on an exact refined taxonomy (Wallace, 1876; Humphries and Parenti, 1986). Using graphical methods to analyse carnivorous plant distribution Croizat proposed a novel close systematic relationship between the families Droseraceae, Lentibulariaceae and Nepenthaceae which are not regarded as being related in traditional plant systematics. Having established the possibility of this relationship on a biogeographic basis Croizat proceeded from a close study of vegetative morphology to propose a new set of homologies for the group. The biogeographic analysis and vegetative characters that support this new group were apparently incongruent with floral morphology. Stressing a reciprocal illumination between biogeography and phylogenetic systematics (see also Hennig, 1950, 1966; Hull, 1967) panbiogeographic analysis was again employed. From this basis a transformation series was erected linking the actinomorphic flowers of Droseraceae and Nepenthaceae with the zygomorphic flowers of Lentibulariaceae via the flowers of the Podostemaceae. Finally a groundplan (hypothesized last common ancestor of the members of a taxon, a form of outgroup analysis; Stevens, 1980) for the carnivorous plants was proposed.

Panbiogeographic analysis can also provide a means of testing whether or not an established taxon is monophyletic. Corner (1963) noted that New Caledonian figs (*Ficus*) in the sections Urostigma and Pharmacosycea shared characters with New World species in the same sections. These similarities were explained by Corner as the result of a close phylogenetic relationship between New Caledonian and New World species established by migration across a trans-Pacific landbridge. Utilizing panbiogeographic analysis Croizat (1968) showed that *Ficus* was not a trans-Pacific group but rather a trans-Indian and trans-Atlantic Oceans group. The similarities between New Caledonian and American figs involved 'archaic characters', i.e. the proposed relationship was based on symplesiomorphy not synapomorphy (Page, 1987). Croizat (1968) also used biogeographic analysis to suggest three alternative hypotheses of relationship between the pigeon *Columba vitiensis* and other *Columba* species that avian systematists could test.

Panbiogeographic analysis can thus not only generate novel phylogenetic hypotheses and test established ones: it can also be used as a means of choosing between alternative hypotheses. Two contradictory phylogenetic hypotheses have been proposed for the interrelationships of the ratite birds based on bone morphology (Cracraft, 1974) and DNA-DNA hybridization (Sibley and Ahlquist, 1981) respectively. Craw (1985) showed that the ratites are biogeographically related by trans-Indian and trans-Atlantic ocean patterns, a biogeographic relationship congruent with the Sibley–Ahlquist hypothesis of ratite interrelationships.

Panbiogeographic method represents an advance beyond vicariance cladistic methods as far as the relationship between phylogenetic systematics and biogeography is concerned. Vicariance cladistic methods are typological (Nelson and Platnick, 1981, Humphries and Parenti, 1986), that is, both taxa and geographic areas are treated as sets or classes and geographic distribution is taken as read and never directly subjected to techniques of spatial analysis (Craw, 1983a). If taxa are spatiotemporally bounded entities then biogeographic analysis must be a necessary pre-requisite for phylogenetic studies and a natural classification can only be established after, not before, a prior biogeographic analysis. This is not to suggest that panbiogeographic analysis is both necessary and sufficient as a basis for a phylogenetic classification if there is to be reciprocal illumination between these separate disciplines. In biogeography and systematics a method of spatiotemporal analysis is needed in order to discover the relationship between space-time and form.

13.3 SPATIAL ANALYSIS IN BIOGEOGRAPHY

13.3.1 A particular pattern of geographic distribution

Lyperobius huttoni is an endemic flightless New Zealand molytine weevil host specific on *Aciphylla* species (Apiaceae). Considered in isolation its geographic distribution seems remarkable. Widely distributed in montane and subalpine habitats on the eastern side of the South Island, an isolated population also occurs at the southern tip of the North Island (Fig 13.1). Almost invariably found above 700 m in the South Island, the North Island population is found on steep cliff faces above deep ravines on the rugged south Wellington coast, only a few metres from the seashore. By examination of the *L. huttoni* distribution map (Fig. 13.1) it is impossible to tell what the weevil's dispersal capabilities are. Generally such an anomalous distribution pattern would be interpreted as the consequence of chance dispersal from the species centre of origin in the eastern South Island mountains across Cook Strait (Davis 1950; Bull, 1967). Such explanations are commonplace in the literature of biogeography but often ignore the possibility that the seemingly anomalous distribution pattern may be shared by other taxa.

Sympatric on the south Wellington cliffs with *L. huttoni* is a large flightless carabid beetle *Mecodema sulcatum*. This species too has a disjunct distribution between the southernmost area of the North Island and parts of the South Island. Over 60 other species, both plants (e.g. *Brachyglottis greyi, Clematis hookeriana, Coriaria sarmentosa, Hymenanthera crassifolia, H. obovata, Myosurus novae-zelandiae, Muehlenbeckia astonii*, and *Pimelea gnidia* (Allan, 1961; Eagle, 1975, 1982; Druce, 1984)) and animals (e.g. *Sharpius sandageri*, a flightless anthribid (Holloway, 1982); and eight species of caddisfly (Henderson, 1985)) share this pattern of disjunct populations at the southern tip of the North Island extending northwards no further than about latitude 40° 30′S and extending across

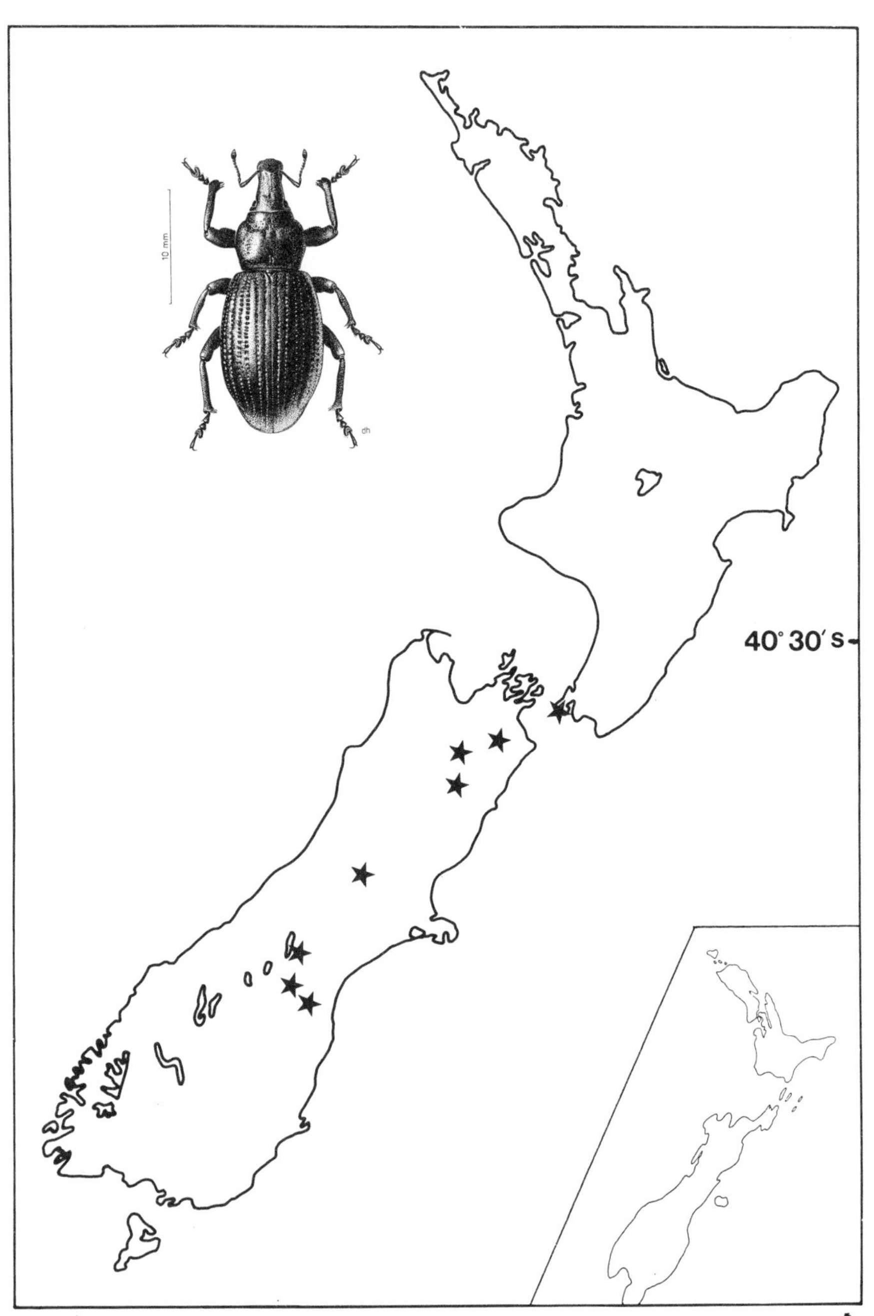

Figure 13.1 Distribution and *habitus* (upper left) of the molytine weevil *Lyperobius huttoni*. Inset (lower right): Paleogeographic reconstruction of New Zealand in the Pliocene (after Suggate *et al.*, 1978, Figs. 11,25).

Cook Strait into the South Island. Means of dispersal of these taxa cover a broad spectrum from no obvious means of overwater dispersal for the three flightless beetles through wind dispersed taxa (*Brachyglottis greyi*) to bird dispersed species. The fact that taxa with widely differing dispersal capabilities exhibit the same disjunct distribution type raises the possibility that one single event might economically explain the southern North Island–South Island disjunction rather than it having arisen through over 60 separate unique dispersal events. Paleogeographic reconstructions of New Zealand in the Pliocene (Suggate *et al.* 1978) show that the South Island and the southern Wellington region of the North Island were one landmass separated by a seaway from the rest of the North Island (Fig. 13.1, inset). The taxa may have once been part of an ancestral biota common to these joined land areas that has subsequently been fragmented by the formation of Cook Strait.

Arguments analogous to the above are presented many times by Croizat (1958) as part of his critique of dispersalist biogeography. Unfortunately what are regarded as the seminal papers and books (D. Rosen, 1975, 1978; Brundin, 1981; Nelson and Platnick, 1981; Nelson and Rosen, 1981) introducing panbiogeography to the biogeographic discipline distorted its nature by maintaining that the comparison of distribution patterns for congruence and their explanation by historical geological events constituted the panbiogeographic method. Such is not the case for arguments based on distributional congruence are a rhetorical device employed by Croizat in order to demonstrate that dispersal capabilities of taxa do not explain the geographic distributions of those taxa.

Approaches to biogeography based on attempts to demonstrate distributional congruence are purely descriptive narrative approaches whether in the form presented above or bolstered by cladistic analyses of the taxa. They are not examples of analytical biogeography because geographic distribution is taken as read and not subjected to the equivalent of character analysis in form systematics to determine what aspects of geographic distributions of taxa are homologies (Craw, 1983a). For spatial analysis of geographic distributions one requires a graphical technique, a vocabulary and a set of symbols (Croizat, 1968, 1982).

13.3.2 Tracks, baselines and nodes

"The track is essentially a graph drawn to render visible and comparable the results of biogeographic investigation . . ." Croizat (1982).

A track is a line graph drawn on a map of the geographic distribution of a particular taxon (be it a species, species-group, genus, or family)

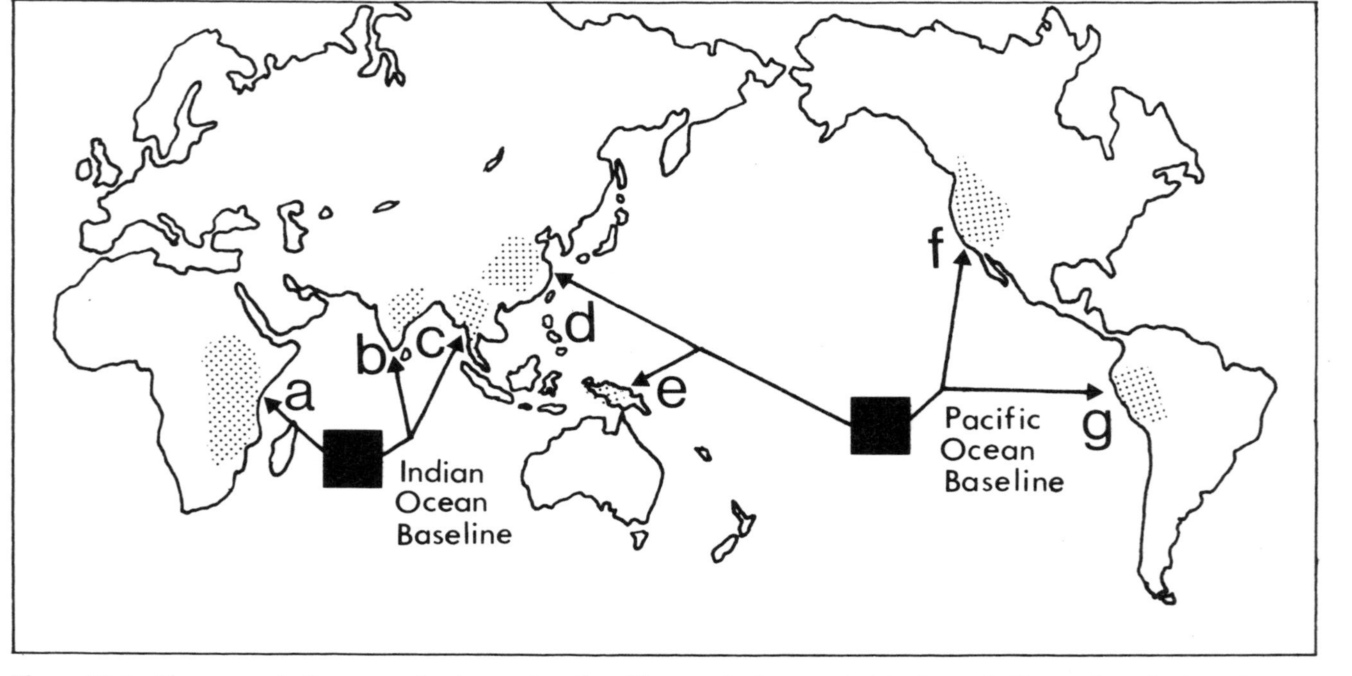

Figure 13.2 The concept of an ocean basin as a baseline (diagnostic characteristic) of a track. Taxa a, b and c (populations or subspecies of a species, species of a genus, genera of a family) are geographically distributed in East Africa, India and Southeast Asia respectively. The track for these taxa is interpreted with respect to the ocean basin that it crosses or circumscribes rather than in terms of the land areas it connects. Their track is described as a trans-Indian Ocean track, and their geographic distribution pattern is characterized by the Indian Ocean as a baseline. Similarly taxa d, e, f and g have a trans-Pacific track and a Pacific Ocean baseline.

that connects the disjunct collection localities of the subordinate taxa belonging to the taxon. This track is interpreted as a graph of the geographic distribution of the taxon representing the 'primary co-ordinates' of that taxon in space (Croizat, 1964). A track is initially constructed by an ordering of collection localities for nearest geographic neighbours within a taxon, making it equivalent to a non-directed minimum spanning graph or tree, i.e. an acyclic graph that connects all the localities occupied by a taxon such that the sum of the lengths of the links connecting each locality is the smallest possible (Craw, 1983a; Page, 1987).

Tracks are oriented in terms of the sea or ocean basins that the track crosses or circumscribes (Fig. 13.2). This allows a hypothesis of the baseline for that track to be proposed. The baseline is a primary biogeographic homology (i.e. diagnostic characteristic) for the group under analysis.

Tracks for many groups can be constructed on the above basis, but more complex patterns of geographic distribution require the use of additional criteria for track construction such as the concept of main massing. A main massing is a numerical, genetical or morphological centre of diversity for a particular taxon or group of taxa (Craw, 1985). The use of this criterion for track construction is best understood through an example. The bird genus *Aegotheles* is distributed in Australia, the Moluccas, New Caledonia and New Guinea (Fig. 13.3a). Track construction for this taxon utilizing solely the minimum distance criterion described earlier would link New Caledonia with Australia for instance and result in a track geometry quite different from that constructed for this group utilizing the additional criteria discussed above. The genus *Aegotheles* has a main massing on New Guinea and the Moluccas (six endemic species) with only one species in Australia and Tasmania, and one in New Caledonia. The closest relative of *Aegotheles* is a species of large fossil owlet-nightjar *Megaegotheles* known from the Pleistocene and Holocene of New Zealand (Rich and Scarlett, 1977). The track for these two genera is therefore drawn straight from Australia/ Tasmania, New Caledonia and New Zealand to link with the New Guinean main massing (Fig. 13.3b). Track orientation in terms of baselines and main massings converts the original non-directed minimal spanning graph (Fig. 13.3a) into a directed graph (Fig. 13.3b) by joining the outlying areas directly to the main massing. Introductions to track construction using several examples are available (Craw and Page, 1988; Grehan and Henderson, 1988).

A suite of individual tracks for various taxonomic groups crossing or circumscribing a particular sea or ocean basin is referred to as a generalized or standard track (Croizat, 1961; Craw, 1983a). Individual

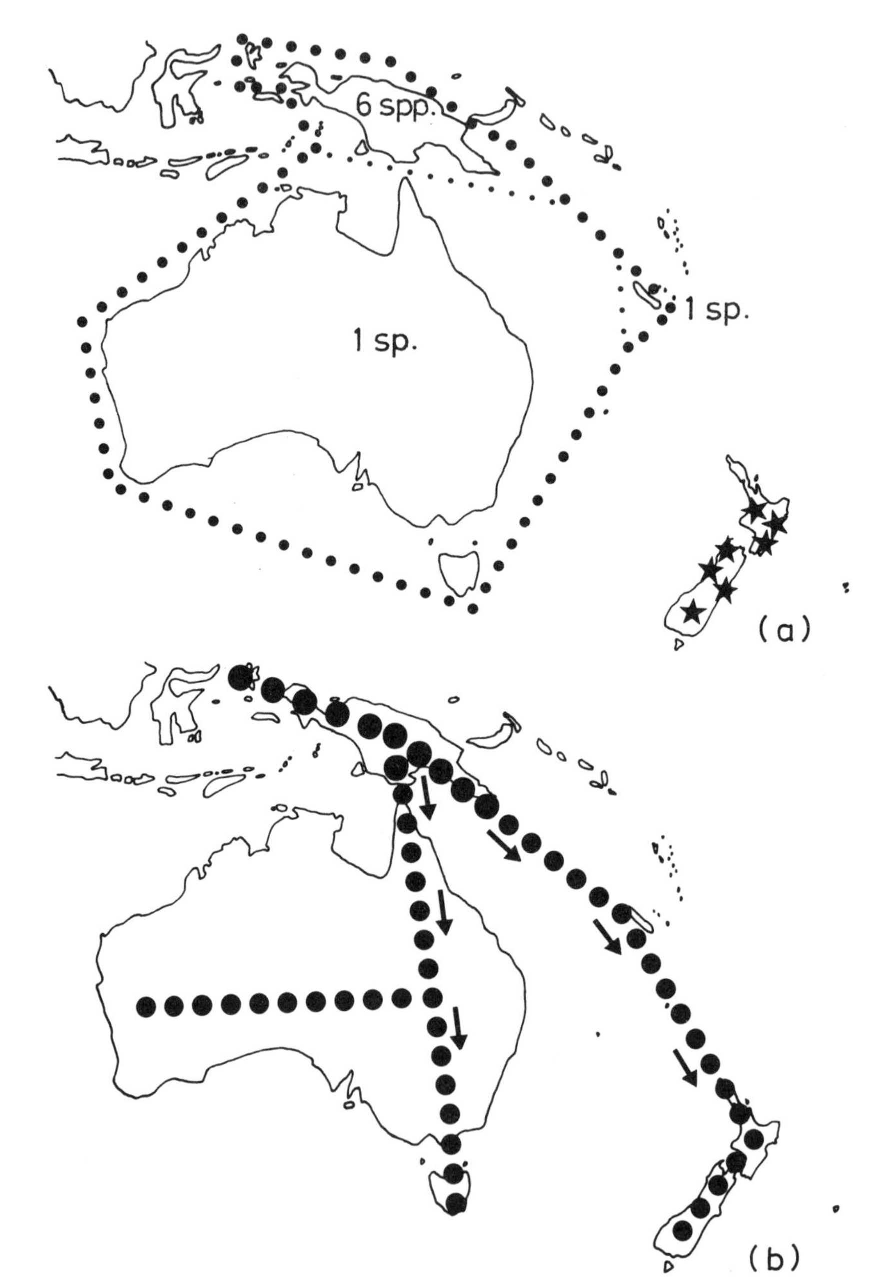

Figure 13.3 (a) Geographic distribution of the bird genus *Aegotheles* (dotted line) and the related fossil genus *Megaegotheles* (stars). Note the main massing of this family in New Guinea. (b) The track (dotted line) analysing the geographic distribution of these birds. The arrows alongside the track indicate the orientation of the track from the New Guinea main massing to Australia, Tasmania, New Caledonia and New Zealand.

tracks are components of a generalized track not so much because of any coincidence or congruence of geographic distributions between the groups analysed but because they are oriented the same way and can be mapped onto the same baseline (Fig. 13.4). The individual tracks belonging to a generalized track share the specialized similarities of coincidence of track orientation and baseline. In terms of graph theory a generalized track is a group of homeomorphic directed graphs mapped onto a common baseline.

An individual track is interpreted as being ". . . a sector of land (and sea) upon which a group – whether plants or animals – has eventually developed . . ." (Croizat, 1961). A generalized track represents the present day pattern of a group of ancestral distributions or biota of which the individual components are relict fragments.

Croizat (1958, 1961) discovered that generalized tracks for terrestrial life bear no apparent relationship to the present day disposition of land masses. They join areas that are now widely separated by oceans and seas. This indicates that portions of continents, larger islands and island groups (e.g. New Caledonia, New Guinea and New Zealand) are related to one another by tracks spanning sea and ocean basins rather than being biogeographic units equivalent to present day geographic areas.

Geographic areas where two or more generalized tracks overlap are recognized as 'gates' or major nodes by Croizat (1958). These major nodes are dynamic biogeographic boundaries where the vicariant remnant fragments of different ancestral biota come into contact (Fig. 13.4). They were interpreted by Croizat (1952, 1958, 1961) as zones of tectonic convergence, regions where fragments of two or more ancestral biological and geological regions meet and combine in space-time (Craw, 1983a, 1984b; Craw and Weston, 1984).

13.3.3 Graph theory and panbiogeographic method

In an important paper Page (1987) has demonstrated that the panbiogeographic method as described above can be rigorously formalized in graph theory terms. Page's reformulation enables baselines and nodes to be recognized through analyses of the connection and incidence matrices constructed from individual tracks. Since tracks can be represented algebraically in connection matrices they are accessible to statistical hypothesis testing for track congruence (Page, 1987). This recent development in panbiogeographic methodology overcomes the frequently raised criticism that such studies are not amenable to statistical testing (McDowall, 1978; Simberloff *et al.*, 1981; Simberloff, 1983).

Panbiogeographic method has been frequently caricatured as being merely the drawing of lines on a map (Rosen, 1975; Platnick and Nelson,

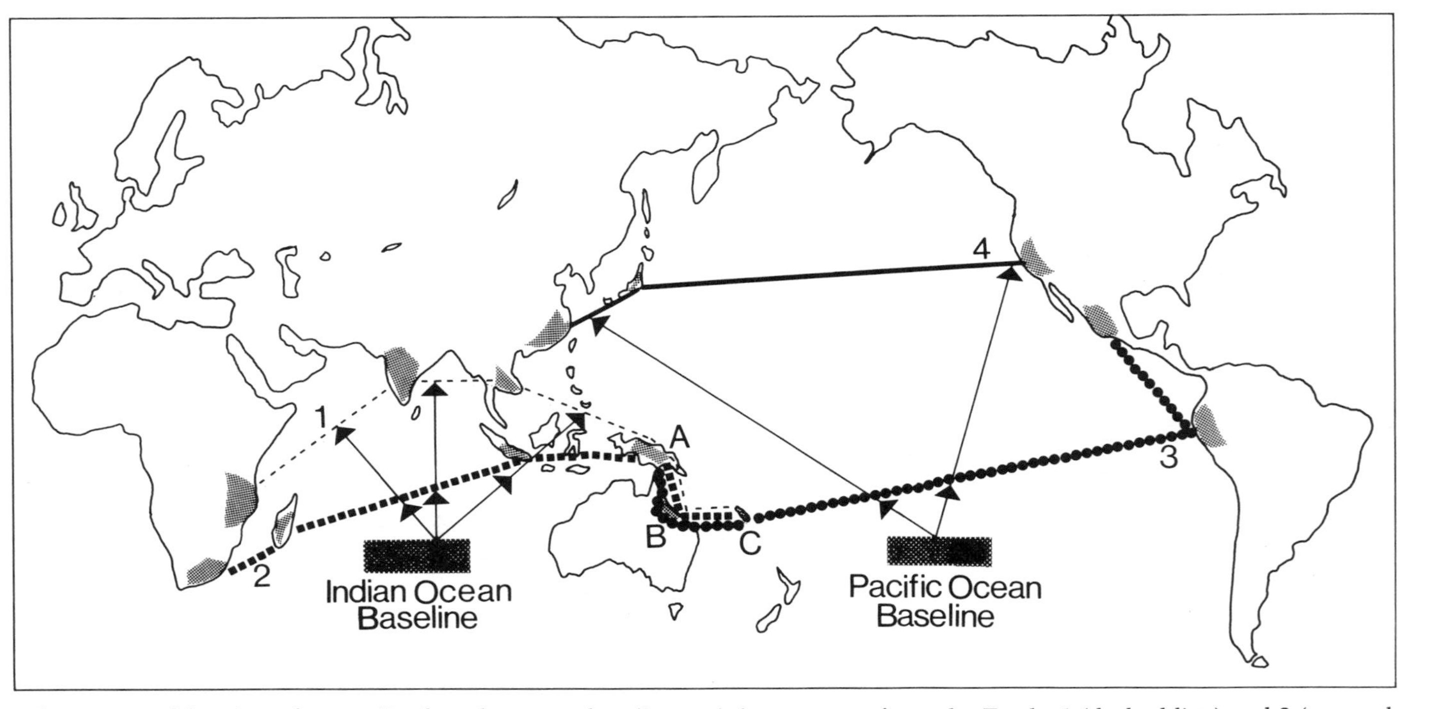

Figure 13.4 Mapping of generalized tracks onto a baseline and the concept of a node. Tracks 1 (dashed line) and 2 (squared line) cross or circumscribe the Indian Ocean and can be mapped onto the same baseline. Similarly tracks 3 (beaded line) and 4 (thick straight line) can be mapped onto a Pacific Ocean baseline. Tracks 1, 2 and 3 intersect in areas A, B and C which are recognized as a node.

1984). Platnick and Nelson (1984) claimed that in a panbiogeographic analysis overlapping lines or tracks in a single region could support a hypothesis of a composite biogeographic area when such was not the case, and that the overlapping tracks were merely components of a larger pattern that could only be detected through employment of cladistic systematic methods. Analysis of examples of overlapping tracks using Page's formulation indicates that this is not the case. Consider tracks 1 and 2 in Fig. 13.5a. These tracks overlap in areas A, B and C although they are geographically incongruent outside these areas. Both tracks share an Indian Ocean baseline and are orientated in the same direction. The connection matrix for these tracks within areas A, B and C (Fig. 13.5b) is asymmetrical and thus no incongruous information on area relationship is implied because all the track links are in the same direction. In contrast consider track 3 in Figure. 13.5a which overlaps with tracks 1 and 2 in areas A, B, and C. The resulting connection matrix for all three tracks is symmetrical with '1' entries on both sides of the diagonal (Fig. 13.5c). Track overlap in this case implies contradictory information on area relationships because the track links are bidirectional. Overlap of tracks 1 and 2 indicates only one set of area interrelationships; overlap of tracks 1 and 2 combined with track 3 indicates more than one set of area interrelationships.

Applications of graph theory to biogeography permit a precise comparison between tracks for individual taxa and allow for detailed analyses of distribution patterns. Clearly the integration of panbiogeographic method with graph theory techniques holds much promise for the future development of Croizat's initial insights into the potential value of graphical analysis of animal and plant geographic distributions.

13.4 DISPERSAL, VICARIANCE AND PANBIOGEOGRAPHIC MODELS OF SOUTHERN HEMISPHERE AND NEW ZEALAND BIOGEOGRAPHY: A COMPARISON

13.4.1 Review of dispersal, vicariance and panbiogeographic models

The recent literature of Southern Hemisphere biogeography has tended to be divided between two apparently competing models, dispersal and vicariance (Fleming, 1979; Cracraft, 1980; Humphries, 1983; Nelson and Platnick, 1984). The dispersal model of Southern Hemisphere biogeography originates with the work of Darwin (1859) and Wallace (1876, 1880). Its more explicit theoretical principles derive from the development of their ideas by Matthew (1915), Simpson (1940), Mayr (1942) and Darlington (1957). The basic concepts of the original form of this model are that species arise at a particular centre of origin and then disperse outwards

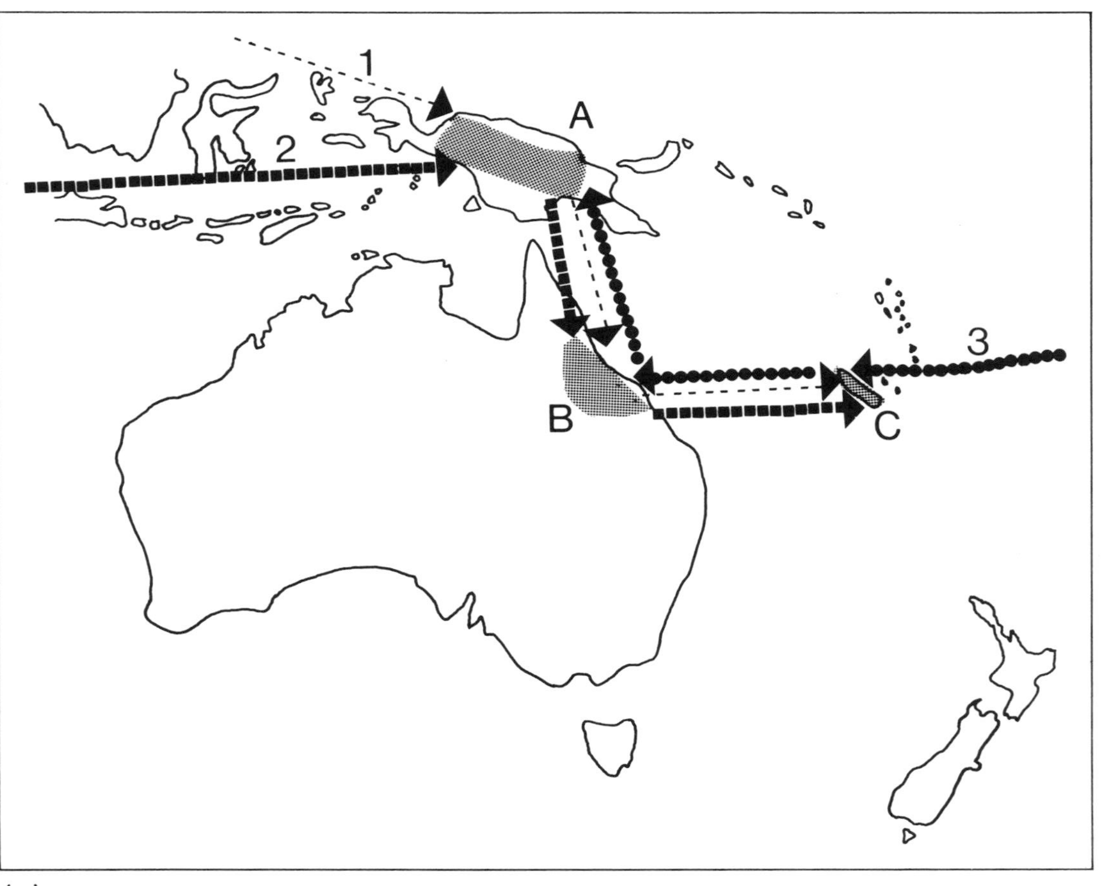

(a)

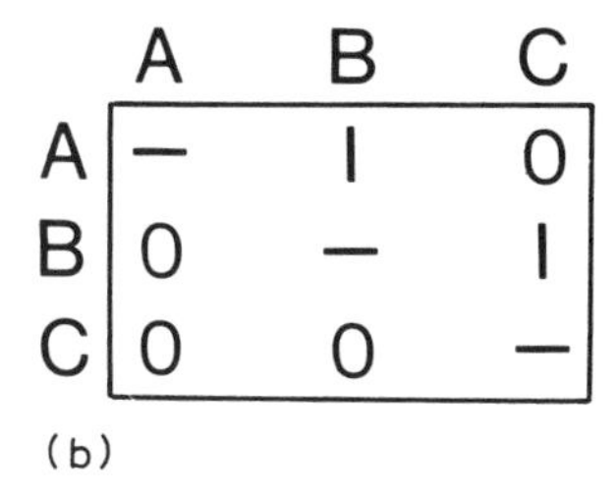

	A	B	C
A	–	1	0
B	0	–	1
C	0	0	–

(b)

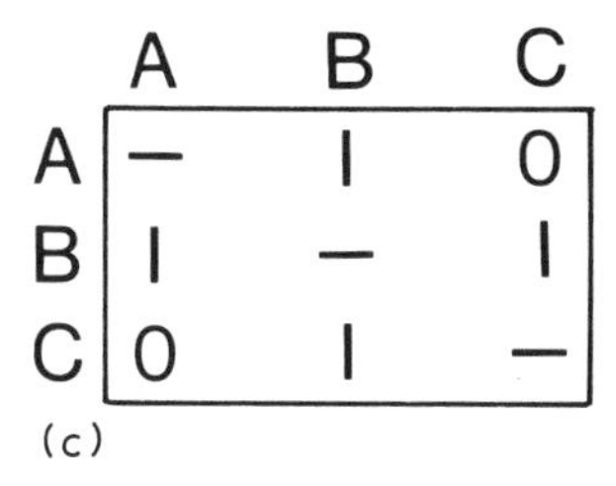

	A	B	C
A	–	1	0
B	1	–	1
C	0	1	–

(c)

Figure 13.5 (a) Details of intersection and overlap of tracks 1, 2 and 3 from Fig. 13.4 in areas A, B and C. Tracks 1 and 2 are oriented (arrows) in the same direction because they have a common baseline. The polarity of the combined link for track 1 and 2 is from A to B to C. Track 3 is oriented in the opposite direction to track 1 and 2 because it has a different baseline. The polarity of the link for track 3 is from C to B to A. (b) Directed connection matrix for tracks 1 and 2 in areas A, B and C. The matrix is constructed so that a '1' is entered for each pair of points when the polarity of the link is from the first to the second point. Note the '1' entries on one side of the diagonal only. (c) Directed connection matrix for tracks 1, 2, and 3 in areas A, B and C. Note the '1' entries on both sides of the diagonal.

from there across a world geography essentially similar to that of today, although land bridges between areas now separated by sea and ocean are sometimes postulated (Chapter 1). These ideas were first applied to the New Zealand biota by Hutton (1873) (reviewed by Craw, 1978). From Hutton (1873) to Fleming (1979, 1982) the dispersal model has involved the concept of an early New Zealand biota that arrived via 'land bridges' up to the end of the Cretaceous, which was subsequently isolated as a relict due to the fragmentation of a southern supercontinent of which New Zealand was once a part.

This relict biota has periodically been supplemented by colonizing species that have dispersed across the oceans from various sources. In the dispersal model, analyses of the raw data of biogeography – the geographic distributions and affinities of taxa – are based on a series of assumptions about the processes of dispersal. For instance, one of the criteria for membership of the relict biota is that a taxon has no obvious

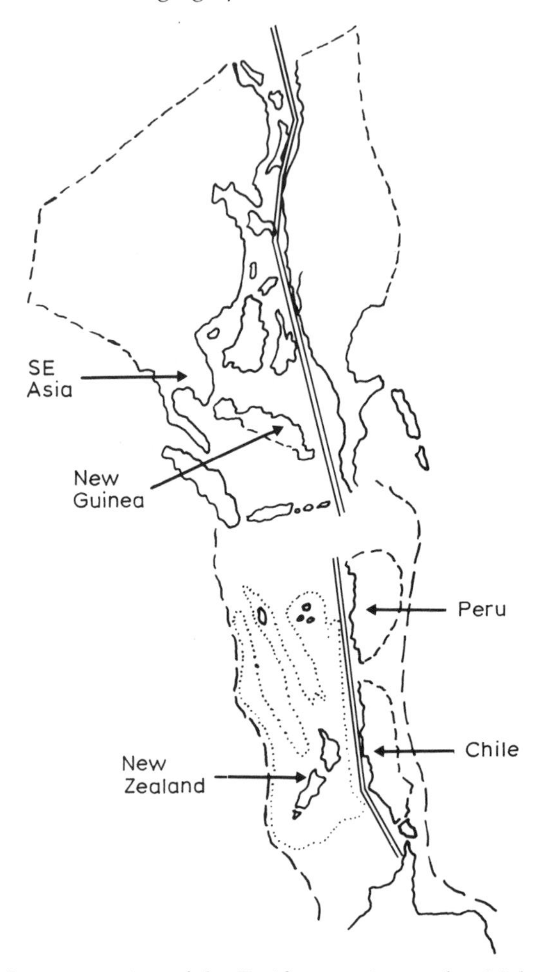

Figure 13.6 Reconstruction of the Pacifica continent after Melville (1981, 1982, pers. comm.). Note the close paleogeographical relationships between part of New Guinea (Wallace's Austro-Malayan subregion) and South-east Asia/East Indonesia (Wallace's Indo-Malayan subregion), and between New Zealand and Chile, also two of Wallace's sub-regions.

'means of dispersal' by which it could cross an ocean barrier to New Zealand.

The vicariance model of Southern Hemisphere biogeography was first explicitly stated by Wegener (1929) in relation to continental drift theory and then extended in detail by a number of European and Southern Hemisphere workers between 1920 and 1940 (Colosi, 1925; Harrison, 1928; Wittmann, 1934; Voss, 1939). It was extensively developed by

Brundin (Chapter 11) and Illies (1965) chiefly under the impetus of Hennig's (1960) phylogenetic speculations. In the modern form of the vicariance model the relict biota is defined from detailed phylogenetic analysis (based on form cladistics), in relation to the break-up sequence of Pangaea, and subsequently Gondwana. Although at first the dispersal and vicariance models seem different they are essentially similar. The criterion for inclusion in the New Zealand relict biota is that a New Zealand taxon is either the closest relative of a South American taxon, or related to a South American – Australian group. Brundin (Chapter 11) terms this a primary transantarctic relationship. Such relationships are assumed to reflect the accepted breakup sequence of Gondwana, and taxa that possess them are regarded as the relict element of the New Zealand biota, upon which is superimposed a dispersal (migration) element with northern or trans-Tasman sea origins (Winterbourn, 1980).

Supporters of the dispersal model generally followed Wallace (1876, 1880) who attributed all direct biogeographic relationships between temperate Australia and New Zealand to "transmission across the sea". Fleming (1962) agreed on geological grounds, stating "at no time is there evidence for direct trans-Tasman connection with Australia". Supporters of the vicariance model also recognize the importance of trans-Tasman dispersal. Brundin (1966, 1970) excluded taxa with direct trans-Tasman ties from the relict biota, attributing their distributions to dispersal from northern centres of origin through New Guinea and then separately down the Inner Melanesian arc to New Caledonia and New Zealand, and down the eastern Australian highlands to Tasmania (Winterbourn, 1980; Craw, 1982). Brundin and subsequent supporters of the vicariance model (Platnick, 1976; Rosen, 1978) have all stressed that the relict New Zealand biota shows no direct relationships to that of Australia. Since Fleming (1979, 1982) accepts the concept of an original Mesozoic, Gondwanan New Zealand relict biota upon which is superimposed an adventitious biota derived from various sources throughout the Tertiary, there is no real conflict between the dispersal model, based on concepts of centres of origin and subsequent dispersal, and the vicariance model, based on analysis of cladograms. Both models deny that taxa which now have direct trans-Tasman relationships could be part of the original relict biota. Yet anomalies remain. Many organisms attributed to the archaic pan-austral Gondwana biota, like the New Zealand ratites, the southern beeches *Nothofagus* and the endemic frog *Leiopelma*, can be interpreted in other ways (Craw, 1985).

A common theme, whether implicit or explicit is evident from Wallace (1876, 1880), Newton (1888), and Guppy (1906) through Brundin (1966) to Fleming (1979, 1982). All agree on a direct relationship of the New Zealand Mesozoic relict biota to that of South America or South

America/Australia combined. They all deny any direct relationship of this biota to the Australian biota without South American taxa also being involved, the explanation being the fragmentation of a Cretaceous Gondwanan biota. Melville (1981, 1982) has presented another explanation with his Pacifica hypothesis. Because the New Zealand relict biota is supposed to have direct affinities with South America, Melville's geological model places the New Zealand region in the mid-Pacific next to Magellanic and Peruvian fragments of South America (Fig. 13.6). In response to criticisms of his model by Haugh (1981), Melville (1981) claimed that the geological positioning of New Zealand against Australia, in many Gondwana reconstructions, "conflicts with a vast amount of biological data".

None of this biological evidence is cited by Melville (1981), but it is clear from an earlier paper (Melville, 1966) that he refers to the apparent absence from the New Zealand flora of characteristic Australian genera such as *Eucalyptus* and *Acacia*. These apparent absences have been cited often as evidence against any direct biogeographic relationships between Australia and New Zealand: ". . . the most conspicuous elements in Australian vegetation are wanting in New Zealand altogether. It has no *Eucalyptus, Acacia* or *Casuarina* or any of the great Australian genera of Proteaceae. It is inconceivable that New Zealand can ever have possessed this assemblage of types and have afterwards lost them" (Thiselton-Dyer, 1878). Thiselton-Dyer's position was reiterated by Wallace (1880) in his influential treatment of New Zealand biogeography, and has generally been uncritically accepted since then. However, fossil *Eucalyptus* leaves were recorded from New Zealand by Ettinghausen (1891), a record recently corroborated by Holden (1983), while *Acacia, Casuarina* and a variety of proteaceous pollens and fossils are known from the New Zealand Tertiary (Mildenhall, 1980; Campbell and Holden, 1984).

Ironically, Melville's Pacifica reconstruction, in which New Zealand is placed against Chile, can be interpreted as a modern restatement of Wallace's (1876) original conception of New Zealand as a transition subregion of his Australian region, related to the Chile (or southern temperate American) subregion of his Neotropical region. The concepts of Wallace can be portrayed as an area cladogram (Fig. 13.7) which shows that, as in Melville's reconstruction (Fig. 13.6), New Zealand is most closely related to Magellanic (Chilean) South America.

Both the dispersal and vicariance models indicate that the relict New Zealand biota originated from a single paleogeographic source: the Cretaceous Gondwana supercontinent. Recently Humphries and Parenti (1986) have proposed a modified vicariance model of Southern Hemisphere biogeography. They suggest that austral and boreal biota

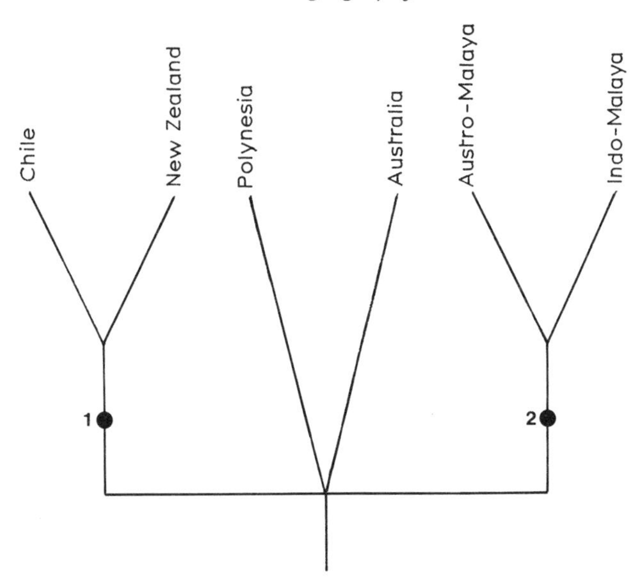

Figure 13.7 Area cladogram portraying Wallace's (1876) conception of the inter-relationships of the four sub-regions of his Australasian region. Components: 1 – transition to Nearctic and Australasian; 2 – transition to Oriental and Australasian.

were once united as a single ancestral biota on a hypothetical Pacific continent. This ancestral biota was fragmented by tectonic processes with the austral and boreal fragments moving to either side of a more distantly related pantropical biota. This is a resurrection of Murray's (1873) hypothesis of a close relation between austral and boreal regions with separate more distantly related tropical regions.

By contrast, Craw (1979, 1982, 1983 a, b, 1984b) and Henderson (1985) suggested an alternative panbiogeographic model in which vicariant fragments of several ancestral biotas have combined and overlapped in space-time to form the present-day New Zealand biota. Independent corroboration of this model has come from studies of the biogeography of southern long-horned caddisflies (Holzenthal, 1986). The geological implications of the panbiogeographic model are that the idea of a geologically monophyletic New Zealand as a unitary marginal fragment of Gondwana (Fig. 13.8) is somewhat anomalous. Croizat (1952) emphasized that the fragmentation of Gondwana and the opening of the Pacific Ocean basin created 'new worlds'. On the basis of biogeographic analysis two opposing lines of 'tectonic stress', one from the direction of the Tasman Sea and Southern Ocean, the other from the mid-Pacific, were identified that converge in the present day New Zealand area

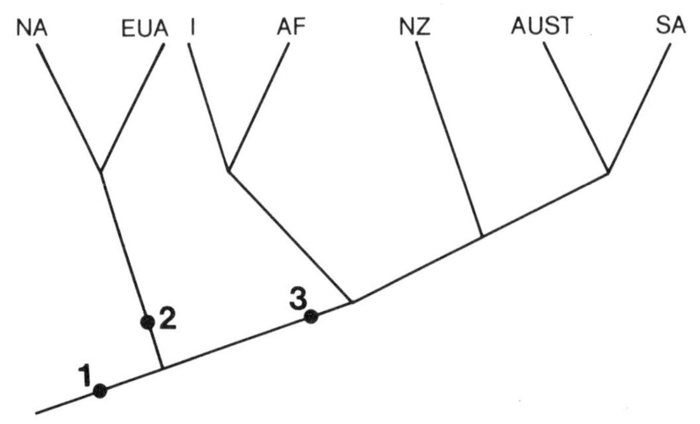

Figure 13.8 Current opinion on the break-up of Pangaea. Note how New Zealand constitutes a distinct, single unit in this summary of orthodox geological theory. Components: 1 – Pangaea, 2 – Laurasia, 3 – Gondwana. Abbreviations: AF – Africa, AUST – Australia, EUA – Eurasia, I – India, NA – North America, NZ – New Zealand, SA – South America (after Craw, 1983a: Fig. 2A).

(Croizat, 1952). A number of recent geological studies have proposed that New Zealand is a geological composite. Craw (1982, 1985) summarized these studies in cladistic form demonstrating the polyphyletic nature of the New Zealand – New Caledonian region as a geological taxon (Fig. 13.9) and remarking on the congruence between the composite geological nature of New Zealand – New Caledonia and the hypothesis of Craw (1979) that fragments of several vicariance biotas overlapped in present day New Zealand as a consequence of tectonic change.

13.4.2 Biogeographic elements, tracks and regions: the problem of biogeographic classification in a New Zealand context

Both dispersal and vicariance models of New Zealand biogeography assume that New Zealand constitutes a single, biogeographical unit or region, the biota of which is a mixture of old relict elements (the vicariance, southern or Gondwana biota) and the more recent trans-oceanic colonists (the dispersal elements). The most detailed and comprehensive classification of New Zealand plants and animals into biogeographic elements is that of Fleming (1963, 1976, 1979; critical review in Craw, 1978), which has gained wide acceptance (Fig. 13.10).

Panbiogeographic analysis provides an alternative biogeographic classification for New Zealand (Fig. 13.11). The generalized tracks that overlap in the present day New Zealand area can be interpreted as

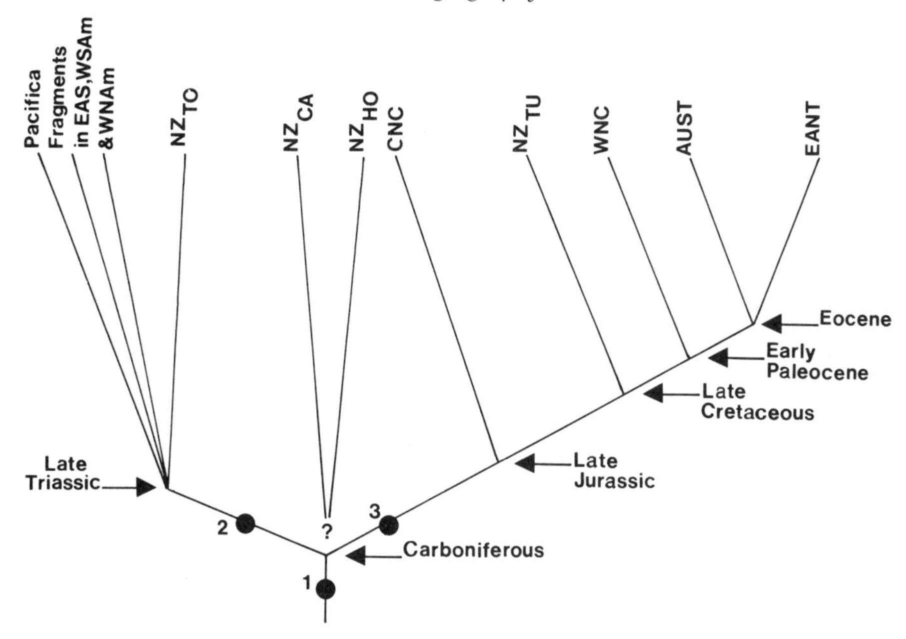

Figure 13.9 Cladistic representation of the composite geological nature of New Zealand and New Caledonia. Indicated are the geological time periods relevant to fragmentation. Components: 1 – Proto-Gondwana; 2 – Pacifica; 3 – Eastern Gondwana. Abbreviations: AUST – Australia; CNC – Central Chain, New Caledonia; EAS – East Asia; EANT – East Antarctica; NZ_{CA} – Caples Ensimatic Arc Terrane, New Zealand; NZ_{HO} – Hokonui Ensimatic Arc Terrane, New Zealand; NZ_{TO} – Torlesse Terrane, New Zealand; NZ_{TU} – Tuhua Ensialic Terrane, New Zealand; WNC – West Coast, New Caledonia; WNAm – Western North America; WSAm – Western South America (after Craw, 1985: Fig. 5).

constituting vicariant fragments of former biogeographic regions or areas of endemism. The individual tracks for the taxa belonging to each generalized track all share a common diagnostic feature, an ocean basin as a baseline. These taxa can thus be classified as belonging to a biogeographic region or area of endemism that corresponds to what is now an ocean basin (Fig. 13.12).

Fleming (1979) writes of the biogeographic elements theory: "Botanists and zoologists on the whole have agreed on the main elements in the biota, classified according to their geographic relationships and probable origin, but have used a variety of names for them". The panbiogeographic classification is completely different from the biogeographic elements classification, as shown by a comparison of a number of New Zealand taxa and their assignment to the appropriate element or panbiogeographic region (Table 13.1).

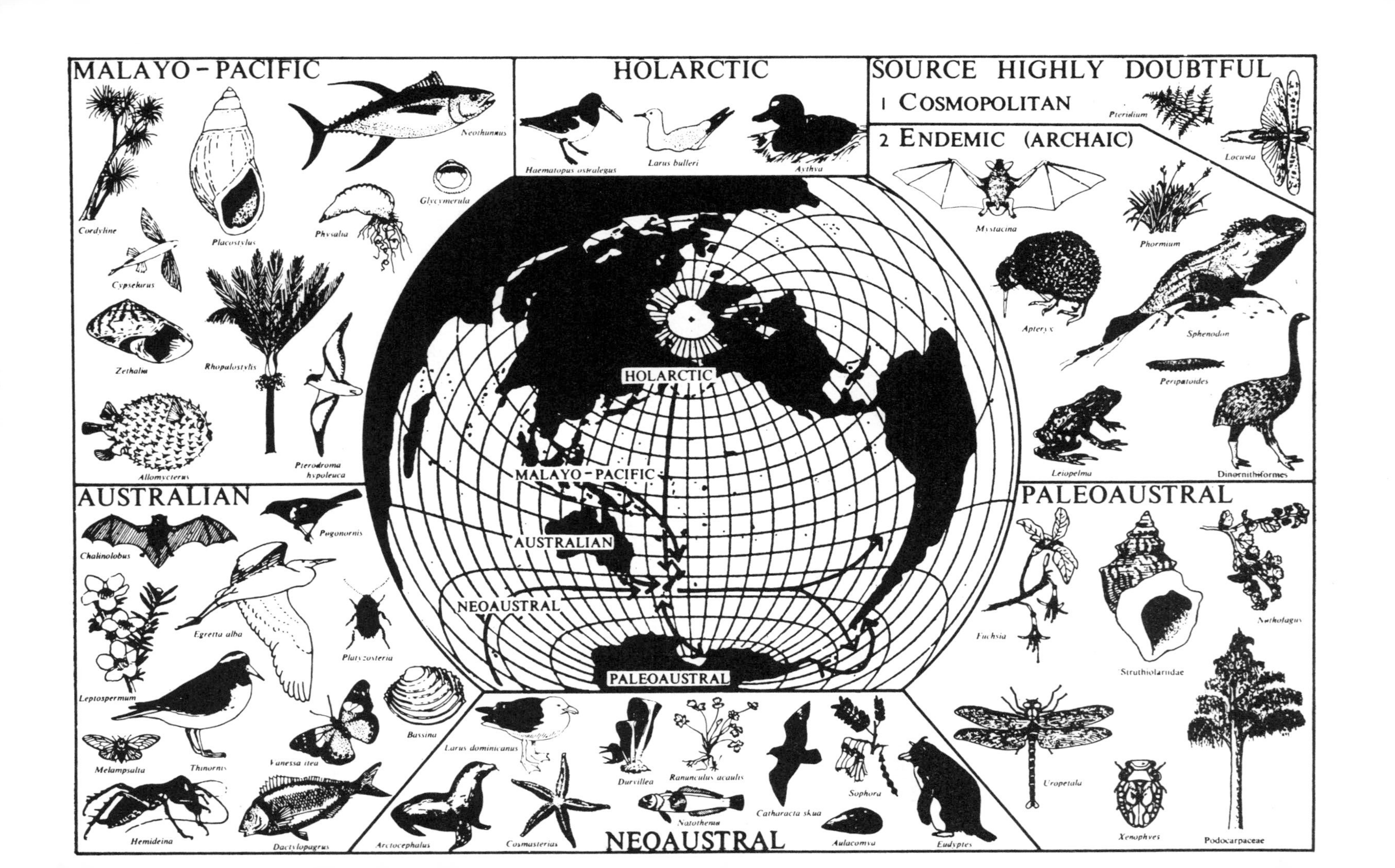
MALAYO-PACIFIC
Cordyline
Placostylus
Neothunnus
Glycymerula
Physalia
Cypselurus
Zethalia
Rhopalostylis
Allomycterus
Pterodroma hypoleuca
HOLARCTIC
Haematopus ostralegus
Larus bulleri
Aythya
SOURCE HIGHLY DOUBTFUL
1 COSMOPOLITAN
Pteridium
Locusta
2 ENDEMIC (ARCHAIC)
Mystacina
Phormium
Apteryx
Sphenodon
Peripatoides
Leiopelma
Dinornithiformes
HOLARCTIC
MALAYO-PACIFIC
AUSTRALIAN
NEOAUSTRAL
PALEOAUSTRAL
AUSTRALIAN
Chalinolobus
Pogonornis
Egretta alba
Platyzosteria
Leptospermum
Bassina
Melampsalta
Thinornis
Vanessa itea
Hemideina
Dactylopagrus
PALEOAUSTRAL
Fuchsia
Struthiolaridae
Nothofagus
Uropetala
Xenophyes
Podocarpaceae
Larus dominicanus
Durvillea
Ranunculus acaulis
Sophora
Catharacta skua
Notothenia
Arctocephalus
Cosmasterias
Aulacomya
Eudyptes
NEOAUSTRAL

There is no association between the biogeographic element and panbiogeographic classification of these taxa. The panbiogeographic classification is different; it is not merely grouping elements in the same way as Fleming's system, and then calling them by different names. For instance, the Pacific Ocean region includes taxa belonging to seven out of Fleming's eight biogeographic elements. The radical difference between the two classifications is best grasped by comparing three well known taxa *Fuchsia, Libocedrus* and *Leiopelma,* which are assigned to three different elements (the Palaeoaustral, Malayo-Pacific and Endemic respectively) in Fleming's classification but which all belong to the Pacific biogeographic region in a panbiogeographic classification. Apart from these qualitative differences it can also be shown statistically that there is no association between the elements and panbiogeographic classifications. Neither a χ^2 test (Sokal and Rohlf, 1981) or a G^2 test adjusted for small sample size (Feinberg, 1980) showed a statistically significant association between the two classifications.

The significant differences between these two classifications arise because the biogeographic elements groupings are based on non-homologous characters such as 'means of dispersal' (e.g. "*Palaeoaustral* . . . consists of . . . species of genera, or endemic genera of southern families, . . . which seem poorly adapted for trans-oceanic dispersal" Fleming, 1963) or on absence characters (e.g. "*Endemic element*: groups that have no close relations in other countries to indicate their place of origin" Fleming, 1979). Groupings on the basis of non-homologous and absence characters are generally recognized as being artificial or phenetic classifications. The panbiogeographic classification is based on graph analyses of biogeographic characters – the geographic distributions of plants and animals – in order to establish shared biogeographic homologies and then group taxa in terms of them.

The philosophical basis for the differences between these two biogeographic classifications is the differing conceptions of space (Section 13.1) underlying each. The elements classification is based on an absolute concept of space and a rigid division between organisms and their environments. The elements classification postulates a separate universe of organisms (the elements) which are moving through absolute geological time into a separate three-dimensional environmental space (the New Zealand subregion). In contrast the panbiogeographic

---

Figure 13.10 New Zealand biogeographic elements. An eighth element, the New Zealand was added in 1976 (from Fleming, 1962: Fig. 12, reproduced with the permission of the Editor, *Tuatara*, Victoria University of Wellington, New Zealand).

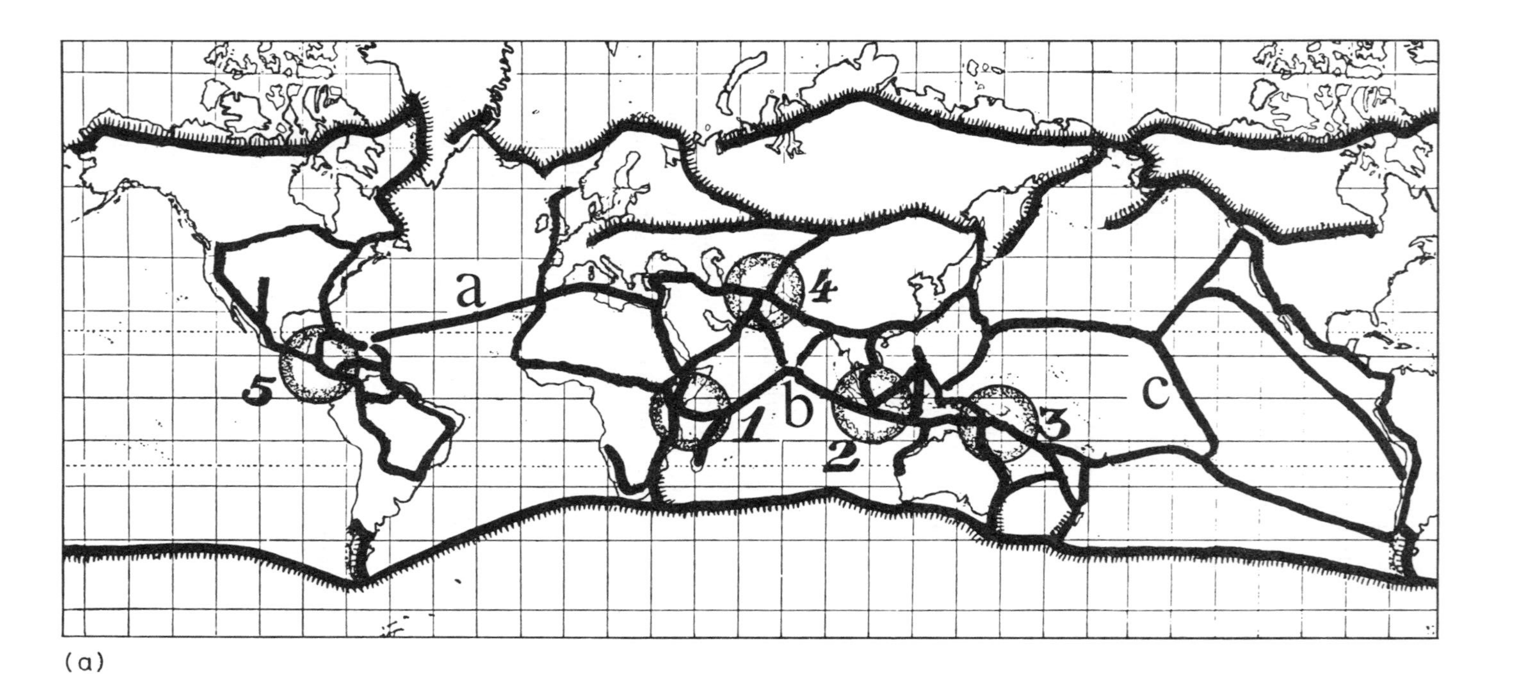

(a)

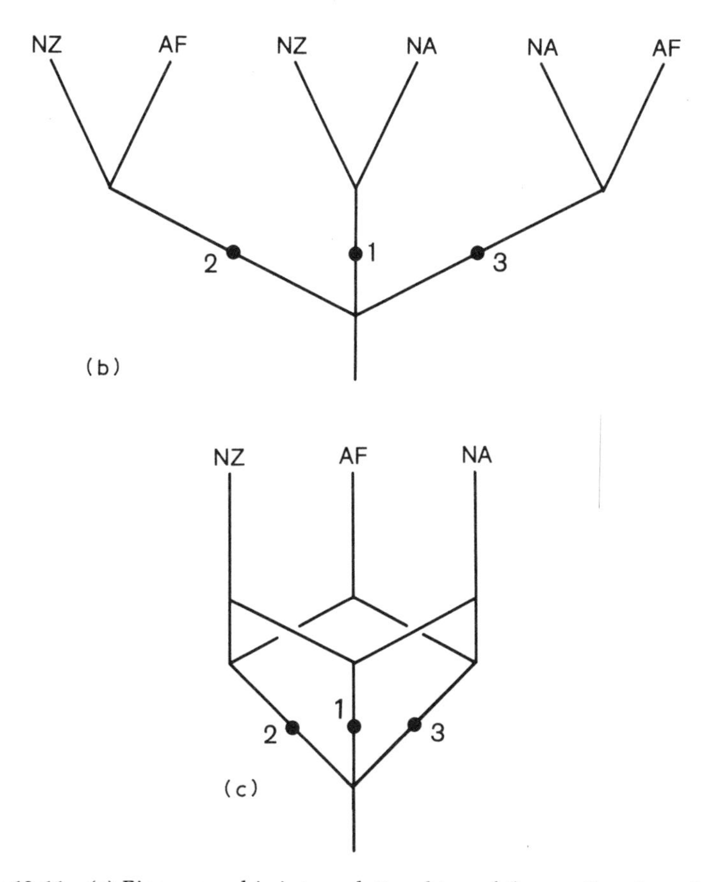

Figure 13.11 (a) Biogeographic inter-relationships of the continents and major islands. The hatched lines at the top and bottom of the map represent trans-Arctic and trans-Antarctic generalized tracks. Proceeding from left to right other generalized tracks are: a – the trans-Atlantic Ocean track connecting the Americas with Africa and Europe; b – the trans-Indian Ocean track linking Africa, Asia and Australasia; c – the trans-Pacific Ocean tracks linking East Asia/Australasia and the Americas. Numbered 1–5 are the major nodes where the generalized tracks intersect and overlap (after Croizat, 1958 [Vol. 2B]: Fig. 259).
(b), (c). Biogeographic relationships of Africa (AF), New Zealand (NZ), and North America (NA) from (a) represented as a cladogram (b) and a reticulate tree (c). Information derived from biogeographic analysis is represented by components 1 (Pacific Ocean), 2 (Indian Ocean) and 3 (Atlantic Ocean) (after Craw, 1983a: Fig. 4).

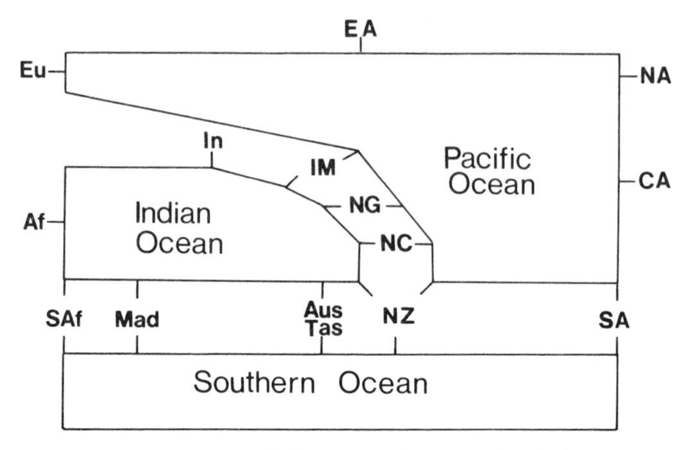

Figure 13.12 Representation of the generalized tracks (linking New Zealand with other land areas) not as lines on the present day map but as networks of biogeographical affinities circumscribing and crossing the ocean basin that each particular generalized track crosses. Abbreviations: AF – Africa; Aus/Tas – Australia/Tasmania; CA – Central America; EA – East Asia; Eu – Europe; In – India; IM – Indo-Malaya; Mad – Madagascar; NA – North America; NC – New Caledonia; NG – New Guinea; NZ – New Zealand; SA – South America; SAf – South Africa.

classification is based upon a relational concept of space. Here the rigid organism/environment dichotomy collapses and the regions and elements are synthesized into one.

De Candolle (1838) introduced the concept of New Zealand as a distinct biogeographic region or 'area of endemism'. Since then New Zealand has generally been regarded as a sub-region of the Australian region (following Wallace 1876), though Huxley (1868) suggested that New Zealand deserved status equal to that of the Australian and the Austro-Columbian (Neotropical) regions (Fig. 13.13). A myriad of phytogeographic and zoogeographic classifications have been proposed in the century since Wallace. Some of the more important or interesting of these are summarized (Fig. 13.14), and bear comparison with the panbiogeographic classification (Figs. 13.11 and 13.12).

The classifications of Engler (van Balgooy, 1971), Skottsberg (1960) and Müeller (1974) are particularly informative since they indicate that the biogeographic relationships between the three southern hemisphere areas of Australia, New Zealand and South America are far more complex than the notion of a unit New Zealand sub-region (common to most biogeographic classifications) would allow. However, these classifications are still based on present day land areas and involve subdividing New Zealand on the basis of this (e.g. in Engler's system the

Table 13.1. Comparison of assignment of examples of plant and animal taxa represented in the New Zealand biota to the appropriate biogeographic element or panbiogeographic region. Assignment to these categories follows Fleming (1963, 1976, 1979) or his criteria for inclusion in a particular element for elements, and Croizat (1952, 1958, 1968) and Craw (1983b, 1985, unpublished) for panbiogeographic regions

Taxon	*Biogeographic element*	*Panbiogeographic region*
Apteryx	Endemic	Indian Ocean
Bulbinella	Neoaustral	Indian Ocean
Chalinolobus	Australian	Indian Ocean
Corokia	Australian	Indian Ocean
Cyanorhamphus	New Zealand	Indian Ocean
Dinornithiformes	Endemic	Indian Ocean
Dodonaea	Malayo-Pacific	Indian Ocean
Hemiphaga	New Zealand	Indian Ocean
Paratrophis	Malayo-Pacific	Indian Ocean
Paryphanta	Palaeoaustral	Indian Ocean
Pelargonium	Neoaustral	Indian Ocean
Peripatoides	Endemic	Indian Ocean
Pittosporum	Malayo-Pacific	Indian Ocean
Podocarpaceae	Palaeoaustral	Indian Ocean
Toronia	Palaeoaustral	Indian Ocean
Chironomid midges	Palaeoaustral	Southern Ocean
Galaxiidae	Neoaustral	Southern Ocean
Leptophlebiid mayflies	Palaeoaustral	Southern Ocean
Restionaceae	Neoaustral	Southern Ocean
Tadorna	Neoaustral	Southern Ocean
Alpine Ranunculi	Holarctic	Pacific Ocean
Casuarina	Australian	Pacific Ocean
Coriaria	Palaeoaustral	Pacific Ocean
Euphrasia	Holarctic	Pacific Ocean
Fuchsia	Palaeoaustral	Pacific Ocean
Gaultheria	Holarctic	Pacific Ocean
Griselinia	Palaeoaustral	Pacific Ocean
Hebe	New Zealand	Pacific Ocean
Leiopelma	Endemic	Pacific Ocean
Libertia	Neoaustral	Pacific Ocean
Libocedrus	Malayo-Pacific	Pacific Ocean
Metrosideros	Malayo-Pacific	Pacific Ocean
Nothofagus	Palaeoaustral	Pacific Ocean
Stylogymnusa	Holarctic	Pacific Ocean

North Island is more closely related to New Caledonia and Queensland than it is to the South Island which has a closer relationship with Southern Australia and Magellanic South America: Fig. 13.14b).

The panbiogeographic classification has a greater empirical content than these earlier predecessors that also stress the complex nature of

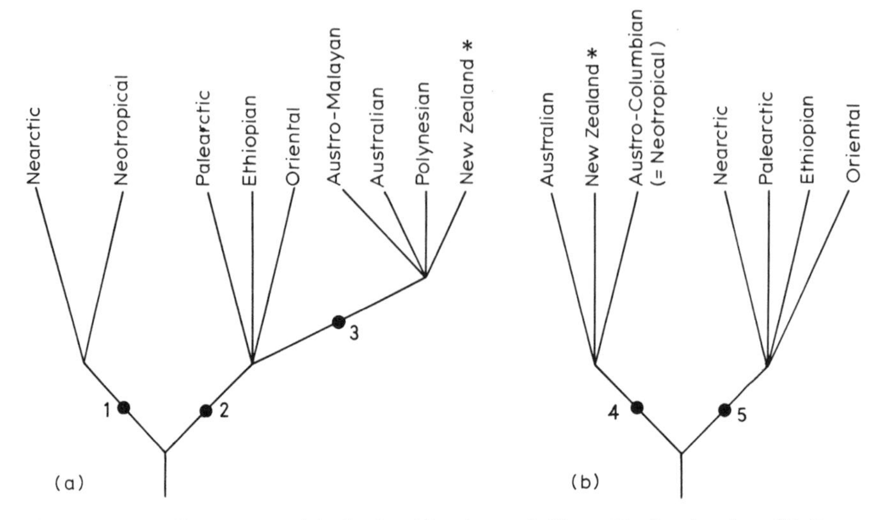

Figure 13.13 Zoogeographical classification of New Zealand after Wallace (1876) (a) and Huxley (1868) (b). Components: (a) = 1 – Neogaea (New World); 2 – Palaeogaea (Old World); 3 – Australian Region; b = 4 – Notogaea; 5 – Arctogaea. In this and Fig. 13.14 the position of New Zealand or parts of New Zealand is indicated by an asterisk.

New Zealand's biogeographic relations, since in it New Zealand is recognized as belonging to three different natural groupings or historical entities. Each of these historical entities is associated with a particular ocean basin. They have been fragmented by tectonic activity (presumably the opening of these basins Croizat, 1952; 1958) and component fragments have combined in space-time to form the present New Zealand biota. The New Zealand area is a dynamic biogeographic fusion zone where fragments of several ancestral biogeographic and geologic worlds meet and fuse into one modern geography in space/time (Craw, 1982, 1983a, b, 1984b).

13.4.3 New Zealand biogeography at the local level

Several important studies have recently demonstrated the potential for the application of panbiogeographic methods to New Zealand

---

Figure 13.14 Phytogeographical and zoogeographical classifications of New Zealand: (a) Grisebach (1872, after review in van Balgooy, 1971). (b) Engler (1882, after review in van Balgooy, 1971). Components: 1 – Araucarient Gebeit; 2 – Altozeanisches Florenreich. (c) Skottsberg (1960). Components: 1 – Subantarctic Zone; 2 – Austral Zone; 3 – Austral Zone B. (d) Thorne (1963). Components: 1 – Oriental; 2 – Australian. (e) Müeller (1974). Components: 1 – Australian Realm; 2 – Archinotic Realm.

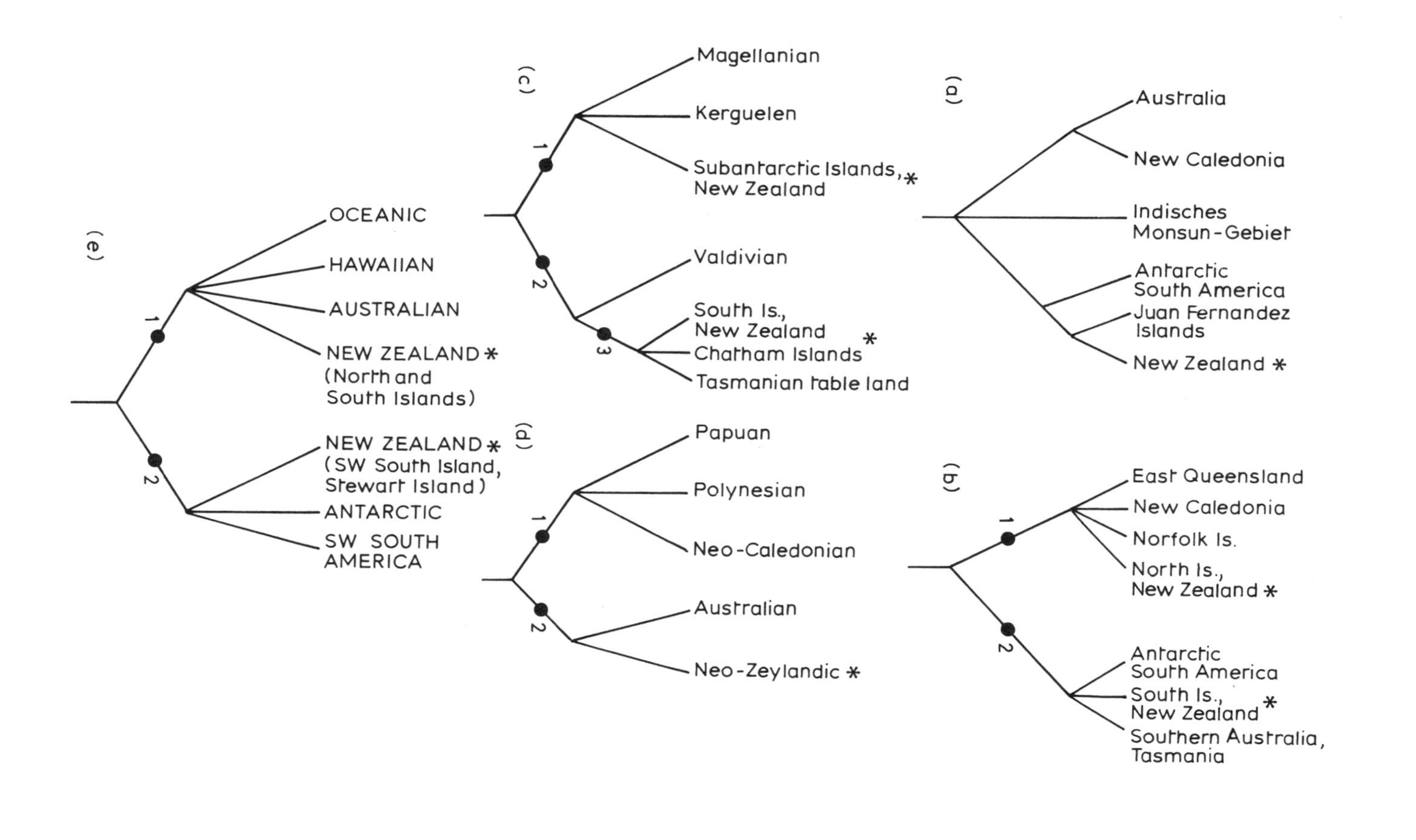
(a)
Australia
New Caledonia
Indisches Monsun-Gebiet
Antarctic South America
Juan Fernandez Islands
New Zealand *
(b)
1
2
East Queensland
New Caledonia
Norfolk Is.
North Is., New Zealand *
Antarctic South America
South Is., New Zealand *
Southern Australia, Tasmania
(c)
1
2
3
Magellanian
Kerguelen
Subantarctic Islands, New Zealand *
Valdivian
South Is., New Zealand
Chatham Islands *
Tasmanian table land
(d)
1
2
Papuan
Polynesian
Neo-Caledonian
Australian
Neo-Zeylandic *
(e)
1
2
OCEANIC
HAWAIIAN
AUSTRALIAN
NEW ZEALAND * (North and South Islands)
NEW ZEALAND * (SW South Island, Stewart Island)
ANTARCTIC
SW SOUTH AMERICA

biogeography. Gibbs (1983) has analysed the biogeography of Australasian micropterigid moths utilizing track analysis. This analysis was the basis from which several hypotheses were proposed relating micropterigid species diversity to the composite geology of the region.

Utilizing a data base of 12 000 distribution records (from within New Zealand and outlying islands) from 800 localities for over 150 caddisfly (Trichoptera) species Henderson (1985) compared and contrasted the merits and defects of dispersalist, vicariance cladistic and panbiogeographic methods of analysis. Panbiogeographic method was found to be the most appropriate and productive approach for analysis of geographic distribution within New Zealand. Heads (1988) has examined distributions of many genera of seed in the most detailed panbiogeographic analysis of New Zealand biogeography yet attempted. This study identifies a number of previously unrecognized distribution patterns and localities of prime importance both to biogeography and conservation biology. Rodgers (1986, 1987) has identified in detail a number of coincident biogeographic patterns in the southern North Island of New Zealand and discussed them in relation to panbiogeographic and dispersal models. This study also demonstrates the importance of panbiogeography in highlighting habitats and landscapes of biological significance.

The above studies clearly indicate the usefulness of panbiogeographic methods at the most local level of biogeographic analysis. These studies demonstrate that the application of panbiogeographic methods has a greater power to generate novel hypotheses about New Zealand biogeography than have either dispersal or vicariance methodologies.

13.5 CONCLUSIONS

Panbiogeography represents a fundamental contribution to the discipline of biogeography. Croizat's vision was that the development of methods of graphical analysis of distribution patterns would form a simple and innovative approach to the problem of biogeographic analysis. Already the application of panbiogeographic methods has laid the foundations for the development of a combined biogeographic history and classification of terrestrial biota that subsumes all the elements of earlier phytogeographical and zoogeographical classifications as well as adding several novel components. Applications of panbiogeography do not stop at biogeographic analyses and classifications. The method has generated novel predictions about historical geology (Craw and Weston, 1984) and provides a biogeographic basis underpinning recent developments in evolutionary theory (Grehan and

Ainsworth, 1985), behavioural ecology (Gray, 1986, 1987) and plant morphology and systematics (Heads, 1984).

Croizat's panbiogeographic methodology and synthesis invite us to investigate with the author the possibilities for the development of biogeography as an independent science. Panbiogeography is a challenge to present biogeographers to be as creative and critical in the context of our time as Croizat was in his time.

ACKNOWLEDGEMENTS

I am grateful to Russell Gray, John Grehan, Michael Heads, Ian Henderson, Rod Page and Geoff Rodgers for making available to me their unpublished manuscripts and theses. Dr Ronald Melville kindly supplied a copy of his Pacifica reconstruction that was used in the production of Fig. 13.6. Thanks to J. S. Dugdale, Russell Gray and G. W. Ramsay for comments on the manuscript. Ian Henderson suggested and conducted the statistical tests comparing biogeographic classifications.

CHAPTER 14

From fossils to earth history: applied historical biogeography

B. R. ROSEN

14.1 RELEVANT PARTS OF THE BIOGEOGRAPHICAL SYSTEM AND OVERVIEW OF METHODS

Fossils have been used to derive earth history through two fundamentally different kinds of evidence: chronological and spatial. The chronological evidence is biostratigraphical; that is, fossils are one of the most important ways of providing relative ages of strata, and this makes it possible to infer sequences of geological events. These biostratigraphical methods are long established, widely accepted both within and outside palaeontology, and well explained in standard texts. There is no need to discuss them any further here.

The purpose of this chapter is to review the ways in which fossils provide spatial information on a geographical scale; that is, reconstructions of the earth's geography in the past (palaeogeography), and the tectonic, eustatic, climatic and oceanographical events (or TECO events; Rosen, 1984) that brought about changes to this geography. I regard pure historical biogeography as being concerned with the history of biotas, and the organic processes that underlie this history. Here, I discuss the way in which these historical patterns are applied to the task of reconstructing earth history, particularly from fossils.

Biogeographers have used both living and fossil organisms to reconstruct earth history from biogeographical distributions. Any biogeographical investigation that has earth history as its objective will probably be based on one or more of the ideas about the biogeographical system (Chapter 2) that link earth history and distributions (Table 14.1). However, they have not all proved equally practical for this purpose: my reasons for eliminating the ideas other than those in the last column of this table are covered in Chapter 2.

The general approach in all methods of reconstructing earth history is

Table 14.1. Summary of the main ideas that biogeographers have believed to be important parts of the biogeographical system (first column; see Chapter 2), showing those that have an interface with earth history (second column). Last column shows those that form the basis of analytical methods, widely or increasingly adopted for reconstructing earth history. Note that for reconstructing earth history, the value of those maintenance ideas having an earth history component is taken into account under Distributional change

Concepts and processes	*Earth history component*	*Practical value for reconstructing earth history*
Maintenance		
Dispersion		
Geoecology	★	
Equilibrium theory and island biogeography		
Barriers	★	
Geoecological assemblages		
Provinces and provinciality	★	
Distributional change		
Range expansion		
Jump dispersal	★	
Equilibrium theory and island biogeography	★	
Earth history (TECO events and changing palaeogeography)	★	★
Environmental tracking as a response to:		
ecological change		
the effect of earth history on ecological change	★	★
Adaptation of a taxon to different conditions		
Originations		
Empirical phylogenetic biogeography	★	★
Vicariance as a response to:		
earth history	★	★
ecological change or the effect of earth history on ecological change	★	★
Fringe isolation as a response to:		
ecological change		
the effect of earth history on ecological change	★	
Jump dispersal	★	
Centres of origin (in part)	★	
Geoecology through time	★	★

to start with taxonomic lists (presence/absence data) based on one or more groups of organisms for a set of different localities or regions. Using methods based on one or more of the assumptions of processes and concepts listed in the final column of Table 14.1, these data are then translated into palaeogeographical reconstructions and TECO events (see Table 14.2). As these tables show, methods that have been used to infer earth history from distributions can be split into two groups. Methods based on distributional change concentrate on analysing the combinations of taxa that occur in different regions, regardless of evolutionary turnover. Conversely, methods based on originations concentrate on geographical patterns of taxonomic appearances (and sometimes extinctions) of taxa, regardless of whether distributional changes have also occurred.

Methods used for reconstructing earth history do not fall neatly into those that use fossils and those that do not. Nevertheless, many of the

Table 14.2. Methods treated in this chapter that are used to reconstruct earth history, and the assumed processes and concepts on which their underlying propositions are based. (★) denotes methods that could be explored

Concepts and assumed processes	*Kinds of methods*			
	Assemblages		*Parsimony analysis of endemicity*	*Phylogenetic methods (panbiogeography and cladistic methods)*
	statistical	*geoecological*		
Distributional change				
Earth history (TECO events and changing palaeogeography)	★		★	
Environmental tracking as a response to the effect of earth history on ecological change		★	★	
Originations				
Empirical phylogenetic biogeography				★
Vicariance as a response to earth history				★
Vicariance as a response to the effect of earth history on ecological change				★
Geoecology through time	★	(★)		(★)

methods based on distributional change are dependent on fossils because changes are inferred by comparing fossil distributions at two or more different geological horizons: a synoptic approach. In many cases, where a single horizon is used, the synoptic element consists of an implicit comparison with the present day. Clearly, if patterns from different horizons are the same, the biotic evidence on its own suggests that their respective palaeogeographies could not have differed very much, or that if the palaeogeography did differ, it was not reflected by analysis of the organisms concerned.

By contrast, there are methods that analyse distributional data for their historical content without recourse to synoptics (historical analysis). These methods can generate a historical hypothesis from just one data set, even one derived from the distributional record of just a single geological horizon (including the present). Here, living and fossil distributions have each been used on their own, as well as in combination. Some of the methods based on historical analysis are treated in other chapters in Part IV, but because they also make use of (or optionally include) fossil data, they must also be mentioned here (Section 14.4).

Interpretations of distributions may be either static or dynamic. A static inference from an analysis of biotas is one that is a reconstruction of geographical and geological features on a single geological horizon, especially features such as mountain ranges, oceanic and continental regions, and land/sea boundaries. A dynamic inference is one that is a reconstruction of geological events and processes (i.e. change) and can only be made by using synoptic comparisons or methods based on historical analysis. Palaeoenvironmental methods can only work synoptically; parsimony analysis of endemicity (PAE) can be used both statically and dynamically; statistical methods work synoptically and just possibly dynamically; and cladistic methods produce dynamic hypotheses through historical analysis.

In all methods, there is great variation in the level of empiricism adopted. Those based on palaeoenvironmental fossil evidence make initial assumptions about the mode of life of the organisms concerned. At the other extreme, cladistic methods and PAE minimize assumptions and emphasize repetition of similar patterns as a criterion of significance in distinguishing fortuitous events from the overall influence of geological events.

The use of fossils introduces constraints and difficulties into methods of inferring earth history that do not arise from the use of living distributions alone, and this (and other constraints) must be discussed before entering into discussion of the methods themselves.

14.2 CONSTRAINTS

14.2.1 The fossil record

In discussing the aims of biogeography in Chapter 2, I noted a divergence in historical biogeography between 'neobiogeographical' and 'palaeobiogeographical' approaches. Though they have shared the same aims (whether 'pure' or 'applied') they have differed in their historical origins, perspectives and emphases, and have adopted different methods of reaching similar goals. Doubt about the value of palaeobiography has been generated by the cladists' well known attack on the phylogenetic and biogeographical value of fossils. I believe that misunderstandings have arisen from this attack which hinder an integration between historical neobiogeography and palaeobiogeography, so it is important for palaeobiogeographers to be clear about the criticisms, while recognizing the constraints of the fossil record. It is therefore worth trying to give a balanced view of the present and potential biogeographical value of fossils.

The origin of criticisms about the biogeographical use of fossils stems from the incompleteness of the fossil record. As a truism, this simply means that any inferences about the past based on fossils must also be incomplete. Methodologically, it also ensures that absence of a taxon from a particular time and place is ambiguous, and cannot be used as constructive or falsifying evidence. It also means that even presence of a fossil has biogeographical ambiguity, since new fossil finds often change our picture of distributions. This happens far less often in the study of living taxa.

The cladistic critique of the value of fossils is in fact no more than a well-reasoned statement showing the limitations of fossil evidence in meeting two of the three aims of historical biogeography: explanation of the distribution of organisms ('pure'), and reconstruction of geological events and palaeogeography from biogeographical evidence ('applied'). In historical analysis in particular, fossils add information, mainly new localities and minimum ages, to that obtained from living distributions, but they cannot contradict the patterns obtained from living organisms alone (Humphries and Parenti, 1986). It is important to recognize two points about these criticisms:

(1) They refer particularly to biogeographical methods based on relationships between living taxa. Thus these criticisms are not directly applicable to palaeobiogeographers who use, for instance, biostratigraphy and palaeoecological evidence to reconstruct geological events.

(2) They are made in the context of using hypothetico-deductive methods (falsifiability), but in an inductive framework (Chapter 2) fossils appear in a better light.

Although it is logically difficult to prove that the fossil record is complete, it does not follow that the biogeographical information content of fossils is so incomplete as to be virtually useless. Admittedly, this may well be nearer the truth for groups whose fossils are often incompletely preserved or that occur relatively rarely, such as many larger vertebrates, terrestrial invertebrates and vascular plants. This is implied by Patterson's (1981a) examples of fossil surprises. But if in Patterson's view, Westoll's (1958) "law of diminishing returns of exploration" does not hold for such groups, it might be more nearly applicable to some abundantly preserved marine groups like foraminifera. Here is the crux of the falsifiability/induction issue. As Rosen and Smith (1988) have argued, whatever the nature of differential preservation, a set of contemporaneous fossil localities with their recorded taxa can be regarded as a set of biogeographical sample points whose taxonomic presences can be analysed rigorously to discover their biogeographical implications. Better or more frequently preserved groups will presumably yield more biogeographical information than poor groups. Biogeographical hypotheses generated on this assumption may still not be falsifiable, but they can be progressively strengthened by independent, rigorous evidence, on an inductive basis (Chapter 2).

Neobiogeographers probably have little choice but to accept the above criticisms about the use of fossils in determining the biogeographical history of living biotas. In any case, many living organisms have little or no fossil record. I would not agree however that fossils are less informative than living organisms in the specific task addressed here of reconstructing geological events. If we use only living distributions, inferences of geological events and biotic history must obviously be increasingly incomplete the greater the relative geological age of the inferred events. I would doubt that living distributions can shed light on more than a fraction of biotic history and geological events before the Mesozoic. By its own standards, neobiogeography can only say less and less (going back in time) with more rigour, while palaeobiogeography can say more, if less rigorously.

Taking an integrated viewpoint, I suggest that the scope for deriving viable hypotheses about geological events and palaeogeography, using all of the available fossil data from throughout the geological (Phanerozoic) record, is fourfold:

(1) through their palaeoenvironmental evidence;

(2) through the data they provide for synoptic 'snapshots' of the biogeography and physical geography for particular slices of geological time;
(3) through inclusion of fossils in cladistic methods;
(4) through extension of cladistic (and possibly other neobiogeographic historical techniques) to entirely fossil data sets.

These approaches are valid in principle (or have not yet been invalidated). It is just unfortunate that the theoretically sound aspects of palaeobiogeographical methods have often been masked by poor practice, especially by the custom of combining fossil evidence with that from other independent sources (notably geophysics) into narratives, and using the hybrid concept of provinces (Chapter 2, Section 2.5.3), instead of comparing results from different kinds of data as potentially conflicting hypotheses in an inductive framework. Thus we should aim to improve the rigour of palaeobiogeographic methods and devise new, more rigorous ones. This will also enable us to assess more effectively the true limitations of fossil data for reconstructing geological events.

14.2.2 Palaeogeographic framework

In synoptic methods in particular, the use of maps of present geography with their modern place names, though a relatively objective approach for compiling separate taxonomic lists for fossil localities, can be distracting when trying to infer palaeogeography. This is because their modern location with respect to one another (distance apart, longitude and latitude, altitude, depth, climate) cannot be assumed to have been the same in the past. If the modern geographical frame of reference cannot be assumed to apply to fossil organisms, what should be used instead?

The solution most widely adopted is to plot fossil distributions (or the processed results derived from them) on palaeogeographic reconstructions based, ideally, on independent evidence like geophysics (e.g. Hughes, 1973). Such reconstructions however are just as hypothetical as any that are based on fossils. They are constantly being amended and updated, and there is therefore no way of knowing whether such independent reconstructions are actually 'correct'. It might be argued that the relevance of this point depends on (1) the chosen scale of resolution, (2) the geological age of a particular reconstruction, (3) consensus and test-durability of particular reconstructions, and (4) level of knowledge of particular sectors of the earth. The continuing, unquelled possibilities of an expanding earth and of a 'Pacifica' landmass (for example) nevertheless suggest that, in the present state of the subject, it would be safer and more rigorous to keep biotic distribution patterns separate from other kinds of reconstructive evidence.

Ideally, palaeogeographical reconstructions derived from different methods and even from different data sets should all be kept independent of each other to facilitate objective comparison between them. Alternative hypotheses can be qualified by stating personal preferences or levels of confidence. Although it could be argued that it is reasonable to bring together mutually consistent elements of different reconstructions, it is also important to highlight anomalies, not to accommodate them within a preferred reconstruction in an *ad hoc* fashion. Unfortunately, some of the theories of distributional change, like jump dispersal, are always at hand for explaining awkward anomalies, because they are independent of geological processes (Chapter 2). The end result of superimposing biotic distributions on independent palaeogeographical reconstructions and accommodating inconsistencies with *ad hoc* accretions, is a confusing morass of facts, ideas and circularity, in which it is difficult to judge precisely what has actually been postulated or discovered.

The alternative is not to use any kind of base map at all, modern or palaeogeographical, except for initial data compilation and for discussion of specific problems, but rather to use other graphical ways of depicting distribution patterns, for example, area cladograms, area dendrograms and area clusters.

Quite different is the problem of generalizing distributions from intitial sample points either by extrapolating them from larger areas, or through interpolating and integrating them within a set of sample points. It is always tempting to replace a cumbersome locality list with visually or politically convenient geographical/geological units, especially states, continents, crustal plates, large islands, archipelagoes, oceans, tectonic units, marine basins, and freshwater systems. The risk of doing this however, is well illustrated by the way in which the Wallace Line (and related biogeographic boundaries) cuts across the otherwise convenient-looking archipelago of Indonesia/New Guinea (Calaby 1972; Fig. 14.1). It is revealing to imagine how, if we were encountering the organisms of this region at just a few localities and for the first time, we would attempt to generalize their distributions. This point applies equally to generalizing distributions by reference to palaeogeographic maps, though the risks are even greater because these maps are only hypothetical.

The accreted terranes of the Pacific margins of the Americas and of New Zealand (Hallam, 1986) similarly illustrate the fallacy of assuming that biogeographical units and apparent physico-geographical entities

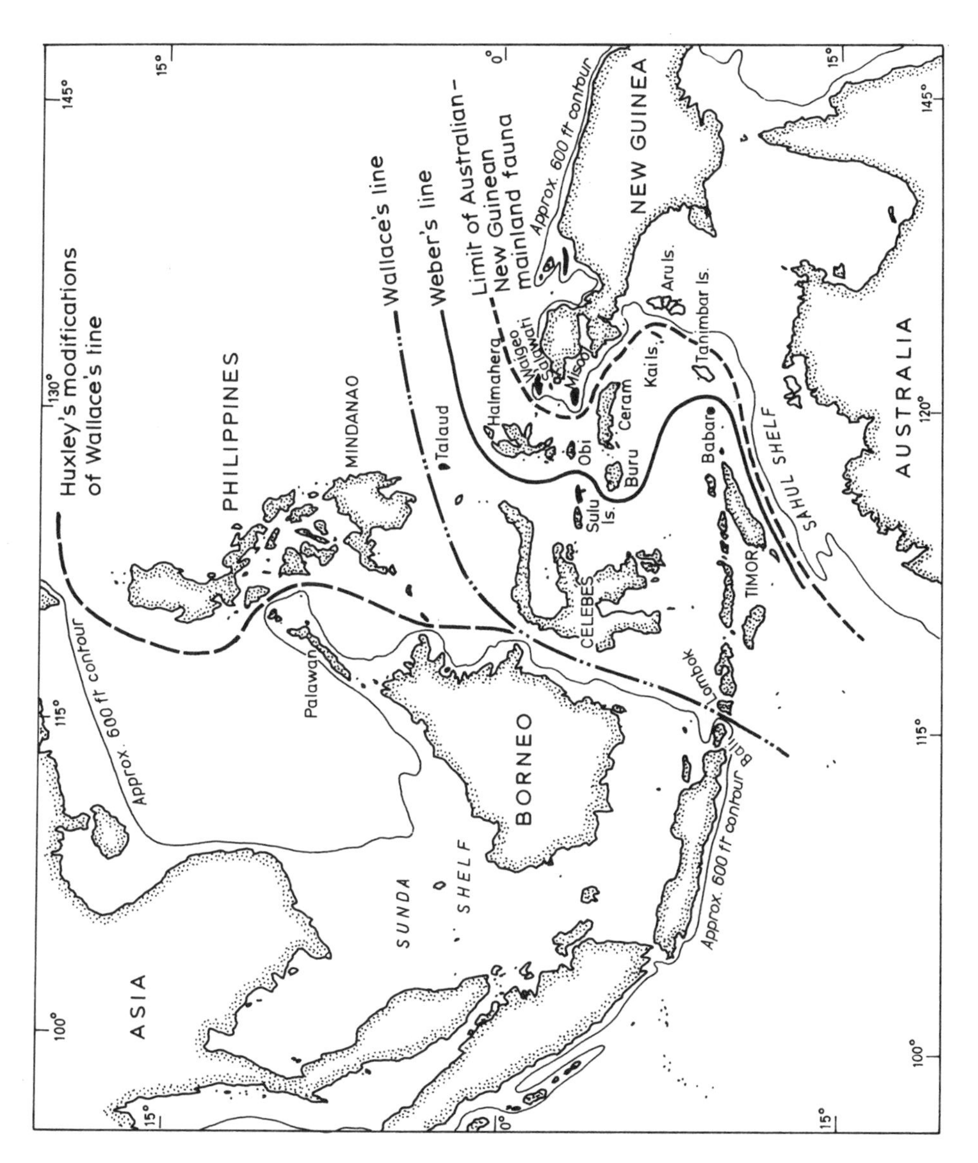

Figure 14.1 Wallace's line and associated biogeographical boundaries in the Indonesian archipelago (from Calaby, 1972). Note how these lines cut across more 'convenient' political boundaries and land mass groupings. They show a much better relationship however with submarine topography.

correspond. Even when a biogeographical method is otherwise rigorous, its results may be affected by locality generalization. Some of the earlier exercises in cladistic biogeography, for example, appear to have encountered this problem. With respect to the problem of South America in Patterson's (1981a) marsupial analysis, Parenti (1981) suggested that this continent may consist not of one region but two: "There is no reason to keep geographical boundaries just because they are familiar, when the objective is the comparison of biotas". Platnick and Nelson (1984) have since argued that cladistic biogeography can recognize composite areas, and does not have an inherent failure to do so.

Nevertheless, some distributional generalization is unavoidable or even desirable, such as when there are very few taxa amongst a few localities in a large region, or where inspection of raw data suggests that several localities share the same biota. This serves to simplify unwieldy data sets, and in some methods like PAE (Section 14.3.3) reduces the otherwise high probability of numerous slightly different alternative results. Geographical resolution in sparse areas can usually be improved by including another group of organisms that is more diverse in these regions. Apart from this, the level of geographical resolution should be kept as fine as is practically possible. It is of course acceptable to adopt a clearly defined set of geographical or geological units for the sake of a particular exercise or for comparison with an independent hypothesis (Rosen and Smith, 1988). In presenting results of a study, it is useful to include relevant comments on locality generalization.

14.2.3 Stratigraphical constraints

The use of fossils raises the question of stratigraphical resolution. In some cases, the same fossils that are used for biogeographical purposes also provide an age for the strata that enclose them. There are of course other non-palaeontological ways of dating rocks.

For synoptic methods in particular it is important to establish that different localities are as closely contemporaneous (same geological age) as possible because the aim is to envisage how the earth looked at particular 'instants' in the past. Clearly, a reconstruction based unwittingly on sampling from rocks of widely different ages is of much less value than a well controlled sample set. On the other hand, some spread of ages is inevitable, although in practice the magnitude of this spread may be difficult to determine because it may be finer than the level of resolution of the best available dating method. Thus a synoptic distribution is actually based on a set of sample points that lies within a hypothetical time slice bounded by two time planes. It is obviously important to ensure that the time-span between these planes is as short as possible for assembling a working data set. The span is not only important in its own right but also provides an estimate of minimum error range for the ages of particular sample localities (this second point being applicable to all fossil data, whether for synoptic or any other purpose). This is because the same factor, maximum stratigraphical resolution, determines both. What is the working minimum for time-spans, and what are the implications of this minimum?

Since fossils are still the predominant means of dating rocks, stratigraphical resolution is usually unlikely to be finer than a biostratigraphical zone. Raup and Stanley (1978) suggest that the approximate accuracy of stratigraphical zonation (based on longevity considerations) is 0.5–5 Myr for species. In the time-scales published by Harland *et al.* (1982), the shortest zone durations include those based on graptolites in the Silurian (Llandovery: *c.* 0.7 Myr), and those based on planktonic foraminifera in the Middle Miocene (*c.* 0.5 Myr). In general, within-basin correlation is sharper in resolution than between-basin correlation (J. B. Richardson, pers. comm.). On a global scale, it may be necessary to think in terms of stages rather than zones, because, according to Harland *et al.*, stages are "essentially the smallest globally recognizable stratigraphic subdivisions." They give figures of about 6 Myr for the average duration of a stage. In the context of geological reconstruction from fossils, this figure must be set against the fact that a portion of the earth's crust can move horizontally tens or even hundreds of kilometres in this time (depending on rates of sea floor spreading). Also, in the time-span of a stratigraphical zone, say 1 Myr, geosynclinal subsidence (or the vertical component of subduction) of 30–200 m, mountain uplift of 500–13 000 m, and strike-slip movement (e.g. along the San Andreas fault in California) of up to 45 km can take place (data derived from Bloom, 1978). These figures give an indication of the minimum possible scales of spatial error due to biostratigraphical resolution in making geological reconstructions from fossil data. Dating by physico-chemical methods however often provides tighter control, especially for the Quaternary (e.g. CLIMAP, 1984).

14.2.4 Taxonomic and phylogenetic considerations

Synonymy is a widespread palaeobiogeographical hazard, especially because it has been so common for fossil systematists to generate synonyms by assuming that the taxa of their own country, region, tectonic unit, or preconceived biogeographic area, are unique ('chauvinotypy'). In the past, this has even been taken to the absurd extreme of using 500 miles (800 km) between areas of fossil collecting as a criterion for naming new species (Ross, 1976b).

Methods that are based on originations also require a sound phylogenetic foundation and, ideally, the taxa used in all biogeographic methods should be monophyletic. To judge from Patterson and Smith's (1987) critique of taxa recently used in statistical analyses of extinctions and diversity, there is little room for complacency; they found that in their particular sample of the fossil record the proportion of monophyletic taxa relative to the total was only 25%.

A problem that arises in the special case of comparing living and fossil distributions is that of unresolved synonymies between living and fossil counterparts. This problem obviously becomes greater as one moves up the geological column. Most living species must have been extant for at least 0.5 Myr (Section 14.2.3). It is also unlikely that fossil taxa can ever be as finely differentiated as their living counterparts.

14.2.5 Ambiguity of absence

Assuming that there is no reason to suspect the taxonomy, the presence of a taxon in a locality list is much less ambiguous than absence, so methods that emphasize presence should be more reliable. For fossil sampling, the possible reasons for absence include all those that might be invoked to explain absence of an organism on the present earth (i.e. with respect to a sampled horizon in the past): poor sampling, poor taxonomy, contemporaneous catastrophe, interim and prevailing ecological conditions and barriers, together with any historical biogeographical (antecedent) factors like extinctions that took place prior to the sampled horizon. But there are also factors that relate only to fossils: absence of rocks of the right age, absence of suitable deposits, loss of fossils through physico-chemical factors like solutioning, diagenesis and metamorphism, together with inadequate or outright failure of preservation. Problems of fossil taxonomy may also underlie apparent absence.

14.2.6 Fact, hypothesis, test or comparison?

The methods discussed below have all yielded patterns that are biotic. They can be interpreted as biotic hypotheses (ecological, historical) and also as hypotheses of physical geology. Unfortunately, our knowledge of the kinds of processes in the biogeographic system that might link earth history to biogeography is still too conjectural, and biotas can only provide absolute constraints or tests for earth history in limited instances (Fortey and Cocks, 1986; and see also Patterson's (1983) suggested biotic test for the idea of an expanding earth). It is more usual to find that one can generate several geological hypotheses from basically the same biotic pattern (e.g. compare Williams, 1969, 1973). Nevertheless, even if geological hypotheses inferred from biotic patterns are refuted by independent geological findings, the biotic patterns and hypotheses remain valid in themselves (Patterson, 1983).

The search for falisifiability, however, does not preclude the usefulness of making careful comparisons between patterns or hypotheses derived from different kinds of evidence. Such comparisons are still preferable to *ad hoc* scenarios, and notable inconsistencies can stimulate further enquiry and improvement of methods. Discovery of repeated patterns can always be construed as significant on an inductive basis.

Whether one believes in absolute tests, or the merit of cumulative inductive evidence, there is always the question of how to bring the biotic and geological sides together. In cladistic biogeography, it has become customary to compare area cladograms derived from biotic data with 'cladograms' derived from geological evidence. These are branching diagrams that represent the history of fragmentation (or sometimes coalescence) of geographical areas through geological time (Young, 1986). They are certainly a convenient graphical way of summarizing geological history and facilitating comparison with biotic cladograms, but they are not true cladograms because they are not derived from explicit analysis or geographical or geological characters in the way that synapomorphies are analysed in systematics. To date, only PAE, discussed below, offers a method of deriving area cladograms that is independent of phylogenetic methods.

14.3 METHODS BASED ON DISTRIBUTIONAL CHANGE

14.3.1 Background

In methods based on distributional change, the underlying assumption is that areas of the earth, now and in the past, can be 'fingerprinted' by the taxa found there, and that these fingerprints are in effect geological features that characterize particular portions of the earth's crust (Crovello, 1981). This biotic fingerprinting can take the form of the presence of the taxa alone, which can be analysed statistically or the taxa can be analysed according to their palaeoenvironmental indications. Although in theory the fingerprint of a region might include all the organisms found there, both living and fossil, in practice it is usual to separate out the patterns of different stratigraphical ages (including the present). All such methods yield patterns that are used to generate palaeogeographical hypotheses.

Use of presence alone is an empirical approach in that there is no need to assume any processes at the outset other than the idea that organisms may change their distributions through geological time, but this is offset at the interpretative stage by the need to introduce the assumption of particular processes, and of attributes of the fossils, in order to relate distributions to earth history. Use of palaeoenvironmental indications on the other hand requires numerous prior ecological assumptions.

14.3.2 Statistical assemblages

Background

Until recently, the dominant empirical methods of identifying biogeographical fingerprints have utilized statistics to group the occurrences of different taxa in different places into assemblages. Strictly, discrete assemblages are a supposition (Durden, 1974), since it is possible that geographical distributions do not occur as entities at all. An alternative model of distributions is that they may be continuously intergradational as in the case of latitudinal gradients (Chapter 3) or they may even be random. (Compare this with the analogous problem of communities in ecology (Whittaker, 1970; Begon *et al.*, 1986)). It is theoretically more sound to proceed by determining if biogeographical assemblages actually exist before deciding what implications they may have for earth history, though there is a risk that the methods one uses may artificially introduce more order than actually exists.

The intuitive origin of the idea of assemblages is probably based on general observations such as the obvious difference between the mammal faunas of Australia and Eurasia, or between the marine biotas of the Atlantic and Indopacific. With respect to living biotas, pioneers of this concept include Sclater, Wallace, and de Candolle (Nelson and Platnick 1981), and for marine organisms, Ekman (1935, 1953). In parallel with neobiogeographers, palaeontologists have for a long time also grouped their fossil biotas into assemblages and biotic regions, early pioneers of which include Lebedeva (1902) and Schuchert (1903).

The broad correlation between present day biotic regions and major physical features of the earth suggests that, in the past, biotic regions probably also reflected ancient geographical and geological features. Ancient biotic regions should therefore make interesting comparisons with independent palaeogeographical hypotheses. The use of statistically defined assemblages for this purpose however is relatively recent, since geologists have generally based their reconstructions more on the palaeoenvironmental evidence of fossils (geoecological assemblages) (Ross, 1976a). Whittington and Hughes (1972), who carried out one of the earliest statistical investigations, trace the use of formal assemblage methods only as far back as 1960. Even now that statistical methods are more widespread, it is still rare for them to be set out independently of other palaeogeographical evidence, possibly, for reasons discussed below, because their implications are often ambiguous.

Methods and patterns

Assemblages can be identified subjectively by assessing how far the biota at one locality corresponds to that at another, and grouping the localities accordingly. The idea of assemblages, however, obviously lends itself to formal statistical treatment. Many methods are based on measures of absolute or percentage similarity or dissimilarity of the biotas at different sample localities, usually in pairwise comparison. Summaries of the formulae used and relevant discussion can be found in Cheetham and Hazel (1969), Sneath and McKenzie (1973), Pielou (1979b), Green (1980), Crovello (1981) and McCoy and Heck (1987). It is invariably the case that some assemblages are closer in taxonomic composition to each other than to others, and statistically this can be expressed in terms of cluster hierarchies or relative biotic distances. Resulting patterns are expressed in numerous ways, not always independently of base maps, the most common being cluster diagrams, dendrograms and graphical representations of biotic distance. For a range of examples, see Williams (1969), Stehli and Wells (1971), Whittington and Hughes (1972), Sneath and McKenzie (1973), Coryndon and Savage (1973), Oliver and Pedder (1979), Savage *et al.* (1979) and Crick (1980). Two examples from this literature, both based on corals, are reproduced in Fig. 14.2.

Figure 14.2 Two examples of statistical analyses of biogeographical assemblages, both using corals. (a) Application of statistical analysis of biotas to palaeogeography. Analysis of Late Emsian (early Devonian) rugose coral genera (upper) compared with an independently derived palaeogeographical reconstruction for the same horizon (lower) (from Oliver and Pedder, 1979). For details of the particular tests intended by the analysis, see authors' original account. Squares enclose number of genera known from each area and the percentages that are endemic to their half of North America. Circles enclose the number of genera that occur in both of the two areas connected by the lines, together with the Otsuka similarity coefficient for the two areas. Shaded areas – land; unshaded areas – seas; circles – "Eastern American Faunas"; squares – "Old World Faunas"; solid symbols – major sample points; open symbols – minor sample points (one or a few genera). (b) Cluster analysis of Recent hermatypic coral general by Stehli and Wells (1971) in Q mode, based on Jaccard's coefficient and unweighted pair groups for generic assemblages. This shows the principle of cluster analysis, although, being Recent, this example is not directly relevant to palaeogeographical reconstruction. Compare the dendrogram with the upper left cladogram in Fig. 14.3 which is based on similar data for some of the same localities. For all other data, see Stehli and Wells' text. A world map showing the distribution of sample points is given on p. 461.

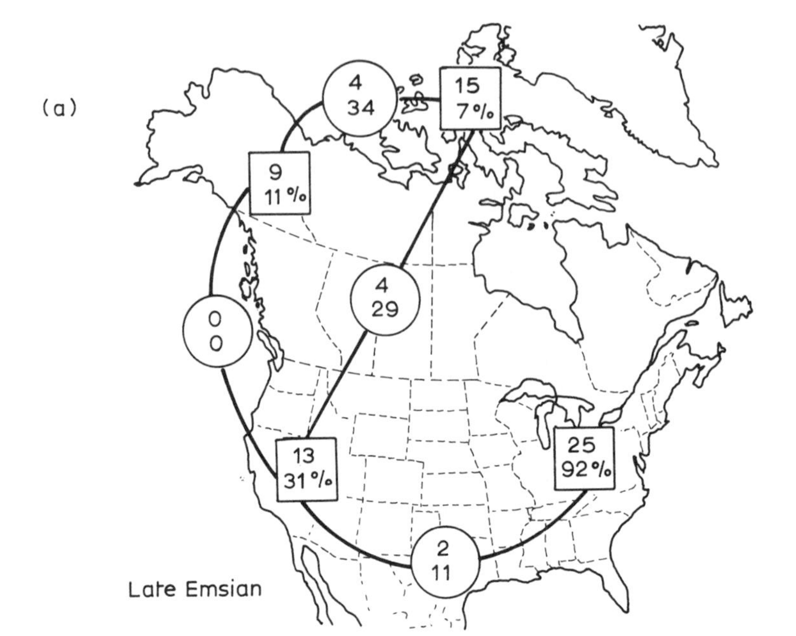

(a)
4
34
15
7%
9
11%
4
29
0
0
13
31%
25
92%
2
11
Late Emsian

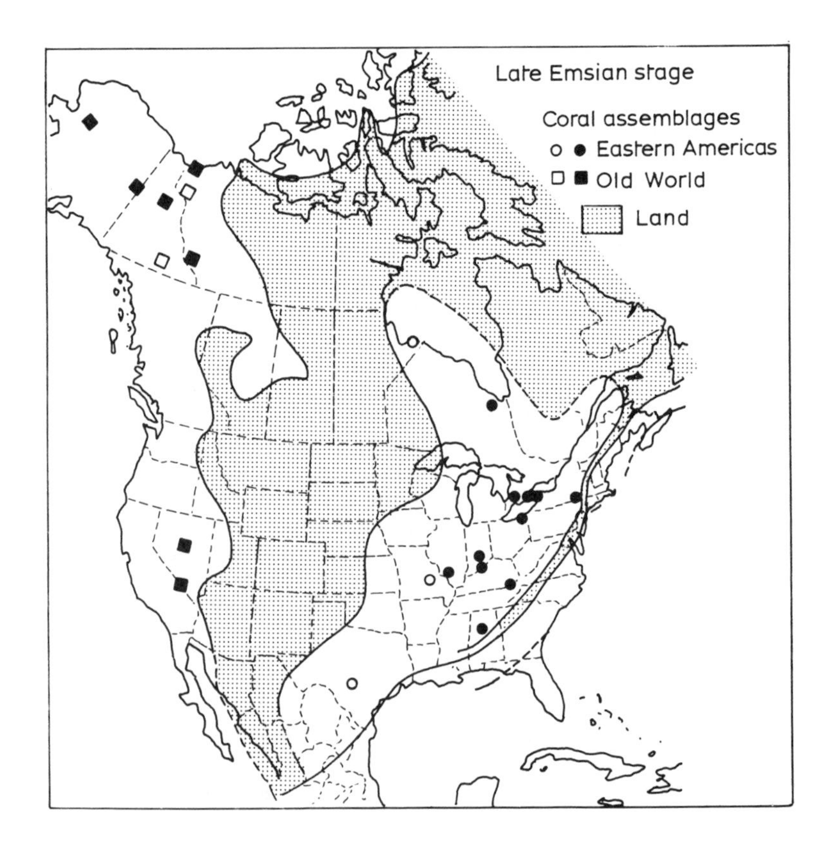

Late Emsian stage
Coral assemblages
Eastern Americas
Old World
Land

(b)

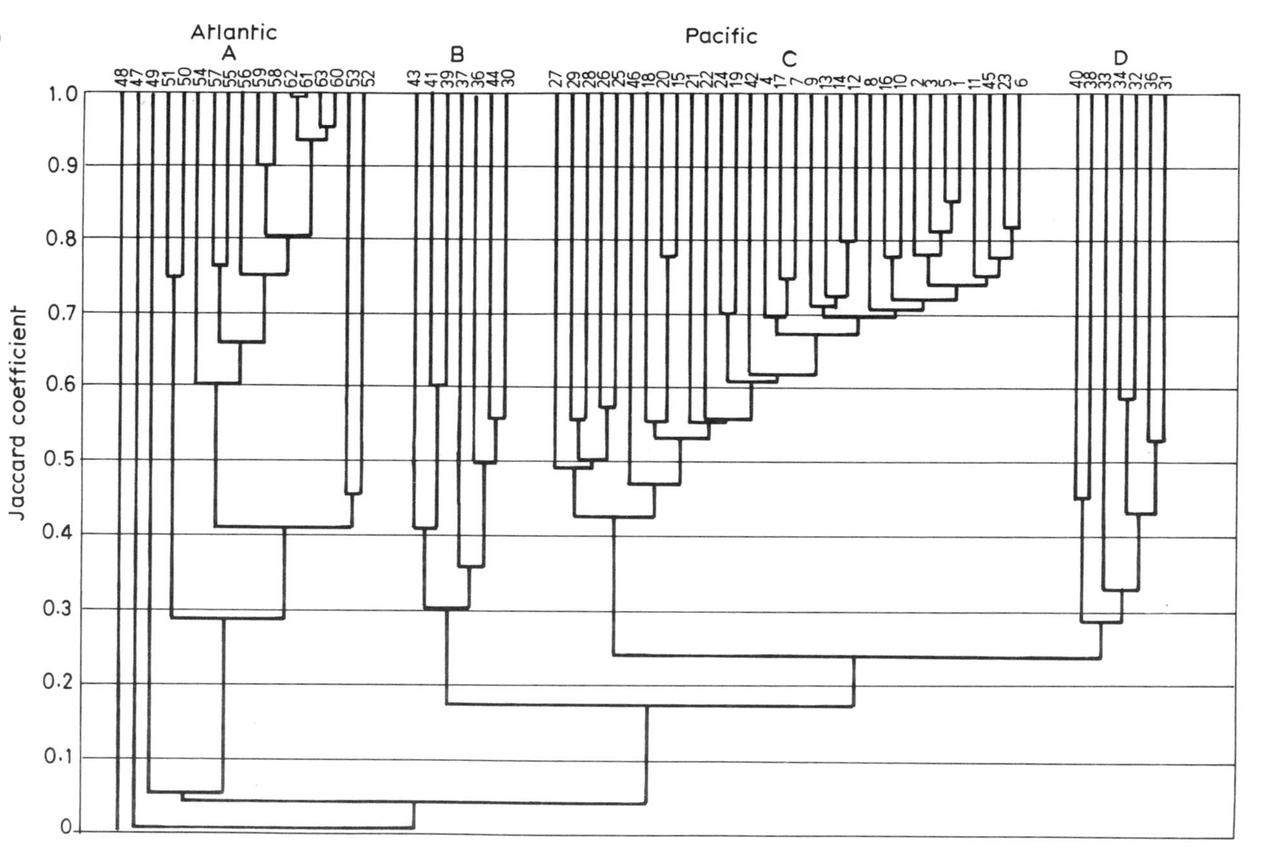
Atlantic
A
B
Pacific
C
D
Jaccard coefficient
1.0
0.9
0.8
0.7
0.6
0.5
0.4
0.3
0.2
0.1
0
48 47 49 51 50 54 57 55 56 59 58 62 61 63 60 53 52
43 41 39 37 36 44 30
27 29 28 26 25 46 18 20 15 21 22 24 19 42 4 17 7 9 13 14 12 8 16 10 2 3 5 1 11 45 23 6
40 38 33 34 32 36 31

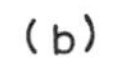

Closely related to the concept of assemblages is the quantification of endemicity – the number of taxa that are unique to an area, a complementary concept to similarity. This is usually expressed in absolute totals or as percentages of the whole biota of an area (Chapter 5; Oliver and Pedder, 1979; Fig. 14.2a). The patterns generated by this method can be interpreted to suggest relative degrees of isolation of different areas, but such figures on their own do not give any indication of relationships between different areas.

The concept of assemblages underlies range or span methods, though range analysis does not necessarily assume that biotas exist as entities. These methods analyse biogeographical data along a line transect of sample localities. For examples of range methods applied to Recent organisms see Valentine (1966) and Rosen (1984). Pielou (1978, 1979b) and Rosen (1981) projected distributions from numerous localities on to an abstract transect, tied only to latitude. The application of these methods to fossil distributions is limited however because it cannot be assumed that a locality set that is approximately linear on a modern map was linear in the past (Section 14.2.2). Nevertheless, there may be particular circumstances which make it acceptable to use fossil data along a transect (e.g. Smith and Xu Juntao, 1988).

There is no agreement about sample size in statistical methods. At one extreme, regional differences between just a few members of a single taxon are used, and at the other, whole biotas are processed using computers. The question of sample size is discussed later.

Static interpretation

The principles used in making palaeogeographical reconstructions from assemblages do not appear to have been set out by any particular author. I have summarized what seems to be the consensus of simple 'rules' for interpreting assemblages in Table 14.3. Four problems arising from these 'rules' can be highlighted:

(1) Though these methods are based on pure presence/absence assemblages, even a most general palaeogeographical interpretation is impossible without also taking into account the fundamentally different palaeogeographical implications of assemblages based on marine and terrestrial organisms.
(2) The rules are only clear if assemblages are either 100% the same, or 100% different. In practice, this is rarely the case, so how does one interpret different levels of similarity (or, equally, clustering levels and gradational measures like biotic distance)?
(3) Where different marine assemblages are found in different places,

Table 14.3. 'Rules' by which statistical assemblages have been interpreted palaeogeographically. Two conditions are shown in the top row for two main kinds of biota in the left column. The matrix of the table shows the five possible interpretations

	Sample localities share the same assemblage	*Sample localities have different assemblages*
Land biotas	Sample localities occurred on the same or connected land masses	Sample localities occurred on different land masses separated by seaways or oceans
Marine biotas	Sample localities occurred in the same or interconnected marine basins or oceans	Sample localities occurred in different marine basins or oceans separated by: *either* land masses *or* great oceanic distance

there are two contrasting interpretations, as shown in Table 14.3.

(4) There are no rules for inferring actual, or even relative, palaeogeographical distances between sample localities, except that if we assume oceanic separation of marine shelf forms, it is usual to invoke 'great' distance. This severely limits the scope offered by assemblages for actually constructing maps (Sneath and McKenzie, 1973).

Dynamic interpretation

The usual way of making dynamic inferences from assemblages is to use a synoptic approach. Static palaeogeographical reconstructions based on interpretation of assemblages from different geological horizons according to the above 'rules' are compared with each other or with present day geography. For example, if assemblages in two sample regions show diminishing similarity through successively younger geological horizons, this can be construed as the breakdown of a pre-existing biogeographical barrier: McKerrow and Cocks (1976) used such a pattern to infer the closure of the Iapetus Ocean through the Lower Palaeozoic. Such an event is convergence, i.e. intermixing or hybridization of biotas (Vermeij 1978; Hallam, 1983), and the converse is divergence of biotas.

Since synoptic methods depend on a prior set of static interpretations, they are subject to whatever errors or problems are inherent in static interpretations. A possible solution to circumvent problems of static interpretation might be to translate relative compositional overlap between assemblages into historical hypotheses. Relative similarity or biotic closeness for example might be taken to be inversely proportional to the age of separation of different biotas on the same time plane, but this is untenable on the grounds that this assumes equal evolutionary rates in all the taxa that make up the different assemblages (C.J. Humphries, pers. comm.). McCoy and Heck (1987) have also pointed out a simple fallacy in making historical inferences directly from similarity measures. Nevertheless, it is generally acknowledged that areas of great biotic dissimilarity have probably been biogeographically distinct for a long period of time, and there is therefore no simple way of distinguishing antecedent (historical) factors from contemporaneous ecological factors. This problem recurs in the interpretation of most other kinds of biogeographical pattern.

14.3.3 Parsimony analysis of endemicity (PAE)

Background

Biogeographers are universally agreed that endemicity (the exclusive occurrence of a taxon in a particular locality or region) is one of the most significant features of biogeographical distribution. In the simplest instance every region would have its own exclusive biota without any overlap with other regions but, in practice, this is rarely the case. It is more usual for some taxa within an area to be unique to it but for the rest to occur elsewhere too. The concept of biotic regions was founded on the delineation of areas according to patterns of occurrence of endemic biotas, but there are invariably intermediate zones and alternative possible schemes, so that decisions about boundaries have usually been arbitrary or generalized. Rather than trying to define such regions at the outset, it is better to work entirely with sample localities (as with the statistical approach to assemblages). This makes it necessary however to distinguish endemicity of an area as it may really exist for a given taxon, and its endemicity as 'seen' through the windows of a sample set of localities within that area. (With extensive sampling, the difference should become negligible.) Note that endemicity of sample localities is relative; a taxon that is shared between two localities but occurs nowhere else is endemic to both of them taken together, but endemic to neither when each locality is considered separately.

This recalls Hennig's (1966) three categories of character distribution in a group of taxa: a character may be unique to one of them, shared by two or more but not all of them, or universal. By this analogy, sample localities can be thought of as 'taxa'– or, more strictly, specimens of taxa, because prior assumptions about the extent of areas should be avoided as far as possible. Subsequent analysis should then be used to work out how sample localities (cf. specimens) might be grouped into areas (cf. taxa). This led Rosen (1984) to suggest that, instead of the more usual phenetic treatments, parsimony analysis as in cladistics might be applied to the distribution of taxa. For this purpose, taxa that are endemic to some but not all of a set of sample localities would, in effect, be geographical/geological synapomorphies, analogous to synapomorphies in a set of taxonomic samples. The method was first attempted by B. Rosen (1985) and realized more fully by Rosen and Smith (1988).

This analogy between systematics and biogeography had already been drawn by implication (Section 14.3). Crovello (1981) has actually formalized this by suggesting that taxa may be analysed as characters of 'OGUs' (Operative Geographic Units), while Young (1986) similarly refers to the idea of areas as 'taxa' that can be recognized by their geological characters.

It is important to recognize the differences between PAE and cladistic biogeography (Chapter 12):

(1) Though they both use the criterion of parsimony, cladistic biogeography applies it to the synapomorphies amongst different taxa in order to obtain relationships between the taxa (phylogenies), and area cladograms are derived from superimposing geographical distributions on these phylogenies. In PAE, parsimony analysis is applied to the 'synapomorphic' (i.e. shared) taxa of different sample localities in order to obtain relationships between the biotas as sampled at these localities. PAE therefore produces area cladograms of sample localities directly from geographical distributions. Cladistic biogeography uses taxonomic characters to fingerprint areas, and PAE uses whole taxa.
(2) Cladistic biogeography can use related taxa from any geological horizon, but PAE currently uses sets of contemporaneous taxa. Until the implications of using taxa as analogues of characters in systematics are worked out, the meaning and scope of using taxa from different horizons within a single sample locality in PAE will not be known.

Like cladistic biogeography however, PAE generates historical (hence geological) hypotheses even from modern distributions, but PAE is still experimental and the theoretical basis for historical inference has yet to be developed satisfactorily.

Methods and patterns

As with phenetic analyses of biogeographical assemblages, the starting point is a data matrix of the occurrence (presence/absence) of taxa with respect to a set of sample localities of the same age. Presence of a taxon is regarded as a derived or 'advanced' character state within a sample locality, and absence as 'primitive'. No attempts have been made so far to use more character states than these or to apply character weighting (e.g. on grounds of abundance).

It is important to minimize at the outset the amount of geographical generalization made from the taxonomic lists of sample localities in order to maximize biogeographical resolution. Nevertheless, a raw data matrix can be modified before analysis according to the following guidelines:

(1) Any taxon common to the whole sample localities set (cf. plesiomorphy), and any that occurs in one locality only (cf. autapomorphy) is omitted as uninformative for finding relationships between the localities.
(2) Localities with very few taxa in relation to most of the other localities should be omitted, or in some cases may be merged. Low diversity is invariably interpreted as 'primitive' by parsimony analysis if Lundberg rooting is used (see below), and such localities can also give rise to numerous equally parsimonious results. There is no rigid guideline on how many taxa constitutes 'very few', but if in doubt, a preliminary analysis usually serves to highlight the problematic localities. However, it is only useful to merge low diversity localities if the merged diversity is then significantly closer to that of other sample localities. At the same time, the accompanying loss of geographical resolution should be justified by merging only those localities that show considerable taxonomic overlap.
(3) Localities with an identical set of taxa should be merged. For the purposes of analysis they constitute a single region. In addition, and especially with large data sets, mergers of localities that are only slightly different taxonomically, and for which there are good grounds for believing that they were neighbours at the time of the sample horizon, might also be reasonable. Sometimes a preliminary analysis points out possible mergers of localities by offering several equally parsimonious solutions amongst the same subset of localities.

In practice, there are no rigid rules about modifying the data set apart from (1) above, but an original data set often has to be modified until a practicable working number of equally parsimonious solutions is obtained. Although this may transgress considerations of objectivity, any detrimental effect on the results should eventually be eliminated by further work. Ideally, all such modifications should be noted in the presentation of results.

The actual methods of analysing sample locality relationships are the same as those used by cladists for analysing taxonomic relationships, and are summarized in Chapter 12. So far however, only the computer program PAUP in several versions (e.g. Swofford, 1984) has actually been used for PAE, but there is scope for trying other methods.

By analogy with systematics, an outgroup locality must also be introduced into the analysis. In using PAUP for systematics, this is conceived as a real or hypothetical taxon that the investigator regards as primitive. In PAE, we therefore have to find a 'primitive area'. Rosen and Smith (1988) decided to use a hypothetical locality that has no taxa at all (Lundberg rooting) because there appears to be no satisfactory way of defining a real 'primitive area'.

Results of PAE consist of nested or hierarchical sets of biotas as represented at the sample localities. They are effectively sample locality cladograms based on the most parsimonious arrangement of the taxa shared by the different localities. Published results of PAE consist of analyses of four data sets for sea urchins (Late Cretaceous, Eocene, Miocene and Recent) and two of reef corals (Miocene and Recent) (Rosen and Smith, 1988; Fig. 14.3). Rosen (1988) also has preliminary results for Recent mangroves and the reef-associated green alga *Halimeda* and Smith and Xu Juntao (1988) have analysed Carboniferous to Triassic data for brachiopods, corals, fusulinid foraminifera, plants and ammonoid cephalopods from Tibet and adjacent areas.

Parsimony analysis of endemicity cladograms based on different data sets can be compared for congruence (Fig. 14.3), and congruent elements can also be sought between cladograms based on different stratigraphical horizons.

Static interpretation

PAE cladograms can be regarded as an alternative to phenetic methods for classifying biotas. However, PAE 'biotas' are taxonomically incomplete in the sense that cosmopolitan (regionally symplesiomorphic) taxa, and taxa found in just one sample locality (regionally autapomorphic taxa) are omitted prior to analysis. Moreover the resulting patterns of endemicity are specific to the actual sample locality set, and some taxa may well occur at places not sampled for the analysis. At least part of a PAE cladogram probably reflects geoecological factors such as barriers,

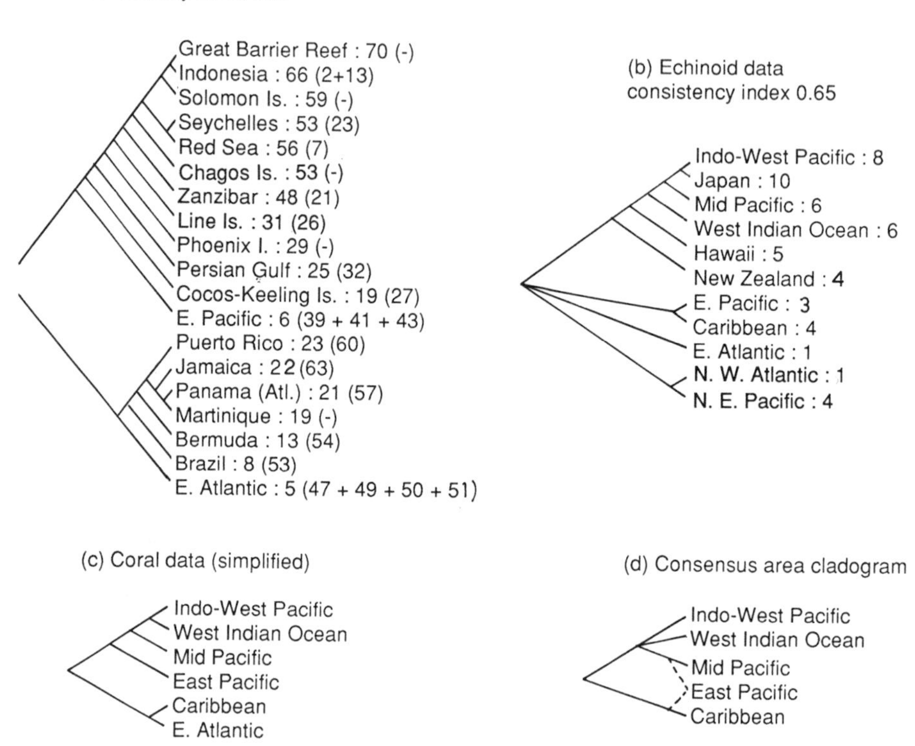

Figure 14.3 Sample locality cladograms derived from PAE for Recent reef coral and clypeasteroid echinoid genera (from Rosen and Smith, 1988). The consensus area cladogram shows only those elements common to both the initial cladograms. Numbers following the locality names refer to the number of genera present that are also found in at least one other sample area. Numbers in brackets in cladogram (a) show the corresponding locality numbers for Stehli and Wells (1971) dendrogram (Fig. 14.2(b)). The world map opposite shows distribution of sample points for which reliable data were available. Solid symbols – Indopacific sample points; open symbols – Atlantic sample points; cross symbols – one genus only.

If Stehli and Wells' dendrogram is simplified to include only these localities, the resulting patterns are found to be very similar, even though they are generated by different methods. They differ mainly in the position of Persian Gulf (32) which here is part of the Indopacific set, but in Stehli and Wells' dendrogram stands out as 'sister' to all those other localities used in both analyses.

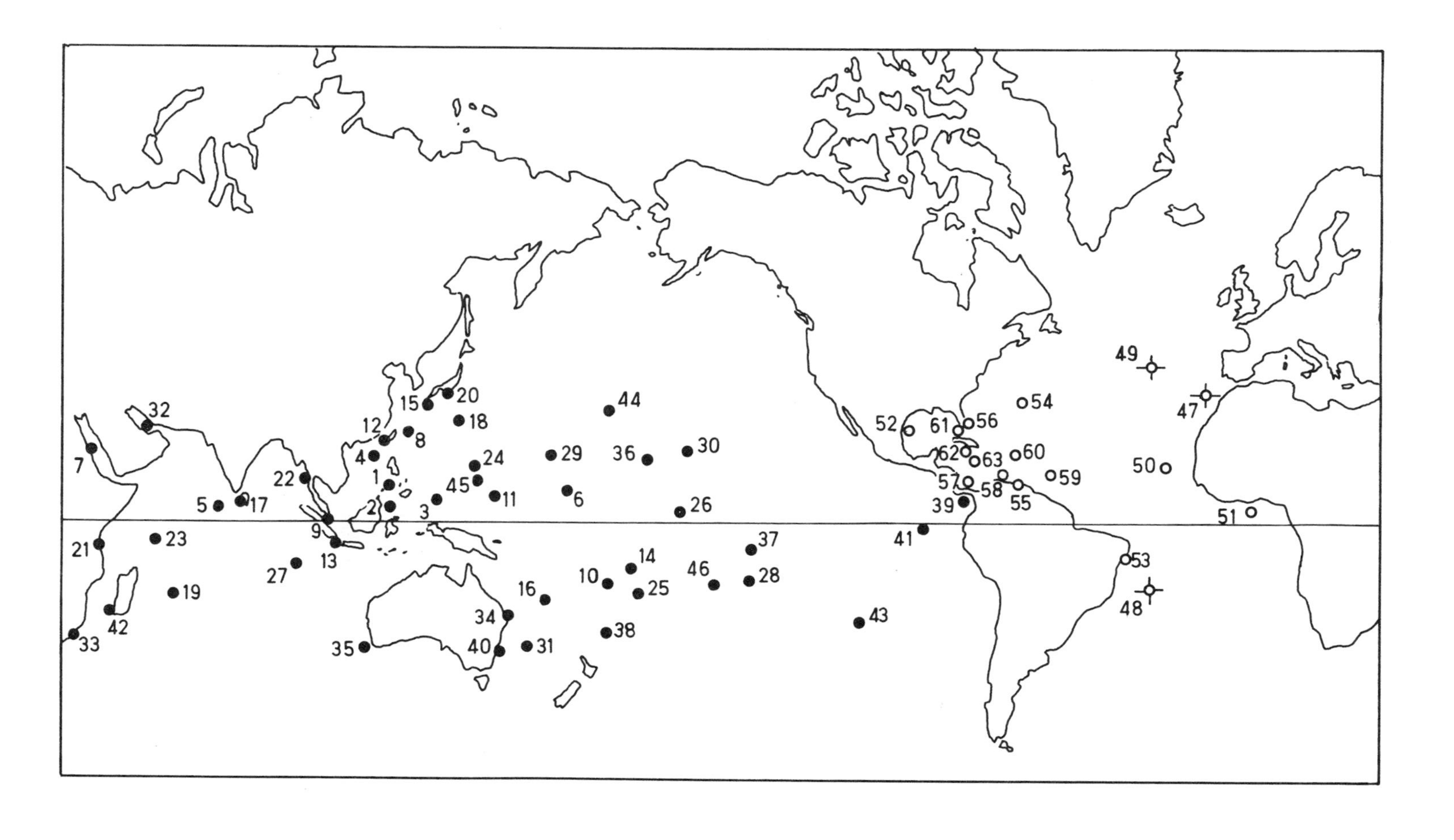

32
7
5
17
22
21
23
19
42
33
12
15
20
18
8
4
1
2
3
45
24
11
9
13
27
44
29
36
30
6
26
35
34
40
31
16
10
14
25
38
46
37
28
43
41
52
61
56
54
62
63
60
57
58
59
55
39
49
47
50
51
53
48

and prevailing ecological conditions. It is not easy to separate such factors from the effects of antecedent events.

Dynamic interpretation

Subject to the following reservations, PAE also provides historical hypotheses. 'Historical' here refers to hypotheses about the history of the analysed biotas previous to the time of the sample horizon, and the secondary hypotheses inferred from these biotic histories about geological events.

At present, the historical value of PAE cladograms is particularly dependent upon how one interprets Lundberg rooting. A Lundberg root locality can be construed either ecologically or historically as: (1) one whose ecological conditions are so unfavourable to all the taxa in the sample locality set that they cannot survive there, (2) one that occurs so far back in geological time relative to the sample horizon that none of the taxa in the sample set had yet evolved.

According to the first interpretation, a PAE cladogram would represent an ecological scheme of relative favourability of the environments for the sampled taxa, localities and horizon, i.e. a hypothesis of contemporaneous ecological conditions. Sister localities would then be those that were most closely related ecologically. According to interpretation (2), PAE cladograms represent schemes of successive emergences of biotas – that is, relative recency of biotic interchange between sister localities or, conversely, historical sequences of biotic divergences and isolations.

It is therefore not yet possible to distinguish how far a single PAE pattern is ecological or historical. A combination of two approaches can be applied to this question. In the first, we can adopt the cladistic biogeography argument that the more a particular area pattern recurs across numerous groups of organisms (congruence, or induction; Chapter 2), the more likely it is to be historical and due to major geological events. This would be further supported by congruent elements between PAE cladograms of different ages. I doubt if all ecological effects would really be eliminated by congruence however, because, as Stehli (1968) and others have shown, ecological patterns like latitudinal diversity gradients can also be recurrent in different groups of organisms through geological time. The second approach is to regard all cladograms simply as hypotheses about the histories of the sample localities, and to see how far these histories correspond with area histories derived from independent geological evidence.

It follows that for a static interpretation of a PAE cladogram, one would need to be able to distinguish the ecological from the historical component in the pattern. This might be possible by elimination of the historical factors inferred on the criterion of generality. Otherwise one would have to start with the hypothesis that it represented an entirely ecological pattern, i.e. relative favourability of contemporaneous environments. On its own however this is not palaeogeographically informative unless one introduces further assumptions about the kinds of geoecological conditions that might be unfavourable, and hence palaeoecological assumptions about the sample organisms. For reef corals, for instance, unfavourable conditions include deep water, high latitude, low temperature, or high primary productivity (Rosen 1977, 1981, 1984). With these reservations, such geoecological interpretations of PAE cladograms might still be useful in comparison with independent palaeogeographical reconstructions.

Although one need not confine geographical parsimony analysis to a single taxonomic group, there is scope for applying analyses to ecologically defined groups of organisms. Admittedly, for fossils this introduces prior palaeoecological assumptions, but it has the attraction of generating hypotheses with more palaeogeographical detail, for example by suggesting the relative position of possible biogeographical barriers and major habitats, and changes of such features through geological time. Note however that PAE cannot directly indicate whether the relative or absolute positions of sample localities differed from the present day. Rosen and Smith (1988) have also suggested that ecological gradients probably generate a pectinated pattern in PAE cladograms (Fig. 14.3 Chagos Is. to E. Pacific inclusive) and therefore sets of such localities should probably be regarded as single historical units.

14.3.4 Geoecological assemblages

Background

Long before palaeontologists turned to numerical analyses of biotas as a palaeogeographical tool, geologists had frequently used fossils to indicate the depositional environment of the rocks in which their fossils occurred. In fact, as most histories of geology point out, the fact that fossils can be found today in places that cannot possibly have been their original habitat (like marine molluscs high up in mountainous terrains) is probably one of the most ancient and fundamental of all recorded geological observations. Provided palaeogeographers have plausible notions about the environment that a particular fossil once inhabited, they only have to overlay a particular set of contemporaneous formations on a geological map with their palaeoenvironmental inferences to obtain an initial hypothesis about the ancient geographical features of that time and region. The basis of the geoecological approach is therefore simple in its logic, and, depending on the amount of detail introduced, it is usually simple to apply – a great deal simpler in fact than producing the prior geological base map and stratigraphy.

In this section, I concentrate on the palaeoenvironmental interpretation of biotas, but there are also numerous important specialized methods of inferring climates and palaeoenvironments from fossils, such as those based on oxygen isotopes and fossil dendrochronology, that I have not covered. All palaeoenvironmental methods can provide a wealth of reconstruction detail, but by definition they ignore the possibility of a direct influence of evolutionary and antecedent geological events in shaping observed distribution patterns.

Palaeoenvironmental interpretations of fossils have an actualistic basis in assuming that the ecology of a particular fossil was the same as its modern counterpart. These methods therefore depend on the current state of knowledge of the ecology and ecological biogeography of living organisms. Advances in marine larval ecology (Jablonski and Lutz, 1983) for example have stimulated new ideas about fossil distributions.

Even if our ecological knowledge is accurate, an obvious source of error is that an extant taxon may not now live in the same conditions as it did in the past. Extinct taxa stretch actualistic assumptions still further, because extra propositions must be introduced based on phylogenetic or adaptive arguments (e.g. that an extinct taxon lived in an environment similar to that of a living sister taxon, or to an unrelated living organisms that has a similar 'adaptive' morphology). If none of these kinds of assumptions is justified, a palaeoenvironmental scheme for a group of entirely extinct organisms can only be built up through the secondary evidence of their patterns of association with other organisms whose indications are better known or more widely accepted, or through independent sedimentary evidence. Such secondary fossils, for example, graptolites, can then be used on their own when more direct palaeoenvironmental indicators are absent (Fig. 14.4).

Geoecological methods are not always tied to strict taxonomic identification, since one can often proceed by recognizing groups of organisms with a common ecological or sedimentological occurrence. A reef fauna, for example, once it has been identified as such, may be taken to indicate tropical marine shallow water, regardless of its geological horizon and taxonomic composition. Circularity or overgeneralization can be a problem however.

For reasons already explained, I have concentrated here on the use of geoecological evidence *per se*, and not discussed the widespread incorporation of this kind of information in schemes of provinces.

Methods and patterns

All geoecological methods start by taking the fossils that occur at different places over the same stratigraphical horizon and grouping them according to similar ecological attributes, generally habitat-preference, hence ecological (or more strictly palaeoecological) assemblages, or sometimes 'communities'. In fact, these assemblages may range from the occurrence of a few indicator fossils to large biotas. In many cases, the principal ecological groupings are well established, and so the initial taxonomic breakdown of the data is often omitted. Obviously, this kind of information has to be on a geographical scale and for this reason I prefer to use 'geoecological' to distinguish such assemblages from those that are recognized on a more detailed, local or palaeoecological scale (though the latter may also be incorporated into biogeographical studies).

Methods based on geoecological assemblages are actually of two main kinds, though a third exists that has not been used very frequently.

(1) Assemblages are recognized principally by informal observation of their ecological/palaeoecological co-occurrence. The ecological groups are used to classify different strata into biofacies. It is difficult to generalize about the kinds of patterns these methods produce. Basically they are maps, profiles or block diagrams depicting reconstructions of contemporaneous geological features and landscapes, as inferred from these biofacies. These may either be independent of, or combined with, reconstructions obtained from other evidence. Fig. 14.4 shows a simple example, but numerous examples can be found in the literature cited below.

(2) A numerical approach can be used, based on geographical plots of data such as diversity (taxonomic richness) and abundance. Diversity gradients in particular have been very widely studied as possible indicators of palaeolatitude (Vine, 1973).

(3) The use of statistical methods, such as those mentioned in Section 14.3.1, does not seem to have been applied to geoecological assemblages, though in practice, it is difficult to draw any real conclusions from statistically defined assemblages without also introducing ecological assumptions. Even their most elementary interpretation depends on the distinction between marine and terrestrial biotas. On the other hand, it is difficult to assess geoecological assemblages as they are rarely based on quantitative methods. The logical conclusion would be to use a combination of geoecological and statistical methods to define geoecological assemblages, even

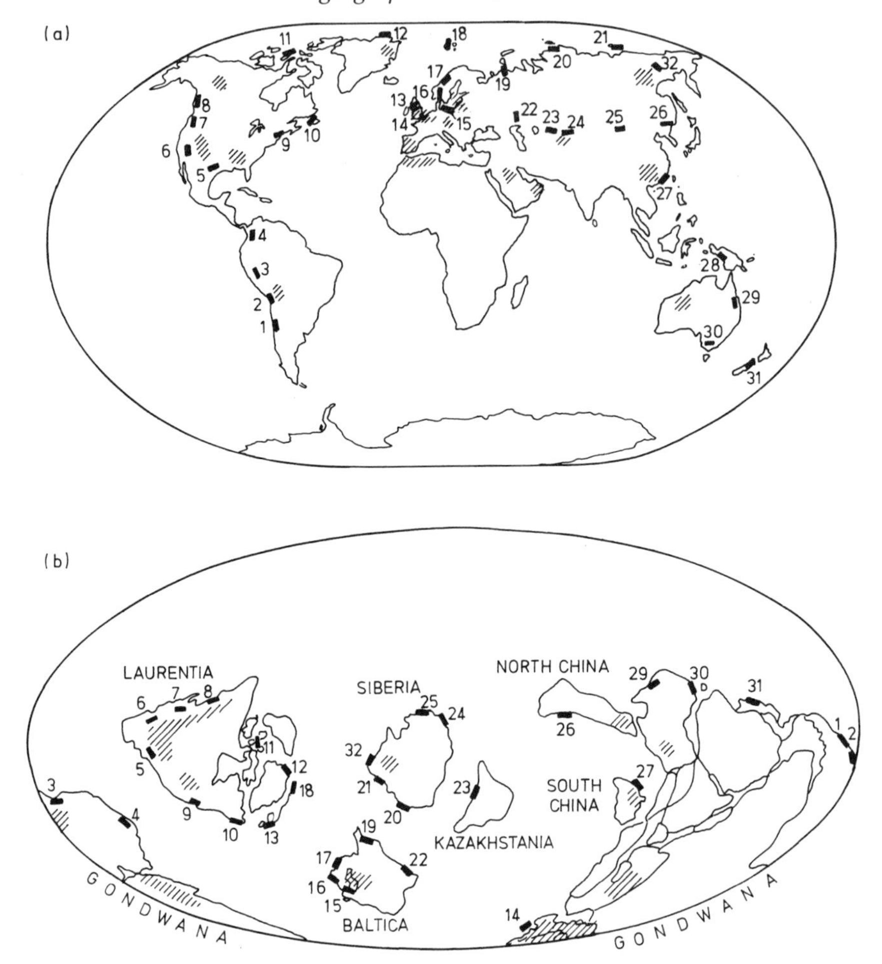

Figure 14.4 Use of biofacies in palaeogeographical reconstruction: a Lower Ordovician example (from Fortey and Cocks, 1986). Black rectangles represent isograptid graptolite facies indicative of oceanic conditions. Shaded areas represent non-isograptid graptolite faunas indicative of continental platforms flooded by seas. (For key to number symbols see Fortey and Cocks, 1986.) (a) Upper map shows modern locations of these facies as found in Lower Ordovician outcrops; (b) the lower map shows them plotted on to a Lower Ordovician palaeogeographical reconstruction derived from a synthesis of numerous lines of evidence. Note that graptolites are an extinct group.

though this introduces numerous actualistic assumptions. (There is also scope for applying parsimony analysis to geoecologically defined biotas.)

Static interpretation

The simplest palaeoecological application of fossils is to use biofacies to infer broad categories of habitat (e.g. lacustrine, marine shelf margin, carbonate platform), and to use these in turn to reconstruct contemporaneous geological and geographical features. This approach, of which the fossil evidence is just one part, is that of facies interpretation of sedimentary rocks. Not all sedimentary rocks contain fossils of course, and other important indications of palaeoenvironments come from inorganic sedimentary structures, grain-size distribution and texture of sediments, their mineralogy and geochemistry, together with stratigraphical details (Hallam, 1981). A great deal of reciprocal (possibly also circular) reasoning is embodied in the facies approach, both between the organisms concerned and other facies evidence.

By far the largest palaeogeographical use of fossils has been through facies interpretation, often in conjunction with statistical assemblages, and other independent evidence. Many palaeobiogeographical provinces are in fact based on such composite evidence (Middlemiss *et al.*, 1971; Hallam, 1973; Hughes, 1973; Gray and Boucot, 1979a). For a small region, facies interpretations are the simplest and most successful way of producing palaeoenvironmental reconstructions. On a larger global scale however, sea floor palaeomagnetic evidence has proved a more potent method over the last 20 years, but because there is no sea floor older than the Mesozoic, facies methods assume especial importance in the pre-Mesozoic. Moreover, although palaeomagnetic and other physical indications, like tectonics, may be the best available means of reconstructing ancient configurations of continents, crustal plate boundaries, orogenic belts and oceans, hypotheses about the distribution of land and sea, together with many detailed regional features, are best derived from facies methods without the broader physical reconstructions.

Cocks and Fortey (1982; Fortey and Cocks, 1986) have provided a useful interpretative scheme for marine geoecological assemblages. In conjunction with Hallam's scheme (1981, 1983), their guidelines can be recast as in Table 14.4. The deep water benthos within a single ocean however may not be as uniform as Cocks and Fortey suggest (Vinogradova, 1979; Paterson, 1986). Since abyssal basins and trenches are often surrounded by much higher relief separating one from another, one might expect them to behave like biogeographical islands. There is obviously scope for further development of such interpretative schemes on a finer geoecological scale, and also for extending this approach to include non-marine environments.

As with statistical assemblages however, a scheme such as Table 14.4

Table 14.4. 'Rules' by which marine biotas can be interpreted palaeogeographically, after Cocks and Fortey (1982) and Hallam (1983). Two conditions are shown in the top row for four main kinds of biotas in the left column. The matrix of the table shows 12 possible interpretations

	Sample localities share the same assemblage	*Sample localities have different assemblages*
Shelf biotas with long distance dispersion	Sample localities occurred in same marine basin or ocean at similar latitude	Sample localities occurred *either* in same marine basin or ocean at different latitudes, *or* in different marine basins or oceans separated by land barriers
Shelf biotas with short distance dispersion	Sample localities occurred on same margin of large ocean at similar latitude, or in same smaller marine basin	*As above, or:* sample localities occurred on different margins of the same ocean
Pelagic biotas	Sample localities occurred in same marine basin or ocean at similar latitude	Sample localities occurred *either* in same marine basin or ocean at different latitudes, *or* in different marine basins or oceans separated by land barriers
Deep water benthos	Sample localities occurred in same marine basin or ocean	Sample localities occurred in different marine basins or oceans, possibly separated by land barriers

raises problems about the significance of intermediate similarities (biotas that partially overlap in composition). Since outright differences in biotas can be attributed to three distinct factors (latitudinal differences, great oceanic distance of separation, or presence of a land barrier), two localities whose shelf biotas (with short range larval dispersion) are 50% the same, might reflect intermediate conditions either of latitude or distance of separation, or a combination of the two. Moreover, this takes no account of possible historical factors (in the sense of antecedent geological and evolutionary events). While latitudinal patterns are often ecological, longitudinal ones may be both historical and ecological in origin. In short, it appears to be impossible to avoid placing patterns of biotas in some kind of prior independent palaeogeographic framework in order to overcome this ambiguity. Where possible, biotas can then be used to provide constraints.

Turning to numerical patterns, areas of similar diversity might be taken to indicate similar latitude, though other interpretations exist (such as the area effect: Chapter 4). Used on their own therefore, diversity patterns seem rather ambivalent or inconclusive. See for example Hallam's (1984) discussion of the Mesozoic boreal/Tethyan question, and Rosen's (1984) historical alternatives to the usual ecological interpretations of diversity patterns.

Dynamic interpretation.

Historical hypotheses can be derived from facies methods by synoptic use of stratigraphical sequences of palaeogeographical reconstructions, including comparison with the present earth. These are hypothetical histories of crustal features. Alternatively, one can omit the palaeogeographical reconstruction stage and use the synoptic sequences of distributional changes in the biotas themselves (i.e. as hypothetical histories of geoecologically defined biotas), either in their own right, or as a means of inferring geological events. In both cases, the link between changing biotas and geological history is based on two assumptions:

(1) Organisms change their distributions through geological time in response to shifting habitat patterns (i.e. environmental tracking). Habitat is interpreted broadly to mean all physico-chemical aspects of the environment. The four sub-ideas embraced by environmental tracking are mentioned in Chapter 2.
(2) Geological change (TECO events) drives ecological change.

Identification of ancient latitude belts has been used to estimate the north-south component of movement of displaced terranes, such as those of the circumpacific belt discussed by Hallam (1986) (Fig. 14.5).

14.4 METHODS BASED ON ORIGINATIONS

14.4.1 Background

Methods based on originations fall into two disparate groups: (1) methods based on statistical analyses of diversity (taxonomic richness), originations and extinctions in space and time as a possible response to changing geological and geoecological conditions; and (2) methods that analyse phylogenetic relationships with respect to the time and place of occurrence of new taxa, as a possible response to geological events.

In phylogenetic methods, a prior phylogenetic scheme of the taxa to be used is necessary. In statistical methods, as with all the previously discussed methods based on distributional change, phylogenetic schemes are not a direct part of the methodology, though analyses

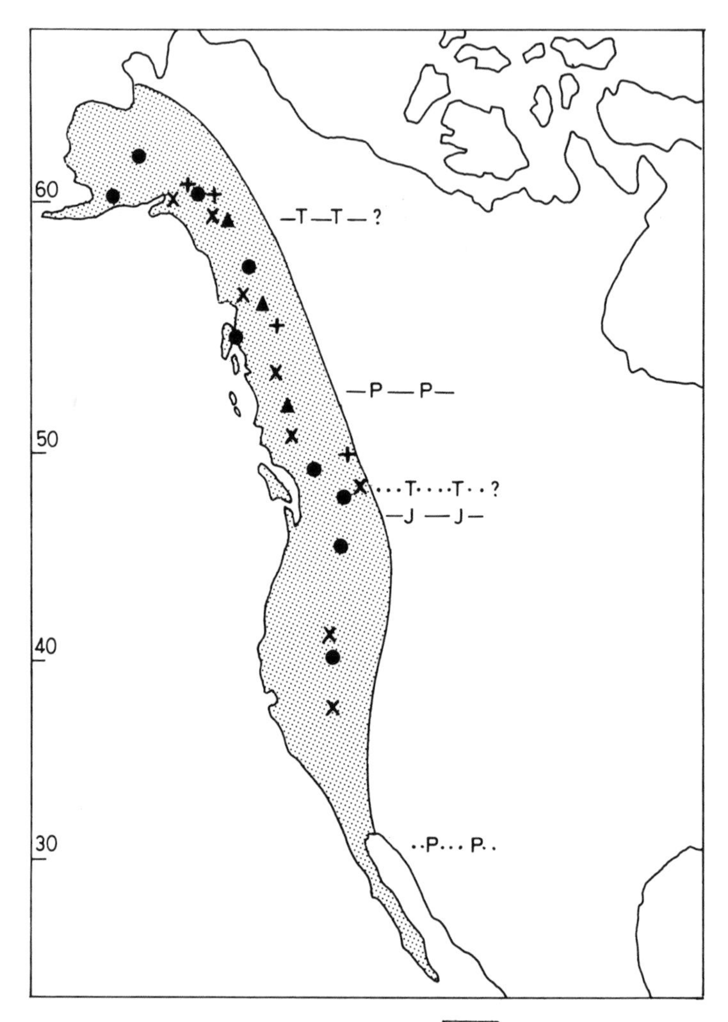

Craton

—J—J— Boundary of Tethyan and Boreal Provinces in Jurassic

—T—T— Boundary of High and Middle Palaeolatitude faunas in
—P—P— Triassic and Permian

···T···T··· Boundary of Middle and Low Palaeolatitude faunas in
···P···P··· Triassic and Permian

Zone of displaced Terranes

Tethyan faunas

▲ Permian

● Triassic

X Jurassic (Pliensbachian)

Boreal faunas

\+ Jurassic (Pliensbachian)

should be based on monophyletic taxa and this presumes sound phylogeny (Section 14.2.4).

14.4.2 Phylogenetics

Phylogenetic methods are unique as a group because they provide direct historical hypotheses even from a single (non-synoptic) data set. This is through the implicit historical component in their assumption that relationships between organisms are evolutionary. Taxa that are least closely related are assumed to have diverged from each other furthest back in geological time. Phylogenetic methods in historical biogeography are discussed in Chapters 11, 12 and 13, and will not therefore be treated here. It is necessary however to comment on their value in the present context of reconstructing geological history from fossils.

In the case of panbiogeography, the method may be limited by its risk of circularity in the concepts of main massing and biogeographical homology and, in any case, it does not appear to have been used by palaeobiogeographers. For cladistic biogeography there are two ways in which fossils can be used: either in combination with living data, or analysing fossils on their own. The problem is not only that of the ambiguity of absence but that the further back in time one goes, the more extinct taxa there are. Patterson and Rosen (1977) have explained the practical difficulties of classifying a group of fossils in the same scheme as their living counterparts, and offered a solution (plesions) that could also be adopted for palaeogeographical purposes. If this results in a biogeographical data base that is too unwieldy however, one would have to break it down into subgroups, or perhaps use only those terminal taxa that were extant at or before a selected geological horizon. (Note that this last suggestion is not a biogeographical extension of Crowson's method (Patterson and Rosen, 1977) or classifying fossils in stratigraphically distinguished groups.) The taxa concerned could then be treated as in Patterson and Rosen's method.

Figure 14.5 Application of faunal indicators of palaeolatitude in palaeogeographical reconstruction: displacement of terranes in the North American cordillera, based on numerous sources (from Hallam, 1986). Map shows location of latitudinally anomalous faunas. The broken lines on the major part of North America east of the zone of displaced terranes indicate known latitudinal outcrop limits of particular faunas for the time shown. The minimum distance of latitudinal movement by the displaced terranes (i.e. relative to the major part of North America) can be obtained by comparing the present extreme latitudinal limits of particular faunas of each age in the terrane zone with their corresponding latitudinal limit in the major part of North America.

In fact only a few cladistic biogeographical authors have made biogeographical analyses of entirely fossil data (Young, 1981, 1984; Milner and Norman, 1984; Grande, 1985). The use of terminal taxa that were extant at or before a particular geological horizon raises the additional interesting possibility of comparing the results of historical analyses based on different geological horizons (including treating the Recent as a geological horizon). In effect, this combines analytical and synoptic approaches. Rosen and Smith (1988) refer to this as 'stratocladistics', and it has only been attempted by Grande (1985). Humphries and Parenti (1986) argue that this approach is not useful because it cannot refute an analysis based on living organisms, but their point is not directly relevant to comparisons between fossil data sets from two or more geological horizons. Although on strict grounds, neither can falsify the other, there should be inductive value if they share some or common patterns (partial congruence). One would expect this if the time interval between any pair of area cladograms is less than the stratigraphical longevity of most of the taxa used. On this and heuristic grounds 'stratocladistics' should therefore be attempted more widely.

Area cladograms derived from phylogenies can be interpreted in terms of either the history of the areas concerned, or the geographical history of the organisms concerned. As with methods based on assemblages and distributional change, there is a whole range of extra assumptions that can then be introduced in order to translate area relationships into earth history. The most stringently empirical approach is simply to compare the most widely corroborated area cladograms derived from phylogenetic analysis with geological theories, minimizing the number of assumptions about why they might correspond. Alternatively, biogeographical theories about how the taxa evolved (e.g. vicariance) can be introduced in order to establish more specific hypotheses. Whether they are based on Recent or on fossil data, the problems of interpreting the resulting area patterns in terms of geology are much the same as those for other methods already discussed. In particular, one cannot produce maps of ancient geography nor specify particular geological events from phylogenetically derived area cladograms. Thus vicariance in a shallow marine biota might have resulted from the emergence of a land barrier in a previously continuous seaway, or from great increase in oceanic distances of separation between different shallow areas.

14.4.3 Geoecology through time

The empirical patterns that result from statistical treatment of biotas through time are usually used to infer earth history through assumptions based on theories of evolutionary ecology (e.g. a time of low global diversity might reflect low provinciality). Although originations, together with diversity patterns and extinctions, are the starting point for data analysis, the underlying propositions derive from the idea that changes in ecological conditions are a major evolutionary driving force (not necessarily through factors affecting isolation of biotas alone). The general approach is synoptic in that data have to be assembled from sets of stratigraphically controlled horizons. For these reasons, I have referred to this part of biogeography as 'geoecology through time'. For the most part, it has been used to make general inferences about the earth, such as its climate or the amount of continental fragmentation at various times in the past, rather than to make specific geological reconstructions. For this reason, further details are not really relevant here. For a summary review, see Jablonski *et al.* (1985).

14.5 DISCUSSION

14.5.1 Relevance of living distributions

Although this chapter is concerned with the use of fossils for geological reconstruction, some comments on the relevance of living organisms are appropriate:

(1) Methods of historical analysis (principally cladistic biogeography, panbiogeography and PAE), based on living distributions, yield hypotheses about earth history in their own right. The view that living distributions provide better hypotheses than fossil distributions has already been discussed, but a final comment is given in the conclusions.
(2) Cladistic biogeographical methods in particular can derive area cladograms from a mixed data set of fossils and living organisms.
(3) Living distributions potentially provide a 'control' for the methods discussed here. How closely do these methods, when applied to modern biotas, generate patterns that correspond to the modern earth? For instance, to what extent does geographical distance of separation correlate with, say, biotic distance, similarity or sister area relationships?

14.5.2 Problems with methods

In the case of statistical methods, there are two problems. Firstly, as is well known, there is a wide choice of methods, formulae and coefficients (or indices). Many are very susceptible to differences in sample size (Cheetham and Hazel, 1969), and different methods often give very different resulting regions, clusters, or estimates of biotic distance. The problem then is to distinguish methodological artifact from biogeographical signal. One solution may be to find common elements in the results of different methods. Although a number of authors (Pielou, 1979b; McCoy and Heck, 1987) have suggested ways in which these methods could be assessed or tested, few palaeobiogeographers seem to have followed this up. Secondly, and overriding the problem of choice and non-standardization, there has been a much more general attack especially from cladists (mostly in the context of systematics but equally in relation to biogeography) concerning the underlying concept of overall similarity.

Two points relevant to most methods concern the assumption of clustering and hierarchies, and the problem of sample size. Whittington and Hughes (1972) have criticized clustering techniques in particular on several grounds including the fact that they yield hierarchies of assemblages. They point out that there is no prior reason for believing that real biotas are hierarchical. With regard to hierarchies generated by cladistic methods and PAE, this may not be the point. In particular, we are dealing with historical entities, and these are inherently hierarchical (Wiley, 1981).

Variation in sample size presents a problem both for phenetics and PAE. On the one hand, there is a possible advantage that a small group of taxa may be well understood by a particular author, and therefore taxonomic, geographical and stratigraphical control will be better. On the other hand small sample areas appear misleadingly 'primitive', and it is difficult to judge whether the pattern derived from study of such a small group on its own is part of a more general pattern, or whether it is unique. Only searching for generality amongst analyses of numerous groups can help to solve this.

The possibility of unique patterns is borne out by studies of unique living organisms' differing ability to cross biogeographical barriers (e.g. the high salinity of the Suez canal system reviewed by Vermeij 1978: see Fig. III.3). In contrast, a large sample biota, especially if it contains different clades or different ecological groups, is more likely to contain different patterns that mask or obscure each other. This is obvious from considering the extreme example of a biota consisting of both terrestrial and marine taxa. A similar problem exists in community ecology (Giller and Gee, 1987). In cladistic biogeography, component analysis is used to find generality amongst different but overlapping patterns (Chapter 12).

14.5.3 Problems of interpretation

Distinguishing stratigraphical and palaeobiogeographical assemblages.

If the time span chosen for an analysis of contemporaneous distributions by assemblage methods and PAE is too long, or if stratigraphical resolution is poor, the results will be confused by effects due to stratigraphical turnover of taxa. This will introduce biogeographically 'false' differences or dissimilarities. Only the tightest stratigraphical control can minimize this, but as already explained, sampling from a true geological 'instant' is impossible. This problem should be less acute if the sample organisms have slow rates of stratigraphical turnover compared with the length of the time span used for sampling. This problem is also apparent when considering extinction rates (Chapter 8).

Static versus dynamic interpretation.

The distinction between static and dynamic interpretations of results has been emphasized throughout. Some methods, like those based on geoecological assemblages, only really provide static reconstructions, unless series of reconstructions are used synoptically. Others, like cladistic biogeography and PAE lead directly to dynamic hypotheses. Static interpretations are essentially palaeoenvironmental, and hence palaeogeographical. Dynamic interpretations are essentially historical in the sense that they provide hypotheses of evolutionary or geological events prior to the time of the sample horizon (in PAE), or prior to the time of the youngest organism in the analysis (cladistic methods). The challenge has been to find ways of distinguishing the effects on distributions of these historical events from those due to the ecological conditions that prevailed at the time of the sample horizon.

There appears to be no ideal single method for solving this, so we must take advantage of the diversity of the methods discussed here. The use of geoecological assemblages (and other methods of facies interpretation) for instance are independent of all but the most general long term evolutionary factors, so they remain unquestionably the best way to date of making static reconstructions of past environments (hence palaeogeography). Unfortunately, the value of these methods has too often been obscured by a narrative approach. Their limitations are that they provide clearer hypotheses about distances and areas within major sedimentary basins or lithological units than between them.

Complementarily to facies methods, cladistic (including phylogenetic) biogeography, panbiogeography and PAE all provide hypotheses about biotas but, objectively, their patterns might be due to one, or a combination of evolutionary, ecological or geological factors. Comparison of these historical patterns with those based on geoecological assemblages should enable us to extract contemporaneous ecological factors. What remains should then be primarily evolutionary or geological, according to the following possibilities: (1) purely evolutionary factors; (2) the effects of geological events on evolution; (3) the effects of ecological factors on evolution; (4) the effects of geological events on ecological conditions and the evolutionary consequences of this; (5) PAE also detects distribution patterns that are historical but not necessarily evolutionary in origin. That is, new combinations of taxa can appear in different regions due to geological events alone, or the effects of geological events on changing ecological conditions ('environmental tracking'), whether or not new taxa have also appeared in the meantime. Of these only (2), (4), and (5) are hypotheses about earth history. Note that possible geological events are not restricted to shifting continents as a cause of vicariance or biotic divergence.

Static implications.

Although the principal evidence of palaeogeography comes from geoecological assemblages, the 'rules' in Tables 14.3 and 14.4 show that even the most basic interpretation of assemblages requires the ecological distinction of terrestrial from marine biotas. In other words, assemblages without ecological information or without an independent palaeogeographical framework are not very informative. This simple ecological distinction however immediately raises the question of the significance of shallow as distinct from pelagic marine organisms. The first should be closely related to terrestrial patterns by occurring in halos around them, while the second should be truly complementary to terrestrial patterns. This generalization however takes no account of the fact that some shallow marine organisms have teleplanic larval dispersion, and their patterns should resemble those of pelagic organisms rather than terrestrial ones. In fact, there is no end to the possible ecological refinements that might be introduced, but the attendant risk is that of extending one's supply lines of ecological assumptions (Section 14.3.4).

A different problem is posed by intergradational assemblages, whether these transitions are used on their own or as a basis for generating clusters, hierarchies or biotic distances. There is certainly no sound reason for translating them into absolute, or even relative, palaeogeographical distances, though it would be interesting to see how far these kinds of analysis carried out on modern biotas reflect modern geographical features. Stehli and Wells' (1971) cluster analysis of Recent reef corals (Fig. 14.2b) would certainly give a distortion of the modern globe if the relative positions of different localities in the cluster hierarchy were translated into present geographical distance. Their analysis in fact makes more sense when interpreted in terms of the relative effectiveness of biogeographical barriers, some of which may be distance-related filter (oceanic) barriers, and some of which are absolute and related to land masses. We should view such patterns as being variously due to a whole range of factors that operated at the time of, and also prior to, a sample horizon.

Of course, the possibilities can be narrowed down if one starts with additional assumptions. If continental drift is assumed for example, increasing similarity of marine assemblages from the same two sample areas through time can be interpreted in terms of a narrowing ocean (McKerrow and Cocks, 1976). Other interpretations of such patterns can invariably be found if continental drift is not assumed. And even if the process of continental drift is assumed by everyone, there are often differences about particular configurations of continents and oceans. Thus, interpretation of statistical assemblages can vary dramatically.

A saving point however is that a certain amount of palaeogeographical control can be obtained by making reference to the lithostratigraphy of the rocks enclosing the sample biotas. Allowing for factors like folding of the strata concerned, the distance apart of sample localities within a single rock body or sequence should not have changed significantly from the time of its deposition, and of the fossilization of its enclosed biotas, to the time of sampling (i.e. the present day). A single rock body is one that has not been significantly disrupted by major faulting, especially strike-slip action and thrusting.

All the foregoing problems must obviously also apply to dynamic interpretations of assemblages based on synoptic comparisons between reconstructions of different horizons.

Conflicting patterns from different groups

Whatever one's assumptions about the nature of distributions (discrete geocommunities, hierarchies or overlapping mosaics), it is common biogeographical experience to find that different groups yield different patterns even when the same biogeographical method is used. Unless it can be shown that the differences are due only to sampling or to inadequate taxonomic, stratigraphical or palaeoecological control, the usual palaeogeographical response has been to claim that only particular groups give the 'correct' pattern, or to accommodate the different patterns by plausible-sounding *ad hoc* accretions. Neither solution is scientifically satisfactory.

There are two other approaches that are more effective. In the interpretation of geoecological assemblages, the different palaeoecological indications of different groups can be carefully and explicitely distinguished and the combined evidence then synthesized to see if a self-consistent reconstruction is possible (Fortey and Barnes, 1977). In historical analytical methods, the solution is to search for common elements (congruence) between the area cladograms of different groups. Note however that all patterns generated by methods of historical analysis can be taken as initial hypotheses, however small the biota on which they are based.

14.6 CONCLUSIONS

It seems that when used on their own, biotic patterns cannot routinely generate palaeogeographical maps or reconstructions of geological events. I agree with Jablonski *et al* (1985) that, "in the light of the multitude of factors influencing the distributional patterns of organisms, it appears that palaeobiogeographic evidence, no matter how intriguing in its own right, is best employed in collaboration with paleogeographic indicators from other fields". This is not surprising since the connection between organisms and geology is not a direct one. Trying to assess precisely what different biogeographical methods can contribute geologically is difficult however, since (biostratigraphy apart) geologists have usually used other ways of deriving earth history, or used biogeography in combination with other evidence rather than alone. Doubts about its resolving power and the consistency of its methods and results have been expressed by Gray and Boucot (1979b) and Jablonski *et al.* (1985). I agree with their criticisms, though I believe that much of the blame can be placed on the conceptual confusion, circularity and pseudo-methodology of 'provinces' (Chapter 2).

Nevertheless, geoecological assemblage methods are important, especially for within-basin reconstructions or for reconstructions based on strata deposited on and immediately around cratonic units. It is significant however that, on their own, these methods have played no critical part in the development of ideas about continental mobility and oceanic closures and openings, because there are always alternative interpretations of their patterns. Thus although Wegener himself used informal comparison between biotas on different continents as evidence of continental drift, the test of continental drift came not from biotas at all, but from the critical evidence of residual magnetic patterns in oceanic crust. But if these methods have not conclusively generated original ideas about global geological processes, they have certainly provided valuable constraints, fine tuning of other kinds of reconstructions, and pointers to uninvestigated possibilities. In this respect it is important to remember that geophysical reconstructions may indicate the relative positions in the past of oceanic and continental crust, but not of land and sea. This is precisely where geoecological methods are important, though their value would be increased if they were used in a more rigorous hypothesis-orientated approach, based on quantitative methods.

More often the process of scientific discovery in this field has been the other way round. Familiar long-established biogeographical patterns have suddenly been perceived to make better sense when made to fit into the new discoveries about processes, or new reconstructions, made by geologists and geophysicists. Examples of this can be found throughout most of the palaeobiogeographical literature of the last 10–15 years, but two that are notable are the reinterpretation of the Lower Palaeozoic biotic patterns of Europe and America, and the history of interpretation of the American Permian and Mesozoic biotas in displaced terranes (but see also Jablonski *et al.*, 1985). Perhaps the closest that geoecological methods have come to contributing to geological theories, albeit in conjunction with all other facies methods, is in the development of the historically important tectonic concept of the geosyncline (now absorbed into plate tectonics models).

The contribution of purely statistical analyses has so far been much more limited than that of geoecological assemblages. Though their potential precision and strict methodology may appeal, it is doubtful if they can yet be used to generate truly independent hypotheses. Unless they are combined with reasonable ecological assumptions, they simply offer ways of classifying sample localities, the exact significance, of which, when taken on their own, remains obscure. There is also no satisfactory way of obtaining a historical interpretation from them, even though their patterns must in part be historical in origin.

Turning to the contribution of the historical methods, PAE is still too experimental to make judgements, and cladistic biogeography and panbiogeography have been applied too little to fossil distributions for their contribution to be compared directly with geoecological methods. It is appropriate however to comment on the geological value of analyses based on living distributions, or a combination of living and fossil organisms.

The strength of cladistic methods lies in their rigour and their explicit methodology (the apparent lack of this in the geoecological approach is not so much inherent as traditional). A disadvantage is that the detailed working out of area cladograms is sometimes laborious, especially when the prior establishment of good phylogenies, as well as component analysis for finding congruence from different groups, is taken into account. Moreover, component analysis seems very much to become a lengthy exercise in its own right through the search for the most parsimonious solution. And if we accept all the limitations of using fossils in these methods, as pointed out by the cladistic biogeographers, it seems that geoecological methods may be better suited to making use of the fossil record. The common criticism that cladistic biogeography can only generate geological hypotheses of area divergence, and cannot therefore be used to infer convergence of areas, has been rejected by cladistic biogeographers but convergence is generally inferred indirectly.

Cladistic methods are tuned to the time and place of appearance of new taxa, and congruence is the means by which more general histories of biotas are distinguished from histories that are unusual or unique to a single taxonomic group. Thus congruence eliminates elements of chance dispersal, and generality is construed to be geologically significant. The remaining general patterns might however be those arising from either vicariance or distributional change (or 'dispersal' in its widest sense). Both of these have geological implications, and are not mutually exclusive. There is no reason to assume, moreover, that vicariance is the result of continental movements and tectonic events. If the process is real (it has not, of course, been directly observed), then it could result from any TECO or ecological event that divides the gene pool of a species population. PAE on the other hand is directly tuned to general patterns of distributional change, specifically biotic divergences and convergences, though these too can result from numerous kinds of TECO event.

Thus although these historical biogeographical methods can delineate biotas, and hence areas, rigorously, they cannot indicate unambiguously whether areas or biotas, or both, have moved. And if either of them has moved, these methods cannot indicate how far. Geoecological methods can sometimes give distance constraints, especially if latitudinal patterns are used, but in general, biogeographical methods cannot generate maps without the addition of tectonic or ecological assumptions. The real geological value of historical methods is that their general patterns demand explanations and so alert us to geological possibilities, such as Humphries and Parenti's (1986) bipolar patterns and their implications for the history of the Pacific basin. In the end, complementary use of different methods seems to be what is needed, rather than belief in a single ideal method. In particular, the present geological potential of biogeography probably lies in further development and application of methods based on geoecological assemblages, cladistics and parsimony analysis of endemicity.

ACKNOWLEDGEMENTS

I have been fortunate in being able to take advantage of the wide range of interests and depth of expertise of my colleagues at the British Museum (Natural History). In particular, I thank Peter Forey, Richard Fortey, Nick Goldman, Chris Humphries, Dick Jefferies, Noel Morris, Gordon Paterson, Colin Patterson, Andrew Smith (who also read and commented on the manuscript), and Dick Vane-Wright (though they do not necessarily share the views expressed). Jill Darrell bore the brunt of the 'Sturm und Drang' generated by the task, and the editors advised helpfully on the manuscript, as well as tolerating my delays with good nature.

CHAPTER 15

Experimental island biogeography

A. SCHOENER

15.1 INTRODUCTION

15.1.1 Islomania

Although it must surely seem that experimental biogeographers have succumbed to islomania – the powerful attraction of islands – they are not alone in doing so. An inkling of the general exuberance for islands can perhaps be glimpsed from the nautical charts of the last century, which designated latitude and longitude of about 200 more islands than are presently known to exist (Stommel, 1984). Many of these were undoubtedly optical illusions, wistfully imagined from ships, but in reality perhaps low cloud banks or icebergs. An unintentional mispositioning of real islands may have occurred at the slip of a pen. Others may have intentionally been given fraudulent identity to satisfy financiers. While additional islands may have subsided, leaving atolls as remembrances, others, such as Surtsey in the Atlantic, occasionally arose abruptly from the ocean floor (Fridriksson, 1975). Nowadays man's activities provide a non-volcanic source of new islands created by dumping dredge sediments or capping dumpsites in nearshore areas.

For whatever reasons, it appears that islands have long had an intrinsic appeal to a variety of investigators. A combination of factors is probably responsible, not least of which is their isolation from events, past and present. In some cases this has protected them from introduced predators or pestilence, maintaining their endemic species or habitats (Chapter 5). Darwin (1845) noted that giant tortoises from each of the dozen or so stark, volcanic islands of the Galapagos were physically distinguishable, although one island was hardly out of sight of the next. These and observations on other Galapagos species were contributing factors to his insights into the process of natural selection (Darwin 1859).

Investigators also appreciated the more tractable nature of islands, particularly their well-defined boundaries, smaller areas, and fewer species, in comparison with continental areas. Collections of representative biotas paved the way for the compilation of species lists from islands and archipelagoes. These lists in turn provided not only considerable information concerning island inhabitants and their distributions but, when combined with palaeontological and geological observations, helped define broad biogeographic patterns and hypotheses about the processes responsible for them.

In addition to the existence of islands classically defined as 'land masses smaller than continents and surrounded by water', many other situations are now viewed as insular. These are termed 'habitat islands', and represent a particular type of habitat, isolated from similar kinds of habitat, by dissimilar habitat. The residents of these habitat islands may be capable of traversing the separating barrier, but are often reluctant to do so and are therefore restricted to their island habitat. The variety of such isolated habitat islands is considerable. Lakes and aquatic beaker microcosms are the mirror image of the classically defined island, and to the freshwater organisms harboured there, dry land is the dispersal barrier (Maguire, 1963; Barbour and Brown, 1974). Other examples of habitat islands range from continental mountain tops (Brown, 1971) to coastal rivers (Sepkoski and Rex, 1974). Caves have also been considered in this light (Culver *et al.*, 1973). Marine habitat islands include submerged surfaces which develop a cover of sessile fouling organisms whose planktonic larvae are incapable of undergoing development to reproductive state without a substrate for attachment (Osman, 1978; Schoener and Schoener, 1981). Isolated plants in deserts have been considered islands for their arthropod fauna (Brown and Kodric-Brown, 1977).

The simplicity of island-like habitats and true islands encourages investigation, and the assumption is made that the generalities discovered may well pertain to more complicated situations. Even were the latter not true, there would still be considerable interest in understanding island characteristics.

15.1.2 Experimentation

Occasionally natural events provide the unique opportunity to follow the course of biogeographic processes under natural conditions. Some examples have been given in Chapter 3. The eruption of Mount St. Helens, in south-western Washington State, is another case in point. At 08.32 on May 18, 1980, the north side of the 3 226 m mountain collapsed (Fig. 15.1). As a plume of steam shot from its summit, an eruption began which eventually reduced one rim of the crater to nearly 1 000 m below its former level. The blast concussion, and debris avalanches and mudflows which followed, brought nearly total devastation to the biota of the subalpine coniferous forest and its lakes.

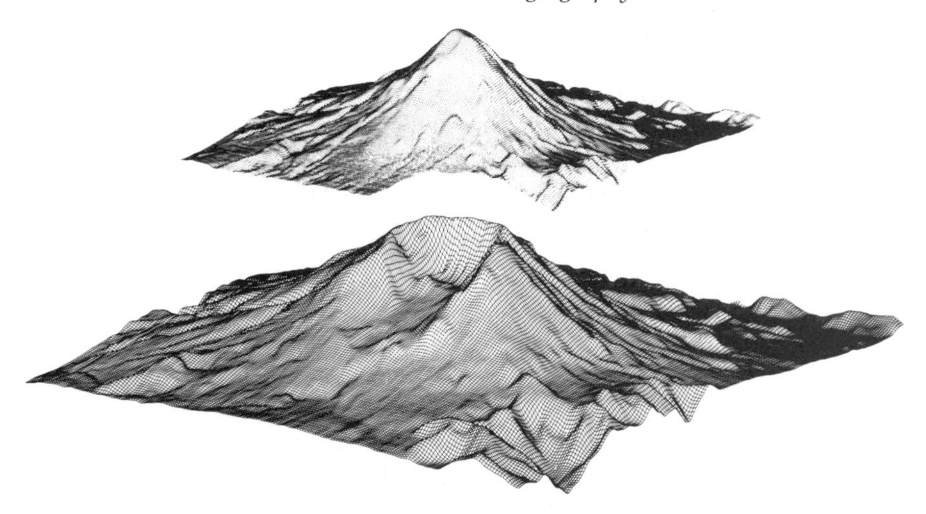

Figure 15.1 Computer generated images of Mount St. Helens, Washington State, before (top) and after (bottom) the 1980 eruption (computer generated using software by Dynamic Graphics, Inc.).

Among the affected blast zone lakes, Spirit Lake was the largest. It received massive amounts of heated debris, superheated magmatic gases, pyroclastic materials and steam-pyrolysed and extracted organic matter from adjacent forested areas (Wissmar *et al.*, 1982). Topographic changes in the dimensions of Spirit Lake were considerable; it became much shallower, enlarging its surface area by 30%. Virtually all lake habitats and food chains were destroyed; changes in lake chemistry, particularly pH and metal enrichment, were documented (Wissmar *et al.*, 1982).

Lake bacteria were sampled several weeks later in those lakes affected by the eruption and those newly-formed lakes created by dammed river tributaries. All lakes yielded such high viable counts of heterotrophic bacteria compared to control lakes unaffected by the blast that they were not accurately assessed by the standard dilution techniques initially used (Staley *et al.*, 1982). Along with other evidence, this suggested that the lakes had undergone rapid transformation from their normal oligotrophic condition to a eutrophic state due to the eruption and its aftermath. Enrichment of these waters may have been due to leaching of organic materials as the indigenous biota began to decompose (Staley *et al.*, 1982), and to the decomposition of unsuccessful post-eruption insect immigrants arriving in late spring and summer (R. Sugg, pers. comm.).

These and other unexpected results recorded during the recolonization process indicate that there is no substitute for direct observation of natural events. However, changes in Spirit Lake's dimensions could have yielded valuable insights into the significance of natural changes in area on species richness, and had a pre-eruption species count been made, subsequent recolonization of other blast zone lakes could have been compared to their prior numbers. Unfortunately this opportunity, unpredictable in its timing and circumstances, could not be anticipated. It is thus towards a more controlled situation that experimenters often turn.

Throughout the past two centuries experimentation has made occasional inroads into the field of biogeography (Darwin, 1859), but a truly concerted effort to incorporate experimentation into biogeography only began in the last quarter century, and much of this deals with islands. Before then the processes involved in determining such phenomena as species' distributions were framed in geological time periods, with the implication that events proceeded slowly, outside the realm of the individual investigator's lifespan. While ecology and its allied fields contributed to biogeography, their contribution was slight. During the previous two centuries, biogeography had been heavily weighted in the past, explaining present-day patterns in the light of long-term processes.

The advent of the MacArthur–Wilson equilibrium model, significantly altered these directions. The model, discussed in Section 15.2 challenged investigators of diverse taxa by providing a set of seemingly simple hypotheses to be examined. And the initial test of these hypotheses, on small mangrove islands in the Florida Keys, indicated that in this case a startlingly short time was involved in the process of island recolonization. Thus biogeographic processes, once viewed only in the long term, may take place over a period of months, and therefore be amenable to short-term experimentation. These short-term processes were thought to work independently of evolution, on a shorter and more local scale, in comparison to either historical evolutionary or phylogenetic biogeography. Such processes should however be viewed as complementary to the longer term ones just as experimentation should be viewed as complementary to direct observations.

For experimental purposes, a variety of habitats have been selected; some workers have preferred to manipulate natural island habitats, while others chose completely artificial ones. Often the choice has been made for utilitarian reasons alone. Good experimental protocol depends upon careful experimental design and sufficient replication to allow a null hypothesis to be tested statistically. Application and validity of such hypotheses are discussed in Chapter 9. Both manipulation of natural situations and use of artificial habitats allows a greater degree of control by the investigator than often exists in the passive observation of natural occurrences. Such variables as size, shape, habitat type, initiation time and extent of replication, are all controllable.

15.2 AN EQUILIBRIUM THEORY

The MacArthur and Wilson (1963, 1967) equilibrium theory of island biogeography incorporates observations on the abundance distribution of individuals among species (Preston, 1962), as well as patterns in the distribution of species among islands. It rests on decades of first-hand biogeographic investigations on islands, yet cuts through the morass of accumulated data, to offer a relatively simple and quantitative theory to explain those distributional patterns at the species level. Both taxonomic and phylogenetic affinities of species are de-emphasized, and species are treated equivalently, species richness instead being the focal point. This suggests interchangeability of species within the taxon investigated, without concern for competitive interactions or predation, which in other situations have significant effects on community composition. Yet in the initial stages of island development, when population densities are low, interactions between species may be thought of as minimal, and these assumptions reasonable. This less taxonomically-oriented emphasis, whose chief proponents were often expert taxonomists themselves, allowed quantitative concepts to be put forward that begged testing.

The main premise of the equilibrium theory is that each island is in dynamic equilibrium in terms of species number between two continuing processes – immigration, adding new species to the island, and extinction – removing already established species. The equilibrium condition ($\hat{S}$) is found at the point of intersection of these immigration and extinction rate curves (Figs. 15.2 and 15.9). Among species, these rates are considered relatively homogeneous, although by curving the rate functions a moderate amount of heterogeneity can be accommodated into the model.

Since the equilibrium is dynamic, change is continual and turnover of species composition is expected. Considerable controversy has accompanied the concept of equilibrium, for what may appear to be at equilibrium over the short term may over the longer term appear to oscillate. Thus selection of an appropriate time-scale for observations could influence the experiment's outcome. Should this be related to the life-span of the organisms, or some other biologically structural criterion? Once the duration of the experiment has been determined, assessment of whether or not equilibrium is being approached can be done in several ways (Section 15.2.6).

The quantitative predictions of the equilibrium theory are mainly

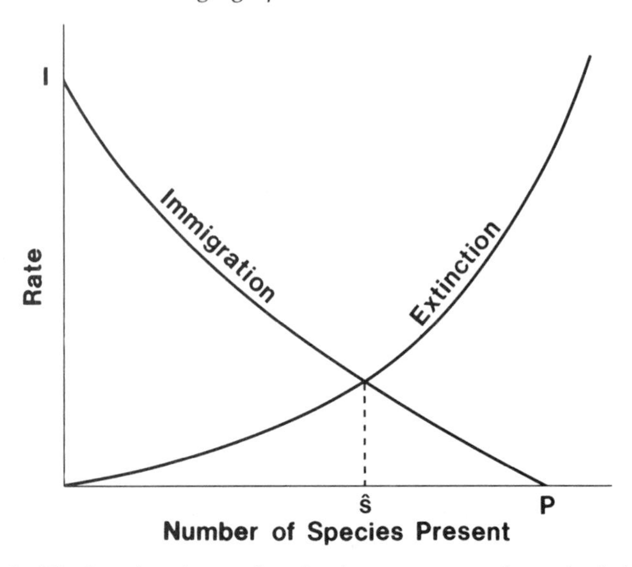

Figure 15.2 The immigration and extinction rate curves for a single island. An equilibrium number of species, $\hat{S}$, occurs at the intersection of these curves. P = total number of species in the species pool; I = the initial immigration rate (modified from MacArthur and Wilson, 1967).

concerned with the variations in island size or distance from a source area of potential colonists, and deal with the colonization of denuded or newly-formed islands, the primal stages in island community development. The novel ways which experimenters have devised to test the model's concepts and expectations are described in the following sections.

15.2.1 Immigration

Predictions from the MacArthur–Wilson equilibrium model include the existence of an island-specific immigration rate curve determined by an island's position relative to its source pool of colonists. This curve is thought to descend from a high point, when no species have yet immigrated, and many are potential new colonists, falling as colonization proceeds and fewer new species from the pool are left to colonize.

In theory immigration occurs when a propagule – that minimal number of individuals of a new species required for successful breeding and population increase under optimal conditions (Simberloff, 1969) – has arrived on an island. In practice, determination of this number of individuals or the sexes of the colonists, is often so difficult that the sole criterion considered instead is the presence of a new species on an island; this is perhaps most justifiable when offspring of island inhabitants will be broadcast from the island in any case, as occurs among reef fishes and fouling organisms with planktonic larval stages.

Comparisons of the taxonomic composition of an island at discrete census periods allow calculation to be made of the number of new species present at the end of each interval that were not present at the beginning of the interval. This number, when divided by the duration of the census interval and plotted against time from initiation, yields a colonization-rate curve, expressed in species per unit time. However, this curve only approximates to the true immigration rate curve, which instead requires continual monitoring to avoid missing immigrations and extinctions between censuses. There are no guidelines to establish the length of time a new species must be present before being considered an immigrant. This is left to the discretion of the investigator, by determining the length of the census interval.

Decreasing colonization-rate curves have been obtained experimentally for a variety of true, habitat or artificial islands – terrestrial (Rey, 1981, 1984), marine (Molles, 1978; Osman, 1978; Smith, 1979) and freshwater (Maguire, 1963; Cairns *et al.*, 1969; Hubbard, 1973). But this support of the theory and the occasional recorded discrepancies (Osman, 1982) are difficult to evaluate because computation of this curve can give erroneous results when species immigrations and extinctions occur between census intervals (Simberloff, 1969; A. Schoener, 1974a, see also discussion of 'cryptoturnover', on p. 494). Although in concept immigration seems attractively simple, the computation of immigration rates actually depends upon how representative the census periods are.

Other experiments suggest that the shape of the island's immigration rate curve fluctuates seasonally in seasonally varying regions: Osman (1978) obtained estimates of monthly species arrival in a different manner from that discussed above, by submerging identical uncolonized substrates monthly. The immigration rate curves constructed from these data indicated that the curves varied in a seasonal pattern. This fluctuation in turn influences the rate at which long-term substrates advance towards the equilibrium state. That these experimental results differ from the expectations suggest that in some cases the attractively simple assumptions of the equilibrium theory require refinement.

15.2.2 Distance effects

The MacArthur–Wilson equilibrium model treats the arrival of island isolation as the decisive factor influencing the rate of arrival of new species on an island, and its descent is steeper for islands nearer a

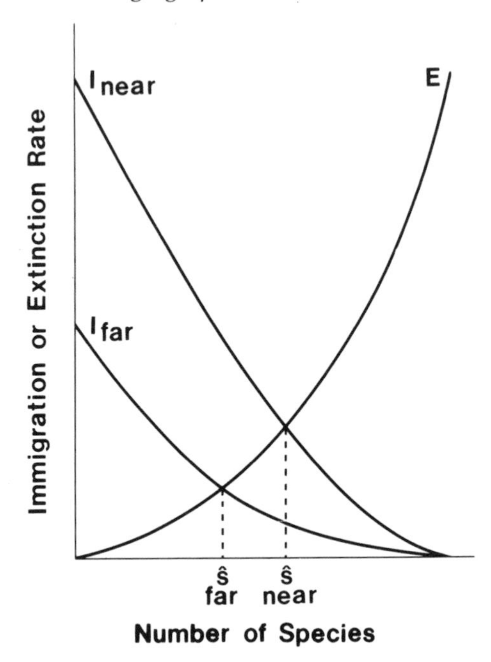

Figure 15.3 The effect of varying island distance is shown for two equally-sized islands either near or far from a source of colonists. The equilibrial numbers are determined by the intersection of the immigration and extinction rate curves as shown (modified from MacArthur and Wilson, 1967).

source of colonists, where initial rate of species accumulation is greater. Islands further from their source area have both a lower number of potential colonizing species and a less frequent arrival of new species; consequently the immigration rate curve for distant islands is one with a lower intercept and a more gradually decreasing slope. For paired islands of similar size at different distances from their source area, the equilibrium model predicts that near islands will maintain a greater number of species at equilibrium, $\hat{S}$, as observed from the intersection of the appropriate immigration rate curves in Fig. 15.3.

Observational data going back over the last two centuries have implied the above relationship, but for more exact tests of the distance premise of this model, both laboratory and field experiments have been devised, although difficulties lie in distinguishing the appropriate source areas. Once this obstacle is overcome, observation of equilibrium numbers is direct.

To circumvent such difficulties, source areas were provided by placement of a fully-colonized synthetic foam sponge in a laboratory tank containing freshwater microscopic organisms and siting uncolonized substrates at distances from the sponge (Cairns and Henebry, 1982): agreement with the equilibrium model's predictions was obtained. Other experiments varied the number of species in the species pool of colonists. Standardized glass microscope slides reached lower equilibrium numbers when exposed to a smaller pool of potential diatom colonists, than they did when the species pool was increased (Patrick, 1967).

Changing rates of immigration have been obtained by Patrick (1967) who exposed glass slides to freshwater diatom colonization under varied rates of flow and observed that greater immigration rates resulted in more species on identical substrates at equilibrium. Transplant experiments allowed fouling panels to be exposed to different immigration rates. Two sets of panels initially exposed in a marine locality with low immigration rates showed close overlap in their colonization curves until one series was moved to a locality with a higher immigration rate and thereafter accumulated a greater number of species (Fig. 15.4).

In other fouling panel experiments, Osman (1982) distinguished the source area (a kelp bed) of sessile epifaunal panel colonists, and positioned panels at distances from the bed. The effects of isolation were in agreement with expectation. More distant panels had lower immigration rates and lower species numbers at equilibrium, but there were indications that the measured extinction rates might also have increased with increasing isolation.

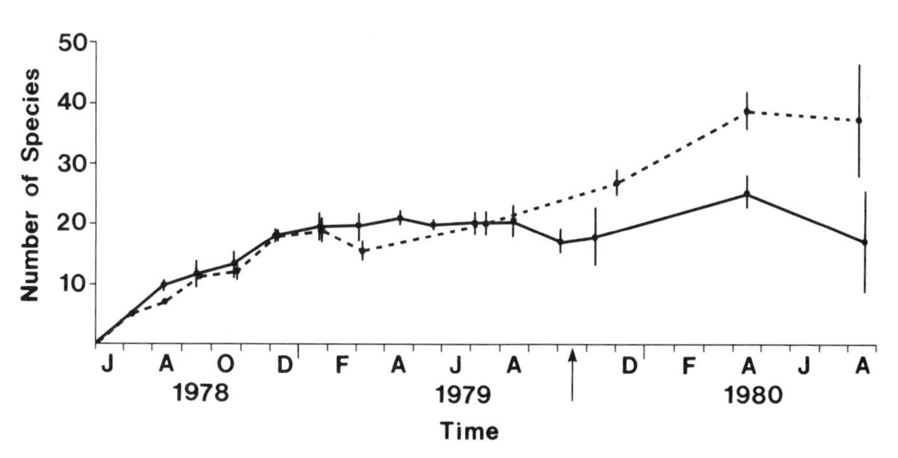

Figure 15.4 Colonization curves experimentally obtained for two marine fouling panel series initially exposed to the same immigration rate until (see arrow) one series (dashed line) was moved to a region with an increased immigration rate. The remaining series (solid line) serves as a control (modified from Osman 1982). Each point is the mean of three replicate panels and the vertical bar is the standard error of the mean.

On soft marine substrates, intertidal plastic trays filled with sand served as islands. Immigration rates to these islands were varied by fitting trays with mesh strips of different widths through which immigration was possible. The meiobenthic species present on the more isolated (smaller width of mesh) islands were fewer than those on the closer islands, an effect attributed to a lowered immigration rate (Hockin, 1982).

Dickerson and Robinson's (1985) laboratory experiments on microscopic freshwater colonists also support the distance predictions, but when they characterized the species as either permanent or transient they noticed that the observed distance effect was the result of a higher number of transients found in laboratory microcosms experiencing a higher immigration rate.

The results of many other experiments run counter to the model's predictions. When Gascon and Miller (1981) created insular and continental artificial reefs, either a few metres from or adjacent to, a nearby source area respectively, the colonization of the continental reef by fishes proceeded more rapidly than did that of the insular reef. Yet a similar equilibrium number of species was reached in both cases. Artificial reefs at Hawaii were observed during their colonization and notable effects of isolation were found; even a relatively slight degree of isolation affected fish species numbers, but isolation increased those numbers, contrary to the prediction (Walsh, 1985). In studies on an archipelago of *Spartina* islands, immigration rates were also not affected by distance (Rey, 1981; Rey and Strong, 1983), and A. Schoener (1974b) was unsuccessful in observing differences in colonization between marine plastic sponge islands thought to be different distances from a source area. While these studies may reflect problems of scale, with either distances too small or time-scales too short relative to the life-histories of the organisms, they may also simply reflect the inability to determine source area or areas accurately. More than one source area may be involved.

Although here we have discussed the effects of distance on individual islands, a further discussion of this effect on the species–area relationship for sets of islands in an archipelago is found in Chapter 4.

15.2.3 Extinction and turnover

Extinction rates

Chapter 8 deals with extinction largely on an evolutionary time scale but here I deal with ecological time exclusively. The MacArthur–Wilson equilibrium model predicts that extinction, the disappearance of a species from an island, will increase as the number of species on the island, and time, increase (Fig. 15.2). This results from consideration both of the greater potential number of candidate species for extinction and the reduced population sizes of each species that occurs once islands accumulate an increased number of species. Although the definition of extinction is clear-cut in theory, its measurement in practice suffers several potential drawbacks both in ecological time and over evolutionary time as discussed in Chapter 8. As is true of monitoring immigration rates, censuses made at discrete intervals may provide an approximation of the extinction rate curve. This is done by dividing the number of species eliminated at the end of each census interval by the interval's duration; however, extinction occurring between censuses may be unrecorded (see also discussion of 'cryptoturnover' below). There is no *a priori* criterion for designating the time interval over which a species must be absent before it is recorded as extinct. Some investigators have been more conservative in this regard than others, recognizing that periodic censuses may miss a species during a census. Therefore, a species was counted extinct when it had been absent for two successive observation periods.

When immigration rates are high, populations can be rescued from the verge of extinction by the continual influx of individuals and new genetic variability (preventing extinction due to genetic accident); this 'rescue effect', observed among defaunated thistle plants in an Arizona desert, essentially lowers the observed extinction rate (Brown and Kodric-Brown, 1977).

Some colonization experiments measure extinction at periodic intervals as discussed above and find agreement with theoretical predictions of the extinction rate curves (Hubbard, 1973; Molles, 1978; Osman, 1978). Sometimes agreement occurs only during a portion of the study: Strong and Rey (1982) report that in general the relationship between island species number and extinction rate curves was in the direction predicted by theory; but these rates were much higher during the first several months of colonization than afterwards when populations became established on islands (Rey, 1985). In other experiments extinction events appeared relatively rare (Osman, 1978; Schoener and Schoener, 1981). It should be noted that these are extinction rates during an ecological time frame, for which short-term experiments are appropriate. The difficulties in extrapolating to evolutionary time and biogeographic patterns may involve other considerations not incorporated into these types of experiments.

Turnover rates

Attainment of equilibrium species numbers on islands, i.e. that point at which both immigration and extinction rates are equal, is thought to be accompanied by continual changes, or turnover, in species composition without alteration of the species equilibrium. The turnover rate is that number of species replaced by new species per unit time at equilibrium. The concept of continual turnover in an island's biota is one of the more unusual assumptions of the MacArthur–Wilson equilibrium model.

For those few island studies where colonization could be followed to provide evidence of approach to equilibrium, determinations of turnover rates have been made. Direct calculation of turnover from censuses made at intervals, however, can suffer pitfalls from 'pseudoturnover', i.e. if local movements of transient species are recorded as extinctions, thus increasing turnover estimates (Lynch and Johnson, 1974) as well as 'cryptoturnover', i.e. species whose extinction may go unnoticed, because of the length of the period between censuses (Simberloff, 1969). Attempts to minimize both these problems were given considerable attention by Rey (1981, 1985) for a set of *Spartina* islands whose recolonization was followed after defaunation. He concluded that there were significant changes in species composition on these islands and demonstrated that turnover was real and not a result of sampling error.

Experimental observations of turnover range from little or no change in species composition at equilibrium to high rates of change. Minimal pre- and post-defaunation turnover was observed in some reef fish studies (Smith, 1979; Gascon and Miller, 1981). At the other end of the spectrum are studies of fish species colonizing both artificial (Walsh, 1985) and natural (Brock *et al.*, 1979) Hawaiian reefs, which report high rates of turnover. In the former study no species was continuously present on the same reef throughout all censuses, and most species had low persistence; in the latter, a comparison of the same reef 11 years later indicated a 40% overlap in species composition. High turnover on the Great Barrier Reef has also been reported and, in this case, attributed to predation (Talbot *et al.*, 1978). Scale factors pertaining to the number of species in the species pool or the size of the observed reef, may also be contributory factors, as turnover might appear low where the number of species on the reef is high relative to that in the source area (Gascon and Miller, 1981).

The MacArthur–Wilson model considers absolute turnover rates to be greater on near than on far islands, and greater on smaller than on larger islands. Williamson (1978) points out that because theory predicts that large, near islands should have more species than small, far islands at equilibrium, the absolute turnover rate will change consistently with species number if distance alone, or area alone, varies but not if both do. Turnover rates can be scaled by the average number of species on an island when relative rates are desired for comparative purposes (Diamond, 1969). This is calculated from periodic censuses, by adding the

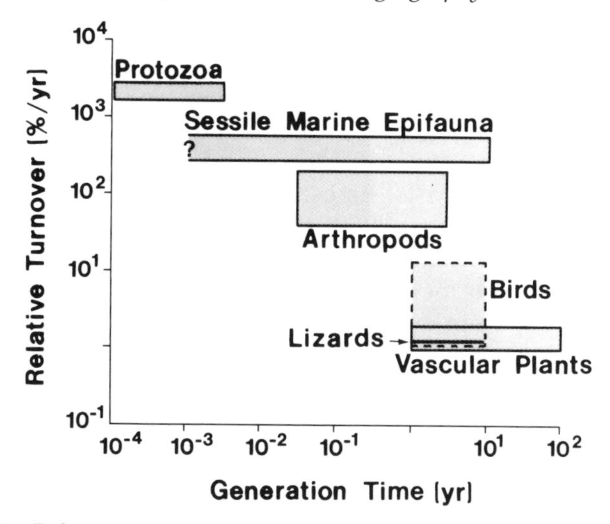

Figure 15.5 Relative species turnover as a function of generation time in six types of organisms (modified from Schoener, 1983b).

number of species at the end of the census interval absent at its start, to the number present at the beginning of a census interval absent at its end. This value is divided by the combined total number of species at the beginning and end of the census interval and multiplied by the time between censuses. A review of published relative turnover rates observed that the more primitive the taxon, the higher was the turnover. Fig. 15.5 indicates that highest rates were observed for protozoans and lowest rates for terrestrial vertebrates and plants. An approximately linear decline with generation time in these organisms was also noted (Schoener, 1983b).

Brown and Kodrick-Brown (1977) point out a further difficulty in predicting turnover rates. On islands distant from a source of colonists the model should correctly predict the relationship between turnover rate and isolation. However, closer to a source, the MacArthur–Wilson model needs be modified to incorporate the rescue effect on extinction rates for islands of varying size and isolation. Brown and Kodric-Brown's (1977) model predicts an effect of isolation on equilibrium turnover rate opposite to the MacArthur–Wilson model, although agreeing in other respects.

15.2.4 Area and its effect

As brought out in several earlier chapters, the effect of area is important in the consideration of speciation, adaptation, extinction, endemism, species richness and community structure. Area is also relevant to the present discussion, as the MacArthur–Wilson equilibrium model considers island area to influence population size, and through it, extinction probabilities. Since a smaller island is expected to maintain a lower population size, a greater probability of extinction due to ecological or genetic accident exists. Its extinction rate curve, therefore, rises more steeply than does that of a larger island. Intersection of this more rapidly rising extinction rate curve with that of the descending immigration rate curve, similar for both large and small islands equidistant from their source, leads to an expected number of species which is lower on the smaller island at equilibrium (Fig. 15.6). Once-connected portions of mainlands, which have become isolated through processes such as rising sea level, are expected to undergo 'relaxation' to a lower species number because of their reduced area (Diamond, 1972): relaxation occurs when recent natural processes displace species numbers from their equilibrial value, leading to a temporary imbalance between

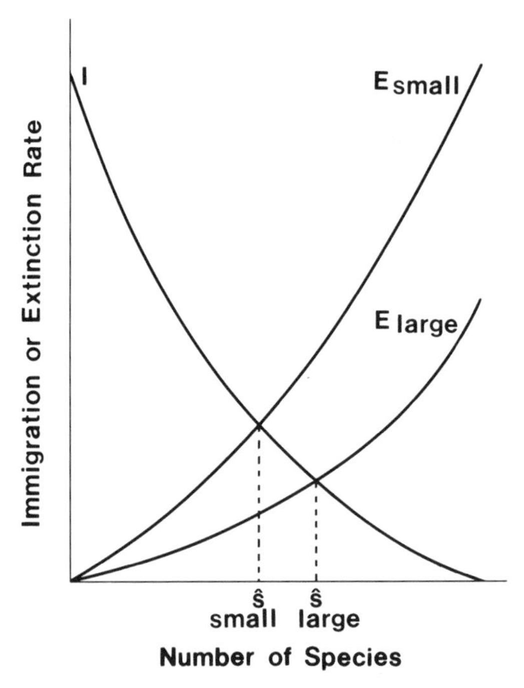

Figure 15.6 The effect of varying island area is shown for two equally-distant islands, one small and the other large. Their equilibrium species numbers are determined by the intersection of the immigration and extinction rate curves (modified from MacArthur and Wilson, 1967).

immigration and extinction and resulting in a lower equilibrium value.

Experimental investigations of the effects of island area on the equilibrium number of species have taken two directions; either colonization has been observed over time for empty islands of varied size, or already-developed island biotas have been modified in ways to induce changes in species number equilibria.

Direct observations of some different sized island systems undergoing colonization have yielded agreement with theoretical predictions. Freshwater diatoms settling on glass slides provide one example (Patrick, 1967). Small plastic dish sponges, submerged singly or in pairs (A. Schoener, 1974b) provide another. Fewer species of marine invertebrate colonists were found on single sponges during most of the colonization period and at equilibrium than were found on two such sponges sewn together, and these larger sponges had a greater total number of individuals. Similarly, the colonization of two sizes of fouling panels by sessile marine epifauna in the Pacific (Osman, 1982) and the Atlantic (Osman, 1978) revealed that larger islands consistently reached a greater equilibrium species number than did smaller islands (Fig. 15.7), although only in the former instance were there significant differences

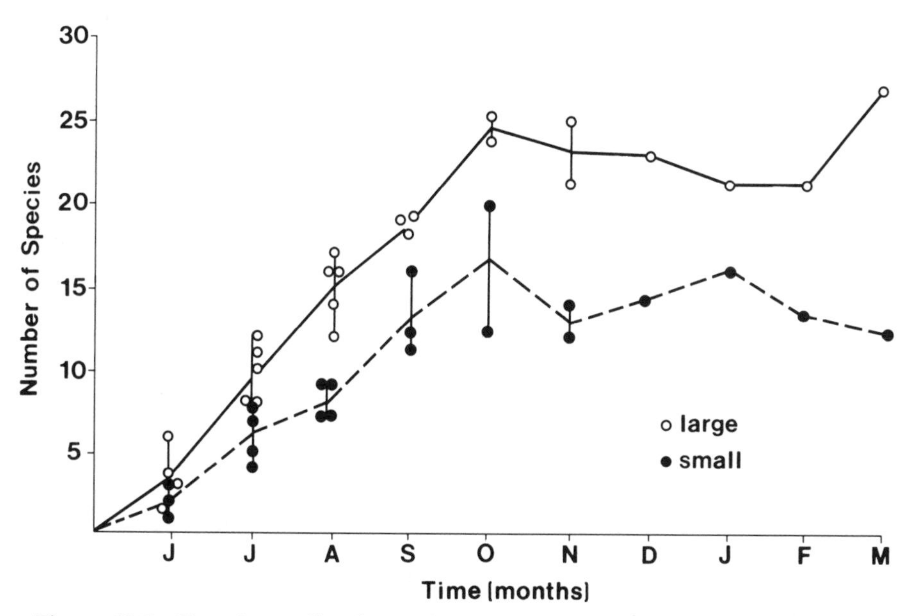

Figure 15.7 Experimentally obtained colonization curves for small and large fouling panel islands equidistant from their marine source area (modified from Osman 1978). Vertical bars indicate the range of values for each month.

between the extinction rate curves for small and large panels. The terrestrial arthropods of defaunated *Spartina* islands (Rey, 1981) showed a significant effect of area on immigration rates early, but not late, in the recolonization process.

Not all studies support the predicted relationships. Plastic tray islands of two sizes showed no appreciable effect of island area on the number of harpacticoid copepod species colonizing the sand (Hockin and Ollason, 1981). Freshwater beakers, insular to their microscopic colonists, likewise failed to show effects of volumetric (equivalent to area) changes on their species' numbers. Maguire (1971) inoculated several algal and protozoan species into beakers with capacities spanning two orders of magnitude and six months later found no differences in the number of species they maintained. Results directly contrary to MacArthur–Wilson expectations were reported for other freshwater beaker microcosms, where smaller beakers had a significantly greater number of species at equilibrium than did larger beakers (Dickerson and Robinson, 1985). These results suggest that there may be something inherently different about laboratory experiments using these microscopic taxa.

Manipulation of island populations and observations on their subsequent survival is another way to investigate island area effects. Introductions of variously sized propagules onto empty islands have been attempted for some taxa. Both their successes and failures are instructive. In a study of mice introduced onto islands in variably sized populations of colonists, larger populations survived longer (Crowell, 1973). Lizard species, introduced into 30 islands spanning four orders of magnitude in area, indicated that time to reproductive extinction for a particular species monotonically increased with island area; above a threshold area rapid colonization ensued, and below that propagules rapidly became extinct. But propagule size itself had no effect (Schoener and Schoener, 1983a).

Experimental manipulation of already developed insular communities is an alternative approach to using empty islands and also allows observation of area effects. Maguire (1971) added a mixture rich in species to beakers of two sizes which had developed an equilibrium species number over a period of months. Reasoning that by the addition of a sufficiently large number of taxa to these islands, this equilibrium number might be temporarily exceeded, differences were expected between the beaker sizes. After additions were made both beaker sizes had similar numbers of species, but no differences were observed between beaker sizes several weeks later and the mean number of species had not decreased in either case. For these relatively short-lived organisms, the time-scale of the experiment appeared adequate.

Small mangrove islands, already saturated with species, provided an opportunity to alter island area without altering habitat diversity (Simberloff, 1976b). The terrestrial arthropod fauna of these islands was censused before sections of the islands were pruned and removed, and after a sufficient interval, these islands of reduced area were again censused. Species numbers had decreased, presumably because of reduced population sizes. This field experiment attempted to determine the response in species number to changes in island area alone, quite apart from any effects of habitat diversity, and it is unique in this respect.

15.2.5 Species – area experiments

The plethora of surveys which indicate a well-defined relationship between species number and island area in a geographic region for a particular taxon, have attracted considerable attention and are reviewed in Chapter 4. The reasons for such a relationship have elicited much speculation. Several explanations have been offered and these are outlined and discussed in Chapter 4. Most data pertinent to this phenomenon are based on survey work, and experimentation on island communities specifically to investigate the species–area relationship is rare indeed.

One experimental study compared the species – area relationship of an archipelago of differently-sized islands once before and twice after experimental defaunation of the islands: this *Spartina* archipelago re-established its species – area relationship within six months (Rey, 1981).

Cairns and Ruthven (1970) computed the species–area relationship for the protozoan colonists of three-dimensional synthetic sponges at several successive times after colonization; even their earliest observations document a strong relationship between the volume of the substrate (corresponding to its area) and the number of freshwater protozoan species.

In order to determine the importance of habitat area to the marine epibenthic organisms on Caribbean fouling panels, Jackson (1977) set out replicates of six panel sizes. Within the first six-month census period, a strong relationship with area had already developed, and a later census indicated an even steeper relationship. The results demonstrated that substrate area was fundamental to the development and maintenance of community structure and its diversity.

The development of the species – area relationship was observed for a series of submerged fouling panels colonized by temperate sessile epifauna (Schoener and Schoener, 1981). By observing colonization in a non-destructive manner at frequent intervals, the ontogeny of this relationship was followed for an archipelago of artificial islands spanning three orders of magnitude in area. During the initial weeks of

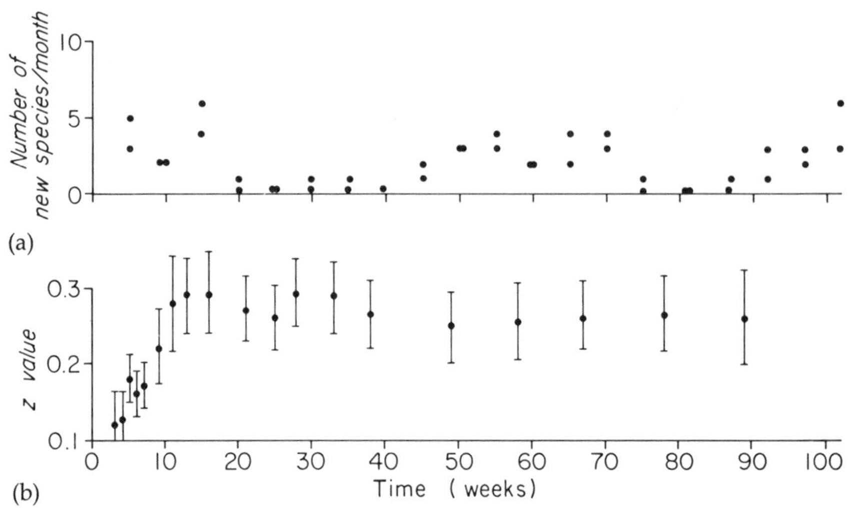

Figure 15.8 (a) immigration rate, defined as the number of new marine sessile species present on clean fouling panels submerged at 5-week intervals. (b) the slope, z, of the species–area relationship with 95% confidence intervals shown (Schoener and Schoener, 1981, with permission of *The American Naturalist*).

colonization the species – area curves had small slopes (small z – values). These gradually increased until by the 10th week values ranged from 0.24–0.29 and during the remaining 100 weeks of the experiment, slopes computed at intervals throughout the course of the experiment remained at these values (Fig. 15.8). Since the colonization curve for each island was also monitored, it was possible simultaneously to determine the temporal approach to equilibrium. When a stable slope was approached in log-log plots of numbers of species present versus area, the colonization curves were generally beginning to approach equilibrium. Even though monthly immigration rates, estimated by submerging an additional series of new surfaces for colonization each month, varied during the course of the year, the slope of the species–area curve fluctuated minimally after the 10th week. Since only a short interval elapsed before a definitive species–area relationship was established, and during this time only slightly less than half of the islands in the archipelago had any observed extinctions, immigration differences were almost entirely responsible for initiating a relatively stable species–area relationship.

These experimental approaches have demonstrated the rapidity with which the species–area relationship is established, usually in less than a year for these taxa. Such information allows further examination of the events during that time frame in order to illuminate causal mechanisms.

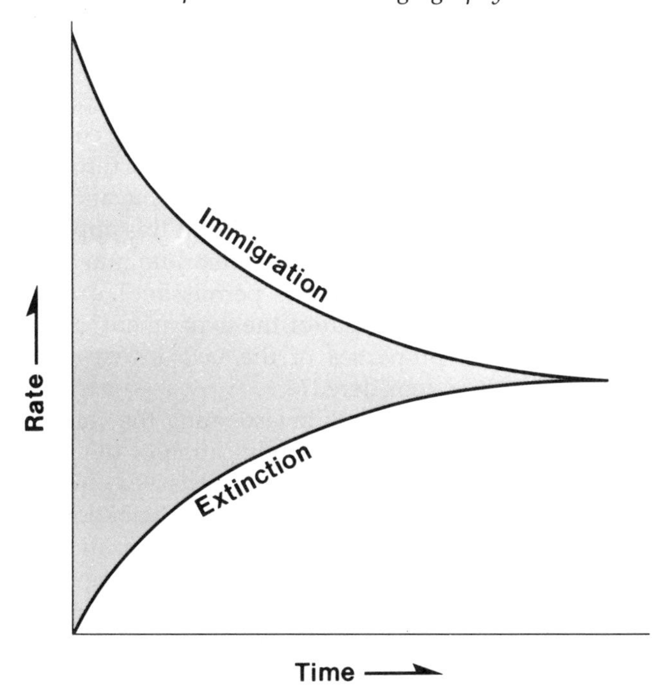

Figure 15.9 As colonization proceeds, the difference between immigration and extinction rate curves (shaded) decreases (modified from MacArthur and Wilson, 1967).

15.2.6 Colonization curves and equilibrium

A colonization curve (Fig. 15.4) plots the quantitative history of species numbers (S) on an island over ecological time. Censusing the island at intervals can provide this information directly, although the colonization curve can also be established indirectly by integrating the difference through time (t) of the immigration rate (I) and extinction rate (E) curves (Fig. 15.9), according to the equation:

$$S_t = \int_0^t (I - E)dt$$

An island undergoing colonization has few species initially. Immigration rates are thus expected to be high while, on the other hand, extinction rates are anticipated to be low. But as colonization proceeds and species numbers increase with time, the difference between these rate curves diminishes. In the equilibrium model the initial rapidly-rising colonization curve becomes asymptotic with time, approaching an equilibrial number of species as the difference between immigration and extinction rates declines further (Fig. 15.9).

Since determination of true immigration and extinction rates often poses difficulties, as discussed earlier, the colonization curve is easier to obtain directly from observations on species numbers through time. The concept of equilibrium, with its dependence on time and spatial scales is, however, a problem to be contended with in this approach. Should this curve indicate oscillation around an equilibrium number, as is often the case, and how much fluctuation is permissible? And what is the appropriate duration for the length of the experiment? Should this be tied to some biological properties of the taxa investigated, such as life-span of the organisms considered?

The experiments geared towards investigating the colonization process and equilibrium directly led to modification of the single equilibrium concept. Recolonization of previously censused, fumigated mangrove islands, was monitored after defaunation (Simberloff and Wilson, 1969; Wilson and Simberloff, 1969). Species recolonization was surprisingly rapid, with species counts already exceeding their original pre-defaunation level after only a few months. This overshoot was considered temporary, representing an early equilibrium state–the non-interactive equilibrium–involving somewhat higher species numbers due to the minimal interaction between species at low population densities. Later, a slightly lower but more lasting equilibrium, the interactive equilibrium, was established (Simberloff and Wilson, 1969, 1970). Wilson (1969) further developed the multi-equilibrial concept to include two additional stages: an assortative equilibrium, occurring during succession when species might coexist more favourably in particular combinations; and a long-term evolutionary equilibrium. To this latter Simberloff (1974) added a taxon cycle equilibrium occurring on an evolutionary time-scale as well. Since the evolutionary time-scale was broached, palaeontologists have applied the dynamic mechanisms of island biogeography theory to questions of regional and global equilibria, postulating, with varying degrees of success, that the number of higher taxa was the result of an equilibrium between speciation and extinction processes. Hoffman (1985a) reviews the status of this research and only considers it plausible on a regional scale, and finds even this unsubstantiated by empirical evidence.

Over ecological time many studies provide experimentally obtained colonization curves for a variety of insular situations. These have been reported for artificial ponds and beakers (Hubbard, 1973; Maguire, 1963, 1971); artificial marine reefs (Walsh, 1985); synthetic sponges in marine and fresh water (Cairns *et al.*, 1969, A. Schoener, 1974b); and marine fouling panels (Fig. 15.7; A. Schoener, 1974a; Schoener and Schoener,

1981; Shin, 1981; Field, 1982). With few exceptions most of the above were considered to be consistent with the MacArthur–Wilson equilibrium model over the observed short-term interval. Natural marine reefs in some cases, particularly where reef spatial scale is sufficiently large, indicate similar trends (Brock *et al.*, 1979; Smith, 1979), but not in other cases where small patch reefs or coral heads are involved (Talbot *et al.*, 1978; Sale, 1980). In the latter studies, fish species exhibit continual change in species composition and substantial variation in species numbers even after several years. Do the community dynamics of small isolated reefs differ fundamentally from larger, less isolated reef systems (Walsh, 1985)?

Plotting the long-term record of colonization of synthetic sponges by freshwater protozoans in a lake, suggested that three of the early equilibrial phases were reached during a five-year colonization period (Cairns and Henebry, 1982). Composite data from several identical experiments initiated during different years in the same lake indicated an initial rise in species numbers towards the non-interactive equilibrium, oscillation during the early interactive equilibrium, and a substantial decrease in those numbers during the assortative equilibrium phase some four to five years after colonization began. Although based on limited replication and composite data, this may represent the first record of colonization through the assortative phase. For short-lived species such as protozoans, half a decade might suffice to incorporate aspects of succession into the colonization process. But if this is the case fewer, rather than more, species, are accommodated during this stage. Time-scales appropriate for detection of similar colonization stages for other taxa might be considerably longer.

Trends relevant to the shape of the colonization curve, but not considered in the equilibrium model, were discovered by comparing identical islands positioned in various seasons and latitudes. The initial rate of rise of the colonization curve varied inversely with northern hemisphere latitude, with colonization curves rising most rapidly for fouling panels in tropical marine regions, probably reflecting the latitudinal gradients in the size of the species pool (Schoener *et al.*, 1978). Fouling panels submerged in one temperate region during seasons with varied immigration rates indicated that panels reached their equilibrium cycle at different times; the oscillating equilibrium that was approached fluctuated within bounds in response to seasonal changes (Osman, 1978).

Determining whether equilibrium is actually being approached is, of course, a critical issue and varying degrees of sophistication have been employed to assess this. Visual observation of the colonization curve can provide some indication, although oscillations resulting from seasonal changes (Osman, 1978) or species interactions (Paine, 1966; Day, 1977) introduce variation, raising the question of how much fluctuation is acceptable. Keough and Butler (1983) propose a test of equilibrium by comparing observed variation in species numbers around the mean with a previously defined level. Another method of establishing whether or not equilibrium is being approached involves arbitrarily dividing the colonization curve into two, and comparing each section to determine whether the earlier portion rises more steeply and is significantly different from the later portion. A still stronger test involves calculation of differences between immigration rate and extinction rate curves over time. Since these curves will converge at equilibrium (Fig. 15.9), the difference between them should decrease with time, yielding a negatively sloping curve (A. Schoener, 1974b); this method raises concerns discussed earlier involving the computation of these rate curves. Comparison of multiple census data over several census periods by an ordination technique, called step-along, has been suggested: this incorporates both species' presence/absence data and their abundances (Williamson, 1983b).

The single most convincing argument regarding the existence of an equilibrium in species numbers is made by comparison of post-defaunation censuses with pre-defaunation species counts. In many experimental investigations only the former data are available, but in a few studies this comparison can be made. In addition to the Simberloff and Wilson (1969, 1970) mangrove studies, the first to suggest an approach to an equilibrium species number, a few other recolonization studies also provide relevant data. Terrestrial arthropods recolonized fumigated *Spartina* islands rapidly and within half a year their species numbers approached their pre-defaunation values (Rey, 1981; Rey and Strong, 1983; Rey, 1984). Collections of reef fishes made on a Hawaiian reef, defaunated twice 11 years apart, yielded 76 species in 1966 and 81 species in 1977 (Brock *et al.*, 1979). Recolonization of a Gulf of Mexico reef attained a species richness of reef fishes similar to its pre-defaunated level after 15 months (Smith, 1979).

15.2.7 Trophic structure

The determination of the existence of definite trends in community trophic structure as island biotas develop has implications for the extent of organization in island communities. According to the MacArthur–Wilson equilibrium model, island biotas comprise chance collections of species with similar physical requirements. Were developing island biotas to move directionally towards a particular trophic structure, more regulation might be suggested than would be expected under the premises of the model. Most experimental studies do not address this particular aspect, but at least two provide relevant analyses.

Heatwole and Levins (1972) analysed Wilson and Simberloff's (1969; Simberloff and Wilson, 1970) mangrove island defaunation data, categorizing the terrestrial arthropod colonists where possible as herbivores, scavengers, detritus feeders, wood borers, ants, predators and parasites. Chi-squared analyses indicated that prior to defaunation species on the six mangrove islands comprised a particular distribution among trophic categories which was indistinguishable from that observed after defaunation, when equilibrium was again approached, and this distribution was gradually converged upon during recolonization. Whether these results implied predictability, as Heatwole and Levins considered to be the case, or whether these were the result of the classificatory scheme (Simberloff, 1976c) has been challenged. Simberloff showed by simulation that increasing numbers of species with time was primarily responsible for the purported approach to a constant trophic distribution. Later reanalysis of these data questioned Simberloff's assumption that all species are available for colonization at one time, and demonstrated a colonization sequence of herbivores, predators and parasites (Glasser, 1982).

A less detailed analysis of recolonization was made by classifying fishes of a Hawaiian coral patch reef into four trophic categories (Brock *et al.*, 1979). This reef was defaunated twice, 11 years apart, and subsequently censused during the two periods at which equilibrium was approached, and twice within the first year after the reef's second

Table 15.1. Trophic structure of a twice-defaunated reef fish assemblage (1966 and 1977 defaunations) observed at defaunation (0) and at two dates thereafter (6, 12 months). Number of species are given in each trophic class. Chi-squared contingency tests are used for comparisons: DF = degrees of freedom; NS = not significant (modified from Brock *et al.*, 1979)

	Months after defaunation						
	1966	*1977*					
Trophic category	*0*	*0*	*6*	*12*	χ^2	*DF*	*Significance*
Carnivores	49	51	16	39			
Omnivores	3	2	2	3			
Planktivores	9	8	3	4			
Herbivores	9	10	10	5			
	———	———			0.35	3	NS
	———	———	———	———	10.87	9	NS
		———	———		5.74	3	NS
		———	———	———	1.57	3	NS

defaunation. During the equilibrial periods nearly all fish species by number and by weight were accounted for in the trophic analyses; censuses conducted during the first year after defaunation were visual censuses. In each of the four observation periods carnivores predominated. The comparisons indicate that significant differences in trophic structure were never found (Table 15.1), suggesting that within six months after defaunation, trophic structure, if altered at all, was re-established.

15.2.8 Actual achievements and potential role of the model

The basic role of any model is not only to depict patterns and explain processes, but to stimulate testing which in turn leads to verification or modification of features of the model which require refinement. In the case of the equilibrium model some modifications have been made in light of experimental results, while in other cases special explanations have been invoked when the model and reality failed to match. Among the actual achievements which can be credited to the model are the introduction of an ecological time frame to a previously evolutionary time-class of problems, a change which has reset our biogeographical clock. The transition between evolutionary and ecological time periods is extensive, however, and although some attempt has been made to interpret aspects of equilibrium theory in an evolutionary light, this transition has so far been only partially illuminated.

The model represents a step forward in understanding island species' richness patterns. By providing a set of simple, testable hypotheses that encouraged input from a variety of taxa rather than focusing primarily on one taxon, it broadened our biogeograhic base. Since taxa may react differently to similar environmental conditions, investigating biological responses through these series of filters increases the power of the analyses.

Finally, the MacArthur–Wilson equilibrium model generated quantitative predictions, necessitating tests of greater precision and more replication, use of null hypotheses, and less dependence on individual island case histories. However, critiques of the rigour of these tests, some of the interpretations, individual aspects of the theory, and some of its suggested applications as discussed next, have emerged, pointing out deficiencies still to be dealt with (Gilbert, 1980; Simberloff, 1976a). As long as such criticism is constructive it will be beneficial to this relatively new research direction.

15.3 IMPLICATIONS OF ISLAND BIOGEOGRAPHY THEORY

Increased fragmentation and isolation of natural landscapes have focused attention on two related problems to which island biogeographic theory has been extended. The first of these is the design of nature reserves, and the second, forest management principles. Before examining the progress made in each area and the ensuing controversy, it is appropriate to determine how, and whether, these areas can validly be considered insular.

In the sense that any habitat separated from similar types of habitat by some intervening, different habitat can be considered insular, both of these situations conform to the general definition of a habitat island. Reserves are enclaves of pristine habitat surrounded by man-made development; present day forests are remnants of larger forested tracts now surrounded by cut areas, reforested areas, or development.

There are ways, however, in which both reserves and forest remnants do now, or will in the future, differ from true islands. At present many species characteristic of these areas can reinvade from not too distant source areas; however, in the future such source areas may well have vanished. In that case, the appropriateness of an equilibrium model considering both immigration and extinction has been questioned (Terborgh, 1976). Immigration of less desirable widespread, fugitive species found in surrounding areas may however increase (Pickett and Thompson, 1978). Although one often considers extinction as the main problem to be dealt with in such situations, immigration of undesirable species may also be troublesome (Janzen, 1983).

Although it is difficult to establish with certainty that mainland habitats are actually becoming increasingly insular, this is definitely the impression, and it is with this assumption in mind that island biogeography theory has been applied to nature reserve design and to forest management.

15.3.1 Nature reserve design

Nature reserves, islands in a sea of development, can potentially span a range of dimensions limited by natural and financial bounds. Can insights regarding the prediction of their optimal dimensions be gleaned from island biogeographic theory?

Although some differences as well as similarities exist between islands and nature reserves, implications of island biogeographic theory have been extended toward the design of such reserves. Whether or not this application is premature, as has been suggested by Simberloff and Abele (1976a), an extensive literature has nonetheless developed along this vein, in part for lack of any other direction. Most pertinent theoretical considerations focus on maximizing species diversity of reserves. But other desirable goals also exist. These include preserving the entire set of interactions that *in toto* characterize a habitat whose protection is desired (Janzen, 1983). Preservation of particular species is, of course, another valid consideration. To the extent that interest focuses on individual species' preservation, or on the complex of interactions in a habitat, the equilibrium theory may not be directly applicable. Some of its precepts, however, particularly those relating to minimum area considerations, may still be useful. Most discussion has so far considered preservation of the biota, and the appropriate size and position of reserves to accomplish this goal.

Wilson and Willis (1975) put forward a series of geometrical 'rules of design' and applied the equilibrium island biogeographic model to nature reserve design. Some of these suggestions have thus far elicited minimal controversy. Among these are, first, the assumption that reserves will have lower extinction rates if they are connected by corridors. Second, because of distance effects, reserves in a clumped grouping are preferable to a similar group of reserves arranged linearly. Shape, another consideration, led to their suggestion of round, rather than elongate area, because of the peninsula effect (Chapter 4 also discusses shape and habitat heterogeneity in relation to species–area relationships). Most controversial has been the question of preferred size of reserves.

Should one advocate the formation of as large as possible a reserve, or if area is limited, instead choose a series of reserves that add up to the same total area? Based on the species–area relationship, indicating a greater number of species present on increasingly large areas, Wilson and Willis (1975) favoured the formation of the largest reserve. Others consider the species–area relationship to be neutral on this matter, arguing that depending upon the gradient of colonizing abilities among species in the species pool, theory does not predict either alternative exclusively (Simberloff and Abele, 1976b). Instead, how population size is related to area, and extinction to population size, determines whether one large or several small reserves will generate the lower extinction rate. Simberloff and Abele (1976b) consider both of these relationships to be species-specific, fundamentally empirical, and poorly known for virutally all species. Under a variety of conditions they advocate the formation of several small areas which total the same as one larger area. Citing several instances where accumulated species richness has been determined for small and large isolates of similar nature, they observe that in a majority of the cases examined, the total number of arboreal arthropods, crustaceans, plants, birds, mammals or lizard species is greater when smaller habitats are combined (Simberloff and Abele, 1982). They suggest on this basis that fragmentation of reserves would be preferable.

Practical considerations in deciding the benefits between a single large reserve and several smaller reserves have been voiced as well. These include larger population sizes to offset genetic inbreeding, as well as management benefits on large reserves; for small reserves there are benefits through edge effects in increasing species richness, as well as reducing potential damage to any one area through natural catastrophes (Higgs, 1981; Simberloff and Abele, 1982). As pointed out in Chapter 4, factors increasing habitat variability or heterogeneity are important considerations both in these questions and in species–area relationships.

Theoretical arguments suggest that the decision to establish small or large reserves is heavily weighted by the number of species in common between a set of smaller reserves. Higgs and Usher (1980) consider the number of species on one large reserve (A) or two smaller reserves $pA + (1 - p)A$ (where $0 < p < 1$). The number of species on one large reserve S_1, is: $S_1 = cA^Z$ where c is a constant usually referred to as the intercept (see Chapter 4) and z is the slope of the species–area relationship.

The number of species on two smaller reserves is:

$$S_2 = c(pA)^z + c[(1 - p)A^z] - V \qquad (15.1)$$

where V is the number of species common to the two smaller areas – the overlap between them.

When the same number of species is conserved, in both cases $S_1 = S_2$ and

$$V = [p^z + (1 - p)^z - 1]cA^z \qquad (15.2)$$

If V exceeds this value, more species will be conserved on one single large reserve, and when V is smaller, more on smaller reserves. The proportional overlap of species on the two smaller areas is:

$$P_v = \frac{V}{S_2}$$

and substituting P_v in equations (15.1) and (15.2) above, gives $P_v = p^z + (1 - p)^z - 1$. Higgs and Usher (1980) have plotted this function for various values of p, and find that the value of p has relatively little effect on the relationship between P_v and z.

Although considerable argument thus favours, in some cases, the formation of several smaller reserves over one large reserve, Gilpin and Diamond (1980) point out that taken to the extreme, numerous small areas prove suboptimal compared with one large reserve; fragmentation into ten small reserves each one tenth the total area, provides fewer species in total than does a single large reserve.

Considering the importance of the question of nature reserve design, the extensive discussion in the literature, and the immediacy of some decisions, one would expect experimentation geared towards this applied question. But so far, this has been minimal. Simberloff and Abele (1982) provide an explanation of an experimental test related to this question. In Simberloff's (1976b) experiments on mangrove islands, two censused islands were each fragmented into archipelagoes of smaller islands by cutting channels through them. For their terrestrial arthropod fauna these minimal water gaps constituted sufficient impediments to dispersal to change one large island into several smaller islands. After allowing sufficient time for these islands fragments to equilibrate, the fragments were censused. Comparisons of the total number of species indicated that the larger of these originally intact islands had a slightly greater number of species on its combined subareas (81) than in the original intact island (77). But the reverse was true of the smaller of the two original islands. As a result, Simberloff and Abele (1982) view the experiment as inconclusive.

Many others have expressed a conservative opinion with regard to the question of optimal reserve size, joining Leopold (1949) in considering wilderness a resource which can shrink but not grow. They advocate a more prudent approach, setting aside the largest possible areas for reserves now, until considerably more experimentation and enlightenment ensues.

15.3.2 Forestry management

Recognizing that a future forest we design should be superior to any we might inherit by default, Harris (1984) advocates a forest management plan whose goals are to maximize biotic diversity and conserve mature forest. Although none of the island biogeographical premises have been validated for the Douglas fir old-growth forest he discusses, Harris integrates ecological principles with those of island biogeography to suggest a long-rotational management scheme that reduces the disastrous effects of cutting the once extensive old-growth stands of the Pacific Northwest of America. By surrounding what is left of this now island-like forest with timber stands of varied age, the smaller central old-growth core island could be transformed into a larger island whose sections are destined to be harvested at various times over three centuries (Fig. 15.10). By alternating timber stands of varying age around this core forest, maximum differences between adjacent stands can be encouraged, and species numbers increased by virtue of increased edge effects. Since the addition of long-rotation management stands will increase the effective size of individual old-growth habitats, Harris advocates that more emphasis be placed on conserving the total number of old-growth cores, than on their increased core size. As these

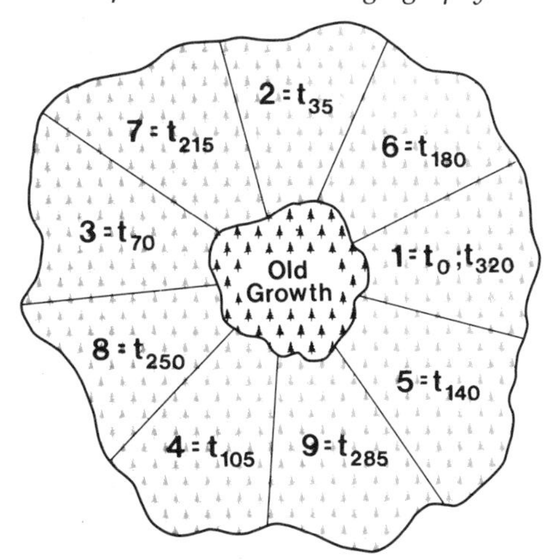

Figure 15.10 Suggested long-rotation management scheme for Douglas fir forests. The old-growth core habitat island would be surrounded by nine tracts of forests in various successional stages. Numbers refer to the sequence of harvesting these tracts over a 320-year cutting cycle. Note that adjacent stands are of considerably different ages (modified from Harris, 1984).

old-growth cores become preserved, inter-island distance between them will become reduced, and their chance of recolonization and genetic exchange should increase.

15.4 SUMMARY

In the past quarter century the emphasis in biogeography has been on an experimental approach and much of this experimentation has been directed towards research on islands and island-like habitats. The MacArthur–Wilson equilibrium theory of island biogeography has set the stage by stimulating a theoretical backdrop, a quantitative basis for explaining species richness patterns observed on islands. This it does in terms of dynamic immigration and extinction rate processes continuing over the short term, and is therefore amenable to experimentation. Some basic predictions of the model, particularly as they relate to island area and distance from source region, have been examined by relevant experimentation. Considering the diversity of island-like situations that have been studied, and the variety of taxa involved, perhaps it is not surprising that not all aspects of the model are supported by these studies. Yet it is not in any one aspect of the model where disagreement occurs. The important contribution made by the model, irrespective of its validation in part or *in toto,* is the impetus that it has given biogeographers to refocus attention in a quantitative way on processes that had only been examined qualitatively previously. Criticism notwithstanding, it continues to encourage experimentation in this field. Extrapolations based on the model have been used in related problems of social and economic consequence, providing further evidence that research on islands still needs to be avidly pursued.

References

Abele, L.G. and Walters, K. (1979) Marine benthic diversity: a critique and alternative explanation. *J. Biogr.*, **6**, 115–26.

Abercrombie, M.L.J. (1960) *The Anatomy of Judgement; an investigation into the process of perception and reasoning*, Hutchinson, London.

Abrams, P. (1983) The theory of limiting similarity. *A. Rev. Ecol. System.*, **14**, 359–76.

Abrams, P.A. (1986) Character displacement and niche shift analyzed using consumer–resource models of competition. *Theoretical Population Biology*, **29**, 107–60.

Abrams, P.A. (1987) On classifying interactions between populations. *Oecologia*, **73**, 272–81.

Abramsky, Z. and Rosenzweig, M.L. (1983) Tilman's predicted productivity–diversity relationship shown by desert rodents. *Nature, Lond.*, **309**, 150–1.

Ab'Saber, A.N. (1977) Espaços ocupados pela expansão dos climas secas na America do Sul, por ocasião dos períodos glacias quaternários. *Paleoclimas*, **3**, 1–19.

Ab'Saber, A.N. (1982) The paleoclimate and paleoecology of Brazilian Amazonia. In *Biological Differentiation in the Tropics* (ed G.T. Prance), Columbia Univ. Press, New York, pp. 41–59.

Adler, G.H. and Wilson, M.L. (1985) Small mammals on Massachusetts islands: the use of probability functions in clarifying biogeographic relationships. *Oecologia*, **66**, 178–86.

Agakhanyants, O.E. (1981) *Arid Mountains of the USSR. Nature and a geographical model of florogenesis*, Mysl. Moscow.

Ahearn, J.N. (1980) Evolution of behavioural reproductive isolation in a laboratory stock of *Drosophila silvestris*. *Experientia*, **36**, 63–4.

Allaby, M. and Lovelock, J. (1983) *The Great Extinction*, Secker and Warburg, London.

Allan, H.H. (1961) *Flora of New Zealand* (Vol. 1), R.E. Owen, Wellington, New Zealand.

Allen, J.C. (1975) Mathematical models of species interactions in time and space. *Am. Nat.* **109**, 319–42.

Alvarez, L.W. (1983) Experimental evidence that an asteroid impact led to the extinction of many species 65 million years ago. *Proc. nat. Acad. Sci.*, **80**, 627–42.

Alvarez, L.W., Alvarez, W., Asaro, F. and Michel, H.V. (1980) Extraterrestrial cause for the Cretaceous–Tertiary extinction. *Science*, **208**, 1095–108.

Alvarez, W. (1986) Towards a theory of impact crisis. EOS, *Trans. Amer. geophys. Un.*, **67**, 649, 653–55, 658.

Alvarez, W., Kauffman, E.G., Surlyk, F. *et al.* (1984) Impact theory of mass extinctions in the invertebrate fossil record. *Science*, **223**, 1135–41.

Alvarez, W. and Muller, R.A. (1984) Evidence from crater ages for periodic impacts of the Earth. *Nature, Lond.*, **308**, 718–21.

Ammermann, A.J. and Cavalli-Sforza, L.L. (1981) *The Neolithic Transition and the Genetics of Populations in Europe*, Princeton Univ. Press, Princeton, N.J.

Anderson, E. (1984) Who's who in the Pleistocene, a mammalian bestiary. In *Quaternary Extinctions: a prehistoric revolution* (eds P.S. Martin and R.C. Klein), University of Arizona Press, Tucson, pp. 40–89.

Andrewartha, H.G. and Birch, L.C. (1954) *The Distribution and Abundance of Animals*, University of Chicago Press, Chicago.

Antonovics, J. (1968) Evolution in closely adjacent plant populations. V. Evolution of self-fertility. *Heredity*, **23**, 219–38.

Aquadro, C.F., Dene, S.F., Bland, M.M. *et al.* (1986) Molecular population genetics of the alcohol dehydrogenase region of *Drosophila melanogaster*. *Genetics*, **114**, 1165–90.

Archibald, J.D. and Clemens, W.A. (1982) Late Cretaceous extinctions. *Am. Scient.*, **70**, 377–85.

Arnold, S.J. (1972) Species densities of predators and their prey. *Am. Nat.*, **106**, 220–36.

Arntzen, J.W. (1978) Some hypotheses on postglacial migrations of the firebellied toad, *Bombina bombina* (L) and the yellowbellied toad, *Bombina variegata* (L). *J. Biogr.*, **5**, 339–95.

Arrhenius, O. (1921) Species and area. *J. Ecol.*, **9**, 95–9.

Atkinson, A. C. (1985) *Plots, Transformations and Regression*, Clarendon Press, Oxford.

Avise, J.C. and Ayala, F.J. (1976) Genetic differentiation in speciose versus depauperate phylads: evidence from Californian minnows. *Evolution*, **30**, 46–58.

Avise, J.C., Giblin-Davidson, C., Laerm, J. *et al.* (1979a) Mitochondrial DNA clones and matriarchal phylogeny within *Geomys pinetis*. *Proc. Nat. Acad. Sci.*, **76**, 6694–8.

Avise, J.C. Lansman, R.A. and Shade, R.O. (1979b) The use of restriction endonucleases to measure mitochondrial DNA sequence relatedness in natural populations. *Genetics*, **92**, 279–95.

Ax, P. (1984) *Das Phylogenetische System*, G. Fischer Verlag, Stuttgart.

Axelrod, D.I. and Bailey, H.P. (1969) Paleotemperature analysis of Tertiary floras. *Paleogr., Paleocl. and Paleoecol.*, **6**, 163–95.

Baker, R.R. (1978) *The Evolutionary Ecology of Animal Migration*, Hodder and Stoughton, London.

Bakker, R.T. (1977) Cycles of diversity and extinction: a plate tectonic/topographic model. In *Patterns of Evolution* (ed A. Hallam), Elsevier, Amsterdam, pp. 431–78.

Ball, I.R. (1975) Nature and formulation of biogeographic hypotheses. *Syst. Zool.*, **24**, 407–30.

Ball, I. (1981) The order of life – towards a comparative biology. *Nature, Lond.*, **294**, 675–76.

Barbour, C.D. and Brown, J.H. (1974) Fish species diversity in lakes. *Am. Nat.*, **108**, 473–89.

Barbour, M.G. and Major, J. (1977) *Terrestrial Vegetation of California*, Wiley, New York.

Barton, N.H. (1979) The dynamics of hybrid zones. *Heredity*, **43**, 341–59.

Barton, N.H. (1980) The fitness of hybrids between two chromosomal races of the grasshopper *Podisma pedestris*. *Heredity*, **45**, 47–59.

Barton, N.H. (1986) The maintenance of polygenic variation through a balance between mutation and stabilising selection. *Genet. Res.*, **47**, 209–16.

Barton, N.H. (1987) The probability of fixation of an advantageous allele in a subdivided population. *Genet. Res*, **50**, 35–40.

Barton, N.H. and Charlesworth, B. (1984) Genetic revolutions, founder effects and speciation. *Ann. Rev. Ecol. System.*, **15**, 133–64.

Barton, N.H. and Hewitt, G.M. (1981) The genetic basis of hybrid inviability between two chromosomal races of the grasshopper *Podisma pedestris. Heredity*, **47**, 367–83.

Barton, N.H. and Hewitt, G.H. (1985) Analysis of hybrid zones. *Ann. Rev. Ecol. System.*, **16**, 113–48.

Baskin, C.C., Baskin, J.M. and Quarterman, E. (1981) Effect of drought stress on transpiration rates and leaf areas of *Astragalus tennesseensis*, a near endemic to cedar glades. *J. Tennessee Acad. Sci.*, **56**, 27–31.

Baskin, J.M. and Baskin, C.C. (1978) The seed bank in a population of an endemic plant species and its ecological significance. *Biol. Conserv.*, **14**, 125–29.

Baskin, J.M. and Baskin, C.C. (1979) The ecological life cycle of the cedar glade endemic *Lobelia gattingeri. Bull. Torrey Botan. Club*, **106**, 176–81.

Baskin, J.M. and Baskin, C.C. (1981) Geographical distribution and notes on the ecology of the rare endemic *Leavenworthia exigua* var. *laciniata. Castanea*, **46**, 243–7.

Bazykin, A.D. (1969) Hypothetical mechanism of speciation. *Evolution*, **23**, 685–7.

Bazzaz, F.A. (1975) Plant species diversity in old-field successional ecosystems in southern Illinois. *Ecology*, **56**, 485–8.

Beever, J.W., III (1979) The niche-variation hypothesis: an examination of assumptions and organisms. *Evol. Theory*, **4**, 181–91.

Begon, M., Harper, J.L. and Townsend, C.R. (1986) *Ecology: individuals, populations and communities*, Blackwell Scientific Publications, Oxford.

Bengtsson, B.O. and Bodmer, W.F. (1976) On the increase of chromosome mutations under random mating. *Theor. Pop. Biol.*, **9**, 260–81.

Bengtsson, B.O. and Christiansen, F.B. (1983) A two-locus mutation-selection model and some of its evolutionary implications. *Theor. Pop. Biol.*, **24**, 59–77.

Benson, W.W. (1982) Alternative models for infrageneric diversification in the humid tropics: tests with passion vine butterflies. In *Biological Diversification in the Tropics* (ed. G.T. Prance), Columbia Univ. Press, New York, pp. 608–40.

Benton, M.J. (1985) Mass extinction among non-marine tetrapods. *Nature, Lond.*, **316**, 811–14.

Berglund, B.E. (1966) Late Quarternary vegetation in eastern Blekinge, south-eastern Sweden, II. Post-glacial time. *Opera Botanica*, **12**, 1–191.

Bernabo, J.C. and Webb, T. (1977) Changing patterns in the Holocene pollen record from northeastern America: a mapped summary. *Quaternary Res.*, **8**, 64–96.

Bernstein, R.J. (1983) What is the difference that makes a difference? Gadamer, Habermas, and Rorty. *PSA, 1982, Volume 2*, 331–59.

Berry, M.V. and Lewis, Z.V. (1980) On the Weierstrass-Mandelbrot fractal function. *Proc. R. Soc.*, A, **370**, 459–84.

Berry, R.J. (1983) Diversity and differentiation: The importance of island biology for general theory. *Oikos*, **41**, 525–9.

Berry, R.J. (1986) Genetics of insular populations of mammals, with particular reference to differentiation and founder effects in British small mammals. *Biol. J. Linn. Soc.*, **28**, 205–30.

Berry, W.B.N. (1968) *Growth of a Prehistoric Time Scale*, W.H. Freeman, San Francisco.

Betancourt, J.L. (1986) Paleoecology of pinyon-juniper woodlands: summary. In *Proceedings, Pinyon-juniper Conference, General Technical Report INT 215*, USDA Forest Service, pp. 129–39.

Beven, S., Connor, E.F. and Beven, K. (1984) Avian biogeography in the Amazon basin and the biological model of diversification. *J. Biogeogr.*, **11**, 383–99.

Bickham, R.J. and Baker, J.W. (1986) Speciation by monobrachial centric fusions. *Proc. Natl. Acad. Sci. (USA)*, **83**, 8245–8.

Birks, H.J.B. (1976) The distribution of European pteridophytes: A numerical analysis. *New Phytol.*, **77**, 257–87.

Bishop, J.A. (1981) A Neodarwinian approach to resistance: examples from mammals. In *Genetic Consequences of Man-Made Change* (eds J.A. Bishop and L.M. Cook), Academic Press, London, pp. 37–52.

Bishop, J.A. and Cook, L.M. (1981) *Genetic Consequences of Man-Made Change*, Academic Press, London.

Blondel, J. (1986) *Biogéographie Évolutive*, Masson, Paris.

Blondel, J., Vuilleumier, F., Marcus, L. and Terouanne, E. (1984) Is there ecomorphological convergence among Mediterranean bird communities of Chile, California and France? In *Evolutionary Biology* (eds M.K. Hecht, B. Wallace and G.T. Prance) (vol. 18), Plenum, New York, pp. 141–214.

Bloom, A.L. (1978) *Geomorphology. A systematic analysis of late Cenozoic landforms*, Prentice-Hall, Englewood Cliffs, New Jersey.

Blouin, M.S. and Connor, E.F. (1985) Is there a best shape for nature reserves? *Biol. Conserv.*, **32**, 277–88.

Boag, P.T. (1983) The heritability of external morphology in Darwin's finches (*Geospiza*) on Isla Daphne Major, Galapagos. *Evolution*, **37**, 877–94.

Boag, P.T. and Grant, P.R. (1981) Intense natural selection in a population of Darwin's finches (Geospizinae) in the Galapagos. *Science*, **214**, 82–5.

Böcher, T.W., Holmen, K. and Jakobsen, K. (1968) *The Flora of Greenland*. P. Haase and Son, Copenhagen.

Bock, I.R. and Parsons, P.A. (1977) Species diversities in *Drosophila* (Diptera): a dependence upon rain forest type of the Queensland (Australian) humid tropics. *J. Biogr.*, **4**, 203–13.

Bohor, B.F., Foord, E.E., Modreski, P.J. and Triplehorn, D.M. (1984) Mineralogic evidence for an impact event at the Cretaceous-Tertiary boundary. *Science*, **224**, 867–9.

Bory de Saint-Vincent, J.B.G.M. (1804) *Voyage dans les quatres iles des mers d'Afrique*, Paris.

Boucher, D.H. (1985) *The Biology of Mutualism: ecology and evolution*, Oxford University Press, New York.

Boucot, A.J. (1975) *Evolution and Extinction Rate Controls*, Elsevier, Amsterdam.

Bowerman, M.L. (1944) *The Flowering Plants and Ferns of Mount Diablo, California*, Gillick Press, Berkeley, California.

Bowers, M.A. and Brown, J.H. (1982) Body size and coexistence in desert rodents: chance or community structure? *Ecology*, **63**, 391–400.

Box, G.E.P. and Cox, D.R. (1964) An analysis of transformations. *J. Rl. Statist. Soc.* B, **26**, 211–46.

Bradshaw, A.D. (1972) Some of the evolutionary consequences of being a plant. *Evol. Biol.*, **5**, 25–47.

Bradshaw, A.D. (1984) The importance of evolutionary ideas in ecology – and *vice versa*. In *Evolutionary Ecology* (ed B. Shorrocks), Blackwell Scientific Pubns, Oxford, pp. 1–25.

Bramwell, D. (ed.) (1979) *Plants and islands*, Academic Press, London.

Bramwell, D. and Bramwell, Z. (1974) *Wild Flowers of the Canary Islands*, S. Thornes, London.

Braun-Blanquet, J. (1923) *L'Origine et le Developpement des Flores dans le Massif Central de France*, Paris.

Braun-Blanquet, J. (1964) *Pflanzensoziologie*, 3rd edn, Springer, Berlin.

Brew, J.S. (1984) An alternative to Lotka-Volterra competition in coarse-grained environments. *Theor. Pop. Biol.*, **25**, 265–88.

Briggs, J.C. (1984) *Centres of Origin in Biogeography*, Biogeographical Monographs 1, University of Leeds.

Brock, R.E., Lewis, C. and Wass, R.C. (1979). Stability and structure of a fish community on a coral patch reef in Hawaii. *Mar. Biol.*, **54**, 281–92.

Brock, T.D. (1985) Life at high temperatures. *Science*, **230**, 132–8.

Brooks, D.R. (1981) Hennig's parasitological method: a proposed solution. *Syst. Zool.*, **30**, 229–49.

Brooks, D.R. (1986) Historical ecology: a new approach to studying the evolution of ecological associations. *Ann. Mo. Bot. Gdn.*, **72**, 660–80.

Brown, G. and Sanders, J.W. (1981) Lognormal genesis. *J. Appl. Probability*, **18**, 542–47.

Brown, J.H. (1981) Two decades of homage to Santa Rosalia: toward a general theory of diversity. *Am. Zool.*, **21**, 877–88.

Brown, J.H. (1971) Mammals on mountain tops: nonequilibrium insular biogeography. *Amer. Nat.*, **105**, 467–78.

Brown, J.H. (1973) Species diversity of seed-eating desert rodents in sand dune habitats. *Ecology*, **54**, 775–87.

Brown, J.H. (1987) Variation in desert rodent guilds: patterns, processes and scales. In *Organization of Communities: past and present* (eds J.H.R. Gee and P.S. Giller), Blackwell Scientific Pubns, Oxford, pp. 185–204.

Brown, J.H. and Davidson, D.W. (1977) Competition between seed-eating rodents and ants in desert ecosystems. *Science*, **196**, 880–2.

Brown, J.H., Davidson, D.W., Munger, J.C. and Inouye, R.S. (1986) Experimental community ecology: the desert granivore system. In *Community Ecology* (eds J. Diamond and T.J. Case), Harper and Row, New York, pp. 41–61.

Brown, J.H. and Gibson, A.C. (1983) *Biogeography*, C.V. Mosby, St. Louis.

Brown, J.H. and Kodric-Brown, A. (1977) Turnover rates in insular biogeography: effect of immigration on extinction. *Ecology*, **58**, 445–9.

Brown, K.S. Jr (1977) Centros de evolucão, refúgios quarternários e conservacãode patrimonios genéticos na região neotropical: padrões de diferencração em Ithomiinae (Lepidoptera: Nymphalidae). *Acta Amazonica*, **7**, 75–137.

Brown, K.S. Jr (1982) Paleoecology and regional patterns of evolution in neotropical forest butterflies. In *Biological Differentiation in the Tropics* (ed G.T. Prance), Columbia University Press, New York, pp. 255–308.

Brown, K.S. and Benson, W.W. (1977) Evolution in modern Amazonian non-forest islands: *Heliconius larmathena*, *Biotropica*, **9**, 95–117.

Brown, V. K. and Southwood, T.R.E. (1983) Trophic diversity, niche breadth and generation times of exopterygote insects in a secondary succession. *Oecologia*, **56**, 220–5.

Brown, W.L. and Wilson, E.O. (1956) Character displacement. *Syst. Zool.*, **5**, 49–64.

Brundin, L. (1966) Transantarctic relationships and their significance as evidenced by chironomid midges. *Svenska Vtenskapsakad. Handl.*, **11**, 1–472.

Brundin, L. (1968) Application of phylogenetic principles in systematics and evolutionary theory. In *Current Problems of Lower Vertebrate Phylogeny* (ed T. Ørvig). Nobel Symposium 4, Stockholm, pp. 473–95.

Brundin, L. (1970) Antarctic land faunas and their history. In *Antarctic Ecology* (ed M.W. Holdgate) Academic Press, London, pp. 41–53.

Brundin, L. (1972a) Evolution, causal biology and classification. *Zoologica Scripta*, **1**, 107–20

Brundin, L. (1972b) Phylogenetics and biogeography. *Syst. Zool.*, **21**, 69–79.

Brundin, L. (1975) Circum-Antarctic distribution patterns and continental drift. In *Biogéographie at Liaisons Intercontinentales au cours de Mésozoique*, 17e Congres International de Zoologie Monaco 1972, *Mém, Mus. natn. Hist. Nat.*, A, Zoologie, **88**, 19–28.

Brundin, L. (1976) A Neocomian chironomid and Podonominae-Aphroteniinae in the light of phylogenetics and biogeography. *Zoological Scripta*, **5**, 139–60.

Brundin, L. (1981) Croizat's panbiogeography versus phylogenetic biogeography. In *Vicariance Biogeography, a critique* (eds G. Nelson and D.E. Rosen), Columbia University Press, New York, pp. 94–138.

Brundin, L. (1986) Evolution by orderly stepwise subordination and largely nonrandom mutations. *Syst. Zool.*, **35**, 602–7.

Buffon, G.L.L., Comte de (1761) *Histoire Naturelle, Generale et Particuliere*, Vol. 9, Imprimerie Royale, Paris.

Bull, R.M. (1967) A Study of the Large New Zealand Weevil, *Lyperobius huttoni* Pascoe, 1876 (Coleoptera: Curculionidae: Molytinae). MSc thesis in zoology, Victoria University of Wellington, New Zealand.

Bush, G.L. (1975) Modes of animal speciation. *Ann. Rev. Ecol. Syst.*, **6**, 339–64.

Bush, G.L., Case, S.M., Wilson, A.C. and Patton, J.L. (1977) Rapid speciation and chromosomal evolution in mammals. *Proc. Natl. Acad. Sci. USA*, **74**, 3942–6.

Buth, D.G. (1979) Empirical tests of the evolutionary protein clock. *13th Ann. Conf. on Numerical Taxonomy*, Harvard University, Cambridge, Mass.

Butlin, R.K. (1987) Speciation by reinforcements. *Trends in Ecol. and Evol.*, **2**, 8–13.

Bykov, B.A. (1979) On a quantitative estimate of endemism. *Botan. Mater. Gerb. Inst. Botan. Akad. Nauk Kazakh. SSR.*, **11**, 3–8.

Bykov, B.A. (1983) *Ecological Dictionary*, Nauka, Kazakhskoi SSR, Alma Ata.

Cairns, J. Jr and Henebry, M.S. (1982) Interactive and noninteractive protozoan colonization processes. In *Artificial Substrates* (ed J. Cairns Jr), Ann Arbor Science, Michigan, pp. 23–70.

Cairns, J. Jr and Ruthven, J.A. (1970) Artificial microhabitat size and the number of colonizing protozoan species. *Trans. Am. Microscopical Soc.*, **89**, 100–9.

Cairns, J. Jr, Dahlberg, M.L., Dickson, K.L. *et al.* (1969) The relationship of fresh-water protozoan communities to the MacArthur–Wilson equilibrium model. *Amer. Nat.*, **103**, 439–54.

Calaby, J.H. (1972) Wallace's Line. In *Encyclopaedia of Papua and New Guinea* (ed P. Ryan), Melbourne University Press, Vol. 1. pp. 1181–2.

Calder, J.A. and Taylor, R.L. (1968) *Flora of the Queen Charlotte Islands Part 1,*

Systematics of the vascular plants, Res. Branch, Canada Dept. Agric. Monogr., **4**(1).

Camarda, I. (1984) Studi sulla flora e sulla vegetazione del Monte Albo (Sardegna centro orientale). 1. La Flora. *Webbia*, **37**, 283–327.

Campbell, J.D. and Holden, A.M. (1984) Miocene casuarinacean fossils from Southland and Central Otago, New Zealand. *N.Z. J. Bot.*, **22**, 159–67.

Candolle, A.-P. de (1820) Geographie botanique. In *Dictionnaire des Sciences Naturelles* Vol. 18, 359–422.

Candolle, A.-P. de (1838) *Statistique de la Famille de Composées*, Treuttel and Wurtz, Paris and Strasbourg.

Cann, R.L., Stoneking, M. and Wilson, A.C. (1987) Mitochondrial DNA and human evolution. *Nature, Lond.*, **325**, 31–5.

Capanna, E., Civitelli, M.V. and Cristaldi, M. (1977) Chromosomal rearrangement, reproductive isolation and speciation in mammals: the case of *Mus musculus*. *Boll. Zool.*, **44**, 213–46.

Carey, S.W. (1958) A tectonic approach to continental drift. In *Continental Drift: a Symposium*, University of Tasmania, Hobart, pp. 177–355.

Carey, S.W. (1975) The expanding earth – an essay review. *Earth Sci. Rev.*, **11**, 105–43.

Carlquist, S. (1974) *Island Biology*, Columbia University Press, New York.

Carlson, S.S., Wilson, A.C. and Maxson, R.D. (1978) Do albumin clocks run on time? *Science*, **200**, 1183–5.

Carr, G.D. and Kyhos, D.W. (1981) Adaptive radiation in the Hawaiian silversword alliance (*Compositae-Madiinae*). I. Cytogenetics of spontaneous hybrids. *Evolution*, **35**, 543–56.

Carr, T.R. and Kitchell, J.A. (1980) Dynamics of taxonomic diversity. *Paleobiology*, **6**, 427–43.

Carson, H.L. (1975) The genetics of speciation at the diploid level. *Amer. Nat.*, **109**, 73–92.

Carson, H.L. (1976) Inference of the time of origin of some *Drosophila* species. *Nature, Lond.*, **259**, 395–6.

Carson, H.L. (1986) Sexual selection and speciation. In *Evolutionary Theory* (eds S. Karlin and E. Nevo) Academic Press, New York, pp. 391–409.

Carson, H.L. and Kaneshiro, K.Y. (1976) Drosophila of Hawaii: systematics and ecological genetics. *Ann. Rev. Ecol. Syst.*, **7**, 11–46.

Carson, H.L. and Yoon, J.S. (1982) Genetics and evolution of Hawaiian *Drosophila*. In *The Genetics and Biology of Drosophila* vol. 3b (eds M. Ashburner, H.L. Carson and J.N. Thomson Jr), pp. 298–344.

Case, T.J. (1975) Species numbers, density compensation and colonizing ability of lizards on the islands in the Gulf of California. *Ecology*, **56**, 3–18.

Case, T.J. (1982) Coevolution in resource-limited competition communities. *Theor. Pop. Biol.*, **21**, 69–91.

Case, T.J. (1983) Sympatry and size similarity in *Cnemidophorus*. In *Lizard ecology: studies on a model organism* (eds R.B. Huey, E.R. Pianka and T. W. Schoener), Harvard University Press, Cambridge, Mass., pp. 297–325.

Case, T.J., Faaborg, J. and Sidell, R. (1983) The role of body size in the assembly of West Indian bird communities. *Evolution*, **37**, 1062–74.

Cassels, R. (1984) The role of prehistoric man in the faunal extinctions of New Zealand and other Pacific Islands. In *Quaternary Extinctions: a Prehistoric Revolution* (eds P.S. Martin and R.G. Klein), University of Arizona Press, Tucson, pp. 741–67.

Caswell, H. (1976) Community structure: a neutral model analysis. *Ecol. Monogr.*, **46**, 327–54.
Caswell, H. (1982) Life history theory and the equilibrium status of populations. *Amer. Nat.*, **120**, 317–39.
Chabot, B.F. and Mooney, H.A., eds (1985) *Physiological ecology of North American Plant Communities*, Chapman and Hall, New York.
Charlesworth, B. (1982) Hopeful monsters cannot fly. *Paleobiology*, **8**, 469–74.
Charlesworth, D. and Charlesworth, B. (1980) Sex differences in fitness and selection for centric fusions between sex chromosomes and autosomes. *Genet. Res.*, **293**.
Cheetham, A.H. and Hazel, J.E. (1969) Binary (presence-absence) similarity coefficients. *J. Paleont.*, **43**, 1130–36.
Chesson, P.L. (1986) Environmental variation and the coexistence of species. In *Community Ecology* (eds J. Diamond and T.J. Case), Harper and Row, New York, pp. 240–256.
Chopik, V.I. (1976) *The High Mountain Flora of the Ukrainian Carpathians*. Naukova Dumka, Kiev.
Cifelli, R. (1969) Radiation of Cenozoic planktonic Foraminifera. *Syst. Zool.*, **18**, 154–68.
Clarke, B.C. (1966) The evolution of morph ratio clines. *Amer. Nat.*, **100**, 389–400
Clausen, J., Keck, D.D. and Heisey, W.W. (1940) Experimental studies on the nature of species. *Carnegie Inst. Washington Pubn.* No. **520**, 1–452.
Clayton, W.D. and Cope, T.A. (1980) The chorology of Old World species of Gramineae. *Kew Bull.*, **35**, 125–71.
Clayton, W.D. and Hepper, F.H. (1974) Computer-aided chorology of West African grasses. *Kew Bull.*, **19**, 213–34.
Clegg, M.T. and Allard, R.W. (1972) Patterns of genetic differentiation in the slender wild oat species *Avena barbata*. *Proc. Natl. Acad. Sci USA*, **69**, 1820–4.
Clemens, W.A. (1982) Patterns of extinction and survival of the terrestrial biota during the Cretaceous-Tertiary transition. *Geol. Soc. Am., Spec. Paper*, **190**, 407–13.
Clemens, W.A., Archibald, J.D. and Hickey, L.J. (1981) Out with a whimper not a bang. *Paleobiology*, **7**, 293–8.
CLIMAP Project Members (1984) The last interglacial ocean. *Quarternary Res.*, **21**, 123–224.
Cocks, L.R.M. and Fortey, R.A. (1982) Faunal evidence for oceanic separations in the Palaeozoic in Britain. *J. Geol. Soc. Lond.*, **139**, 465–78.
Cody, M.L. (1974) *Competition and the Structure of Bird Communities*, Princetown University Press, Princetown, New Jersey.
Cody, M.L. (1975) Towards a theory of continental species diversity. In *Ecology and Evolution of Communities* (eds M.L. Cody and J.M. Diamond) Harvard University Press, Cambridge, Mass., pp. 214–357.
Cody, M.L., Fuentes, E.R. Glanz, W. *et al.* (1977) Convergent evolution in the consumer organisms of Mediterranean Chile and California. In *Convergent Evolution in Chile and California: Mediterranean climate ecosystems* (ed H.A. Mooney), Dowden, Hutchinson and Ross, Stroudsburg, Penn., pp. 144–92.
Cody, M.L. and Mooney, H.A. (1978) Convergence versus nonconvergence in Mediterranean-climate ecosystems. *Ann. Rev. Ecol. Syst.*, **9**, 265–321.
Cohan, F.M. (1984) Can uniform selection retard random genetic divergence between isolated conspecific populations. *Evolution*, **38**, 495–504.
Cohen, J.E. (1970) A Markov contingency table model for replicated Lotka-Volterra systems near equilibrium. *Amer. Nat.*, **104**, 547–60.

Cole, B.J. (1983a) Assembly of mangrove ant communities: patterns of geographic distribution. *J. Anim. Ecol.*, **52**, 339–48.
Cole, B.J. (1983b) Assembly of mangrove ant communities: colonization abilities. *J. Anim. Ecol.*, **52**, 349–55.
Cole, K. (1985) Past rates of change, species richness and a model of vegetational inertia in the Grand Canyon, Arizona. *Amer. Nat.*, **125**, 289–303.
Coleman, B.D. (1981) On random placement and species-area relations. *Mathematical Biosci.*, **54**, 191–215.
Coleman, B.D., Mares, M.A., Willig, M.R. and Hsieh, Y.H. (1982) Randomness, area and species richness. *Ecology*, **63**, 1121–33.
Colinvaux, P. (1986) *Ecology*, Wiley, New York.
Collinson, M.E. and Scott, A.C. (1987) Factors controlling the organization and evolution of ancient plant communities. In *Organization of Communities, Past and Present* (eds J.H.R. Gee and P.S. Giller), Blackwell Scientific Pubns, Oxford, pp. 367–88.
Colosi, G. (1925) La teoria della trabżione dei continenti e le dottrine biogeografiche. *L'Universo*, **6**, 179–86.
Colwell, R.K. and Winkler, D.W. (1984) A null model for null models in biogeography. In *Ecological Communities: Conceptual Issues and the Evidence* (eds D.R. Strong Jr, D. Simberloff, L.G. Abele and A.B. Thistle), Princetown University Press, Princetown, New Jersey, pp. 344–59.
Comfort, A. (1987) What real world? (Review of *Philosophy and the brain* by J.Z. Young.) *The Guardian*, (January 23) 11.
Connell, J.H. (1975) Some mechanisms producing structure in natural communities: a model and evidence from field experiments. In *Ecology and Evolution of Communities* (eds M.L. Cody and J.M. Diamond), Harvard University Press, Cambridge, Mass., pp. 460–90.
Connell, J.H. (1978) Diversity in tropical rain forests and coral reefs. *Science*, **199**, 1302–10.
Connell, J.H. (1983) On the prevalence and relative importance of interspecific competition: evidence from field experiments. *Amer. Nat.*, **122**, 661–96.
Connor, E.F. (1986) The role of Pleistocene forest refugia in the evolution and biogeography of tropical biotas. *Trends in Ecology and Evolution*, **1**, 161–5.
Connor, E.F. and McCoy, E.D. (1979) The statistics and biology of the species–area relationship. *Amer. Nat.*, **113**, 791–833.
Connor, E.F., McCoy, E.D. and Cosby, B.J. (1983) Model discrimination and expected slope values in species–area studies. *Amer. Nat.*, **122**, 789–96.
Connor, E.F. and Simberloff, D. (1978) Species number and compositional similarity of the Galapagos flora and avifauna. *Ecol. Monogr.*, **48**, 219–48.
Connor, E.F. and Simberloff, D. (1979) The assembly of species communities: chance or competition? *Ecology*, **60**, 1132–40.
Contandriopoulos, J. (1962) *Recherches sur la Flore endemique de la Corse et sur ses origines*. Thèse presentée à la Faculté des Sciences de Montpellier, France.
Contandriopoulos, J. and Cardona, M.A. (1984) Caractere original de la flore endemique des Baleares. *Botanica Helv.*, **94**, 101–31.
Contandriopoulos, J. and Gamisans, J. (1974) A propos de l'element artico-alpin de la flore corse. *Bull. Soc. bot. Fr.*, **121**, 175–204.
Cook, R.E. (1969) Variation in species density of North American birds. *Syst. Zool.*, **18**, 63–84.
Coope, G.R. (1979) Late Cenozoic fossil Coleoptera: evolution, biogeography and ecology. *Ann. Rev. Ecol. Syst.*, **10**, 247–67.
Coope, G.R. (1987) The response of late Quarternary insect communities to

sudden climatic change. In *Organization of Communities, Past and Present* (eds J.H.R. Gee and P.S. Giller), Blackwell Scientific Pubns, Oxford, pp. 389–406.

Corner, E.J.H. (1963) *Ficus* in the Pacific region. In *Biogeography* (ed. J.L. Gressitt) Bishop Museum, Honolulu, Hawaii, pp. 233–45.

Coryndon, S.C. and Savage, R.J.G. (1973) The origin and affinities of African mammal faunas. In *Organisms and Continents through Time* (ed N.F. Hughes) *Spec. Papers in Palaeontol.*, **12**, 121–35.

Cowen, R. and Stockton, W.L. (1978) Testing for evolutionary equilibria. *Paleobiology*, **4**, 195–200.

Coyne, J.A. (1984) Correlation between heterozygosity and rate of chromosome evolution in animals. *Amer. Nat.*, **123**, 725–9.

Coyne, J.A. and Lande, R. (1985) The genetic basis of species differences in plants. *Amer. Nat.*, **126**, 141–5.

Cracraft, J. (1974) Phylogeny and evolution of the ratite birds. *Ibis*, **116**, 494–521.

Cracraft, J. (1975) Historical biogeography and earth history: perspectives for a future synthesis. *Ann. Missouri bot. Gdn.*, **62**, 227–50.

Cracraft, J. (1980) Biogeographic patterns of terrestrial vertebrates in the southwest Pacific. *Palaeogeogr. Palaeoclimatol. Palaeoecol.*, **31**, 353–69.

Cracraft, J. (1981) Pattern and process in paleobiology: the role of cladistic analysis in systematic paleontology. *Paleobiology*, **7**, 456–68.

Cracraft, J. (1982) Geographic differentiation, Cladistics and Vicariance Biogeography: reconstructing the Tempo and Mode of Evolution. *Am. Zool.*, **22**, 411–24.

Craddock, E.M. and Johnson, W.E. (1979) Genetic variation in Hawaiian *Drosophila*. V. Chromosomal and allozymic diversity in *Drosophila silvestris* and its homosequential species. *Evolution*, **33**, 137–55.

Crane, P.R. (1985) Phylogenetic analysis of Seed Plants and the origin of Angiosperms. *Ann. Missouri bot. Gdn.*, **72**, 716–93.

Craw, R.C. (1978) Two biogeographical frameworks: implications for the biogeography of New Zealand. *Tuatara*, **23**, 81–114.

Craw, R.C. (1979) Generalized tracks and dispersal in biogeography: a response to R.M. McDowall. *Syst. Zool.*, **28**, 99–107.

Craw, R.C. (1982) Phylogenetics, areas, geology and the biogeography of Croizat: a radical view. *Syst. Zool.*, **31**, 304–16.

Craw, R.C. (1983a) Panbiogeography and vicariance cladistics: are they truly different? *Syst. Zool.*, **32**, 431–8.

Craw, R.C. (1983b) *Biogeography of New Zealand: a methodological and conceptual approach*, PhD thesis in Zoology, Victoria University of Wellington, New Zealand.

Craw, R.C. (1984a) Leon Croizat's biogeographic work: a personal appreciation. In *Croizat's Pangeography and Principia Botanica* (eds R.C. Craw and G.W. Gibbs) Victoria University Press, Wellington, as *Tuatara*, **27**, 8–13.

Craw, R.C. (1984b) Biogeography and biogeographic principles. *N.Z. Ent.*, **8**, 49–52.

Craw, R.C. (1985) Classic problems of Southern Hemisphere Biogeography Re-examined: Panbiogeographic analysis of the New Zealand frog *Leiopelma*, the ratite birds and *Nothofagus*. *Z. zool. Syst. & Evolutionsforsch.*, **23**, 1–10.

Craw, R.C. and Gibbs, G.W. (1984) Croizat's Panbiogeography and Principia Botanica: search for a novel synthesis. *Tuatara*, **27**, 1–75.

Craw, R.C. and Page, R. (1988) Panbiogeography: method and metaphor in the new biogeography. In *Process and Metaphors in the New Evolutionary Paradigm* (eds M.W. Ho and S. Fox) Wiley & Sons, Chichester.

Craw, R.C. and Weston, P. (1984) Panbiogeography: a progressive research program? *Syst. Zool.*, **33**, 1–13.

Crick, R.E. (1980) Integration of paleobiogeography and paleogeography: evidence from Arenigian nautiloid biogeography? *J. Paleontol.*, **54**, 1218–36.

Crisp, D.J. (1958) The spread of *Elminus modestus* Darwin in north-west Europe. *J. Mar. Biol. Ass. UK*, **37**, 483–520.

Crisp, D.J. (1960) The northern limits of *Elminius modestus* in Britain. *Nature, Lond.*, **188**, 681.

Croizat, L. (1952) *Manual of Phytogeography*, Junk, The Hague.

Croizat, L. (1958) *Panbiogeography*, vol. 1,2a,2b, published by the author, Caracas, Venezuela.

Croizat, L. (1961) *Principia Botanica*, published by the author, Caracas, Venezuela.

Croizat, L. (1964) *Space, Time, Form: the biological synthesis*, published by the author, Caracas, Venezuela.

Croizat, L. (1967) An introduction to the subgeneric classification of *Euphorbia* L. with stress on the South African and Malagasy species. *Webbia*, **22**, 83–202.

Croizat, L. (1968) The biogeography of the tropical lands and islands east of Suez-Madagascar: with particular reference to the dispersal and form-making of *Ficus* L., and different other vegetal and animal groups. *Atti Ist. Univ. Lab. Crittogam. Pavia*, **4**, 1–400.

Croizat, L. (1978) Hennig (1966) entre Rosa (1918) y Løtrup (1977): medio singlo de ''sistemática filogenética''. *Boln. Acad. Cienc. fis. Mat. Nat.*, (*Caracas*), **38**, 59–147.

Croizat, L. (1982) Vicariance/vicariism, panbiogeography, ''vicariance biogeography'' etc.: a clarification. *Syst. Zool.*, **31**, 291–304.

Croizat, L., Nelson, G. and Rosen, D.E. (1974) Centres of origin and related concepts. *Syst. Zool.*, **23**, 265–87.

Crovello, T.J. (1981) Quantitative biogeography: an overview. *Taxon*, **30**, 563–75.

Crowder, L.B. (1980) Ecological convergence of community structure: a neutral model analysis. *Ecology*, **61**, 194–8.

Crowell, K.L. (1962) Reduced interspecific competition among the birds of Bermuda. *Ecology*, **43**, 75–88.

Crowell, K.L. (1973) Experimental zoogeography: introduction of mice to small islands. *Amer. Nat.*, **107**, 535–58.

Culver, D.C., Holsinger, J.R. and Baroody, R. (1973) Toward a predictive cave biogeography: the Greenbrier Valley as a case study. *Evolution*, **27**, 689–95.

Dammermann, K.W. (1948) The fauna of Krakatau, 1883–1933. *K. Ned. Akad. Wetenschappen Verhandelingen*, **44**, 1–594.

Darlington, P.J. Jr. (1957) *Zoogeography, the geographical distribution of Animals*. J. Wiley and Sons, New York.

Darlington, P.J. Jr (1965) *Biogeography of the Southern End of the World: Distribution and History of Far Southern Life and Land, with an assessment of continental drift*, Harvard University Press, Cambridge, Mass.

Darlington, P.J. Jr (1970) A practical criticism of Hennig – Brundin ''phylogenetic systematics'' and Antarctic biogeography. *Syst. Zool*, **19**, 1–18.

Darwin, C.R. (1845) *A Journal of Researches in to the Natural History and Geology of the Countries Visited during the Voyage of HMS Beagle round the World, under the command of Capt. Fitzroy, R.N.*, Harper, New York.

Darwin, C. (1859) *On the Origin of Species by Means of Natural Selection or the Preservation of Favoured Races in the Struggle for Life*, John Murray, London.

Davis, J.H. (1950) Evidences of trans-oceanic dispersal of plants to New Zealand. *Tuatara*, **3**, 87–97.
Davis, M., Hut, P. and Muller, R.A. (1985) Terrestrial catastrophism: Nemesis or Galaxy. *Nature, Lond.*, **313**, 503.
Davis, M.B. (1986) Climatic instability, time lags and community disequilibrium. In *Community Ecology* (eds J. Diamond and T.J. Case) Harper and Row, New York, 269–84.
Day, R.W. (1977) Two contrasting effects of predation on species richness in coral reef habitats. *Mar. Biol.*, **44**, 1–5.
Diamond, J.M. (1969) Avifaunal equilibria and species turnover rates on the Channel Islands of California. *Proc. Natl. Acad. Sci. USA*, **64**, 57–63.
Diamond, J.M. (1970) Ecological consequences of island colonization by southwest Pacific birds. *Proc. Natl. Acad. Sci. USA*, **67**, 529–63.
Diamond, J.M. (1972) Biogeographic kinetics: estimation of relaxation times for avifaunas of Southwest Pacific Islands. *Proc. Natl. Acad. Sci. USA*, **69**, 3199–203.
Diamond, J.M. (1973) Distributional ecology of New Guinea birds. *Science*, **179**, 759–69.
Diamond, J.M. (1975) Assembly of species communities. In *Ecology and Evolution of Communities* (eds M.L. Cody and J.M. Diamond) Harvard University Press, Cambridge, Mass., pp. 343–444.
Diamond, J.M. (1984a) "Normal" extinctions of isolated populations. In *Extinctions* (ed M.N. Nitecki), University of Chicago Press, Chicago, pp. 191–246.
Diamond, J.M. (1984b) Biogeographic mosaics in the Pacific. In *Biogeography of the Tropical Pacific* (eds F.J. Radovsky, P.H. Raven, S.H. Sohmer), Association of Systematic Collections, Lawrence, Kansas, pp. 1–14.
Diamond, J.M. (1984c) Historic extinctions: a rosetta stone for understanding Prehistoric extinctions. In *Quarternary Extinctions: a prehistoric revolution* (eds P.S. Martin and R.G. Klein), University of Arizona Press, Tucson, pp. 824–62.
Diamond, J.M. (1985) Rats as agents of extermination. *Nature, Lond.*, **318**, 602–3.
Diamond, J.M. (1986a) Overview: laboratory experiments, field experiments and natural experiments. In *Community Ecology* (eds J. Diamond and T.J. Case) Harper and Row, New York, pp. 3–22.
Diamond, J.M. (1986b) Evolution of ecological segregation in the New Guinea montane avifauna. In *Community Ecology* (eds J. Diamond and T.J. Case), Harper and Row, New York, pp. 98–125.
Diamond, J.M. and Case, T.J. (1986) Overview: Introductions, Extinctions and Invasions. In *Community Ecology* (eds J.M. Diamond and T.J. Case) Harper and Row, New York, pp. 65–79.
Diamond, J.M. and Gilpin, M.E. (1982) Examination of the "null" model of Connor and Simberloff for species co-occurrences on islands. *Oecologia*, **52**, 64–74.
Diamond, J.M. and Marshall, A.G. (1977) Niche shifts in New Hebridean birds. *Emu*, **77**, 61–72.
Dickerson, J.E. Jr and Robinson, J.V. (1985) Microcosms as islands: a test of MacArthur–Wilson equilibrium theory. *Ecology*, **66**, 966–80.
Digerfeldt, G. (1974) The post-glacial development of the Ranviken Bay in Lake Immeln. *Geol. För. Stockh. Förh.*, **96**, 3–32.
Dingus, L. (1984) Effects of stratigraphic completeness on interpretations of extinction rates across the Cretaceous–Tertiary boundary. *Paleobiology*, **10**, 420–38.

Dixon, J.R. (1979) Origin and distribution of reptiles in lowland tropical rainforests of South America. In *The South American Herpetofauna: its origin, evolution and dispersal. Mus. Nat. Hist. Univ. Kansas Monogr.*, **7**, 1–485.

Dobzhansky, T. (1940) Speciation as a stage in evolutionary divergence. *Amer. Nat.*, **74**, 312–21.

Dobzhansky, T. (1950) Evolution in the tropics. *Am. Scient.*, **38**, 209–21.

Dony, J.G. (1970) *Species–area Relationships*, unpublished report to the National Environment Research Council, London.

Doronina, Y.A. (1973) *Flora and Vegetation of the Uda Rivero Basin*, Nauka, Sibir. Otd., Novosibirsk.

Dressler, R.L. (1981) *The Orchids, Natural History and Classification*, Harvard University Press, Cambridge, Mass.

Druce, A.P. (1984) Distribution of Indigenous Higher Plants in North Island and Northern South Island, New Zealand, unpublished report, Botany Division, DSIR.

Duellman, W.E. (1972) South American frogs of the *Hyla rostrata* group (Amphibia, Anura, Hylidae). *Zoöl. Meded. Rijksmus. Nat. Hist.*, **47**, 177–92.

Duellman, W.E. (1979) The South American Herpetofauna: its origin, evolution and dispersal. *Mus. Nat. Hist. Univ. Kansas Monogr.*, **7**, 1–485.

Duellman, W.E. (1982) Quarternary climatic ecological fluctuations in the lowland tropics: frogs and forests. In *Biological Differentiation in the Tropics* (ed G.T. Prance), Columbia University Press, New York, pp. 389–402.

Duellman, W.E. and Crump, M.L. (1974) Speciation in frogs of the *Hyla parviceps* group in the upper Amazon Basin. *Occ. Pap. Mus. Nat. Hist. Univ. Kansas*, **23**, 1–40.

Dunham, A.E., Smith, G.R. and Taylor, J.N. (1979) Evidence for ecological character displacement in western American catostomid fishes. *Evolution*, **33**, 877–96.

Durden, C.J. (1974) Biomerization: an ecologic theory of provincial differentiation. In *Paleogeographical Provinces and Provinciality* (ed C.A. Ross), *Soc. Econ. Paleontologists and Mineralogists Spec. Pubns.*, **21**, pp. 18–53.

Eagle, A. (1975) *Eagle's Tree and Shrubs of New Zealand in Colour*, Collins, Aukland and London.

Eagle, A. (1982) *Eagle's Tree and Shrubs of New Zealand Second Series*, Collins, Aukland, Sydney and London.

Eberhard, W.G. (1986) *Sexual Selection and Animal Genitalia*, Harvard University Press, Cambridge, Mass.

Edmondson, W.T. (1944) Ecological studies of sessile Rotatoria. *Ecol. Monogr.*, **14**, 32–66.

Edmunds, G.F. Jr (1975) Phylogenetic biogeography of mayflies. *Ann. Missouri Bot. Gdn.*, **62**, 251–63.

Edmunds, G.F. Jr (1981) Discussion. In *Vicariance Biogeography, a Critique* (eds G. Nelson, D.E. Rosen) Columbia University Press, New York, pp. 287–97.

Ehrendorfer, F. (1962) Cytotaxonomische Beiträge zur Genese der mitteleuropäischen Flora und Vegetation. *Ber. Deutsch. Botan. Ges.*, **75**, 137–52.

Ehrlich, P. and Ehrlich, A. (1981) *Extinction: the Causes and Consequences of the Disappearance of Species*, Random House, New York.

Ehrlich, P.R., Murphy, D.D., Singer, M.C. *et al.* (1980) Extinction, reduction, stability and increase: the response of checkerspot butterflies (*Euphydras*) populations to Californian drought. *Oecologia*, **46**, 101–5.

Ehrlich, P.R. and Raven, P.H. (1969) Differentiation of populations. *Science*, **165**, 1228–32.

Ehrlich, P., Sagan, C., Kennedy, D. and Roberts, W.O. (1984) *The Cold and the Dark: the world after nuclear war*, W.W. Norton, New York.
Ekman, S. (1935) *Tiergeographie des Meeres*, Akademische Verlagsgesellschaft, Leipzig.
Ekman, S. (1953) *Zoogeography of the Sea*, Sidgwick and Jackson, London.
Eldredge, N. (1981) Discussion of M.D.F. Uvardy's paper: The riddle of dispersal: dispersal theories and how they affect biogeography. In *Vicariance Biogeography; a Critique* (eds G. Nelson and D.E. Rosen), Columbia University Press, New York, pp. 34–8.
Eldredge, N. and Cracraft, J. (1980) *Phylogenetic Patterns and the Evolutionary Process: method and theory in comparative biology*, Columbia University Press, New York.
Eldridge, J.L. and Johnson, D.H. (1988) Size Differences in Migrant Sandpiper Flocks: ghosts in ephemeral guilds, *Oecologia*, in press.
Elenevsky, A.G. (1966) On some floristic peculiarities of inland Daghestan. *Byull. Moskov. Obshch. Ispitateli Prirody, Otd. Biol.*, **71**, 107–17.
Ellenberg, H. (1978) *Vegetation Mitteleuropas mit den Alpen in ökologischer Sicht*, 2nd edn, Ulmer, Stuttgart.
Endler, J.A. (1977) *Geographic Variation, Speciation, and Clines*, Princeton University Press, Princeton, New Jersey.
Endler, J.A. (1982a) Pleistocene forest refuges: fact or fancy? In *Biological Diversification in the Tropics*, (ed G.T. Prance), Columbia University Press, New York, pp. 641–57.
Endler, J.A. (1982b) Problems in distinguishing historical from ecological factors in biogeography. *Amer. Zool.*, **22**, 441–52.
Endler, J.A. (1982c) Alternative hypotheses in biogeography: introduction and synopsis of the symposium. *Amer. Zool.* **22**, 349–54.
Endler, J.A. (1986) *Natural Selection in the Wild*, Princeton University Press, Princeton, New Jersey.
Engler, A. (1879–82) *Versuch einer Entwicklungsgeschichte der extratropischen Florengebiete* (2 vols), Englemann, Leipzig.
Ernst, W. (1978) Discrepancy between ecological and physiological optima of plant species. A re-interpretation. *Oecol. Plant.*, **13**, 175–88.
Erwin, T. L. (1981) Taxon pulses, and dispersal: an evolutionary synthesis illustrated by carabid beetles. In *Vicariance Biogeography; a critique* (eds G. Nelson and D.E. Rosen), Columbia University Press, New York, pp. 159–83.
Estabrook, G.F., Johnson, C.S. and McMorris, F.R. (1976) A mathematical foundation for analysis of cladistic character compatibility. *Math. Biosci.*, **29**, 181–87.
Ettingshausen, C. von (1981) Contributions to the knowledge of the fossil flora of New Zealand. *Trans. Proc. N.Z. Inst.*, **23**, 237–310.
Faeth, S.H. (1984) Density compensation in vertebrates and invertebrates: a review and an experiment. In *Ecological Communities: conceptual issues and the evidence* (eds D.R. Strong, Jr., D. Simberloff, L.G. Abele and A.B. Thistle) Princeton University Press, Princeton, New Jersey, pp. 491–509.
Fallow, W.C. and Dromgoogle, E.L. (1980) Faunal similarities across the south Atlantic among Mesozoic and Cenozoic invertebrates correlated with widening of the ocean basin. *J. Geol.*, **88**, 723–7.
Farris, J.S. (1970) Methods for computing Wagner trees. *Syst. Zool.*, **19**, 83–92.
Farris, J.S. (1983) The empirical basis of Phylogenetic Systematics. In *Advances in Cladistics*, **1** (eds N. Platnick and V.A. Funk), Columbia University Press, New York, pp. 7–36.

Farris, J.S. (1986) On the boundaries of phylogenetic systematics. *Cladistics*, **2**, 14–27.

Favarger, C. and Contandriopoulos, J. (1961) Essai sur léndemisme. *Bull. Soc. Botan. Suisse*, **71**, 383–408.

Feinberg, S.E. (1980) *The Analysis of Cross Classified Categorical Data*, 2nd edn, MIT Press, Cambridge, Mass.

Felsenstein, J. (1978) The number of evolutionary trees. *Syst. Zool.*, **27**, 27–33.

Felsenstein, J. (1982) How can we infer geography and history from gene frequencies? *J. Theor. Biol.*, **96**, 9–20.

Felsenstein, J. (1985) Confidence limits on phylogenesis: an approach using the bootstrap. *Evolution*, **39**, 783–91.

Fenchel, T. (1975) Character displacement and coexistence in mud snails. *Oecologia*, **20**, 19–32.

Ferlatte, W.J. (1974) *A Flora of the Trinity Alps of Northern California*, University of California Press, Berkeley, California.

Fernald, M.L. (1950) *Gray's manual of Botany*, 8th edn, American Book Company, New York.

Ferris, V.R. (1980) A science in search of a paradigm? – Review of the symposium "Vicariance biogeography: a critique". *Syst. Zool.*, **29**, 67–76.

Ferson, S., Downey, P., Klerks, P. *et al.* (1986) Competing reviews, or why do Connell and Schoener disagree? *Amer. Nat.*, **127**, 571–6.

Feyerabend, P. (1978) *Science in a Free Society*, New Left Books, London.

Ficken, R.W., Ficken, M.S. and Morse, D.H. (1968) Competition and character displacement in two sympatric pine-dwelling warblers. *Evolution*, **22**, 307–14.

Field, B. (1982) Structural analysis of fouling community development in the Damariscotta River estuary. *J. Exp. Mar. Biol. Ecol.*, **57**, 25–33.

Fischer, A.G. (1960) Latitudinal variation in organic diversity. *Evolution*, **14**, 64–81.

Fischer, A.G. and Arthur, M.A. (1977) Secular variations in the pelagic realm. *Soc. Econ. Paleontol. Minerol., Spec. Pubn.*, **25**, 19–50.

Fisher, R.A. (1930) *The Genetical Theory of Natural Selection*, Clarendon Press, Oxford.

Fisher, R.A. (1937) The wave of advance of advantageous genes. *Ann. Eugenics*, **7**, 355–69.

Fjeldsa, J. (1983) Ecological character displacement and character release in grebes Podicipedidae. *Ibis*, **125**, 463–81.

Fleming, C.A. (1962) New Zealand biogeography: a palaeontologist's approach. *Tuatara*, **10**, 53–108.

Fleming, C.A. (1963) The nomenclature of Biogeographic elements in the New Zealand biota. *Trans., R. Soc. N.Z.*, **1**, 13–22.

Fleming, C.A. (1976) New Zealand as a minor source of terrestrial plants and animals in the Pacific. *Tuatara*, **22**, 30–7.

Fleming, C.A. (1979) *The Geological History of New Zealand and its Life*, Aukland University Press/Oxford University Press, Aukland, N.Z.

Fleming, C.A. (1982) *George Edward Lodge: the unpublished New Zealand Bird Paintings*, Nova Pacifica, Wellington, N.Z.

Flessa, K.W. (1975) Area, continental drift and mammalian diversity. *Paleobiology*, **1**, 189–94.

Flessa, K.W., Bennett, S.G., Cornue, D.B. *et al.* (1979) Geological implications of the relationship between mammalian faunal similarities and geographic distance. *Geology*, **7**, 15–18.

Flessa, K.W. and Imbrie, J. (1973) Evolutionary pulsations: evidence from Phanerozoic diversity patterns. In *Implications of Continental Drift to the Earth Sciences* (eds D.H. Tarling and S.K. Runcorn), Academic Press, New York, pp. 247–85.

Flessa, K.W. and Jablonski, D. (1983) Extinction is here to stay. *Paleobiology*, **9**, 315–21.

Flessa, K.W. and Jablonski, D. (1985) Declining Phanerozoic background extinction rates: effect of taxonomic structure. *Nature, Lond.*, **313**, 216–18.

Flessa, K.W. and Sepkoski, J.J. Jr (1978) On the relationship between phanerozoic diversity and changes in habitable area. *Paleobiology*, **4**, 359–66.

Forman, R.T.T. (1964) Growth under controlled conditions to explain the hierarchical distributions of a moss, *Tetraphis pellucida*. *Ecol. Monogr.*, **34**, 1–25.

Fortey, R.A. and Barnes, C.R. (1977) Early Ordovician conodont and trilobite communities of Spitsbergen: influence on biogeography. *Alcheringa*, **1**, 297–309.

Fortey, R.A. and Cocks, L.R.M. (1986) Marginal faunal belts and their structural implications, with examples from the Lower Palaeozoic. *J. Geol. Soc. Lond.*, **143**, 151–60.

Fougere, P.F. (1985) On the accuracy of spectrum analysis of red noise processes using maximum entropy and periodogram methods: simulation studies and application to geophysical data. *J. Geophys. Res.*, **90**(A5), 4355–66.

Fridriksson, S. (1975) *Surtsey: Evolution of Life on a Volcanic Island*, Butterworths, London.

Frost, D.R. (1985) *Amphibian Species of the World: a taxonomic and geographical reference*, Allen Press, Inc. and Association Systematic Collections.

Fryer, G., Greenwood, P.H. and Peake, J.F. (1983) Punctuated equilibria, morphological stasis and the paleontological documentation of speciation: a biological appraisal of a case history in an African lake. *Biol. J. Linn. Soc.*, **20**, 195–205.

Fryer, G. and Iles, T.D. (1972) *The Cichlid Fishes of the Great Lakes of Africa: their biology and evolution*, Oliver and Boyd, Edinburgh.

Fuentes, E.R. (1976) Ecological convergence of lizard communities in Chile and California. *Ecology*, **57**, 3–17.

Fuentes, E.R. (1980) Convergence of community structure: neutral model vs field data. *Ecology*, **61**, 198–200.

Fujimori, T.J. (1977) Stem biomass and structure of a mature *Sequoia semper vivens* stand on the Pacific coast of northern Carolina. *J. Jap. For. Soc.*, **59**, 435–41.

Fujimori, T., Kawanabe, S., Saito, H. *et al.* (1976) Biomass and primary production in forests of 3 major vegetation zones of the northwestern United States. *J. Jap. For. Soc.*, **58**, 360–73.

Furley, P.A. and Newey, W.W. (1983) *Geography of the Biosphere; an introduction to the nature, distribution and evolution of the world's life zones*, Butterworths, London.

Futuyma, D.J. and Mayer, G.D. (1980) Non-allopatric speciation in animals. *Syst. Zool.*, **29**, 254–71.

Gadzhiev, V.D., Kulieva, K.G. and Bagabov, Z.V. (1979) *Flora and Vegetation of the High Mountains of Talysh*, Elm, Baku.

Gagnidze, R.I. and Kemularia-Natadze, L.M. (1985) *Botanical Geography and Flora of Racha-Lechkhumi (Western Georgia)*, Metsniereba, Tbilisi.

Galland, P. (1982) *Etude de la Vegetation des pelouses alpines au Parc National Suisse*, Thesis, Univ. Neuchatel.

Gamisans, J., Aboucaya, A. and Antoine, C. (1985) Quelques donnees numeriques et chorologiques sur la flora vasculaire de la Corse. *Candollea*, **40**, 571–82.

Gankin, R. and Major, J. (1964) *Arctostaphylos myrtifolia*, its biology and relationship to the problem of endemism. *Ecology*, **45**, 793–808.

Gascon, D. and Miller, R.A. (1981) Colonization by near shore fish on small artificial reefs in Barkley Sound, British Columbia. *Can. J. Zool.*, **59**, 1635–46.

Gatrell (1983) *Distance and Space. A geographical perspective*, Clarendon Press, Oxford.

Gee, J.H.R. and Giller, P.S., eds. (1987) *Organization of Communities: past and present*, Blackwell Scientific Pubns, Oxford.

Gentry, A.H. (1986) In *Conservation Biology, the Science of Scarcity and Diversity* (ed M.E. Soule), Sinauer Assoc., Sunderland, Mass., pp. 153–81.

Georghiou, G.P. (1972) The evolution of resistance to pesticides. *Ann. Rev. Ecol. Syst.*, **3**, 133–68.

Ghiselin, M. (1974) A radical solution to the species problem. *Syst. Zool.*, **23**, 536–44.

Ghiselin, M. (1981) Categories, life and thinking. *Behavioural Brain Sci.*, **4**, 269–313.

Gibbs, G. W. (1983) Evolution of Micropterigidae (Lepidoptera) in the S.W. Pacific. *J. Geo.*, **7**, 505–10.

Gilbert, F.S. (1980) The equilibrium theory of island biogeography: fact or fiction. *J. Biogeography*, **7**, 209–35.

Giller, P.S. (1984) *Community Structure and the Niche*, Chapman and Hall, London.

Giller, P.S. and Gee, J.H.R. (1987) Analysis of community organization: the influence of equilibrium, scale and terminology. In *Organization of Communities: Past and Present* (eds J.H. Gee and P.S. Giller), Blackwell Scientific Pubns, Oxford, pp. 514–42.

Gillespie, J.H. (1984) Molecular evolution over the mutational landscape. *Evolution*, **38**, 1116–29.

Gilpin, M.E. (1975) Limit cycles in competition communities. *Amer. Nat.*, **109**, 51–60.

Gilpin, M.E. and Diamond, J.M. (1976) Calculation of immigration and extinction curves from the species-area-distance relation. *Proc. Natl. Acad. Sci. USA*, **73**, 4130–4.

Gilpin, M.E. and Diamond, J.M. (1980) Subdivision of nature reserves and the maintenance of species diversity. *Nature, Lond.*, **285**, 567–8.

Gilpin, M.E. and Diamond, J.M. (1982) Factors contributing to non-randomness in species co-occurrences on islands. *Oecologia*, **52**, 75–84.

Gingerich, P.D. (1984) Pleistocene extinctions in the context of origination and extinction equilibria in Cenozoic mammals. In *Quaternary Extinctions: a prehistoric revolution* (eds P.S. Martin and R.G. Klein) University of Arizona Press, Tucson, pp. 211–22.

Glasser, J.W. (1982) On the causes of temporal change in communities: modification of the biotic environment. *Amer. Nat.*, **119**, 375–90.

Gleason, H.A. (1922) On the relation between species and area. *Ecology*, **3**, 158–62.

Gold, J.R. (1980) Chromosomal change and rectangular evolution in North American cyprinid fishes. *Genet. Res.*, **35**, 157–64.

Goldblatt, P. (1978) An analysis of the flora of southern Africa: its characteristics, relationships and origins. *Ann. Missouri Bot. Gdn.*, **63**, 369–436.

Goldschmidt, R.B. (1940) *The Material Basis of Evolution* (repr. 1982), Yale University Press, New Haven.

Golenberg, M. *et al.* (1990) Chloroplast DNA sequence from a Miocene *Magnolia* species. *Nature*, **344**, 656–8.

Goloskokov, V.P. (1984) *Flora of the Dzhungarsk Alatau*, Izd. Nauka, Kazakhskoi SSR, Alma Ata.

Goodall, D.W. (1952) Quantitative aspects of plant distribution. *Biol. Rev.*, **27**, 194–245.

Goodrich, S. and Neese, E. (1986) *Uinta Basin Flora*, U.S. Forest Service, Ashley National Forest and Bureau of Land Mgt.

Gorchakovskii, P.L. (1975) *The Plant World of the High Ural Mountains.*, Nauka, Moscow.

Gorchakovskii, P.L. and Zueva, V.N. (1982) Inter- and intrapopulation variability of Ural endemic *Astragalus* spp. *Ekologiya Transl.*, **4**, 231–6.

Gorchakovskii, P.L. and Zueva, V.N. (1984) Age structure and dynamics of small isolated populations of Ural endemic milk vetches. *Ekologiya Transl.*, **3**, 103–10.

Gottlieb, L.D. (1973) Enzyme differentiation and phylogeny in *Clarkia franciscana*, *C. rubicunda* and *C. amoena*. *Evolution*, **27**, 205–14.

Gottlieb, L.D. (1974) Genetic confirmation of the origin of *Clarkia lingulata*. *Evolution*, **28**, 244–50.

Gottlieb, L.D. (1977a) Phenotypic variation in *Stephanomeria exigua* ssp. *coronaria* (Compositae) and its recent derivative species *'malheurensis'*. *Amer. J. Bot.*, **64**, 873–80.

Gottlieb, L.D. (1977b) Genotypic similarity of large and small individuals in a natural population of the annual plant *Stephanomeria exigua* ssp. *coronaria* (Compositae). *J. Ecol.*, **65**, 127–34.

Gottlieb, L.D. (1978a) *Stephanomeria malheurensis* (Compositae), a new species from Oregon. *Madrono*, **25**, 44–6.

Gottlieb, L.D. (1978b) Allocation, growth rates and gas exchange in seedlings of *Stephanomeria* ssp. *coronaria* and its recent derivative *S. malheurensis*. *Amer. J. Bot.*, **65**, 970–7.

Gottlieb, L.D. (1979) The origin of phenotype in a recently evolved species. In *Topics in Plant Population Biology* (eds O.T. Solbrig, S. Jain, G.B. Johnson and P.H. Raven), Columbia University Press, New York. pp. 265–86.

Gottlieb, L.D. (1984) Genetics and morphological evolution in plants. *Amer. Nat.*, **123**, 681–709.

Gould, S.J. (1981) Palaeontology plus ecology as Palaeobiology. In *Theoretical Ecology, Principles and Applications* (ed R.M. May) Blackwell Scientific Pubns, Oxford, pp. 295–317.

Gould, S.J. (1983) *Hen's Teeth and Horse's Toes*, W.W. Norton and Co., New York.

Gould, S.J. (1985) All the news that's fit to print and some opinions that aren't. *Discover*, November, pp. 86–91.

Graham, A. (1982) Diversification beyond the Amazon Basin. In *Biological Differentiation in the Tropics* (ed. G.T. Prance), Columbia University Press, New York, pp. 78–90.

Graham, R.W. and Lundelius, E.L. (1984) Coevolutionary disequilibrium and Pleistocene extinctions. In *Quarternary Extinctions: a prehistoric revolution* (eds P.S. Martin and R.G. Klein), University of Arizona Press, Tucson, pp. 223–49.

Grande, L. (1985) The use of paleontology in systematics and biogeography and a time control refinement for historical biogeography. *Paleobiology*, **11**, 234–43.

Grant, P.R. (1972) Convergent and divergent character displacement. *Biol. J. Linn. Soc.*, **4**, 39–68.

Grant, P.R. (1975) The classical case of character displacement. In *Evolutionary Biology*, Vol. 8 (eds T. Dobzhansky, M.K. Hecht and W.C. Steere), Plenum Press, New York, pp. 237–7.

Grant, P.P.R. (1986) *The Evolution and Ecology of Darwin's Finches*, Princeton University Press, Princeton.

Grant, P.R. and Schluter, D. (1984) Interspecific competition inferred from patterns of guild structure. In *Ecological Communities: conceptual issues and the evidence* (eds D.R. Strong, Jr, D. Simberloff, L.G. Abele and A.B. Thistle), Princetown University Press, Princetown, New Jersey, pp. 201–33.

Grant, V. (1980) Gene flow and the homogeneity of species populations. *Biol. Zentralbl.*, **99**, 157–69.

Granville, J.J. de, (1982) Rain forest and xeric flora refuges in French Guiana. In *Biological Differentiation in the Tropics* (ed G.T. Prance), Columbia University Press, New York, pp. 159–81.

Grassle, J.F. (1985) Hydrothermal vent animals: distribution and biology. *Science*, **229**, 713–17.

Gray, J. and Boucot, A.J. (1979a) *Historical biogeography, plate Tectonics, and the Changing Environment*, Oregon State University Press, Corvallis.

Gray, J. and Boucot, A.J. (1979b) Editors' disclaimer, pp. vii–ix. In *Historical Biogeography, Plate Tectonics, and the Changing Environment* (eds J. Gray and A.J. Boucot), Oregon State University Press, Corvallis.

Gray, R. (1986) Faith and foraging: a critique of the paradigm argument from design. In *Foraging Behaviour* (eds A.C. Kamil, J.R. Krebs and H.R. Pulliam). Plenum Press, New York.

Gray, R. (1987) Metaphors and methods: behavioural ecology panbiogeography and the evolving synthesis. In *Process and Metaphors in the New Evolutionary Paradigm*, J. Wiley and Sons, New York.

Graybosch, R.A. and Buchanan, H. (1983) Vegetative types and endemic plants of the Bryce Canyon breaks. *Gt. Basin Nat.*, **43**, 701–12.

Grayson, D.K. (1984) Explaining Pleistocene extinctions. In *Quaternary Extinctions: a prehistoric revolution* (eds P.S. Martin and R.G. Klein), University of Arizona Press, Tucson, pp. 807–23.

Green, R.H. (1980) Multivariate analysis in ecology: the assessment of ecologic similarity. *Ann. Rev. Ecol. Syst.*, **11**, 1–14.

Gregory, R.L. (1974) *Concepts and Mechanisms of Perception*, Duckworth, London.

Gregory, R.L. (1981) *Mind in Science; a history of explanations in Psychology and Physics*, Weidenfeld and Nicholson, London.

Grehan, J.R. and Ainsworth, R. (1985) Orthogenesis and Evolution. *Syst. Zool.*, **34**, 174–92.

Grehan, J.R. and Henderson, I. (1988) Panbiogeography. In *Vertebrate Zoogeography and Evolution in Australasia* (eds M. Archer, G. Clayton and J. Long), Hesperian Press, Marrickville, Australia.

Gressitt, J.L. (1961) Problems in the zoogeography of Pacific and Antarctic insects. *Pac. Insects Monogr.*, **2**, 1–94.

Gressitt, J.L. (1970) Subantarctic entomology and biogeography. *Pac. Insects Monogr.*, **23**, 295–374.

Grieve, R.A., Sharpton, V.L., Goodacre, A.K. and Garvin, J.B. (1985) Periodic

cometary impacts and the terrestrial cratering record. EOS, *Trans. Amer. Geophys. Un.*, **66**, 813.
Grubb, P. (1977) The maintenance of species richness in plant communities: the importance of the regeneration niche. *Biol. Rev.*, **52**, 107–45.
Grubov, V.I. (1982) *Key to the Vascular Plants of Mongolia*, Nauka, Leningrad.
Guilday, J.E. (1967) Differential extinction during late-Pleistocene and Recent times. In *Pleistocene Extinctions: the search for a Cause* (eds P.S. Martin and H.E. Wright), Yale University Press, New Haven, pp. 121–40.
Guilday, J.E. (1984) Pleistocene extinctions and environmental change. In *Quaternary Extinctions: a prehistoric Revolution* (eds P.S. Martin and R.G. Klein), University of Arizona Press, Tucson, pp. 250–8.
Guppy, H.B. (1906) *Observations of a Naturalist in the Pacific between 1896 and 1899*, vol. II, MacMillan and Co., London.
Guthrie, R.D. (1984) Mosaics, allelochemics and nutrients: an ecological theory of Late Pleistocene megafaunal extinctions. In *Quaternary Extinctions: a prehistoric revolution* (eds P.S. Martin and R.G. Klein), University of Arizona Press, Tucson, pp. 259–98.
Gyllenstein, U. and Wilson, A.C. (1987) Interspecific mitochondrial DNA transfer and the colonization of Scandinavia by mice. *Genet. Res.*, **49**, 25–9.
Haeupler, H. (1983) Die Mikroarealophyten der Balearen. Ein Beitrag zum Endemismus–Begriff und zur Inselbiogeographie. *Tüxenia*, **3**, 271–88.
Haffer, J. (1969) Speciation in Amazonian forest birds. *Science*, **165**, 131–7.
Haffer, J. (1974) Avian speciation in tropical South America. *Publ. Nutall Ornith. Club*, **No. 14**, 1–390.
Haffer, J. (1979) Quaternary biogeography of tropical lowland South America. In *The South American Herpetofauna; its origin evolution and dispersal.* (ed W.E. Duellmann) *Mus. Nat. Hist. Univ. Kans. Monogr.*,**7**, pp. 107–40.
Haffer, J. (1981) Aspects of neotropical bird speciation during the Cenozoic. In *Vicariance Biogeography; a Critique* (eds G. Nelson, D.E. Rosen), Columbia University Press, New York, pp. 371–94.
Haffer, J. (1982) General aspects of the refuge theory. In *Biological Differentiation in the Tropics* (ed. G.T. Prance), Columbia University Press, New York, pp. 6–24.
Haines-Young, R.W. and Petch, J.R. (1980) The challenge of critical rationalism for methodology in physical geometry. *Prog. Phys. Geogr.*, **4**, 63–78.
Hairston, N.G., Smith, F.E. and Slobodkin, L.B. (1960) Community structure, population control, and competition. *Amer. Nat.*, **94**, 421–5.
Hallam, A. (ed.) (1973) *Atlas of Palaeobiogeography*, Elsevier Scientific Publishing Company, Amsterdam.
Hallam, A. (1981) *Facies Interpretation and the Stratigraphic Record*, W.H. Freeman and Company, Oxford.
Hallam, A. (1983) Plate tectonics and evolution. In *Evolution from Molecules to Men* (ed. D.S. Bendall) Cambridge University Press, Cambridge, U.K. pp. 367–86.
Hallam, A. (1984) Distribution of fossil marine invertebrates in relation to climate. In *Fossils and Climate* (ed. P.J. Brenchley), *Geol. J., Spec.* Issue **11**, 107–25.
Hallam, A. (1986) Evidence of displaced terranes from Permian to Jurassic faunas found around the Pacific margins. *J. Geol. Soc. Lond.*, **143**, 209–16.
Halliday, R.B., Barton, N.H. and Hewitt, G.H. (1983) Electrophoretic analysis of a chromosomal hybrid zone in the grasshopper *Podisma pedestris*. *Biol. J. Linn. Soc.*, **19**, 51–62.

Hamill, D.N. and Wright, S.J. (1988) Interspecific interaction and similarity in species composition. *Amer. Nat.*, **131**, 412–23.
Hamilton, W.D. and May, R.M. (1977) Dispersal in stable habitats. *Nature, Lond.*, **269**, 378–81.
Hammond, A.L. (1976) Paleoceanography: sea floor clues to earlier environments. *Science*, **191**, 168–70, 208.
Hamrick, J.L., Linhart, Y.B. and Mitton, J.B. (1979) Relationships between life history characteristics and electrophoretically detectable genetic variation in plants. *Ann. Rev. Ecol. Syst.*, **10**, 173–200.
Hanski, I. (1981) Exploitative competition in transient habitat patches. In *Quantitative Population Dynamics* (eds D.G. Chapman and V.F. Gallucci), International Co-operative Publishing House, Fairland, Maryland, pp. 25–38.
Hanski, I. (1982) Dynamics of regional distribution: the core and satellite species hypothesis. *Oikos*, **38**, 210–21.
Hanski, I. (1983) Coexistence of competitors in patchy environment. *Ecology*, **64**, 493–500.
Harland, W.B., Cox, A.V., Llewellyn, P.G. *et al.* (1982) *A Geologic Time Scale*, Cambridge University Press, Cambridge.
Harper, J.L. (1968) The regulation of numbers and mass in plant populations. *Population Biology and Evolution*, (ed. R.C. Lewontin) Syracuse Univ. Press, New York, pp. 139–58.
Harper, J.L. (1977) *Population Biology of Plants*, Academic Press, New York.
Harper, J.L. and White, J. (1974) The demography of plants. *Ann. Rev. Ecol. Syst.*, **4**, 419–63.
Harris, L.D. (1984) *The Fragmented Forest*, University of Chicago Press, Chicago.
Harris, R.P. (1987) Spatial and temporal organization in marine plankton communities. In *Organization of Communities, Past and Present* (eds J.H.R. Gee and P.S. Giller), Blackwell Scientific Pubns, Oxford, pp. 327–46.
Harrison, L. (1928) The composition and origins of the Australian fauna, with special reference to the Wegener hypothesis. *Rep. Australas. Ass. Adv. Sci.*, **18**, 332–96.
Harvey, P.H., Colwell, R.K., Silvertown, J.W. and May, R.M. (1983) Null models in ecology. *Ann. Rev. Ecol. Syst.*, **14**, 189–211.
Hassell, M.P. and Comins, H.N. (1976) Discrete-time models for two-species competition. *Theor. Pop. Biol.*, **9**, 202–21.
Hastings, A. (1978) Spatial heterogeneity and the stability of predator–prey systems: predator-mediated coexistence. *Theor. Pop. Biol.*, **14**, 380–95.
Hastings, A. (1987) Can competition be detected using species co-occurrence data? *Ecology.*, **68**, 117–23.
Haugh, B.N. (1981) Discussion. In *Vicariance Biogeography: a critique* (eds G. Nelson and D.E. Rosen), Columbia University Press, New York, pp. 275–86.
Hayward, T.L. and McGowan, J.A. (1979) Pattern and structure in an oceanic zooplankton community. *Amer. Zool.*, **19**, 1045–55.
Heads, M. (1984) Principia Botanica: Croizat's contribution to botany. In *Croizat's Panbiogeography and Principia Botanica* (eds R.C. Craw and G.W. Gibbs), Victoria University Press, Wellington, as *Tuatara*, **27**, 26–48.
Heads, M. (1985) Biogeographic analysis of *Nothofagus* (Fagaceae). *Taxon*, **34**, 474–80.
Heads, M. (1988) *Biogeography and taxonomy of some New Zealand plants*, PhD thesis in Botany, University of Otago, New Zealand.
Heads, M. and Craw, R.C. (1984) Bibliography of the scientific work of Leon

Croizat 1932–82. In *Croizat's Panbiogeography and Principia Botanica* (eds R.C. Craw and G.W. Gibbs), Victoria University Press, Wellington, as *Tuatara*, **27**, 67–75.

Heaney, L.R. (1986) Biogeography of mammals in southeast Asia: estimates of rates of colonization, extinction and speciation. *Biol. J. Linn. Soc.*, **29**, 127–65.

Heatwole, H. and Levins, R. (1972) Trophic structure stability and faunal change during recolonization. *Ecology*, **53**, 531–4.

Hecht, M.K. (1975) The morphology and relationships of the largest known terrestrial lizard, *Megalania prisca* Owen, from the Pleistocene of Australia. *Proc. R. Soc. Vict.*, **87**, 239–49.

Hedrick, P.W. (1986) Genetic polymorphism in heterogeneous environments: a decade later. *Ann. Rev. Ecol. Syst.*, **17**, 535–66.

Hedrick, P.W., Ginevan, M.E. and Ewing, E.P. (1976) Genetic polymorphism in heterogeneous environments. *Ann. Rev. Ecol. Syst.*, **7**, 1–32.

Henderson, I. (1985) *Systematic studies of New Zealand Trichoptera and critical analysis of systematic methods*, PhD thesis in Zoology, Victoria University of Wellington, New Zealand.

Hendrickson, J.A. Jr (1981) Community-wide character displacement re-examined. *Evolution*, **35**, 794–810.

Hendrych, R. (1981) Bemerkungen zum Endemismus in der Flora der Tschechoslowakei. *Preslia*, **53**, 97–120.

Hendy, M.D. and Penny, D. (1982) Branch and Bound algorithms to determine minimum evolutionary trees. *Math. Biosci.*, **59**, 277–90.

Hennig, W. (1950) *Grundzüge einer Theorie der phylogenetischen Systematik*, Deutscher Zentralverlag, Berlin.

Hennig, W. (1960) Die Dipteren-fauna von Neuseeland als systematisches und tiergeographishes problem. *Beitr. Ent.*, **10**, 221–329.

Hennig, W. (1966) *Phylogenetic Systematics*, University of Illinois, Urbana.

Hennig, W. (1984) *Aufgaben und Probleme stammesgeschichtler Forschung*, Parey, Berlin.

Hertz, P.E. (1983) Eurythermy and niche breadth in West Indian *Anolis* lizards: a reappraisal. In *Advances in Herpetology and Evolutionary Biology* (eds A.G.J. Rhodin and K. Miyata) Museum of Comparative Zoology, Harvard University, Cambridge, Mass., pp. 472–83.

Hewitt, G.M. (1975) A sex chromosome hybrid zone in the grasshoper *Podisma pedestris* (Orthoptera: Acrididae). *Heredity*, **35**, 375.

Hewitt, G.M. (1979) *Animal Cytogenetics III Orthoptera*, Gebruder Borntraeger, Stuttgart.

Heyer, W.R. (1979) Systematics of the *pentadactylus* species group of the frog genus *Leptodactylus* (Amphibia: Leptodactylidae). *Smithsonian Contr. Zool.*, **301**, 1–43.

Heyer, W.R. and Maxson, L.R. (1982) Distributions, relationships, and zoogeography of lowland frogs/the *Leptodactylus* complex in South America, with special reference to Amazonia. In *Biological Differentiation in the Tropics* (ed. G.T. Prance), Columbia University Press, New York, pp. 375–88.

Higgs, A.J. (1981) Island biogeography theory and nature reserve design. *J. Biogr.*, **8**, 117–24.

Higgs, A.J. and Usher, M.B. (1980) Should nature reserves be large or small. *Nature, Lond.*, **285**, 568–9.

Hilborn, R. (1975) The effect of spatial heterogeneity on the persistence of predator–prey interactions. *Theor. Pop. Biol.*, **8**, 346–55.

Hilu, K.W. (1983) The role of single gene mutations in the evolution of flowering plants. *Evol. Biol.*, **16**, 97–128.

Hockin, D.C. (1982) Experimental insular zoogeography: some tests of the equilibrium theory using meiobenthic harpacticoid copepods. *J. Biogeogr.*, **9**, 487–97.

Hockin, D.C. and Ollason, J.G. (1981) The colonization of artificially isolated volumes of intertidal estuarine sand by harpacticoid copepods. *J. exp. Mar. Biol. Ecol.*, **53**, 9–29.

Hoffman, A. (1985a) Island biogeography and palaeobiology: in search for evolutionary equilibria. *Biol. Rev.*, **60**, 455–71.

Hoffman, A. (1985b) Patterns of family extinction depend on definition and geological timescale. *Nature, Lond.*, **315**, 659–62.

Holden, A.M. (1983) *Eucalyptus*-like leaves from Miocene rocks in Central Otago. *Pac. Sci. Ass. 15th Congr. Abstr.*, **1**, 103.

Holloway, B. (1982) Anthribidae (Insecta: Coleoptera). *Fauna of New Zealand*, **3**.

Holt, R.D. (1977) Predation, apparent competition and the structure of prey communities. *Theor. Pop. Biol.*, **12**, 197–229.

Holzenthal, R. (1986) Studies in neo-tropical Leptoceridae (Trichoptera), VI: Immature stages of *Hudsonema flaminii* (Navás) and the evolution and historical biogeography of Hudsonemini. *Proc. Ent. Soc. Washington*, **88**, 268–79.

Hooker, J.D. (1844–60) *The Botany of the Antarctic Voyage of H.M. Discovery Ships Erebus and Terror in the Years 1839–1843*. I. *Flora Antarctica* (1844–47), London. 2 pt. II. *Flora Novae Zelandiae* (1853–55), London, 2 pt. III. *Flora Tasmaniae* (1855–60), London, 2 pt.

Hooker, J.D. (1882) *On Geographical Distribution, Presidential Address*, Report of the 51st Meeting of the British Association for the Advancement of Science (York, 1881), London.

Hoover, R.F. (1970) *The Vascular Plants of San Luis Obispo County, California*, University of California Press, Berkeley, California.

Hopper, S.D. (1979) Biogeographical aspects of speciation in southwest Australian flora. *Ann. Rev. Ecol. Syst.*, **10**, 399–422.

Horn, M.H. and Allen, L.G. (1978) A distributional analysis of California coastal marine fishes. *J. Biogeogr.*, **5**, 23–42.

Horn, H.S. and MacArthur, R.H. (1972) Competition among fugative species in a harlequin environment. *Ecology*, **53**, 749–52.

Horton, D.R. (1984) Red kangaroos: last of the Australian megafauna. In *Quaternary Extinctions: a prehistoric revolution* (eds P.S. Martin and R.G. Klein), University of Arizona Press, Tucson, pp. 639–80.

Horvat, I., Glavac, V. and Ellenberg, H. (1974) *Vegetation Südosteuropas*, Gustav Fischer, Jena.

Howell, J.T. (1970) *Marine Flora*, 2nd edn, University of California Press, Berkeley, California.

Hsü, K.J. (1980) Terrestrial catastrophe caused by cometary impact at the end of the Cretaceous. *Nature, Lond.*, **285**, 201–3.

Hubbard, M.D. (1973) Experimental insular biogeography. *Florida Scient.*, **36**, 132–41.

Hubbell, S.P. (1979) Tree dispersion, abundance and diversity in a dry tropical forest. *Science*, **203**, 1299–309.

Hubbell, S.P. (1980) Seed predation and coexistence of tree species in tropical forests. *Oikos*, **35**, 214–29.

Hubbell, S.P. and Foster, R.B. (1983) Diversity of canopy trees in a neotropical forest and implications for conservation. *Brit. Ecol. Soc. Spec. Publication*, **2**, 25–41.

Hubbell, S.P. and Foster, R.B. (1986) Biology, chance and history, and the structure of tropical rain forest tree communities. In *Community Ecology* (eds J. Diamond and T.J. Case), Harper and Row, New York, pp. 314–30.

Huber, O. (1982) Significance of savanna vegetation in the Amazon Territory of Venezuela. In *Biological Differentiation in the Tropics* (ed G.T. Prance), Columbia University Press, New York, pp. 221–44.

Hueck, K. (1966) *Die Wälder Südamerikas*, Fischer Verlag, Stuttgart.

Huey, R.B. and Pianka, E.R. (1974) Ecological character displacement in a lizard. *Amer. Zool.*, **14**, 1127–36.

Hughes, N.F. ed. (1973) Organisms and continents through time. *Spec. Publns. Palaeontol.*, **12**.

Hull, D.L. (1967) Certainty and circularity in evolutionary taxonomy. *Evolution*, **21**, 174–89.

Hull, D.L. (1976) Are species really individuals? *Syst. Zool.*, **25**, 174–91.

Hull, D.L. (1978) A matter of individuality. *Philosophy Sci.*, **45**, 335–60.

Hull, D.L. (1979) The limits of cladism. *Syst. Zool.*, **28**, 416–40.

Hull, D.L. (1980) Cladism gets sorted out. (Review of *Phylogenetic analysis and paleontology* (1979) by J. Cracraft and N. Eldredge (eds), Columbia University Press, New York.) *Paleobiology*, **6**, 131–6.

Hulten, E. (1968) *Flora of Alaska and Neighboring Territories*, Stanford University Press, Stanford, California.

Hulten, E. (1970) *Atlas över växternas Utbredning i Norden*, 2nd edn, Stockholm.

Humphrey, R.R. (1974) *The Boojum Tree and its Home*, University of Arizona Press, Tucson.

Humphries, C.J. (1983) Biogeographical explanations and the southern beeches. In *Evolution, Time and Space: the emergence of the biosphere* (eds R.W. Sims, J.H. Price and P.E.S. Whalley), Academic Press, London and New York, pp. 336–65.

Humphries, C.J. and Funk, V.A. (1984) Cladistic Methodology. In *Current Concepts in Plant Taxonomy* (Eds V.H. Heywood and D.M. Moore), Academic Press, London and Orlando, pp. 323–62.

Humphries, C.J. and Parenti, L. (1986) *Cladistic Biogeography*, Clarendon Press, Oxford.

Huston, M. (1979) A general hypothesis of species diversity. *Amer. Nat.*, **113**, 81–101.

Hut, P., Alvarez, W., Elder, W. *et al.* (1985) Comet showers as possible causes of stepwise mass extinctions. EOS, *Trans. Am. Geophys. Un.*, **66**, 813.

Hutchinson, G.E. (1957) Concluding remarks. *Cold Spring Harbour Symposium in Quantitative Biol.*, **22**, 415–27.

Hutchinson, G.E. (1959) Homage to Santa Rosalia, or 'Why are there so many kinds of animals?' *Amer. Nat.*, **93**, 145–59.

Hutchinson, G.E. (1961) The paradox of the plankton. *Amer. Nat.*, **95**, 137–45.

Hutton, F.W. (1873) On the geographical relations of the New Zealand fauna. *Trans. Proc. N.Z. Inst.*, **5**, 227–56.

Huxley, T.H. (1868) On the classification and distribution of the Alectoromorphae and Heteromorphae. *Proc. Zool. Soc. Lond.*, 249–319.

Iatrou, G. and Georgiadis, T. (1985) *Minuartia favargeri*: a new species from Peloponnesus (Greece). *Candollea*, **40**, 129–38.

Illies, J. (1965) Die Wegnersche Kontinentalverschiebungstheorie im lichte der modernen biogeographie. *Naturwissenschaften*, **18**, 505–11.

Iturralde, R.B. (1986) Algunas aspectos fitogeograficos de plantas serpentinicolos cubanas. *Feddes Repert.*, **97**, 49–58.

Ivashyn, D.S., Ya, R., Isayeva, P.I. *et al.* (1981) Relict and endemic plants of the Donets River valley in its lower course. *Ukrayen. Botan. Zhurnal*, **38**, 60–4.

Jablonski, D. (1986a) Background and mass extinctions: the alternation of macroevolutionary regimes. *Science*, **231**, 129–33.

Jablonski, D. (1986b) Causes and consequences of mass extinctions: a comparative approach. In *Dynamics of Extinction* (ed D.K. Elliott), Wiley, New York, pp. 183–229.

Jablonski, D., Flessa, K.W. and Valentine, J.W. (1985) Biogeography and paleobiology. *Paleobiology*, **11**, 75–90.

Jablonski, D. and Lutz, R.A. (1983) Larval ecology of marine benthic invertebrates: paleobiological implications. *Biol. Rev.*, **58**, 21–89.

Jackson, J.B.C. (1977) Habitat area, colonization and development of epibenthic community structure. In *Biology of Benthic Organisms* (eds B.F. Keegan, P.O. Ceidigh and P.J.S. Boaden), Pergamon Press, New York, pp. 349–58.

Jalas, J. and Suominen, J. (1972) *Atlas Florae Europaeae* (6 vols to date), Helsinki.

Jannasch, H.W. and Mottl, M.J. (1985) Geomicrobiology of deep-sea hydrothermal vents. *Science*, **229**, 717–25.

Janzen, D.H. (1967) Why mountain passes are higher in the tropics. *Amer. Nat.*, **101**, 233–49.

Janzen, D.H. (1970) Herbivores and the number of tree species in tropical forests. *Amer. Nat.*, **104**, 501–28.

Janzen, D.H. (1981) The peak in North American ichneumoid species richness lies between 38 and 42°N. *Ecology*, **62**, 532–7.

Janzen, D.H. (1983) No park is an island: increase in interference from outside as park size decreases. *Oikos*, **41**, 402–10.

Jenny, H. (1980) *The Soil Resource. Origin and Behavior*. Ecological Studies **37**, Springer, New York.

Jenny, H., Arkley, R.J. and Schultz, A.M. (1969) The pygmy forest-podsol ecosystem and its dune associates of the Mendocino Coast. *Madrono*, **20**, 60–74.

Johanson, D. and Edey, M. (1981) *Lucy: the Beginnings of Humankind*, Warner Books, New York.

Johnson, M.S. (1982) Polymorphism for direction of coil in *Partula suturalis*: behavioural isolation and positive frequency-dependent selection. *Heredity*, **49**, 145–51.

Johnson, M.S., Clarke, B.C. and Murray, J. (1977) Genetic variation and reproductive isolation in *Partula*. *Evolution*, **31**, 116–26.

Johnson, M.S., Murray, J. and Clarke, B. (1986) Allozymic similarities among species of *Partula* in Moorea. *Heredity*, **56**, 319–28.

Johnston, D.W. and Odum, E.P. (1956) Breeding bird populations in relation to plant succession in the piedmont of Georgia. *Ecology*, **37**, 50–62.

Jones, J.S., Bryant, S.H., Lewontin, R.C. *et al.* (1981) Gene flow and the geographic distribution of a molecular polymorphism in *Drosophila pseudoobscura*. *Genetics*, **98**, 157–79.

Jones, J.S., Leith, B.H. and Rawlings, P. (1977) Polymorphism in *Cepaea*: problem with too many solutions. *Ann. Rev. Ecol. Syst.*, **8**, 109–43.

Jones, M.G. (1933) Grassland management and its influence on the sward. *Empire J. exp. Agric.*, **1**, 43–367.

Kamelin, R.V. (1973) *Florogenetic Analysis of the Natural Flora of the Mountains of Middle Asia*, Nauka, Leningrad.

Karavaev, M.N. (1958) *Conspectus of the Flora of Yakutia*, Nauka, Moscow and Leningrad.

Karmysheva, N.K. (1982) *Flora and Vegetation of the Western Spur Ranges of the Talasski Alatau*, Izd. Nauka, Kazakhskoi SSR, Alma-Ata.

Keast, A. (1970) Adaptive evolution and shifts in niche occupation in island birds. *Biotropica*, **2**, 61–75.

Keeler-Wolf, T. (1986) The barred antshrike (*Thamnophilus doliatus*) on Trinidad and Tobago: habit niche expansion of a generalist forager. *Oecologia*, **30**.

Keener, C.S. (1983) Distribution and biohistory of the endemic flora of the mid-Appalachian shale barrens. *Botan. Rev.*, **49**, 65–115.

Kemp, M. (1981) Tertiary palaeogeography and the evolution of Australian climate. In *Ecological Biogeography of Australia* (ed. A. Keast) Junk, The Hague, pp. 31–50.

Kennett, J.P., Houtz, R.E., Andrews, P.B., *et al.* (1974) Development of the circum-Antarctic current. *Science*, **186**, 144–7.

Keough, M.J. and Butler, A.J. (1983) Temporal changes in species number in an assemblage of sessile marine invertebrates. *J. Biogr.*, **10**, 317–30.

Key, K.H.L. (1982) Species, parapatry and the morabine grasshoppers. *Syst. Zool.*, **30**, 425–58.

Khanminchun, V.M. (1980) *Flora of the Eastern Tannu-Ola (Southern Tuva)*, Nauka Sibitskoe Ordelenie, Novosibirsk.

Kiester, A.R. (1971) Species density of North American amphibians and reptiles. *Syst. Zool.*, **20**, 127–37.

Kiester, A.R. (1983) Zoogeography of the skinks (Sauria: Scincidae) of Arno Atoll, Marshall Islands. In *Advances in Herpetology and Evolutionary Biology* (eds A.G.J. Rhodin and K. Miyata), Museum of Comparative Zoology, Cambridge, Mass., pp. 359–64.

Kikkawa, J. and Williams, E.E. (1971) Altitudinal distribution of land birds in New Guinea. *Search*, **2**, 64–9.

Kilias, G. and Alahiotis, S.N. (1982) Genetic studies on sexual isolation and hybrid sterility in long-term cage populations of *Drosophila melanogaster*. *Evolution*, **36**, 121–31.

Kinne, O. (1971) Salinity–invertebrates. In *Marine Ecology* (Vol. 1), Environmental Factors, Part 2 (ed O. Kinne) MacLehose, Glasgow, pp. 821–995.

Kinzel, H. (1982) *Pflanzenökologie und Mineralstoffwechsel*, Ulmer, Stuttgart.

Kirkpatrick, M. (1982) Sexual selection and the evolution of female choice. *Evolution*, **36**, 1–12.

Kitchell, J.A. and Pena, D. (1984) Periodicity of extinctions in the geologic past: deterministic versus stochastic explanations. *Science*, **226**, 686–92.

Kluge, A.G. and Farris, J.S. (1969) Quantitative Phyletics and the evolution of anurans. *Syst. Zool.*, **18**, 1–32.

Knaben, G.S. (1982) Om arts- og rasedannelse i Europa under kvartærtiden. 1. Endemiske arter i Nord-Atlanteren. *Blyttia*, **40**, 229–35.

Knoll, A.H. (1984) Patterns of extinction in the fossil record of vascular plants. In *Extinctions* (ed M.N. Nitecki), University of Chicago Press, Chicago, pp. 21–68.

Knoll, A.H. (1986) Patterns of change in plant communities through geological times. In *Community Ecology* (eds J. Diamond and T.J. Case), Harper and Row, New York, pp. 126–42.

Koehn, R.K. (1978) Physiology and biochemistry of enzyme variation: the interface of ecology and population genetics. In *Ecological Genetics: the interface* (ed. P.F. Brussard), Springer-Verlag, New York, pp. 51–72.

Kohn, A.J. (1978) Ecological shift and release in an isolated population: *Conus miliaris* at Easter Island. *Ecol. Monogr.*, **48**, 323–36.

Koopman, K.F. (1981) Discussion. In *Vicariance Biogeography: a critique* (eds D. Nelson and D.E. Rosen), Columbia Press, New York, pp. 184–7.

Kornfeld, I., Smith, D.C., Gagnan, P.S. and Taylor, J.N. (1982) The cichlid fish of Cuetro Cieniges, Mexico: direct evidence of conspecificity among distinct trophic morphs. *Evolution*, **36**, 658–64.

Kostitzin, V.A. (1939) *Mathematical Biology*, Harrap, London.

Krasnoborov, I.M. (1976) *The High Mountain Flora of the Western Sayan*, Nauka, Sibirskoe Otd., Novosibirsk.

Krasnoborov, I.M. and Khanminchun, V.M. (1985) Review of Grubov, I.M. (1982) Key to the vascular plants of Mongolia, Nauka, Leningrad. *Botan. Zhurnal*, **70**, 563–4.

Krasnoborov, I.M., Lomonosova, M.N., Khanminchun, V.M. *et al.* (1980) Fifth addition to the flora of the Tuva ASSR. *Botan. Zhurnal*, **65**, 1024–8.

Krebs, D.J., Keller, B.L. and Tamarin, R.H. (1969) *Microtus* population biology: demographic changes in fluctuating populations of *M. ochrogaster* and *M. pennsylvanicus* in Southern India. *Ecology*, **50**, 587–607,

Kruckeberg, A.R. (1984) California serpentines: flora, vegetation, geology, soils and management problems. *Univ. California Publ. Botany*, **78**.

Kubitski, K. ed. (1983) *Dispersal and Distribution. An international symposium. Sonderbände Naturwiss. Vereins in Hamburg 7*. Parey, Hamburg.

Kuhn, T.S. (1970) *The Structure of Scientific Education*, 2nd edn, Chicago University Press, Chicago.

Kuminova, A.V. (1960) *The Vegetation of the Altai*, Akad. Nauk SSSR, Sibir. Otd., Novosibirsk.

Kunkel, G. (1980) *Die Kanarischen Inseln und Ihre Pflanzenwelt*, Gustav Fischer, Stuttgart.

Ladiges, P.Y. and Humphries, C.J. (1983) A cladistic study of *Arillastrum, Angophora* and *Eucalyptus* (Myrtaceae). *Bot. J. Linn. Soc.*, **86**, 105–34.

Ladiges, P.Y. and Humphries, C.J. (1986) Relationships in the stringy barks. *Eucalyptus* L Herit. informal subgenus *Monocalyptus* series Capitellatae and Olsenianae: phylogenetic hypotheses, biogeography and classification. *Aust. J. Bot.*, **34**, 603–32.

Ladiges, P.Y., Humphries, C.J. and Brooker, M.I.H. (1987) Cladistic and biogeographic analysis of Western Australian species of *Eucalyptus* L Herit. informal subgenus *Monocalyptus* Pryor & Johnson. *Aust. J. Bot*, **35**, 251–81.

Lande, R. (1979) Effective deme sizes during long-term evolution estimated from rates of chromosomal rearrangement. *Evolution*, **33**, 234–51.

Lande, R. (1981) Models of speciation by sexual selection on polygenic traits. *Proc. Nat. Acad. Sci., USA*, **78**, 3721–5.

Lande, R. (1985) The fixation of chromosomal rearrangement in a subdivided population with local extinction and recolonization. *Heredity*, **54**, 323–32.

Lane, R.P. and Marshall, J.E. (1981) Geographical variation, races and subspecies. In *The Evolving Biosphere. Chance Change and Challenge* (Ed P.L. Forey), British Museum (Natural History) Cambridge University Press, Cambridge, pp. 9–19.

Langley, C.H., Montgomery, F. and Quattlebaum, W.F. (1982) Restriction map variation in the *Adh* region of *Drosophila*. *Proc. Nat. Acad. Sci. USA*, **79**, 5631–5.

Langridge, J. and Griffing, B. (1959) A study of high temperature lesions in *Arabidopsis thaliana*. *Aust. J. Biol. Sci.*, **12**, 117–35.

Larcher, W. (1980) *Oekologie der Pflanzen, auf physiologischer Grundlage*, 3rd edn, Uni-Taschenbücher, Ulmer, Stuttgart.

Larson, A. and Highton, R. (1978) Geographic protein variation and divergence in populations of the salamander *Plethodon welleri* group (Amphibia, Plethodontidae). *Syst. Zool.*, **27**, 431–48.

Lasker, H.R. (1978) The measurement of taxonomic evolution preservational consequences. *Paleobiology*, **4**, 135–49.

Lassen, H.H. (1975) The diversity of fresh water snails in view of the equilibrium theory of island biogeography. *Oecologia*, **19**, 1–8.

Lawlor, T.E. (1983) The mammals. In *Island Biogeography in the Sea of Cortez* (eds T.J. Case and M.L. Cody), University of California Press, Berkeley, pp. 265–89, 482–500.

Lawton, J.H. and Strong, D.R. (1981) Community patterns and competition in foliovorous insects. *Amer. Nat.*, **118**, 317–38.

Lebedeva, N. (1902) Rol korallov u devonskikh otlozheniyaka rossii (and) Die Bedeutung der korallen in den devonische Ablagerung en Russlands. *Mém. Com. Géol.*, **17**, i–x, 1–130, 137–80.

Ledig, F.T. and Conkle, M.T. (1983) Gene diversity and genetic structure in a narrow endemic, Torrey Pine (*Pinus torreyana* Parry ex Carr.) *Evolution*, **37**, 79–85.

Lein, M.R. (1972) A trophic comparison of avifaunas. *Syst. Zool.*, **21**, 135–50.

Leopold, A. (1949) *A Sand County Almanac*, Oxford University Press, Oxford.

Levin, S.A. (1974) Dispersion and population interaction. *Amer. Nat.*, **108**, 207–28.

Levin, S.A. (1978) Population models and community structure in heterogeneous environments. In *Studies in Mathematical Biology, Part II, Populations and Communities. Studies in Mathematics, Vol. 16* (ed S.A. Levin), Mathematical Assoc. of America, Providence, RI pp. 439–76.

Levins, R. and Culver, D.C. (1971) Regional coexistence of species and competition between rare species. *Proc. Nat. Acad. Sci. USA*, **68**, 1246–8.

Levinton, J.S. (1987) The body size–prey size hypothesis and *Hydrobia*. *Ecology*, **68**, 229–31.

Lewin, R. (1983) Origin of species in stressed environments. *Science*, **222**, 1112.

Lewin, R. (1985a) Pattern and process in life's history. *Science*, **229**, 151–3.

Lewin, R. (1985b) Catastrophism not yet dead. *Science*, **229**, 640.

Lewin, R. (1986) Mass extinctions select different victims. *Science*, **231**, 219–20.

Lewontin, R.C. and Birch, L.C. (1966) Hybridization as a source of variation for adaptation to new environments. *Evolution*, **20**, 315–36.

Lillegraven, J.A. (1972) Ordinal and familial diversity in Cenozoic mammals. *Taxon*, **21**, 261–74.

Linnaeus, C. (1781) *Selected Dissertations from the Amoenitates Academicae*, a supplement to Mr. Stillingfleet's Tracts relating to Natural History 2. On the increase of the Habitable earth, London.

Lister, B.C. (1976) The nature of niche expansion in West Indian *Anolis* lizards. I: Ecological consequences of reduced competition. *Evolution*, **30**, 659–76.

Little, J.W. and Mount, D.W. (1982) The SOS regulatory system of *Escherichia coli*. *Cell*, **29**, 11–22.

Lomolino, M.V. (1984) Immigrant selection, predation, and the distributions of *Micotus pennsylvanicus* and *Blarina brevicauda* on islands. *Amer. Nat.*, **123**, 468–83.

Löve, A. and Löve, D. eds, (1963) *North Atlantic Biota and their History*, Pergamon Press, London.

Lovejoy, T.E., Rankin, J.M., Bierregaard Jr., *et al.* (1984) Ecosystem decay of Amazon forest remnants. In *Extinctions* (ed. M.N. Nitecki), University of Chicago Press, Chicago, pp. 295–325.

Løvtrup, S. (1982) The four theories of evolution. *Rev. Biol.*, **75**, 53–66, 231–72, 385–409.

Lubchenco, J. (1978) Plant species diversity in a marine intertidal community: importance of herbivore food preference and algal competitive abilities. *Amer. Nat.*, **112**, 23–39.

Lubchenco, J. (1986) Relative importance of competition and predation: early colonization by seaweeds in New England. In *Community Ecology* (eds J. Diamond and T.J. Case). Harper and Row, New York, pp. 537–55.

Ludwig, D., Jones, D.D. and Holling, C.S. (1978) Qualitative analysis of insect outbreak systems: the spruce budworm and forest. *J. Anim. Ecol.*, **47**, 315–32.

Lutz, T.M. (1985) The magnetic reversal record is not periodic. *Nature, Lond.*, **317**, 404–7.

Lynch, J.D. (1979) The amphibians of the lowland tropical forests. In *The South American Herpetofauna; its origin, evolution and dispersal.* (ed W.E. Duellmann) *Mus. Nat. Hist. Univ. Kans. Monogr.*, **7**, pp. 189–215.

Lynch, J.D. (1981) The systematic status of *Amblyphrynus ingeri* (Amphibia: Leptodactylidae) with the description of an allied species in western Colombia. *Caldasia*, **13**, 313–32.

Lynch, J.D. (1982) Relationships of the frogs of the genus *Ceratophrys* (Leptodactylidae) and their bearing on hypotheses of Pleistocene forest refugia in South America and punctuated equilibria. *Syst. Zool.*, **31**, 166–79.

Lynch, J.D. (1986) The definition of the Middle American clade of *Eleutherodactylus* based on jaw musculature (Amphibia: Leptodactylidae). *Herpetologica*, **42**, 248–58.

Lynch, J.D. and Duellman, W.E. (1980) The *Eleutherodactylus* of the Amazonian slopes of the Ecuadorian Andes (Anura: Leptodactylidae). *Mus. Nat. Hist. Univ. Kansas Misc. Publ.*, **69**, 1–86.

Lynch, J.F. and Johnson, M.K. (1974) Turnover and equilibrium in insular avifaunas, with special reference to the California Channel Islands. *Condor*, **76**, 370–84.

MacArthur, J.W. (1975) Environmental fluctuations and species diversity. In *Community Ecology* (eds M.L. Cody and J.M. Diamond) Harvard University Press, Cambridge, Mass., pp. 74–80.

MacArthur, R.H. (1955) Fluctuations of animal populations and a measure of community stability. *Ecology*, **36**, 533–6.

MacArthur, R.H. (1965) Patterns of species diversity. *Biol. Rev.*, **40**, 510–33.

MacArthur, R.H. (1969) Patterns of communities in the tropics. *Biol. J. Linn. Soc.*, **1**, 19–30.

MacArthur, R.H. (1972) *Geographical Ecology/patterns in the distribution of species.* Harper and Row, New York.

MacArthur, R.H., Diamond, J.M. and Karr, J. (1972) Density compensation in island faunas. *Ecology*, **53**, 330–42.

MacArthur, R.H. and Levins, R. (1967) The limiting similarity, convergence and divergence of coexisting species. *Amer. Nat.*, **101**, 377–85.

MacArthur, R.H. and MacArthur, J.W. (1961) On bird species diversity. *Ecology*, **42**, 594–8.

MacArthur, R.H., Recher, H. and Cody, M. (1966) On the relation between habitat selection and species diversity. *Amer. Nat.*, **100**, 319–22.

MacArthur, R.H. and Wilson, E.O. (1963) An equilibrium theory of insular zoogeography. *Evolution*, **17**, 373–87.

MacArthur, R.H. and Wilson, E.O. (1967) *The Theory of Island Biogeography*, Princeton University Press, Princeton.

McCoy, E.D. and Heck, K.L. Jr (1976) Biogeography of corals, sea grasses and mangroves: an alternative to the center of origin concept. *Syst. Zool.*, **25**, 201–10.

McCoy, E.D. and Heck, K.L. (1983) Centers of origin revisited. *Paleobiology*, **9**, 17–19.

McCoy, E.D. and Heck, K.L. Jr (1987) Some observations on the use of taxonomic similarity in large-scale biogeography. *J. Biogeogr.*, **14**, 79–87.

McCulloch, G.E. (1985) Variance tests for species association. *Ecology*, **66**, 1676–81.

McDowall, R.M. (1978) Generalized tracks and dispersal in biogeography. *Syst. Zool.*, **27**, 88–104.

McGowan, J.A. and Walker, P.W. (1979) Structure in the copepod community of the North Pacific central gyre. *Ecol. Monogr.*, **49**, 193–226.

McGowan, J.A. and Walker, P.W. (1985) Dominance and diversity maintenance in an oceanic ecosystem. *Ecol. Monogr.*, **55**, 103–18.

McGuinness, K.A. (1984) Equations and explanations in the study of species–area curves. *Biol. Rev.*, **59**, 423–40.

McKenna, M.C. (1973) Sweepstakes, filters, corridors, Noah's Arks and beached Viking funeral ships in palaeogeography. In *Implications of Continental Drift to the Earth Sciences* (Vol. 1) (eds D.H. Tarling and S.K. Runcorn), Academic Press, London.

McKenzie, J.A. and Parsons, P.A. (1974) Microdifferentiation in a natural population of *Drosophila melanogaster* to alcohol in the environment. *Genetics*, **77**, 385–94.

McKerrow, W.S. and Cocks, L.R.M. (1976) Progressive faunal migration across the Iapetus Ocean. *Nature, Lond.*, **263**, 304–6.

MacLaren, D.J. (1970) Time, life and boundaries. *J. Paleontol.*, **44**, 801–15.

McLean, D.M. (1978) A terminal Mesozoic 'greenhouse': lessons from the past. *Science*, **101**, 401–6.

McLean, D.M. (1981) Terminal Cretaceous extinctions and volcanism: a link. *Amer. Ass. Adv. Sci.*, Abstracts, Torongo, p. 128.

McLean, D.M. (1985) Mantle degassing unification of the Trans-K-T beobiological record. *Evol. Biol.*, **19**, 287–313.

McManus, J.W. (1985) Marine speciation, tectonics and sea-level changes in southeast Asia. *Proc. 5th Int. Coral Reef Congr., Tahiti*, **4**, 133–38.

McMurtrie, R. (1978) Persistence and stability of single-species and prey-predator systems in spatially heterogeneous environments. *Math. Biosci.*, **39**, 11–51.

McNab, B.K. (1971) On the ecological significance of Bergmann's rule. *Ecology*, **52**, 845–54.

MacNair, M.R. (1981) Tolerance of higher plants to toxic materials. In *Genetic Consequences of Man-Made Change* (eds J.A. Bishop and L.M. Cook) Academic Press, London, pp. 177–208.

Maguire, B. Jr (1963) The passive dispersal of small aquatic organisms and their colonization of isolated bodies of water. *Ecol. Monogr.*, **33**, 161–85.
Maguire, B. (1970) On the flora of the Guayana Highland. *Biotropica*, **2**, 85–100.
Maguire, B. Jr (1971) Phytotelmata: Biota and community structure determination in plant-held waters. *Ann. Rev. Ecol. Syst.*, **2**, 439–64.
Main, J.L. (1981) Magnesium and calcium nutrition of a serpentine endemic grass. *Amer. Midl. Nat.*, **105**, 196–9.
Major, J. (1974) Kinds and rates of changes in vegetation and chronofunctions. In *Handbook of Vegetation Science*, Part VIII, Vegetation Dynamics (ed. R. Knapp) pp. 9–18.
Major, J. (1977) California climate in relation to vegetation. In *Terrestrial Vegetation of California* (eds M.G. Barbour and J. Major) Wiley, New York, pp. 11–74.
Malecot, G. (1948) *The mathematics of heredity*. Masson, Paris.
Mallet, J.L.B. (1986) Hybrid zones of *Heliconius* butterflies in Panama and the stability and movement of warning colour clines. *Heredity*, **56**, 191–202.
Malyshev, L.I. (1965) *High Mountain Flora of the Eastern Sayan*. Nauka, Moscow and Leningrad.
Malyshev, L.I. (ed.) (1972) Properties and genesis of the flora. In *High Mountain Flora of the Stanovoi Uplands* Nauka, Sibir. Otd., Novosibirsk, pp. 150–89.
Malyshev, L.I. (ed.) (1976) *Flora Putorana*, Izd. Nauka, Sibirskoe Otd., Novosibirsk.
Mandelbrot, B. (1983) *The Fractal Geometry of Nature*, Freeman, New York.
Mares, M.A., Blair, W.F., Enders, F.A. *et al.* (1977) The strategies and community patterns of desert animals. In *Convergent Evolution in Warm Deserts: an examination of strategies and pattern in deserts of Argentina and the United States* (eds G.H. Orians and O.T. Solbrig), Dowden, Hutchinson and Ross, Stroudsburg, Penn., pp. 107–63.
Mark, G.A. and Flessa, K.W. (1977) A test for evolutionary equilibria: Phanerozoic branchipods and Cenozoic mammals. *Paleobiology*, **3**, 17–22.
Mark, G.A. and Flessa, K.W. (1978) A test for evolutionary equilibrium revisited. *Paleobiology*, **4**, 201–2.
Marshall, L.G. (1977) Evolution of the carnivorous adaptive zone in South America. In *Major Patterns in Vertebrate Evolution* (eds M.K. Hecht, P.C. Goody, B.M. Hecht), Plenum Press, New York, pp. 709–21.
Marshall, L.G. (1981) The great American interchange: an invasion induced crisis for South American mammals. In *Biotic Crises in Ecological and Evolutionary Time* (ed. M.H. Nitecki), Academic Press, New York, pp. 133–229.
Marshall, L.G. (1984) Who killed cock robin?: an investigation of the extinction controversy. In *Quaternary Extinctions: a prehistoric revolution* (eds P.S. Martin and R.G. Klein), University of Arizona, Tucson, pp. 785–806.
Marshall, L.G. (1985) Geochronology and land mammal biochronology of the transamerican faunal interchange. In *The Great American Biotic Interchange* (eds F.G. Stehli and S.D. Webb), Plenum, New York, pp. 49–85.
Marshall, L.G. and Corruccini, R.S. (1978) Variability, evolutionary rate, and allometry in dwarfing lineages. *Paleobiology*, **4**, 101–19.
Marshall, L.G., Webb, S.D., Sepkoski, J.J. and Raup, D.M. (1982) Mammalian evolution and the Great American Interchange. *Science*, **215**, 1351–7.
Martin, T.E. (1980) Diversity and abundance of spring migratory birds using habitat islands on the great plains. *Condor*, **82**, 430–39.
Martin, P.S. (1984a) Prehistoric overkill: the global model. In *Quaternary*

Extinctions: a prehistoric revolution (eds P.S. Martin and R.G. Klein), University of Arizona Press, Tucson, pp. 354–403.
Martin, P.S. (1984b) Catastrophic extinctions and late Pleistocene blitzkrieg: two radiocarbon tests. In *Extinctions* (ed. M.H. Nitecki), University of Chicago Press, Chicago, pp. 153–189.
Martin, P.S. and Klein, R.G. (eds) (1984) *Quaternary Extinctions: a prehistoric revolution*, University of Arizona Press, Tucson.
Martin, T.E. (1981a) Species-area slopes and coefficients: a caution on their interpretation. *Amer. Nat.*, **118**, 823–37.
Martin, T.E. (1981b) Limitation in small habitat islands: chance or competition? *Auk*, **98**, 715–34.
Matessi, C. and Jayakar, S.D. (1981) Coevolution of species in competition: a theoretical study. *Proc. Nat. Acad. Sci. USA*, **78**, 1081–4.
Matthew, W.D. (1915) Climate and evolution. *Ann. N.Y. Acad. Sci.*, **24**, 171–318.
May, R.M. (1973) *Stability and Complexity in Model Ecosystems*, Princeton University Press, Princeton, New Jersey.
May, R.M. (1975) Patterns of species abundance and diversity. In *Ecology and Evolution of Communities* (eds M.L. Cody and J.M. Diamond), Harvard University Press, Cambridge, Mass., pp. 81–120.
May, R.M. (1981) Patterns in multi-species communities. In *Theoretical Ecology: principles and application*, 2nd edn (ed. R.M. May), Blackwell Scientific Pubns, Oxford, pp. 197–227.
May, R.M. (1984) Real and apparent patterns in community structure. In *Ecological Communities: conceptual issues and the evidence* (eds D.R. Strong, D. Simberloff, L. Abele and A. Thistle), Princeton University Press, Princeton, pp. 3–16.
May, R.M. (1986) The search for patterns in the balance of nature: advances and retreats. *Ecology*, **67**, 1115–26.
May, R.M. and MacArthur, R.H. (1972) Niche overlap as a function of environmental variability. *Proc. Nat. Acad. Sci. USA*, **69**, 1109–13.
Maynard Smith, J. (1966) Sympatric speciation. *Amer. Nat.*, **100**, 637–50.
Mayr, E. (1942) *Systematics and the Origin of Species*, Columbia University Press, New York.
Mayr, E. (1954) Change of genetic environment and evolution. In *Evolution as a Process* (eds J. Huxley, A.C. Hardy and E.B. Ford), Allen and Unwin, London.
Mayr, E. (1963) *Animal Species and Evolution*, Belknap, Harvard.
Mayr, E. (1982) *The Growth of Biological Thought: diversity, evolution and inheritance*, Belknap, Harvard.
Mayr, E. and O'Hara, R.J. (1986) The biogeographic evidence supporting the Pleistocene forest refuge hypothesis. *Evolution*, **40**, 55–67.
Mazaud, A., Laj, C., Sèze, L. de, and Verosub, K.L. (1983) 15-Myr periodicity in the frequency of geomagnetic reversals since 100 Myr. *Nature, Lond.*, **304**, 328–30.
Means, D.B. (1975) Competitive exclusion along a habitat gradient between two species of salamanders (*Desmognathus*) in western Florida. *J. Biogeogr.*, **2**, 253–63.
Mears, J.A. (1980) Flavonoid diversity and geographic endemism in *Parthenium*. *Biochem. Systematics and Ecology*, **8**, 361–70.
Melville, R.M. (1966) Continental drift, Mesozoic continents and the migrations of angiosperms. *Nature, Lond.*, **211**, 116–120.

Melville, R.M. (1981) Vicarious plant distributions and paleography of the Pacific region. In *Vicariance Biogeography: a critique* (eds G. Nelson and D. Rosen), Columbia University Press, New York, pp. 238–74, 298–302.

Melville, R.M. (1982) The biogeography of *Nothofagus* and *Trigonobalanus* and the origin of the Fagaceae. *Bot. J. Linn. Soc.*, **85**, 75–88.

Menge, B.A. and Sutherland, J.P. (1976) Species diversity gradients: synthesis of the roles of predation, competition and temporal heterogeneity. *Amer. Nat.*, **110**, 351–69.

Mickevich, M.F. (1978) Taxonomic Congruence. *Syst. Zool.*, **27**, 143–58.

Mickevich, M.F. (1981) Quantitative phylogenetic biogeography. In *Advances in Cladistics*, **1** (eds V.A. Funk and D.R. Brooks), Columbia University Press, New York, pp. 209–22.

Middlemiss, T.A., Rawson, P.F., and Newall, G. eds (1971) Faunal provinces in space and time. *Geol. J. Spec. Issue*, **4**, 1–236.

Mildenhall, D.C. (1980) New Zealand late Cretaceous and Cenozoic plant biogeography: a contribution. *Palaeogeogr., Palaeoclimatol., Palaeoecol.*, **31**, 197–233.

Milligan, B.G. (1985) Evolutionary divergence and character displacement in two phenotypically variable, competing species. *Evolution*, **39**, 1207–22.

Milner, A.R. and Norman, D.B. (1984) The biogeography of advanced ornithopod dinosaurs (Archosauria: Ornithischia) – a cladistic–vicariance model. In *Third Symposium on Mesozoic Terrestrial Ecosystems, Short Papers* (eds W.-E. Reif and F. Westphal), Attempto Verlag, Tübingen, pp. 145–50.

Miyamoto, M.M. (1985) Consensus cladograms and general classifications. *Cladistics*. **1**, 186–9.

Molles, M.C. (1978) Fish species diversity on model and natural reef patches: experimental insular biogeography. *Ecol. Monogr.*, **48**, 289–305.

Mooney, H.A. and Cody, M.L. (1977) Summary and conclusions. In *Convergent Evolution in Chile and California: Mediterranean climate ecosystems* (ed. H.A. Mooney), Dowden, Hutchinson and Ross, Stroudsburg, Prnn., pp. 193–9.

Moore, W.S. (1977) An evaluation of narrow hybrid zones in vertebrates. *Q. Rev. Biol.*, **52**, 263–78.

Moran, C. (1981) Spermatogenesis in natural and experimental hybrids between chromosomally differentiated taxa of *Caledia captiva*. *Chromosoma*, **81**, 579–91.

Morse, D.R., Lawton, J.H., Dodson, M.M. and Williamson, M.H. (1985) Fractal dimension of vegetation and the distribution of arthropod body lengths. *Nature, Lond.*, **314**, 731–2.

Muller, H.J. (1939) Reversibility in evolution considered from the standpoint of genetics. *Biol. Rev. Camb. Phil. Soc.*, **14**, 261–80.

Muller, H.J. (1942) Isolating mechanisms, evolution and temperature. *Biol. Symp.*, **6**, 71–125.

Müller, P. (1973) *Dispersal Centres of Terrestrial Vertebrates in the Neotropical Realm, Biogeographica 2*, Junk, The Hague.

Müller, P. (1974) *Aspects of Zoogeography*, Junk, The Hague.

Murray, A. (1873) On the geographical relations of the chief coleopterous faunae. *J. Linn. Soc. (Zoology)*, **11**, 1–89.

Murray, D.F. (1980) *Threatened and Endangered Plants of Alaska*. U.S. Forest Service and Bureau of land Mgt, vii and 59.

Murray, J. and Clarke, B.C. (1980) The genus *Partula*: speciation in progress. *Proc. Roy. Soc. Lond.* B, **211**, 83–117.

Myers, A.A. (1977) The effect of temperature during pupal actiphase on

imaginal colouration in *Mythimna loreyi* (Duponechl) (Lep. Noctuidae). *Entomologist's Gaz.*, **28**, 75–79.
Myers, A.A. (1988) A cladistic and biogeographic analysis of the Aorinae sub-family nov. In *VIth International Colloquium on Amphipoda* (ed. J. Stock), *Crustaceana Supplement*, **13**, 167–92.
Nagylaki, T. (1975) Conditions for the existence of clines. *Genetics*, **80**, 595–615.
Negi, J.G. and Tiwari, R.K. (1983) Matching long-term periodicities of geomagnetic reversals and galactic motions of the solar system. *Geophys. Res. Letters*, **10**, 713–16.
Nei, M. Maruyama, T. and Wu, C.I. (1983) Models of evolution of reproductive isolation. *Genetics*, **103**, 557–79.
Nelson, C.E. (1981) Phytogeography of Southern Australia. In *Ecological Biogeography of Australia* (ed. A. Keast), Junk, The Hague, pp. 735–57.
Nelson, G. (1974) Historical biogeography. An alternative formalization. *Syst. Zool.*, **23**, 555–8.
Nelson, G. (1978) From Candolle to Croizat: comments on the history of biogeography. *J. Hist. Biol.*, **11**, 269–305.
Nelson, G. (1979) Cladistic analysis and synthesis: principles and definitions, with a historical note on Adanson's *Familles des Plantes (1763–64). Syst. Zool.*, **28**, 1–21.
Nelson, G. (1984) Cladistics and biogeography. In *Cladistics: perspectives on the reconstruction of evolutionary history* (eds T. Duncan and T.F. Stuessy), Columbia University Press, New York, pp. 273–93.
Nelson, G. (1985) A decade of challenge, the future of biogeography. *J. Hist. Earth Sci. Soc.*, **4**, 187–96.
Nelson, G. and Platnick, N. (1981) *Systematics and Biogeography/cladistics and vicariance*, Columbia University Press, New York.
Nelson, G. and Platnick, N. (1984) *Biogeography*, Carolina Biology Reader No. 119.
Nelson, G. and Rosen, D.E. (eds) (1981) *Vicariance Biogeography; a critique*, Columbia University Press, New York.
Nelson, J.S. (1984) *Fishes of the World*, Oxford University Press, Oxford.
Nevo, E. (1985) Speciation in action and adaptation in subterranean mole rats: patterns and theory. *Boll. Zool.*, **52**, 65–95.
Nevo, E. (1986) Mechanisms of adaptive speciation at the molecular and organismal levels. In *Evolutionary Processes and Theory* (eds S. Karlin and E. Nevo), Academic Press, New York, pp. 439–74.
Nevo, E., Gorman, G.C., Soule, M.F. *et al.* (1972) Competitive exclusion between insular *Lacerta* species (Sauria: Lacertidae). *Oecologia*, **10**, 183–90.
Nevo, E., Beiles, A. and Ben-Shlomo, R. (1984) The evolutionary significance of genetic diversity: ecological, demographic and life-history correlates. In *Evolutionary Dynamics of Genetic Diversity* (ed. G.S. Mani), Springer-Verlag, Berlin, pp. 13–213.
Newell, N.D. (1967) Revolutions in the history of life. *Geol. Soc. Am. Spec. Paper*, **89**, 63–91.
Newell, N.D. (1971) An outline history of tropical organic reefs. *Am. Mus. Novit.*, **2465**, 1–37.
Newton, A. (1888) Presidential Address. Section D. Biology. *Rep. 57th Meeting Br. Ass. Adv. Sci.*, 726–33.
Nichols, R.A. (1984) *Ecological and genetical studies of a hybrid zone in the grasshopper Podisma pedestris*. PhD thesis, University of East Anglia, Norwich.

Nichols, R.A. and Hewitt, G.M. (1986) Population structure and the shape of a chromosomal cline between two races of *Podisma pedestris* (Orthoptera: Acrididae). *Biol. J. Linn. Soc.*, **29**, 301–16.

Niklas, K.J., Tiffney, B.H. and Knoll, A.H. (1983) Patterns in vascular land plant diversification. *Nature, Lond.*, **303**, 614–16.

Nix, H.A. (1981) The environment of *Terra Australis*. In *Ecological Biogeography of Australia* (ed. A. Keast), Junk, The Hague, pp. 103–33.

Noodt, W. (1965) Natürliches System und Biogeographie der Syncarida (Crustacea, Malacostraca). *Gewäss, Abwäss.*, **37–38**, 77–186.

Odin, G.S. (ed) (1982) *Numerical Dating in Stratigraphy*, Wiley, New York.

Odum, E.P. (1969) The strategy of ecosystem development. *Science*, **164**, 262–70.

Officer, C.B. and Drake, C.L. (1985) Terminal Cretaceous environmental events. *Science*, **277**, 1161–7.

Oliver, W.A. and Pedder, A.E.H. (1979) Biogeography of late Silurian rugose corals in North America. In *Historical Biogeography, Plate Tectonics, and the Changing Environment* (eds J. Gray and A.J. Boucot), Oregon State University Press, Corvallis.

Opdam, P. van Dorp, D. and ter Braak, C.J.F. (1984) The effect of isolation on the number of woodland birds in small woods in the Netherlands. *J. Biogeogr.*, **11**, 473–8.

Orians, G.H. and Solbrig, O.T. (1977) Degree of convergence of ecosystem characteristics. In *Convergent Evolution in Warm Deserts: an examination of strategies and patterns in deserts of Argentina and the United States* (eds G.H. Orians and O.T. Solbrig). Dowden, Hutchinson and Ross, Stroudsburg, Penn., pp. 225–55.

Osman, R.W. (1978) The influence of seasonality and stability on the species equilibrium. *Ecology*, **59**, 383–99.

Osman, R.W. (1982) Artificial substrates as ecological islands. In *Artificial Substrates* (ed. J. Cairns, Jr.), Ann Arbor, pp. 71–114.

Osman, R.W. and Whitlatch, R.B. (1978) Patterns of diversity: fact or artifact? *Paleobiology*, **4**, 41–54.

Ozenda, P. (1977) *Flore du Sahara*, 2nd edn, Centre National de la Recherche Scientifique, Paris.

Pacala, S.W. and Roughgarden, J. (1982) Spatial heterogeneity and interspecific competition. *Theor. Pop. Biol.*, **21**, 92–113.

Pacala, S.W. and Roughgarden, J. (1985) Population experiments with *Anolis* lizards of St Maarten and St Eustatius. *Ecology*, **66**, 129–41.

Page, R. (1987) Graphs and generalized tracks: quantifying Croizat's panbiogeography. *Syst. Zool.*, **36**(1), 1–17.

Paik, Y.K. and Sung, K.C. (1969) Behaviour of lethals in *Drosophila melanogaster* populations. *Jap. J. Genet.*, **44**, Suppl. 1, 180–92.

Paine, R.T. (1966) Food web complexity and species diversity. *Amer. Nat.*, **100**, 65–75.

Pal, P.C. and Creer, K.M. (1986) Geomagnetic reversal spurts and episodes of extra terrestrial catastrophisms. *Nature, Lond.*, **320**, 148–50.

Papanicolaou, K., Babalonas, D. and Kokkini, S. (1983) Distribution patterns of some Greek mountain endemic plants in relation to geological substrate. *Flora*, **174**, 405–37.

Parenti, L.R. (1981) Discussion of C. Patterson's paper: Methods of paleobiogeography. In *Vicariance Biogeography: a critique* (eds G. Nelson and D.E. Rosen) Columbia University Press, New York, pp. 490–7.

Park, T. (1962) Beetles, competition and populations. *Science*, **138**, 1369–75.

Parry, M.L. (1978) *Climatic Change, Agriculture and Settlement*, Dawson, Folkestone, England.

Parsons, P.A. (1980) Adaptive strategies in natural populations of *Drosophila*: ethanol tolerance, desiccation resistance and development times in climatically optimal and extreme environments. *Theor. Appl. Genet.*, **57**, 257–66.

Parsons, P.A. (1982) Evolutionary ecology of Australian *Drosophila*: a species analysis. *Evol. Biol.*, **14**, 297–350.

Parsons, P.A. (1983) *The Evolutionary Biology of Colonizing Species*, Cambridge University Press, New York.

Parsons, P.A. (1986) Evolutionary rates under environmental stress. *Evol. Biol.*

Parsons, P.A. (1987) Features of colonising animals: genotypes and phenotypes. In *Colonization, Succession and Stability, 26th Symp. British Ecological Soc.*, Blackwell Scientific Pubns, Oxford, pp. 133–54.

Paterson, G.L.J. (1986) Aspects of the zoogeography of some benthic animals in Rockall Tough (abstract). *Proc. R. Soc. Edin.* Ser. B, **88**, 316–18.

Paterson, H.E.H. (1982) Perspective on speciation by reinforcement? *S. Afr. J. Sci.*, **78**, 53–7.

Patrick, R. (1961) A study of the numbers and kinds of species found in rivers in eastern United States. *Proc. Acad. Nat. Sci. Philad.*, **113**, 215–58.

Patrick, R. (1967) The effect of invasion rate, species pool, and size of area on the structure of the diatom community. *Proc. Nat. Acad. Sci. USA*, **58**, 1335–42.

Patterson, B.D. (1984) Mammalian extinction and biogeography in the southern Rocky Mountains. In *Extinction* (ed. M.H. Nitecki), University of Chicago Press, Chicago, pp. 247–93.

Patterson, C. (1981a) Methods in Paleobiogeography. In *Vicariance Biogeography: a Critique* (eds G. Nelson and D.E. Rosen), Columbia University Press, New York, pp. 446–89.

Patterson, C. (1981b) Significance of fossils in determining evolutionary relationships. *Ann. Rev. Ecol. Syst.*, **12**, 195–223.

Patterson, C. (1982) Classes and cladists or individuals and evolution. *Syst. Zool*, **31**, 284–6.

Patterson, C. (1983) Aims and methods in biogeography. In *Evolution, Time and Space: the emergence of the Biosphere*, Systematics Assoc. Spec. Vol. **23**, 1–28.

Patterson, C. and Rosen, D.E. (1977) Review of ichthyodectiform and other Mesozoic teleost fishes and the theory and practice of classifying fossils. *Bull. Am. Must. Nat. Hist.*, **158**, 81–172.

Patterson, C. and Smith, A.B. (1987) Periodicity of extinctions: a taxonomic artefact. *Nature, Lond.* (in press)

Peake, J.F. (1981) The land snails of islands – a dispersalist's viewpoint. In *Chance, Change and Challenge: the evolving biosphere* (eds P.H. Greenwood and P.L. Forey), British Museum (Natural History) and Cambridge University Press, Cambridge, pp. 247–63.

Peet, R.K. (1974) Measurement of species diversity. *Ann. Rev. Ecol. Syst.*, **5**, 285–307.

Phillips, D.L. (1982) Life forms of granite outcrop plants. *Amer. Midl., Nat.*, **107**, 206–8.

Pianka, E.R. (1966) Latitudinal gradients in species diversity: a review of concepts. *Amer. Nat.*, **100**, 33–46.

Pianka, E.R. (1967) On lizard species diversity: North American flatland deserts. *Ecology*, **48**, 331–51.

Pianka, E.R. (1983) *Evolutionary Ecology*, 3rd edn, Harper and Row, New York.
Pianka, E.R. (1986) *Ecology and Natural History of Desert Lizards: analyses of the ecological niche and community structure*, Princeton University Press, Princeton, New Jersey.
Pickett, S.T.A. and Thompson, J.N. (1978) Patch dynamics and the design of nature reserves. *Biol. Conserv.*, **13**, 27–37.
Pielou, E.C. (1975) *Ecological Diversity*, John Wiley and Sons, New York.
Pielou, E.C. (1978) Latitudinal overlap of seaweed species: evidence for quasi-sympatric speciation. *J. Biogeogr.*, **5**, 227–38.
Pielou, E.C. (1979a) *Biogeography*, John Wiley and Sons, New York.
Pielou, E.C. (1979b) Interpretation of paleoecological similarity matrices. *Paleobiology*, **5**, 435–43.
Platnick, N. (1976) Drifting spiders or continents? Vicariance biogeography of the spider subfamily Laroniinae (Araneae: Gnaphosidae). *Syst. Zool.*, **25**, 101–9.
Platnick, N.I. and Nelson, G. (1978) A method of analysis for historical biogeography. *Syst. Zool.*, **27**, 1–16.
Platnick, N. and Nelson, G. (1984) Composite areas in vicariance biogeography. *Syst. Zool.*, **33**, 328–35.
Playford, P.E., McLaren, D.J., Orth, C.J. *et al.* (1984) Iridium anomaly in the Upper Devonian of the Canning Basin, Western Australia. *Science*, **226**, 437–9.
Popper, K. (1959) *The Logic of Scientific Discovery*, Hutchinson, London.
Popper, K. (1972) *Objective Knowledge: an evolutionary approach*, Clarendon Press, Oxford.
Por, F.D. (1978) *Lessepsian Migration. The influx of Red Sea biota into the Mediterranean by way of the Suez Canal*, Springer Verlag, Berlin.
Potts, D.C. (1985) Sea-level fluctuations and speciation in Scleractinia. *Proc. 5th Internat. Coral Reef Congr. Tahiti*, **4**, 127–32.
Pounds, J.A. and Jackson, J.F. (1981) Riverine barriers to gene flow and the differentiation of fence lizard populations. *Evolution*, **35**, 516–26.
Powell, J.R. (1978) The founder-flush speciation theory – an experimental approach. *Evolution*, **32**, 465–74.
Prance, G.T. (1978) The origin and evolution of the Amazon flora. *Interscientia*, **3**, 207–22.
Prance, G.T. (1982a) (ed) *Biological Diversification in the Tropics*. Columbia University Press, New York, pp. 714.
Prance, G.T. (1982b) Forest refuges: evidence from woody angiosperms. In *Biological Diversification in the Tropics* (ed G.T. Prance), Columbia University Press, New York, pp. 137–57.
Prentice, H.C. (1976) A study of endemism: *Silene diclinis*. *Biol. Conserv.*, **10**, 15–30.
Prentice, H.C. (1984) Enzyme polymorphism, morphometric variation and population structure in a restricted endemic, *Silene diclinis* (Caryophyllaceae). *Biol. J. Linn. Soc.*, **22**, 125–44.
Preston, F.W. (1962) The canonical distribution of commonness and rarity. *Ecology*, **43**, 185–215 and 410–32.
Prothero, D.R. (1985) Mid-Oligocene extinction event in North American land mammals. *Science*, **229**, 550–51.
Provine, W. (1986) *Sewall Wright and Evolutionary Biology*, University of Chicago Press, Chicago.

Pryor, L.D. and Johnson, L.A.S. (1971) *A Classification of the Eucalypts*, A.N.U. Press, Canberra.

Quezel, P. (1978) Analysis of the flora of Mediterranean and Saharan Africa. *Ann. Missouri Bot. Gdn.*, **65**, 479–534.

Radinsky, L. (1978) Do albumin clocks run on time? *Science*, **200**, 1182–3.

Rafe, R.W., Usher, M.B. and Jefferson, R.G. (1985) Birds on reserves: the influence of area and habitat on species richness. *J. Appl. Ecol.*, **22**, 327–35.

Rampino, M.R. and Strothers, R.B. (1984) Terrestrial mass extinctions cometary impacts and the sun's motion perpendicular to the galactic plane. *Nature, Lond.*, **308**, 709–12.

Rand, A.S. and Williams, E.E. (1969) The anoles of La Palma: aspects of their ecological relationships. *Breviora*, **327**, 1–18.

Raup, D.M. (1978) Approaches to the extinction problem. *J. Paleontol.*, **52**, 517–23.

Raup, D.M. (1979a) Biases in the fossil record of species and genera. *Bull. Carnegie Mus. Nat. Hist.*, **13**, 85–91.

Raup, D.M. (1979b) Size of the Permo-Triassic bottleneck and its evolutionary implications. *Science*, **206**, 217–18.

Raup, D.M. (1984) Death of species. In *Extinctions* (ed. M.H. Nitecki), University of Chicago Press, Chicago, pp. 1–19.

Raup, D.M. (1985a) Magnetic reversals and mass extinctions. *Nature, Lond.*, **314**, 341–3.

Raup, D.M. (1985b) Rise and fall of periodicity. *Nature, Lond.*, **317**, 384–5.

Raup, D.M. (1986) Biological extinction in earth history. *Science*, **231**, 1528–33.

Raup, D.M. and Jablonski, D. (eds) (1986) *Patterns and Processes in the History of Life*, Springer-Verlag, Berlin.

Raup, D.M. and Marshall, L.G. (1980) Variation between groups in evolutionary rates: a statistical test of significance. *Paleobiology*, **6**, 9–23.

Raup, D.M. and Sepkoski, J.J. Jr (1982) Mass extinctions in the marine fossil record. *Science*, **215**, 1501–3.

Raup, D.M. and Sepkoski, J.J. Jr (1984) Periodicity of extinctions in the geologic past. *Proc. Nat. Acad. Sci.*, **81**, 801–15.

Raup, D.M. and Sepkoski, J.J. Jr (1986) Periodic extinction of families and genera. *Science*, **231**, 833–6.

Raup, D.M. and Stanley, S.M. (1978) *Principles of Paleontology*, 2nd edn, W.H. Freeman and Company, San Francisco.

Raven, P.H. and Axelrod, D.I. (1978) Origin and relationships of the California flora. *Univ. Calif. Publ. Botany*, **72**, pp. 1–134.

Rebristaya, O.V. (1977) *Flora of the Eastern Bolshezemelsk Tundra*, Nauka, Leningrad.

Reeves, R.D., Macfarlane, R.M. and Brooks, R.R. (1983) Accumulation of nickel and zinc by western North American genera containing serpentine-tolerant species. *Amer. J. Botan.*, **70**, 1297–1303.

Rex, M.A. (1981) Community structure in the deep-sea benthos. *Ann. Rev. Ecol. Syst.*, **12**, 331–54.

Rey, J.R. (1981) Ecological biogeography of arthropods on *Spartina* islands. *Ecol. Monogr.*, **51**, 181–92.

Rey, J.R. (1984) Experimental tests of island biogeographic theory. In *Ecological Communities: conceptual issues and the evidence* (eds D. Strong, D. Simberloff, L.G. Abele and A.B. Thistle) Princeton University Press, pp. 101–12.

Rey, J.R. (1985) Insular ecology of salt marsh arthropods: species level patterns. *J. Biogeogr.*, **12**, 97–107.

Rey, J.R. and Strong, D.R. Jr (1983) Immigration and extinction of salt marsh arthropods on islands: an experimental study. *Oikos*, **41**, 396–401.

Reynolds, C.S. (1987) Community organization in the freshwater plankton. In *Organization of Communities Past and Present*. (Eds J.H.R. Gee and P.S. Giller) Blackwell Scientific Publications, Oxford, pp. 297–326.

Rich, P.V. and Scarlett, S.R.J. (1977) Another look at *Megaegotheles,* a large owlet-nightjar from New Zealand. *Emu*, **77**, 1–6.

Richards, P.W. (1952) *The Tropical Rain Forest*, Cambridge University Press.

Ricklefs, R.E. and Cox, G.W. (1972) Taxon cycles in the West Indian avifauna. *Amer. Nat.*, **106**, 195–219.

Ricklefs, R.E. and Cox, G.W. (1978) Stage of taxon cycle, habitat distribution and population density in the avifauna of the West Indies. *Amer. Nat.*, **112**, 875–95.

Ricklefs, R.E. and O'Rourke, K. (1975) Aspect diversity in moths: a temperate–tropical comparison. *Evolution*, **29**, 313–24.

Riedl, R. (1978) *Order in Living Organisms*, translated by R.P.S. Jefferies, John Wiley and Sons, Chichester [Die Ordnung der Lebendigen, Verlag Paul Parey, Hamburg].

Robbins, L.W., Moulton, M.P. and Baker, R.J. (1983) Extent of geographic range and magnitude of chromosomal evolution. *J. Biogeogr.*, **10**, 533–41.

Roberts, J.D. and Maxson, L.R. (1985) Tertiary speciation models in Australian anurans: molecular data challenge Pleistocene scenario. *Evolution*, **39**, 325–34.

Rodgers, G. (1986) Moawhango ecological district: a biogeographic hot spot. *Forest and Bird*, **17**, 24–7.

Rodgers, G. (1987) *A historical interpretation of landscapes of the Moawhango Ecological District*. PhD thesis in Botany, Victoria University of Wellington, New Zealand.

Rodin, L.E. and Bazilevich, N.I. (1967) *Dynamics of the Organic Matter and Biological Cycling of Ash Elements and Nitrogen in the Main Types of the World's Vegetation*, Oliver and Boyd, London.

Roose, M.L. and Gottlieb, L.D. (1976) Genetic and biochemical consequences of polyploidy in *Tragopogon*. *Evolution*, **30**, 818–30.

Root, R.B. (1967) The niche exploitation pattern of the Blue–grey Gnatcatcher. *Ecol. Monogr.*, **37**, 317–50.

Rorty, R. (1979) *Philosophy and the Mirror of Nature*, Princetown University Press, Princetown, New Jersey.

Rosa, D. (1918) *Ologenesi, nuova Tooria del Evoluzione e della Distribuzione Geografica dei Viventi*, R. Bemporad and Figlio, Firenze, Italy.

Rose, R.K. and Birney, E.C. (1985) Community ecology. In *Biology of New World Microtus* (ed. R.H. Tamarin), Am. Soc. Mammalogists Spec. Pubn. No. 10, pp. 310–39.

Rose, S. (1976) *The Conscious Brain*, rev. edn, Penguin Books Ltd, Harmondsworth.

Rosen, B.R. (1975) The distribution of reef corals. *Rep. Underwater Ass.* (*new series*), **1**, 1–16.

Rosen, B.R. (1977) The depth distribution of Recent hermatypic corals and its palaeontological significance. *Mém. Bur. Rech. Géol. Min.*, **89**, 507–17.

Rosen, B.R. (1981) The tropical high diversity enigma – the coral's eye view. In *Chance, Change and Challenge: the emerging biosphere* (eds P.H. Greenwood and P.L. Forey), British Museum (Natural History) and Cambridge University Press, Cambridge, 103–29.

Rosen, B.R. (1984) Reef coral biogeography and climate through the late Cainozoic: just islands in the sun or a critical pattern of islands. In *Fossils and Climate* (ed. P.J. Brenchley), *Geol. J. Spec. Issue*, **11**, 201–62.
Rosen, B.R. (1985) Long-term geographical controls on regional diversity. *The Open University Geological Soc., Journal*, **6**, 25–30.
Rosen, B.R. (1988) The Cainozoic history of tropical marine benthic distribution patterns as exemplified by reef corals in comparison with other reef-associated organisms. *Helgoländer Meeresuntersuchung*. (in press).
Rosen, B.R. and Smith, A.B. (1988) Tectonics from fossils? Analysis of reef coral and sea urchin distributions from late Cretaceous to Recent, using a new method. In *Gondwana and Tethys* (eds M.G. Audley-Charles and A. Hallam), Geol. Soc. Lond. Spec. Pubn (in press).
Rosen, D.E. (1975) A vicariance model of Caribbean biogeography. *Syst. Zool.*, **24**, 431–64.
Rosen, D.E. (1978) Vicariant patterns and historical explanation in biogeography. *Syst. Zool.*, **27**, 159–88.
Rosen, D.E. (1979) Fishes from the uplands and intermontane basins of Guatemala: revisionary studies and comparative geography. *Bull. Am. Mus. Nat. Hist.*, **162**, 267–376.
Rosen, D.E. (1985) Geological hierarchies and biogeographic congruence in the Caribbean. *Ann. Missouri Bot. Gdn.*, **72**, 636–59.
Rosen, D.E. and Buth, D.G. (1980) Empirical evolutionary research versus neo-Darwinian speculation. *Syst. Zool.*, **29**, 300–8.
Rosenzweig, M.L. (1971) Paradox of enrichment: destabilization of exploitation ecosystems in ecological time. *Science*, **171**, 385–7.
Rosenzweig, M.L. (1973) Habitat selection experiments with a pair of coexisting heteromyid rodent species. *Ecology*, **54**, 111–17.
Rosenzweig, M.L. (1975) On continental steady states of species diversity. In *Ecology and Evolution of Communities* (eds M.L. Cody and J.M. Diamond), Harvard University Press, Cambridge, Mass., pp. 121–41.
Rosenzweig, M.L. (1979) Optimal habitat selection in two-species competition systems. In *Population Ecology* (eds U. Halbach and J. Jacobs), Fischer Verlag, Stuttgart, pp. 283–93.
Rosenzweig, M.L. and Winakur, J. (1969) Population ecology of desert rodent communities: habitats and environmental complexity. *Ecology*, **50**, 558–72.
Ross, C.A. (ed) (1976a) *Paleobiogeography, Benchmark Papers in Geology*, **31**, Dowden, Hutchinson and Ross, Stroudsburg, Penn.
Ross, C.A. (1976b) Editor's comments on papers 6 and 7. In *Paleobiogeography, Benchmark Papers in Geology 31* (ed C.A. Ross), Dowden, Hutchinson and Ross, Stroudsburg, Penn., p. 66.
Ross, H.H. (1974) *Biological Systematics*, Addison Wesley, Reading, Penn.
Rotondo, G.M., Springer, V.G., Scott, G.A.J. and Schlanger, S.O. (1981) Plate movement and island integration – A possible mechanism in the formation of endemic biotas, with special reference to the Hawaiian islands. *Syst. Zool.*, **30**, 12–21.
Roughgarden, J. (1972) Evolution of niche width. *Amer. Nat.*, **106**, 683–719.
Roughgarden, J. (1974a) Species packing and the competition function with illustrations from coral reef fish. *Theor. Pop. Biol.*, **5**, 163–86.
Roughgarden, J. (1974b) Niche width: biogeographic patterns among *Anolis* lizard populations. *Amer. Nat.*, **108**, 429–42.
Roughgarden, J. (1976) Resource partitioning among competing species – a coevolutionary approach. *Theor. Pop. Biol.*, **9**, 388–424.

Roughgarden, J. (1979) *Theory of Population Genetics and Evolutionary Ecology: an introduction*, Macmillan, New York.
Roughgarden, J. (1983) Coevolution between competitors. In *Coevolution* (eds D.J. Futuyma and M. Slatkin), pp. 383–403.
Roughgarden, J. and Feldman, M. (1975) Species packing and predation procedure. *Ecology*, **56**, 489–92.
Rouhani, S. and Barton, N.H. (1987a) The probability of peak shifts in a founder population. *J. Theor. Biol.*, **26**, 51–62.
Rouhani, S. and Barton, N.H. (1987b) Speciation and the 'shifting balance' in a continuous population. *Theor. Pop. Biol.*, **31**, 465–92.
Rummel, J.D. and Roughgarden, J. (1983) Some differences between invasion-structured and coevolution-structured competitive communities: a preliminary theoretical analysis. *Oikos*, **41**, 477–86.
Rune, O. (1953) Plant life on serpentine and related rocks in the north of Sweden. *Acta Phytogeogr. Suec.*, **31**, 1–139.
Runemark, H. (1969) Reproductive drift, a neglected principle in reproductive biology. *Botan. Notiser*, **122**, 90–129.
Runemark, H. (1971) The phytogeography of the central Aegean. In *Evolution of the Aegean* (ed. A. Strid). *Opera Botan.*, **30**, 20–8.
Rusterholz, K.A. and Howe, R.W. (1979) Species–area relations of birds on small islands in a Minnesota lake. *Evolution*, **33**, 468–77.
Ryvarden, L. (1975a) Studies in seed dispersal. I. Trapping of diaspores in the alpine zone at Finse Norway. *Norw. J. Botan.*, **22**, 21–4.
Ryvarden, L. (1975b) Studies in seed dispersal. II. Trapping of winter-dispersed diaspores in the alpine zone at Finse, Norway. *Norw. J. Botan.*, **22**, 21–4.
Säether, O.A. (1983) The canalized evolutionary potential: Inconsistencies in phylogenetic reasoning. *Syst. Zool.*, **32**, 343–59.
Sainz Ollero, H. and Hernandez Bermejo, J.E. (1985) Sectorizacion fitogeografica dela Peninsula Iberica e islas Baleares: la contribucion de su endemoflora coma criterio de semejanza. *Candollea*, **40**, 485–508.
Sale, P.F. (1979) Recruitment, loss and coexistence in a guild of territorial coral reef fishes. *Oecologia*, **42**, 159–77.
Sale, P.F. (1980) Assemblages of fish on patch reefs – predictable or unpredictable? *Environmental Bull. Fish.*, **5**, 243–9.
Sale, P.F. and Dybdahl, R. (1975) Determinants of community structure for coral reef fishes in an experimental habitat. *Ecology*, **56**, 1343–55.
Sale, P.F. and Dybdahl, R. (1978) Determinants of community structure for coral reef fishes in isolated coral heads at lagoonal and reef slope sites. *Oecologia*, **34**, 57–74.
Sanders, H.L. (1968) Marine benthic diversity: a comparative study. *Amer. Nat.*, **111**, 337–159.
Sanders, H.L. (1969) Marine benthic diversity and the stability–time hypothesis. In *Diversity and Stability in Ecological Systems* (eds G.M. Woodwell and H.H. Smith) *Brookhaven Symp. Biol.*, **22**, pp. 71–81.
Sauer, J. (1969) Oceanic islands and biogeographic theory: a review. *Geogr. Rev.*, **59**, 582–93.
Sauer, W. and Sauer, G. (1980) *Paraskevia* gen. nov. mit. *P. cesatiana* comb. nov. (*Boraginaceae*), eine endemische Gattung Griechenlands. *Phyton* (Austria), **20**, 285–306.
Savage, N.M., Perry, D.G. and Boucot, A.J. (1979) A quantitative analysis of Lower Devonian brachiopod distribution. In *Historical Biogeography, Plate*

Tectonics, and the Changing Environment (eds J. Gray and A.J. Boucot), Oregon State University Press, Corvallis, pp. 169–200.

Sayles, R.S. and Thomas, T.R. (1978) Surface topography as a non-stationary random process. *Nature, Lond.*, **271**, 431–4 and **273**, 573.

Schaffer, C.W. (1977) On dispersal. *Syst. Zool.*, **26**, 446.

Schall, J.R. and Pianka, E.R. (1978) Geographical trends in numbers of species. *Science*, **201**, 679–86.

Schindewolf, O.H. (1950) *Grundfragen der Paläontologie*, Schweizerbart, Stuttgart.

Schlee, D. and Dietrich, H.G. (1970) Insekter führender Bernstein aus der Unterkreide des Libanon. *Neues J. Geol. Paläontol.*, M. **1**, 40–50.

Schluter, D. (1984) A variance test for detecting species associations with some example applications. *Ecology*, **65**, 998–1005.

Schluter, D., Price, T.D. and Grant, P.R. (1985) Ecological character displacement in Darwin's finches. *Science*, **227**, 1056–9.

Schluter, D. (1986) Tests for similarity and convergence of finch communities. *Ecology*, **67**, 1073–83.

Schluter, D. and Grant, P.R. (1982) The distribution of *Geospiza difficilis* in relation to *G. fuliginosa* in the Galapagos Islands: tests of three hypotheses. *Evolution*, **36**, 1213–26.

Schmid, M. (1982) Endemisme et speciation en Nouvelle Caledonie. *C.R. Soc. Biogeogr.*, **58**, 52–60.

Schminke, H.K. (1973) Evolution, System und Verbeitungs geschichte der Famile Parabathynellidae (Bathynellaea, Malacostraca). *Mikrofauna des Meeresbodens*, **24**, 1–192.

Schminke, H.K. (1974) Mesozoic intercontinental relationships as evidenced by Bathynellid Crustacea (Syncarida: Malacostraca). *Syst. Zool.*, **23**, 157–64.

Schmithusen, J. (1968) *Allgemeine Vegetationsgeographie*, de Gruyter.

Schoener, A. (1974a) Colonization curves for planar marine islands. *Ecology*, **55**, 818–27.

Schoener, A. (1974b) Experimental zoogeography: colonization of marine mini-islands. *Amer. Nat.*, **108**, 715–38.

Schoener, A. and Schoener, T.W. (1981) The dynamics of the species–area relation in marine fouling systems. I. Biological correlates of changes in the species–area slope. *Amer. Nat.*, **118**, 339–60.

Schoener, A., Lond, E.R. and DePalma, J.R. (1978) Geographic variation in artificial island colonization curves. *Ecology*, **59**, 367–82.

Schoener, T.W. (1969) Models of optimal size for solitary predators. *Amer. Nat.*, **103**, 277–313.

Schoener, T.W. (1970) Size patterns in West Indian *Anolis* lizards II. Correlations with the sizes of particular species–displacement and convergence. *Amer. Nat.*, **104**, 155–74.

Schoener, T.W. (1971) Large-billed insectivorous birds: a precipitous diversity gradient. *Condor*, **73**, 154–61.

Schoener, T.W. (1974a) Competition and the forum of habitat shift. *Theor. Pop. Biol.*, **6**, 265–307.

Schoener, T.W. (1974b) The compression hypothesis and temporal resource partitioning. *Proc. Nat. Acad. Sci. USA*, **71**, 4169–72.

Schoener, T.W. (1974c) Resource partitioning in ecological communities. *Science*, **185**, 27–39.

Schoener, T.W. (1975) Presence and absence of habitat shift in some widespread lizard species. *Ecol. Monogr.*, **45**, 233–58.

Schoener, T.W. (1976a) Alternatives to Lotka–Volterra competition: models of intermediate complexity. *Theor. Pop. Biol.*, **10**, 309–33.
Schoener, T.W. (1976b) The species–area relation within archipelagos: models and evidence from island land birds. *Proc. 16th Int. Ornithol. Congr.*, 629–42.
Schoener, T.W. (1977) Competition and the niche. In *Biology of the Reptilia* (eds C. Gans and D.W. Tinkle), Academic Press, London, pp. 35–136.
Schoener, T.W. (1982) The controversy over interspecific competition. *Amer. Scient.*, **70**, 586–95.
Schoener, T.W. (1983a) Field experiments on interspecific competition. *Amer. Nat.*, **122**, 240–85.
Schoener, T.W. (1983b) Rate of species turnover decreases from lower to higher organisms. *Oikos*, **41**, 372–7.
Schoener, T.W. (1984) Size differences among sympatric bird-eating hawks: a worldwide survey. In *Ecological Communities: conceptual issues and the evidence* (eds D.R. Strong Jr., D. Simberloff, L.G. Abele and A.B. Thistle), Princeton University Press, Princeton, New Jersey, pp. 254–81.
Schoener, T.W. (1985) Some comments on Connell's and my reviews of field experiments on interspecific competition. *Amer. Nat.*, **125**, 730–40.
Schoener, T.W. (1986a) Resource partitioning. In *Community Ecology* (eds J. Diamond and T.J. Case), Harper and Row, New York.
Schoener, T.W. (1986b) Kinds of ecological communities – ecology becomes pluralistic. In *Community Ecology* (eds J. Diamond and T.J. Case), Harper and Row, New York, pp. 467–79.
Schoener, T.W. (1986c) Patterns in terrestrial vertebrate versus arthropod communities: do systematic differences in regularity exist? In *Community Ecology* (eds J. Diamond and T.J. Case), Harper and Row, New York, pp. 556–86.
Schoener, T.W. (1986d) When should a field experiment be counted?: a reply to Galindo and Krebs. *Oikos*, **46**, 119–21.
Schoener, T.W. (1987) Axes of controversy in community ecology. In *Stream-fish Community Ecology* (ed W. Matthews), University of Oklahoma Press.
Schoener, T.W. and Janzen, D.H. (1968) Some notes on tropical versus temperate insect size patterns. *Amer. Nat.*, **101**, 207–24.
Schoener, T.W., Huey, R.B. and Pianka, E.R. (1979). A biogeographic extension of the compression hypothesis: competitors in narrow sympatry. *Amer. Nat.*, **113**, 295–8.
Schoener, T.W., Roughgarden, J. and Fenchel, T. (1986) The body-size-prey-size hypothesis: a defence. *Ecology*, **67**, 260–1.
Schoener, T.W. and Schoener, A. (1983a) The time to extinction of a colonizing propagule of lizards increases with island area. *Nature, Lond.*, **302**, 332–4.
Schoener, T.W. and Schoener, A. (1983b) Distribution of vertebrates on some very small islands: I. Occurrence sequences of individual species. *J. Anim. Ecol.*, **52**, 209–35.
Schoener, T.W., Slade, J. and Stinson, C. (1982) Diet and sexual dimorphism in the very catholic lizard genus, *Leiocephalus* of the Bahamas. *Oecologia*, **53**, 160–9.
Schreiber, H. (1978) Dispersal centres of Sphingidae (Lepidoptera) in the Neotropical region. *Biogeographica*, **10**, Junk, The Hague.
Schuchert, C. (1903) On the faunal provinces of the middle Devonic of America and the Devonic sub-provinces of Russia, with two paleogeographic maps. *Am. Geol.*, **32**, 137–62.

Schuh, R.T. (1981) Discussion (of A. Solem's paper: Land-snail biogeography: a true snail's pace of change). In *Vicariance Biogeography; a critique* (eds G. Nelson and D.E. Rosen), Columbia University Press, New York, pp. 231–4.

Schwartz, R.D. and James, P.B. (1984) Periodic mass extinctions and the sun's oscillation about the galactic plane. *Nature, Lond.*, **309**, 712–13.

Scott, J.A. (1972) Biogeography of Antillean butterflies. *Biotropica*, **4**, 32–45.

Scriber, J.M. (1973) Latitudinal gradients in larval feeding specialization of the world Papilionidae (Lepidoptera). *Psyche*, **80**, 355–73.

Searle, J.B. (1986) Factors responsible for a karyotypic polymorphism in the common shrew, *Sorex araneus*. *Proc. Roy. Soc. (London)*, **229**, 277–98.

Seberg, O. (1986) A critique of the theory and methods of panbiogeography. *Syst. Zool.*, **35**, 369–80.

Sedelnikov, V.P. (1979) *Flora and Vegetation of the High Mountains of the Kuznetski Alatau*, Nauka, Sibirskoe Otd., Novosibirsk.

Segerstrale, S.G. (1957) Baltic Sea. In *Treatise on Marine Ecology and Paleoecology* (ed. J.W. Hedgpeth), *Geol. Soc. Am. Mem.*, **67**, pp. 751–800.

Sene, F.M. and Carson, H.L. (1977) Genetic variation in Hawaiian *Drosophila*. IV. Allozymic similarity between *D. silvestris* and *D. heteroneura* from the island of Hawaii. *Genetics*, **86**, 187–98.

Sepkoski, J.J. Jr (1984) A kinetic model of Phanerozoic taxonomic diversity. III. Post-Paleozoic families and mass extinctions. *Paleobiology*, **10**, 246–67.

Sepkoski, J.J. Jr, Bambach, R.K., Raup, D.M. and Valentine, J.W. (1981) Phanerozoic marine diversity and the fossil record. *Nature, Lond.*, **293**, 435–7.

Sepkoski, J.J. and Rex, M.A. (1974) The distribution of freshwater mussels: coastal rivers as biogeographic islands. *Syst. Zool.*, **23**, 165–88.

Serra-Tosio, B. (1986) Un nouveau chironomide Antarctique des Iles Crozet, *Parochlus crozetensis* n. sp. (Diptera, Nematocera). *Nouv. Rev. Entomol.*N.S., **3**(2).

Sharrock, J.T.R. (1976) *The Atlas of Breeding Birds in Britain and Ireland*, Poyser, Berkhamsted.

Shaw, D.D. and Wilkinson, P. (1980) Chromosome differentiation, hybrid breakdown and the maintenance of a narrow hybrid zone in *Caledia*. *Chromosoma*, **80**, 1–31.

Shaw, D.D., Wilkinson, P. and Coates, D.J. (1982) The chromosomal component of reproductive isolation in the grasshopper *Caledia captiva*. The relative viabilities of recombinant and non-recombinant chromosomes during embryogenesis. *Chromosoma*, **86**, 533–47.

Sheppard, P.M. (1958) *Natural Selection and Heredity*, Hutchinson, London.

Shields, A. (1979) Evidence for initial opening of the Pacific Ocean in the Jurassic. *Palaeogeogr., Palaeoclimatol., Palaeoecol.*, **26**, 181–220.

Shigesada, N., Kawasaki, K. and Teramoto, E. (1979) Spatial segregation of interacting species. *J. Theor. Biol.*, **79**, 83–99.

Shigesada, N. and Roughgarden, J. (1982) The role of rapid dispersal in the population dynamics of competition. *Theor. Pop. Biol.*, **21**, 353–72.

Shin, P.K.S. (1981) The development of sessile epifaunal communities in Kylsalia, Kilkieran Bay (West Coast of Ireland). *J. exp. Mar. Biol. Ecol.*, **54**, 97–111.

Schlotgauer, S.D. (1978) *Flora and Vegetation of Western pre-Okhotiya*, Nauka, Moscow.

Shmida, A. (1984) Endemism in the flora of Israel. *Bot. Jahrb. Syst.*, **104**, 537–67.

Shorrocks, B., Rosewell, J., Edwards, K. and Atkinson, W. (1984) Interspecific competition is not a major organizing force in many insect communities. *Nature, Lond.*, **310**, 310–12.

Shreve, F. (1951) *Vegetation of the Sonoran Desert*, Carnegie Inst. of Washington Publ. 591.

Shy, E. (1984) Habitat shift and geographical variation in North American tanagers (Thraupinae. *Piranga*). *Oecologia*, **63**, 281–5.

Sibley, C.G. and Ahlquist, J.E. (1981). The phylogeny and relationships of the ratite birds as indicated by DNA–DNA hybridization. In *Evolution Today, Proceedings of the 2nd International Congress on Systematic and Evolutionary Biology* (eds G.G.E. Scudder and J.L. Reveal), Hunt Institute for Botanical Documentation, Pittsburgh.

Signor, P.W. III and Lipps, J.H. (1982) Sampling bias, gradual extinction patterns and catastrophes in the fossil record. *Geol. Soc. Amer. Spec. Paper*, **190**, 291–6.

Sih, A., Crowley, P., McPeek, M. *et al.* (1985) Predation, competition and prey communities: a review of field experiments. *Ann. Rev. Ecol. Syst.*, **16**, 269–311.

Silver, L.T. and Schultz, P.H., eds (1982) Geological implications of impacts of large asteroids and comets on the earth. *Geol. Soc. Amer. Spec.*, **190**, 1–528.

Silvertown, J. (1985) History of a latitudinal diversity gradient: woody plants in Europe. *J. Biogeogr.*, **12**, 519–25.

Simberloff, D.S. (1969) Experimental zoogeography of islands: A model for insular colonization. *Ecology*, **50**, 296–314.

Simberloff, D.S. (1974) Equilibrium theory of island biogeography and ecology. *Ann. Rev. Ecol. Syst.*, **5**, 161–82.

Simberloff, D. (1976a) Species turnover and equilibrium island biogeography. *Science*, **194**, 572–8.

Simberloff, D. (1976b) Experimental zoogeography of islands: effects of island size. *Ecology*, **57**, 629–48.

Simberloff, D. (1976c) Trophic structure determination and equilibrium in an arthropod community. *Ecology*, **57**, 395–8.

Simberloff, D. (1983) Biogeography: the unification and maturation of a science. In *Perspectives in Ornithology* (eds A.M. Brush and G.A. Clark, Jr), Cambridge University Press, Cambridge, pp. 411–55.

Simberloff, D. (1984) Properties of coexisting bird species in two archipelagos. In *Ecological Communities: conceptual Issues and the Evidence* (eds D.R. Strong, Jr, D. Simberloff, L.G. Abele and A.B. Thistle), Princetown University Press, Princetown, New Jersey, pp. 234–53.

Simberloff, D. and Abele, L.G. (1976a) Island biogeographic theory and conservation practice. *Science*, **191**, 285–6.

Simberloff, D. and Abele, L.G. (1976b) Island biogeography and conservation: strategy and limitations. *Science*, **193**, 1032.

Simberloff, D. and Abele, L.G. (1982) Refuge design and island biogeography theory: effects of fragmentation. *Amer. Nat.*, **120**, 41–50.

Simberloff, D. and Boecklen, W. (1981) Santa Rosalia reconsidered: size ratios and competition. *Evolution*, **35**, 1206–28.

Simberloff, D. and Connor, E.F. (1984) Inferring competition from biogeographic data: a reply to Wright and Biehl. *Amer. Nat.*, **124**, 429–36.

Simberloff, D., Heck, K.L., McCoy, E.D. and Connor, E.F. (1981) There have been no statistical tests of cladistic biogeographical hypotheses. In *Vicariance*

Biogeography; a critique (eds G. Nelson and D.E. Rosen). Columbia University Press, New York, pp. 40–63.
Simberloff, D.S. and Wilson, E.O. (1969) Experimental zoogeography of islands: The colonization of empty islands. *Ecology*, **50**, 278–96.
Simberloff, D.S. and Wilson, E.O. (1970) Experimental zoogeography of islands: A two-year record of colonization. *Ecology*, **51**, 934–7.
Simon, J.L. and Dauer, D.M. (1977) Reestablishment of a benthic community following natural defaunation. In *Ecology of Marine Benthos* (ed. B.C. Coull) University of South Carolina Press, Columbia, South Carolina, pp.139–54.
Simpson, B.B. and Haffer, J. (1978) Speciation patterns in the Amazonian forest biota. *Ann. Rev. Ecol. Syst.*, **9**, 497–518.
Simpson, G.G. (1940) Antarctica as a faunal migration route. *Proc. 5th Pacif. Sci. Congr.*, **2**, 755–66.
Simpson, G.G. (1952) Probabilities of dispersal in geologic time. *Bull. Am. Mus. Nat. Hist.*, **99**, 163–76.
Simpson, G.G. (1953) *The Major Features of Evolution*, Columbia University Press, New York.
Simpson, G.G. (1964) Species density of North American recent mammals. *Syst. Zool.*, **13**, 57–73.
Singer, M.J. and Nkedi, P. (1980) Properties and history of an exhumed Tertiary oxisol in California. *Soil Sci. Soc. Amer. Proc.*, **14**, 587–90.
Sjörs, H. (1956) *Norkisk Växtgeografi*, Svenska Bokiolaget, Stockholm.
Skellam, J.G. (1973) The formulation and interpretation of mathematical models of diffusionary processes in population biology. In *Formulation and Interpretation of Mathematical Models of Diffusionary Processes in Population Biology* (eds M.S. Bartlett and R.W. Hiorns), pp. 63–85.
Skottsberg, C. (1956) Derivation of the flora and fauna of Juan Fernandez and Easter Island. In *The Natural History of Juan Fernandez and Easter Island* (ed. C. Skottsberg) Uppsala, Sweden, pp. 193–438.
Skottsberg, C. (1960) Remarks on the plant geography of the southern cold temperate zone. *Proc. Roy. Soc. Lond. B*, **152**, 447–57.
Slatkin, M.W. (1974) Competition and regional coexistence. *Ecology*, **55**, 128–34.
Slatkin, M.W. (1980) Ecological character displacement. *Ecology*, **61**, 163–77.
Slatkin, M. (1985) Gene flow in natural populations. *Ann. Rev. Ecol. Syst.*, **16**, 393–430.
Sloan, R.E., Rigby, J.K. Jr, Van Valen, L.M. and Gabriel, D. (1986) Gradual dinosaur extinction and simultaneous ungulate radiation in the Hell Creek Formation. *Science*, **232**, 629–33.
Slud, P. (1976) Geographic and climatic relationships of avifaunas with special reference to comparative distribution in the neotropics. *Smithson, Contr. Zool.*, **212**, 1–149.
Smit, J. and Hertogen, J. (1980) An extra terrestrial event at the Cretaceous–Tertiary boundary. *Nature, Lond.*, **285**, 198–200.
Smith, A.B. and Xu Juntao (1988) Palaeontology of the 1985 geotraverse of Tibet, Lhasa to Golmud. In *The Geology of Tibet* (eds R.M. Shackleton and P. Wallace) Phil. Trans. R. Soc. A (in press).
Smith, G.B. (1979) Relationship of eastern Gulf of Mexico reef-fish communities to the species equilibrium theory of insular biogeography. *J. Biogeogr.*, **6**, 49–61.
Smith, A.G. and Hallam, A. (1970) The fit of the southern continents. *Nature (Lond.)*, **225**, 139–44.

Sneath, P.H.A. and McKenzie, K.G. (1973) Statistical methods for the study of biogeography. In *Organisms and Continents through Time* (ed. N.F. Hughes) *Spec. Pap. Palaeontol.*, **12**, 45–60.

Snider-Pellegrini, A. (1858) *La Création et ses Mystères Dévoilees*, A. Franck, Paris.

Sobolevskaya, K.A. (1953) *Conspectus of the Flora of Tuva*, Zapadno Sibiro Fil., Botan. Sad. Novosibirsk.

Sokal, R.R. (1983a) A phylogenetic study of the Caminalcules. II. Estimating the true cladogram. *Syst. Zool.*, **32**, 185–201.

Sokal, R.R. (1983b) A phylogenetic study of the Caminalcules. III. Fossils and classification. *Syst. Zool.*, **32**, 248–58.

Sokal, R.R. and Rohlf, F.J. (1981) *Biometry: the principles and practice of statistics in biological research*, 2nd edn, W.H. Freeman and Co., San Francisco.

Solbrig, O.T. and Rollins, R.C. (1977) The evolution of autogamy in species of the mustard genus *Leavenworthia*. *Evolution*, **31**, 265–81.

Somero, G. (1986) Protein adaptation and biogeography: threshold effects on molecular evolution. *Trends in Ecol. and Evol.*, **1**, 124–7.

Sousa, W.P. (1984) The role of disturbance in natural communities. *Ann. Rev. Ecol. Syst.*, **15**, 353–91.

Southwood, T.R.E. (1977) Habitat, the templet for ecological strategies. *J. Anim. Ecol.*, **46**, 337–65.

Southwood, T.R.E. (1987) The concept and nature of the community. In *Organization of Communities: past and present* (eds J.H.R. Gee and P.S. Giller) Blackwell Scientific Pubns, Oxford, pp. 3–17.

Staley, J.T., Lehmicke, L.G., Palmer, F.E. *et al.* (1982) Impact of Mount St Helens eruption on bacteriology of lakes in the blast zone. *Appl. Environmental Microbiol.*, **43**, 664–70.

Stanley, S.M. (1979) *Macroevolution. Pattern and process*, W.H. Freeman and Co., San Francisco.

Stanley, S.M. (1984) Marine mass extinctions: a dominant role for temperature. In *Extinctions* (ed. M.H. Nitecki), University of Chicago Press, Chicago, pp. 69–117.

Stanley, S.M. (1986) Anatomy of a regional mass extinction: Plio-Pleistocene decimation of the western Atlantic bivalve fauna. *Palaios*, **1**, 17–36.

Stanley, S.M., Parsons, P.A., Spence, G.E. and Weber, L. (1980) Resistance of species of the *Drosophila melanogaster* subgroup to environmental extremes. *Aust. J. Zool.*, **28**, 413–21.

Stebbins, G.L. (1950) *Variation and Evolution in Plants*, Columbia University Press, New York.

Stebbins, G.L. (1952) Aridity as a stimulus to plant evolution. *Amer. Nat.*, **86**, 33–44.

Stebbins, G.L. (1974) *Flowering Plants: evolution above the species level*, Harvard University Press, Cambridge, Mass.

Stebbins, G.L. (1978a) Why are there so many rare plants in California? 1. Environmental factors. *Fremontia*, **5**, 6–10.

Stebbins, G.L. (1978b) Why are there so many rare plants in California? 2. Youth and age of species. *Fremontia*, **6**, 17–20.

Stebbins, G.L. (1980) Rarity of plant species: A synthetic viewpoint. *Rhodora*, **82**, 77–86.

Stebbins, G.L. and Major, J. (1965) Endemism and speciation in the California flora. *Ecol. Monogr.*, **35**, 1–35.

Steele, J.H. (1985) A comparison of terrestrial and marine ecological systems. *Nature, Lond.*, **313**, 355–8.

Stehli, F.G. (1968) Taxonomic diversity gradients in pole locations: the recent model. In *Evolution and Environment* (ed. E.T. Drake), Peabody Museum Centennial Symposium, Yale University Press, New Haven, Conn., pp. 163–227.
Stehli, F.G., Douglas, R.G. and Newell, N.D. (1969) Generation and maintenance of gradients in taxonomic diversity. *Science*, **164**, 947–9.
Stehli, F.G. and Wells, J.W. (1971) Diversity and age patterns in hermatypic corals. *Syst. Zool.*, **20**, 115–26.
Stenseth, N.C. (1979) Where have all the species gone? On the nature of extinction and the Red Queen hypothesis. *Oikos*, **33**, 196–227.
Stepanova, E.F. (1962) *Vegetation and Flora of the Tarbagatai Range*, Izd. Akad. Nauk Kazakhskoi SSR, Alma Ata.
Stetter, K.O. (1986) Diversity of extremely thermophilic archaebacteria. In *Thermophiles: general, molecular and applied microbiology* (ed. T.D. Brock), Wiley, New York.
Stevens, P. (1980) Evolutionary polarity of character states. *Ann. Rev. Ecol. Syst.*, **11**, 333–58.
Stommel, H. (1984) *Lost Islands*, University of British Columbia Press, Victoria.
Strawson, P.F. (1959) *Individuals: an essay in descriptive metaphysics*, Methuen and Co., London.
Street, H.E., ed (1978) *Essays in Plant Taxonomy*, Academic Press, London.
Strobeck, C. (1973) N-species competition. *Ecology*, **54**, 650–4.
Strong, D.R. Jr (1982) Harmonius coexistence of hispine beetles on *Heliconia* in experimental and natural communities. *Ecology*, **63**, 1039–49.
Strong, D.R. Jr (1983) Natural variability and the manifold mechanisms of ecological communities. *Amer. Nat.*, **122**, 636–60.
Strong, D.R. Jr, Lawton, J. and Southwood, T.R.E. (1984a) *Insects on Plants. Community patterns and mechanisms*, Blackwell Scientific Pubns, Oxford.
Strong, D.R. Jr and Rey, J.R. (1982) Testing for MacArthur–Wilson equilibrium with the arthropods of the miniature *Spartina* archipelago at Oyster Bay. *Amer. Zool.*, **22**, 355–60.
Strong, D.R. Jr, Simberloff, D., Abele, L.G. and Thistle, A.B. (1984b) *Ecological Communities: conceptual issues and the evidence*, Princeton University Press. New Jersey.
Strong, D.R. Jr, Szyska, L.A. and Simberloff, D.S. (1979) Tests of community-wide character displacement against null hypotheses. *Evolution*, **33**, 897–913.
Suggate, R.P., Stevens, G.R. and Te Punga, M.T., eds (1978) *The Geology of New Zealand*, Volume 2, Government Printer, Wellington.
Sugihara, G. (1981) $S = CAz$, $z = 1/4$: A reply to Connor and McCoy. *Amer. Nat.*, **117**, 790–93.
Sutter, R. (1969) Ein Beitrag zur kenntnis der soziologischen Bindung süd-südostalpiner Reliktendemismen. *Acta Bot. Croat.*, **28**, 349–66.
Swofford, D.L. (1984) *PAUP: phylogenetic analysis using parsimony. Version 2.4.* Illinois Natural History Survey, Champaign, Illinois.
Szafer, W., ed. (1966) *The Vegetation of Poland*, Pergamon Press, Oxford and Polish Scientific Pubns, Warszawa.
Szymura, J.M. and Barton, N.H. (1986) Genetic analysis of a hybrid zone between the fire-bellied toads *Bombina bombina* and *B. variegata*, near Cracow in Southern Poland. *Evolution*, **40**, 1141–59.
Szymura, J.M., Spolsky, C. and Uzzell, T. (1985) Concordant change in mitochondrial and nuclear genes across a hybrid zone between two frog species (genus *Bombina*). *Experientia*, **41**, 1469–70.

Takhtajan, A.L. (1978) *Floristic Regions of the World*, Nauka, Leningrad, English edn., Transl. T.J. Crovello, 1986, University of California Press, Berkeley.

Talbot, F.H., Russell, B.C. and Anderson, G.R.V. (1978) Coral reef fish communities: unstable high-diversity systems? *Ecol. Monogr.*, **48**, 425–40.

Tanksley, S.D., Medina-Filho, H. and Rick, C.M. (1982) Use of naturally occurring enzyme variation to detect and map genes controlling quantitative traits in an interspecific backcross of tomato. *Heredity*, **49**, 11–26.

Taper, M.L. and Case, T.J. (1985) Quantitative genetic models for the coevolution of character displacement. *Ecology*, **66**, 355–71.

Tarling, D.H. (1972) Another Gondwanaland. *Nature, Lond.*, **238**, 92–3.

Tattersall, I. (1981) Discussion. In *Vicariance Biogeography; a critique* (eds G. Nelson and D.E. Rosen) Columbia University Press, New York, pp. 406–10.

Tauber, C.A. and Tauber, M.J. (1977a) A genetic model for sympatric speciation through habitat diversification and seasonal isolation. *Nature, Lond.*, **268**, 701–5.

Tauber, C.A. and Tauber, M.J. (1977b) Sympatric speciation based on allelic changes at three loci: evidence from natural populations in two habitats. *Science*, **197**, 1298–9.

Tauber, F. (1984) Endemische Phytoassoziationen aus den Rumànischen Karpater. *Folia Geobot. Phytotax.*, **19**, 337–48.

Templeton, A.R. (1980) The theory of speciation via the founder principle. *Genetics*, **94**, 1011–38.

Templeton, A.R. (1981) Mechanisms of speciation – a population genetic approach. *Ann. Rev. Ecol. Syst.*, **12**, 23–48.

Templeton, A.R. and Gilbert, L.E. (1985) Population genetics and the coevolution of mutualism. In *The Biology of Mutualism: Ecology and Evolution* (ed D.H. Boucher), Oxford University Press, New York, pp. 128–44.

Terborgh, J. (1973) On the notion of favourableness in plant ecology. *Amer. Nat.*, **107**, 481–501.

Terborgh, J. (1976) Island biogeography and conservation: strategy and limitations. *Science*, **193**, 1029–30.

Terborgh, J. (1977) Bird species diversity on an Andean elevational gradient. *Ecology*, **58**, 1007–19.

Terborgh, J. (1980) Vertical stratification of a Neotropical forest bird community. *Proc. Int. Ornithol. Congr.*, **XVII**, 1005–12.

Terborgh, J. (1985) The vertical component of plant species diversity in temperate and tropical forests. *Amer. Nat.*, **126**, 760–76.

Thiselton-Dyer, W.T. (1878) Lecture on plant distribution as a field for geographical research. *Proc. R. Geogr. Soc.*, **22**, 412–55.

Thoday, J.M. (1972) Disruptive selection. *Proc. Roy. Soc. Lond.* B, **182**, 109–43.

Thomas, J.H. (1961) *Flora of the Santa Cruz Mountains of California*, Stanford University Press, Stanford, California.

Thompson, E. (1975) *Human Evolutionary Trees*, Cambridge University Press, Cambridge.

Thorne, R.F. (1963) Biotic Distribution Patterns in the Tropical Pacific. In *Pacific Basin Biogeography* (ed J.L. Gressitt), Bishop Museum Press, Honolulu, pp. 311–50.

Thornton, I.W.B. (ed.) (1986) *1985 Zoological Expedition to the Krakataus: preliminary report*, La Trobe University, Bundoora, Victoria.

Thorpe, R.S. (1984) Primary and secondary transition zones in speciation and population differentiation: a phylogenetic analysis of range expansion. *Evolution*, **38**, 233–43.

Tilman, D. (1982) *Resource Competition and Community Structure*, Princeton, University Press, Princeton, New Jersey.

Toft, C.A. (1986) Communities of species with parasitic life-styles. In *Community Ecology* (eds J. Diamond and T.J. Case), Harper and Row, New York, pp. 445–63.

Toft, C.A. and Schoener, T.W. (1983) Abundance and diversity of orb spiders on 106 Bahamian islands: biogeography at an intermediate trophic level. *Oikos*, **41**, 411–26.

Toft, C.A., Trauger, D.L. and Murdy, H.W. (1982) Tests for species interactions: breeding phenology and habitat use in subarctic ducks. *Amer. Nat.*, **120**, 586–613.

Toledo, V.M. (1982) Pleistocene changes of vegetation in tropical Mexico. In *Biological Differentiation in the Tropics* (ed G.T. Prance), Columbia University Press, New York, pp. 93–111.

Tolmachev, A.I. (1957) On the alpine flora of the Hissar Range. *Zemlevedenie*, **4**, 142–67.

Tolmachev, A.I. (1974a) *Introduction to the Geography of Plants*, Izd. Leningrad University.

Tolmachev, A.I. (ed) (1974b) *The Plant World of High Mountains and its Mastering. Problems of Botany 12*, Izd. Nauka, Leningrad.

Tolmachev, A.I. (ed) (1974c) *Determining the Higher Plants of Yakutia*, Nauka, Sibir. Otd., Novosibirsk.

Tonkyn, D.W. and Cole, B.J. (1986) The statistical analysis of size ratios. *Amer. Nat.*, **128**, 66–81.

Townsend, C.H.T. (1913) A new application of taxonomic principles. *Ann. entomological Soc. Am.*, **6**, 226–32.

Tsvelev, N.N. (1976) *Grasses of the USSR*, Nauka, Leningrad.

Tukey, J.W. (1977) *Exploratory Data Analysis*, Addison-Wesley, Reading, Mass.

Turelli, M. (1981) Niche overlap and invasion of competitors in random environments. I. Models without demographic stochasticity. *Theor. Pop. Biol.*, **20**, 1–56.

Turner, J.R.G. (1971) Mullerian mimicry in burnet moths and Heliconid butterflies. In *Ecological Genetics and Evolution* (ed E.R. Creed), pp. 224–60.

Turner, J.R.G. (1976) Mullerian mimicry: classical beanbag evolution and the role of ecological islands in adaptive race formation. In *Evolutionary Theory* (eds S. Karlin and E. Nevo), pp. 185–218.

Turner, J.R.G. (1982) How do refuges produce biological diversity? Allopatry and parapatry, extinction and gene flow in mimetic butterflies. In *Biological Diversification in the Tropics* (ed G.T. Prance), Columbia University Press, New York, pp. 309–35.

Turrill, W.B. (1953) *Pioneer Plant Biogeography: The phytogeographical Researches of Sir Joseph Dalton Hooker*, M. Nijhoff, The Hague.

Turrill, W.B. (1958) The evolution of floras with special reference to those of the Balkan peninsula. *J. Linn. Soc. (Botany)*, **56**, 136–52.

Twisselmann, E.C. (1967) A flora of Kern County, California. *Wasmann J. Biol.*, **25**, 1–395.

Valentine, D.H. (1972) *Taxonomy, Phytogeography and Evolution*, Academic Press, London.

Valentine, J.W. (1966) Numerical analysis of marine molluscan ranges on the extratropical northeastern Pacific shelf. *Limnol. Oceanogr.*, **11**, 198–211.

Valentine, J.W. (1973) *Evolutionary Paleoecology of the Marine Biosphere*, Prentice-Hall, Englewood Cliffs.

Valentine, J.W. (1984a) Neogene marine climate trends: implications for biogeography and evolution of the shallow-sea biota. *Geology*, **12**, 647–50.

Valentine, J.W. (1984b) Climate and evolution in the shallow sea. In *Fossils and Climate* (ed P.J. Brenchley), *Geol. J., Spec. Issue*, **11**, 265–77.

Valentine, J.W. and Jablonski, D. (1982) Major determinants of the biogeographic pattern of the shallow-sea fauna. *Bull. Soc. Geol. Fr.* (series 7), **24**, 893–99.

van Balgooy, M.M.J. (1971) Plant geography of the Pacific. *Blumea* (supplement) Vol. **6**.

Van Bruggen, T. (1976) *The Vascular Plants of South Dakota*, Iowa State University Press, Ames.

Vandermeer, J.H. (1980) Indirect mutualism: variations on a theme by Stephen Levine. *Amer. Nat.*, **116**, 441–8.

Van Valen, L. (1971) Adaptive zones and the orders of mammals. *Evolution*, **25**, 420–8.

Van Valen, L. (1973) A new evolutionary law. *Evol. Theory*, **1**, 1–33.

Van Valen, L. (1984a) Catastrophes, expectations and the evidence. *Paleobiology*, **10**, 121–37.

Van Valen, L. (1984b) A resetting of Phanerozoic community evolution. *Nature, Lond.*, **307**, 660–2.

Van Valen, L. (1985a) How constant is extinction. *Evol. Theory*, **7**, 93–106.

Van Valen, L. (1985b) Patterns of origination. *Evol. Theory*, **7**, 107–25.

Van Valen, L. (1985c) A theory of origination and extinction. *Evol. Theory*, **7**, 133–42.

Van Valen, L. and Sloan, R.E. (1966) The extinction of the multituberculates. *Syst. Zool.*, **15**, 261–78.

Vanzolini, P.E. and Williams, E.E. (1970) South American anoles; the geographic differentiation and evolution of the *Anolis chrysoepis* species group (Sauria; Iguanidae). *Arq. Zool. Sao Paulo*, **19**, 1–298.

Vasek, F.C. (1980) Creosote bush: long-lived clones in the Mohave Desert. *Amer. J. Bot.*, **67(2)**, 246–55.

Vermeij, G.J. (1978) *Biogeography and Adaptation; patterns of marine life*, Harvard University Press, Cambridge, Mass.

Viereck, L.A. (1970) Forest succession and soil development adjacent to the Chena River in interior Alaska. *Arctic and Alpine Res.*, **2**, 1–26.

Vincent, P. (1981) From theory to practice – a cautionary tale of island biogeography. *Area*, **13**, 115–18.

Vine, F.J. (1973) Organic diversity, palaeomagnetism and Permian palaeogeography. In *Organisms and Continents through Time* (ed N.F. Hughes), *Spec. Pap. Palaeontol.*, **12**, 61–77.

Vinogradova, N.G. (1959) The zoogeographical distribution of the deep-water bottom fauna in the abyssal zone of the ocean. *Deep Sea Res.*, **5**, 205–8.

Vinogradova, N.G. (1979) The geographical distribution of the abyssal and hadal (ultra-abyssal) fauna in relation to the vertical zonation of the ocean. *Sarsia*, **64**, 41–50.

Vogel, F. and Motulsky, A.G. (1986) *Human Genetics: problems and Approaches*, Springer Verlag, Berlin.

Voss, E. (1939) Bemerkenswers inter kontinentale Zusammenhange in den Unterfamilien Rynchitinae, Attelabinae und Apoderinae. *Proc. 7th Int. Congr. Ent.*, **vol. 1**, 444–60.

Vuilleumier, F. and Simberloff, D. (1980) Ecology versus history as determinants of patchy and insular distribution in high Andean birds. In *Evolutionary*

Biology, **12**, (eds M.K. Hecht, W.C. Steere and B. Wallace), Plenum Press, New York, pp. 235–379.

Wagner, W.H. (1961) Problems in the classification of ferns. In *Recent Advances in Botany*, University of Toronto Press, Toronto, pp. 841–4.

Wagner, W.H. (1980) Origin and philosophy of the groundplan–divergence method of Cladistics. *Syst. Bot.*, **5**, 173–93.

Wake, D.B., Yanev, K.P. and Brown, C.W. (1986) Intraspecific sympatry in a 'ring species', the plethodontid salamander *Ensatina escholitzii* in Southern California. *Evolution*, **40**, 866–88.

Wallace, A.R. (1860) On the zoological geography of the Malay Archipelago. *J. Linn. Soc.*, **4**, 172–84.

Wallace, A.R. (1876) *The Geographical Distribution of Animals*, II, MacMillan, London.

Wallace, A.R. (1880) *Island Life*, MacMillan, London.

Wallace, B. (1981) *Basic Population Genetics*, Columbia University Press, New York.

Walsh, J.B. (1982) Rate of accumulation of reproductive isolation by chromosome rearrangements. *Amer. Nat.*, **120**, 510–32.

Walsh, W.J. (1985) Reef fish community dynamics on small artificial reefs: The influence of isolation. *Bull. mar. Sci.*, **36**, 357–6.

Walter, H. (1977) The oligotrophic peatlands of western Siberia – the largest peino-heliobiome in the world. *Vegetatio*, **34**, 167–78.

Walter, H. (1981) Weber Höchstwerte der Producktion von natürlichen Pflanzenbeständen in Nordost Asien. *Vegetatio*, **44**, 37–41.

Walter, H. and Breckle, S.W. (1984) *Oekologie der Erde, 1. Spezielle Oekologie der tropischen und subtropischen Zonen*, Fischer, Stuttgart.

Walter, H. and Straka, H. (1970) *Arealkunde, floristisch–historische Geobotanik*, Ulmer, Stuttgart.

Waring, R.H. and Franklin, J.F. (1979) Evergreen coniferous forests of the Pacific Northwest. *Science*, **204**, 1380–86.

Watson, H.C. (1835) *Remarks on the Geographical Distribution of British Plants*, Longman, Rees, Orme, Brown, Green and Longman, London.

Webb, D.A. (1954) Is the classification of plant communities either possible or desirable? *Bot. Tidsskr.*, **51**, 362–70.

Webb, D.A. (1978) Flora Europaea – a retrospect. *Taxon*, **27**, 3–14.

Webb, L.J. and Tracey, J.G. (1981) The rain forests of northern Australia. In *Australian Vegetation* (ed R.H. Groves), Cambridge University Press, New York, pp. 67–101.

Webb, S.D. (1969) Extinction-origination equilibria in Late Cenozoic land mammals of North America. *Evolution*, **23**, 688–702.

Webb, S.D. (1977) A history of savanna vertebrates in the New World. Part I: North America. *Ann. Rev. Ecol. Syst.*, **8**, 355–80.

Webb, S.D. (1978) A history of savanna vertebrates in the New World. Part II: South America and the Great Interchange. *Ann. Rev. Ecol. Syst.*, **9**, 393–426.

Webb, S.D. (1984a) On two kinds of rapid faunal turnover. In *Catastrophes and Earth History* (eds W.A. Berggren and J.A. Van Couvering), Princeton University Press, Princeton, pp. 417–36.

Webb, S.D. (1984b) Ten million years of mammal extinctions in North America. In *Quaternary Extinctions: a prehistoric revolution* (eds P.S. Martin and R.G. Klein), University of Arizona Press, Tucson, pp. 189–210.

Webb, S.D. (1987) Community patterns in extinct terrestrial vertebrates. Organization of communities, past and present. In *Organization of Communities,*

Past and Present (eds J.H.R. Gee and P.S. Giller) Blackwell Scientific Pubns, Oxford, pp. 439–67.
Wegener, A. (1929) *The Origin of Continents and Oceans*, Dover Pubns, New York, 1966 reprint.
Weitzman, S.H. (1978) Three new species of fishes of the genus *Nannostomus* from the Brazilian states of Para and Amazonas (Teleostéi: Lebiasinidae). *Smithson. Contr. Zool.* **263**, 1–14.
Weitzman, S.H. and Weitzman, M. (1982) Biogeography and evolutionary diversification in neotropical fresh water fishes, with comments on the refuge theory. In *Biological Differentiation in the Tropics* (eds G.T. Prance), Columbia University Press, New York, pp. 403–22.
Welsh, S.L. and Chatterly, L.M. (1985) Utah's rare plants revisited. *Gt Basin Nat.*, **45**, 173–236.
Westermann, M., Barton, N.H. and Hewitt, G.M. (1987) DNA content variation between chromosomal races of *Podisma pedestris*. *Heredity*, **58**, 221–8.
Westman, W.E. (1983) Island biogeography: studies on the xeric shrublands of the Inner Channel Islands, California. *J. Biogeogr.*, **10**, 97–118.
Westoll, T.S. (1958) The origins of continental vertebrate faunas. *Trans. Geol. Soc. Glasgow*, **23**, 79–105.
Whitaker, A.H. (1973) Lizard populations on islands with and without Polynesian rats *Rattus exulans* (Peale). *Proc. N.Z. ecol. Soc.*, **20**, 121–30.
White, M.J.D. (1978) *Modes of Speciation*, W.H. Freeman, San Francisco.
Whiteside, M.C. and Harmsworth, R.V. (1967) Species diversity in chydorid (Cladocera) communities. *Ecology*, **48**, 664–7.
Whitmire, D.P. and Jackson, A.A. (1984) Are periodic mass extinctions driven by a distant solar companion? *Nature, Lond.*, **309**, 713–15.
Whitmire, D.P. and Matese, J.J. (1985) Periodic comet showers and Planet X. *Nature, Lond.*, **313**, 36–8.
Whittaker, R.H. (1970) *Communities and Ecosystems*, MacMillan, London.
Whittaker, R.H. (1977) Evolution of species diversity in land communities. *Evol. Biol.*, **10**, 1–67.
Whittaker, R.H. and Niering, W.A. (1975) Vegetation of the Santa Catalina Mountains, Arizona. V. Biomass, production and diversity along the elevation gradient. *Ecology*, **56**, 771–90.
Whittaker, R.J., Richards, K., Wiridamata, W., and Flenley, J.R. (1984) Krakatau 1883–1983: a biogeographical assessment. *Progr. Phys. Geogr.*, **8**, 61–8.
Whittam, T.S. and Siegel-Causey, D. (1981) Species interactions and community structure in Alaskan seabird colonies. *Ecology*, **62**, 1515–24.
Whittington, H.B. and Hughes, C.P. (1972) Ordovician geography and faunal provinces deduced from trilobite distribution. *Phil. Trans. Roy. Soc.* B, **263**, 235–78.
Wiebe, P.H. (1982) Rings of the Gulf Stream. *Scient. Am.*, **246**, 69–76.
Wiens, J.A. (1977) On competition and variable environments. *Am. Scient.*, **65**, 590–7.
Wiggins, I.L. and Thomas, J.H. (1962) *A Flora of the Alaskan Arctic Slope, Arctic Inst. of North America Spec. Publ. 4*, University of Toronto Press, Toronto.
Wijsman, E. and Cavalli-Sforza, L.L. (1984) Migration and genetic population structure with special reference to man. *Ann. Rev. Ecol. Syst.*, **15**, 279–302.
Wild, H. and Bradshaw, A.D. (1977) The evolutionary effects of metalliferous and other anomalous soils in south central Africa. *Evolution*, **31**, 282–93.
Wiley, E.D. (1981) *Phylogenetics: the theory and practice of Phylogenetic Systematics*, Wiley-Interscience, New York.

Williams, A. (1969) Ordovician faunal provinces with reference to brachiopod distribution. In *The Pre-Cambrian and Lower Palaeozoic Rocks of Wales* (ed A. Wood), University of Wales Press, Cardiff.

Williams, A. (1973) Distribution of brachiopod assemblages in relation to Ordovician palaeogeography. In *Organisms and Continents through Time* (ed N.F. Hughes), *Spec. Pap. Palaeontol.*, **12**, 241–69.

Williams, C.B. (1964) *Patterns in the Balance of Nature and Related Problems in Quantitative Ecology*, Academic Press, New York.

Williams, E.E. (1972) The origin of faunas. Evolution of lizard congeners in a complex island fauna. A trial analysis. *Evol. Biol.*, **4**, 47–89.

Williams, E.E. (1983) Ecomorphs, faunas, island size and diverse end points in island radiations of *Anolis*. In *Lizard Ecology: studies of a model organism* (eds R.B. Huey, E.R. Pianka and T.W. Schoener), Harvard University Press, Cambridge, Mass., pp. 326–70.

Williamson, G.B. (1978) A comment on equilibrium turnover rates for islands. *Amer. Nat.*, **112**, 241–3.

Williamson, M. (1972) *The Analysis of Biological Populations*, E. Arnold, London.

Williamson, M. (1981) *Island Populations*, Oxford University Press, Oxford.

Williamson, M. (1983a) A century of islands: from Darwin to the Hawaiian Drosophilidae. *Biol. J. Linn. Soc.*, **20**, 3–10.

Williamson, M. (1983b) The land-bird community of Skockholm, ordination and turnover. *Oikos*, **41**, 378–84.

Williamson, M. (1985) Apparent systematic effects on species-area curves under isolation and evolution. In *Statistics in Ornithology* (eds B.J.T. Morgan and P.M. North), Springer-Verlag, Berlin, pp. 171–8.

Williamson, M. (1987) Are communities ever stable? In *Colonisation, Succession and Stability*, (eds A.J. Gray, M.J. Crawley and P.J. Edwards), *26th Symp. Brit. Ecol. Soc.*, Blackwell Scientific Pubns, 353–72.

Wills, C.J. (1970) A mechanism for rapid allopatric speciation. *Amer. Nat.*, **111**, 603–5.

Wilson, A.C., Cann, R.L., Carr, S.M. *et al.* (1985) Mitochondrial DNA and two perspectives on evolutionary genetics. *Biol. J. Linn. Soc.*, **26**, 375–400.

Wilson, A.C., Carlson, S.S. and White, T.J. (1977) Biochemical evolution. *Ann. Rev. Biochem.*, **46**, 573–639.

Wilson, A.C., Sarich, V.M. and Maxon, L.R. (1974) The importance of gene rearrangement in evolution: evidence from studies on rates of chromosomal, protein and anatomical evolution. *Proc. Nat. Acad. Sci. USA*, **72**, 5061–5.

Wilson, E.O. (1961) The nature of the taxon cycle in the Melanesian ant fauna. *Amer. Nat.*, **95**, 169–93.

Wilson, E.O. (1969) The species equilibrium. In *Diversity and Stability in Ecological Systems* (eds G.M. Woodwell and H.H. Smith), Brookhaven Symp. in Biology No. 22, Associated Universities, Inc. Upton, New York, pp. 38–47.

Wilson, E.O. and Simberloff, D.S. (1969) Experimental zoogeography of islands: Defaunation and monitoring techniques. *Ecology*, **50**, 267–78.

Wilson, E.O. and Willis, E.O. (1975) Applied biogeography. In *Ecology and Evolution of Communities* (eds M.L. Cody and J.M. Diamond), Harvard University Press, Cambridge, Mass., pp. 522–34.

Wilson, J.W. (1974) Analytical zoogeography of North American mammals. *Evolution*, **28**, 124–40.

Wilson, M.V. and Shmida, A. (1984) Measuring beta diversity with presence–absence data. *J. Ecol.*, **72**, 1055–64.

Winterbourn, M.J. (1980) The freshwater insects of Australasia and their affinities. *Paleogr., Paleoclimatol., Paleoecol.*, **31**, 235–49.

Wissmar, R.C., Devol, A.H., Nevissi, A.E. and Sedell, J.R. (1982) Chemical changes of lakes within the Mount St Helens blast zone. *Science*, **216**, 175–8.

Wittmann, O. (1934) Die biogeographischen Beziehungen der Südkontinente: Die antarktischen Beziehungen. *Zoogeographica*, **2**, 246–304.

Woodger, J.H. (1952) From biology to mathematics. *Br. J. Phil. Sci.*, **2**, 193–216.

Wright, D.H. (1983) Species–energy theory: an extension of species–area theory. *Oikos*, **41**, 496–506.

Wright, S.J. (1932) The roles of mutation, inbreeding, crossbreeding and selection in evolution. *Proc. sixth Int. Congr. Genet.*, **1**, 356–66.

Wright, S.J. (1940) Breeding structure of populations in relation to speciation. *Amer. Nat.*, **74**, 232.

Wright, S.J. (1941) On the probability of fixation of reciprocal translocations. *Amer. Nat.*, **74**, 513–22.

Wright, S.J. (1980) Density compensation in island avifaunas. *Oecologia*, **45**, 385–9.

Wright, S.J. (1981) Intra-archipelago vertebrate distributions: the slope of the species–area relation. *Amer. Nat.*, **118**, 726–48.

Wright, S.J. (1982) The shifting balance theory and micromutation. *Ann. Rev. Genet.*, **16**, 1–20.

Wright, S.J. and Biehl, C.C. (1982) Island biogeographic distributions: testing for random, regular and aggregated patterns of species occurrence. *Amer. Nat.*, **119**, 345–57.

Wright, S.J. Dobzhansky, T. and Hovanitz, W. (1942) Genetics of natural populations. VII. The allelism of lethals in the third chromosome of *Drosophila pseudoobscura*. *Genetics*, **27**, 363–94.

Wu, L., Bradshaw, A.D. and Thurman, D.A. (1975) The potential for evolution of heavy metal tolerance in plants. III. The rapid evolution of copper tolerance in *Agrostis stolonifera*. *Heredity*, **34**, 165–87.

Yanev, K.P. and Wake, D.B. (1981) Genetic differentiation in a relict desert salamander, *Batrachoceps campi*. *Herpetologia*, **37**, 16–28.

Yoda, K. (1967) A preliminary survey of the forest vegetation of eastern Nepal. II. General description, structure and floristic composition of sample plots chosen from different vegetation zones. *J. Coll. Arts Sci. Chiba Univ. Nat. Sci.*, **ser. 5**, 99–140.

Yodzis, P. (1986) Competition, mortality and community structure. In *Community Ecology* (eds J. Diamond and T.J. Case), Harper and Row, New York, pp. 480–91.

Young, G.C. (1981) Biogeography of Devonian fishes. *Alcheringa*, **5**, 225–43.

Young, G.C. (1984) Comments on the phylogeny and biogeography of antiarchs (Devonian placoderm fishes) and the use of fossils in biogeography. *Proc. Linn. Soc. N.S.W.*, **107**, 443–73.

Young, G.C. (1986) Cladistic methods in Paleozoic continental reconstruction. *J. Geol.*, **94**, 523–37.

Young, J.Z. (1971) *An Introduction to the Study of Man*, Oxford University Press, Oxford.

Yurtsev, B.A. (1968) *Flora of the Suntar-Khayata Mountains: problems of the history of the high mountain landscapes of northeastern Siberia*, Nauka, Leningrad.

Yurtsev, B.A., Tolmachev, A.I. and Rebristaya, O.V. (1978) Floristic boundaries and divisions of the Arctic. In *The Arctic Floristic Region* (ed B.A. Yurtsev), Nauka, Leningrad, pp. 9–104.

Zach, L.W. (1950) A northern climax, forest or muskeg? *Ecology*, **31**, 304–6.
Zakirov, K.Z. (1961) *Flora and Vegetation of the Zeravshan River Basin, Part 2, Conspectus of the flora*, Akad. Nauk Uzbekskoi SSR, Tashkent.
Zakirov, P.K. (1969) *The Plant Cover of the Nuratinsk Mountains*, Izd. Fan, Uzbekskoi SSR, Tashkent.
Zakirov, P.K. (1971) *Botanical Geography of the Low Mountains of the Kyzl Kum and the Nuratau Range*, Izd. Fan, Uzbekskoi SSR, Tashkent.
Zandee, M. (1985) *C.A.F.C.A.: a collection of APL functions for cladistic analysis.*
Zandee, M. and Roos, M. (1987) Component compatibility in historical biogeography. *Cladistics*, **3**, (4), 305–32.
Zaverukha, B.V. (1980) Some theoretical questions related to the study of the phenomenon of endemism in the flora of Volyn-Podolia. *Ukrayinsky Botan. Zhurnal*, **37**, 15–19.
Zedler, P.H. (1977) Life history attributes of plants and the fire cycle. In *Proc. Symp. on the Environmental Consequences of Fire and Fuel Management in Mediterranean Ecosystems, 1–5 Aug., 1977, Palo Alto, California* (eds H.A. Mooney and C.E. Conrad), US Forest Service, Gen. Techn. Report WO-3.
Ziman, J. (1968) *Public Knowledge*, Cambridge University Press, Cambridge.
Zoller, W.H., Parrington, J.P. and Plelan Kotra, J.M. (1984) Iridium enrichment in airborne particles from Kilavea volcano: January, 1983. *Science*, **222**, 1118–21.
Zouros, E. and d'Entremont, C. (1980) Sexual isolation among populations of *Drosophila mojavensis*: response to pressure from a related species. *Evolution*, **34**, 421–30.

Index